Applications of Supercomputers in Engineering II

SECOND INTERNATIONAL CONFERENCE
ON
APPLICATIONS OF SUPERCOMPUTERS IN ENGINEERING
ASE/91

SCIENTIFIC COMMITTEE

Applications of Supercomputers in Engineering II

Editors: C.A. Brebbia, Wessex Institute of Technology, U.K.
D. Howard, Oxford University, U.K.
A. Peters, IBM Scientific Center, Germany.

Compuational Mechanics Publications
Southampton Boston

Co-published with

Elsevier Applied Science
London New York

C.A. Brebbia
Wessex Institute of Technology
Computational Mechanics Institute
Ashurst Lodge, Ashurst
Southampton SO4 2AA
UK

D. Howard
Computing Laboratory
Oxford University
11 Keble Road
Oxford OX1 3QD
UK

A. Peters
IBM Scientific Center
Institute for Supercomputing
Tiergartenstr 15
PO Box 103068
D-6900 Heidelberg
Germany

Co-published by

Computational Mechanics Publications
Ashurst Lodge, Ashurst, Southampton, UK

Computational Mechanics Publications Ltd
Sole Distributor in the USA and Canada:

Computational Mechanics Inc.
25 Bridge Street, Billerica, MA 01821, USA

and

Elsevier Science Publishers Ltd
Crown House, Linton Road, Barking, Essex IG11 8JU, UK

Elsevier's Sole Distributor in the USA and Canada:

Elsevier Science Publishing Company Inc.
655 Avenue of the Americas, New York, NY 10010, USA

British Library Cataloguing-in-Publication Data

A Catalogue record for this book is available
from the British Library

ISBN 1-85166-695-8 Elsevier Applied Science, London, New York
ISBN 1-85312-118-5 Computational Mechanics Publications, Southampton
ISBN 1-56252-052-0 Computational Mechanics Publications, Boston, USA

Library of Congress Catalog Card Number 91-72946

PREFACE

This book comprises an edited version of the Proceedings of the 2nd International Conference on Applications of Supercomputers in Engineering which took place at the Massachusetts Institute of Technology, Cambridge, USA during August 1991. The Conference was organized by the Wessex Institute of Technology, Southampton, UK with the support of the International Society for Boundary Elements. The first International Conference on Applications of Supercomputers in Engineering held in Southampton, UK in September 1989 was a very successful meeting and the resulting Conference Proceedings are now widely distributed throughout the world.

The revolutionary aspects of the next generation of computers are now fully recognised by many engineers and scientists. Vector and parallel computers form the basis of the computing power needed to address the complex problems with which engineers are faced. The new machines not only increase the size of the problems which can be solved, but also require a different computational approach to obtain the most efficient results.

The book includes papers on topics related to modern computer architecture and their relevance to engineering calculations, development of new computer algorithms, their application to solve engineering problems and further trends in engineering software, particularly as related to new hardware. The Volume is divided into the following sections:

1. Parallel Processing
2. Visualization and Graphics
3. Performance
4. Applications in Fluid Flow
5. Applications in Structural Mechanics

The Editors are grateful to all the authors for their excellent contributions and in particular to the members of the International Scientific Advisory Committee for their help. They hope that this volume will increase the awareness of the potential of new supercomputers amongst engineers and scientists.

The Editors
Southampton, August 1991

CONTENTS

SECTION 1: PARALLEL PROCESSING

SECTION 2: VISUALIZATION AND GRAPHICS

SECTION 3: PERFORMANCE

SECTION 4: APPLICATIONS IN FLUID FLOW

SECTION 5: APPLICATIONS IN STRUCTURAL MECHANICS

SECTION 1: PARALLEL PROCESSING

Multi Block Flow Solvers for Use on Parallel Computers

J.B. Vos (*), C.M. Bergman (**)

(*) Hydraulic Machines and Fluid Mechanics Institute (IMHEF), Swiss Federal Institute of Technology - Lausanne, CH-1015 Lausanne, Switzerland

(**) Centre for Computational Mathematics & Mechanics (C2M2), The Royal Institute of Technology (KTH), S-100 44 Stockholm, Sweden

ABSTRACT.

Multi block solvers for the 2 and 3 Dimensional Euler Equations have been developed. The spatial discretization scheme is that of Jameson [1] using central differencing with added artificial dissipation. The resulting system of ODE's is integrated in time using the explicit Runge Kutta procedure. The computational domain is split into multiple blocks, first of all to facilitate the grid generation for complex geometries, and secondly to allow the utilization of different parallel computers, on which this paper places the emphasis. The spectrum of parallel computers envisaged is broad, ranging from computers having only a relatively small number of processors to massively parallel computers. Results discussed in this paper includes the CRAY 2, ALLIANT FX/80 and the Intel iPSC/2 SX, while the implementation on the Connection Machine is briefly discussed. The multi block approach proved to be easily ported between these different computers, and - most important - attain high parallel efficiencies.

INTRODUCTION.

The field of Computational Fluid Dynamics (CFD) is rapidly developing. An example of this is in the area of Hypersonic Flows where ambitious projects like the European Space Shuttle HERMES and the German Space Plane SANGER have stimulated the research in CFD for high speed flows including air chemistry. New developments in the areas of CFD and Computational Chemistry, together with the rapid increase in computer speed and central memory capacity of computers has made numerical simulations as important as the sometimes cumbersome experimental simulations. However, numerical simulation has its limitations too. For example, solving the complete set of equations describing the flow around an spacecraft during the re-entry, i.e. the Compressible Navier Stokes Equations together with Non-Equilibrium Chemistry models, is far beyond the capacity of the present generation of computers. The resolution of the small length scales in the boundary layer require very dense computational meshes, which could be in excess of 10^7 grid points. Moreover, high Reynolds number flows are unsteady and the time must be added as a fourth dimension to obtain an appropriate representation of the flow problem. Recent estimates [2] indicate that for the industrial design of military or civil aircrafts at least 20 Million grid points are necessary together with an computational speed of 100 Gflops (10^{11} floating point operations per second).

Today, state of the art supercomputers have sustained processing speeds in the order of 0.5 - 1 Gflops for optimized CFD codes, which is far from the speed required for the industrial design of complete aircrafts. Moreover, it became clear in the last years that the maximum speed of a single processor is reaching a plateau, which is in the order of 1 Gflops. In 1985 the CRAY 2 came on the market, which had a clock cycle of 4.1 ns. Today, the fastest computers available have clock cycles of 6 ns (CRAY Y-MP) and 2.9 ns (NEC SX3). To increase the speed of a single computer system, the attention is now focused on parallel architectures. For example, CRAY Research has announced their Teraflop project in which 1000 processors having each a speed of 1 Gflop are put in a single machine. Moreover, computer systems having up to 65384 processors are available on the market today. To use these parallel computers efficiently new programming concepts need to be developed.

Here the multi block approach using structured meshes is used to explore the inherent parallelism in the underlying numerical algorithm. The implementation of this approach on MIMD machines is straightforward, as will be shown further on. Implementation of the multi block solvers on SIMD machines requires another approach, and is only briefly addressed. Lately research in CFD is more and more focusing on the use of unstructured meshes because its data structure facilitates both the mesh generation and adaptation. The implementation of solvers using unstructured meshes on parallel computers is less straightforward than for structured meshes due to the indirect addressing required in these solvers. However, results have already been published in the literature. Moreover, much progress has been made in recent years in implementing solvers using unstructured meshes on SIMD computers, by using pre-compilers which identify the data that can be exchanged concurrently. The reason for adopting a structured multi block approach in the work described here was for its simple data structure which facilitates the implementation of this approach on different parallel computers.

GOVERNING EQUATIONS.

The multi block flow solver described here solves the 2- and 3D Euler equations. In 2D cartesian coordinates these equations can be written as

$$\frac{\partial}{\partial t}\mathbf{w} + \frac{\partial}{\partial x}\mathbf{F}(\mathbf{w}) + \frac{\partial}{\partial y}\mathbf{G}(\mathbf{w}) = 0 \qquad (1)$$

where the state vector **w** and the flux vectors **F** and **G** are given by

$$\mathbf{w} = \begin{pmatrix} \rho \\ \rho u \\ \rho v \\ \rho E \end{pmatrix}, \mathbf{F}(\mathbf{w}) = \begin{pmatrix} \rho u \\ \rho u^2 + p \\ \rho u v \\ u(\rho E + p) \end{pmatrix}, \mathbf{G}(\mathbf{w}) = \begin{pmatrix} \rho v \\ \rho u v \\ \rho v^2 + p \\ v(\rho E + p) \end{pmatrix} \qquad (2)$$

In these equations ρ denotes the density, u,v the cartesian velocity components, E the total energy and p the pressure. The system of equations is completed by specifying the relation between the pressure p and the state vector **w**. For a caloric perfect gas, this equation states

$$p = (\gamma - 1)\,\rho\left(E - \frac{1}{2}\left(u^2 + v^2\right)\right) \tag{3}$$

where γ is the ratio of specific heats, equal to 1.4 for air. The extension of this system of equations to 3D is straightforward.

NUMERICAL METHOD.

Following the concept of the finite volı me method Eq. (1) is integrated over a domain Ω. Then by using the theorem of Gauss one obtains

$$\int_{\Omega} \frac{\partial}{\partial t} \mathbf{w}\, dS + \int_{\partial\Omega} \mathbf{H}\cdot\mathbf{n}\, ds = 0 \tag{4}$$

where $\mathbf{H} = (\mathbf{F}, \mathbf{G})$ is the flux tensor, see Eq. (2), and $\mathbf{n}$ the normal at the boundary of Ω, pointing in the outward direction. Assume that the domain Ω is a Finite Volume cell with index (i,j), and consider $\mathbf{w}_{i,j}$ as an approximation of the state vector in this cell. Then Eq. (4) can be approximated as

$$\frac{d}{dt}\left(S_{i,j}\mathbf{w}_{i,j}\right) + \mathbf{Q}_{i,j} = 0 \tag{5}$$

where $S_{i,j}$ is the volume of the cell (i,j) and $\mathbf{Q}_{i,j}$ represents the net convective flux leaving and entering the cell, calculated from

$$\mathbf{Q}_{i,j} = \mathbf{h}_{i+1/2,\,j} + \mathbf{h}_{i-1/2,\,j} + \mathbf{h}_{i,\,j+1/2} + \mathbf{h}_{i,\,j-1/2} \tag{6}$$

where

$$\mathbf{h}_{i-1/2,j} = \int_{i-1/2,j} \mathbf{H}\cdot\mathbf{n}\ ds = \mathbf{H}_{i-1/2,j} \cdot \int_{i-1/2,j} \mathbf{n}\, ds \tag{7}$$

The value of the flux tensor, $\mathbf{H}_{i-1/2,j}$ situated at the surface is calculated from Eq. (2) using the average of the state vectors having the surface at i-1/2,j in common. On a cartesian grid this results in a centered scheme which is second order accurate in space.

Adaptive Dissipation

The Finite Volume scheme is augmented by the addition of a second order artificial viscosity term to be used near discontinuities, and of a fourth order dissipation term to suppress odd/even oscillations allowed for by centered schemes. The dissipation terms are formed from the second and fourth order differences of the state vector multiplied by a scaling factor and a weight, the latter usually referred to as a switch [1]. This switch is constructed from the absolute value of the normalized second difference of the pressure, implying that the second order dissipation term is small except in regions of large pressure gradients, such as in the neighborhood of a shock wave or a stagnation point. The fourth order difference is used everywhere except in regions where the second order dissipation is strong in order to prevent the generation of oscillations in these regions. After addition of the dissipative terms, the following numerical scheme results,

$$\frac{d}{dt}(S_{i,j}w_{i,j}) + Q_{i,j} - D_{i,j} = 0 \qquad (8)$$

where $D_{i,j}$ is the dissipation operator. The conservative form of the discretized equations is preserved by, analogous to the convective fluxes, introducing dissipative fluxes for each equation. The operator $D_{i,j}$ is then split as

$$D_{i,j} = d_{i+1/2,j} + d_{i-1/2,j} + d_{i,j+1/2} + d_{i,j-1/2} \qquad (9)$$

where each dissipative flux is evaluated at the same location as the corresponding convective flux.

Explicit time integration

Equation (8) is integrated in time using the explicit Runge Kutta scheme [1]. This type of schemes is normally used for their high accuracy in time, but here properties as stability and damping are of greater interest. Furthermore the explicit integration scheme does not require a matrix inversion which is difficult to parallelize, as for example is the case for implicit schemes. The largest difference stencil is that of the dissipation operator, requiring for the switch 7 points in each direction.

Boundary conditions

At all the sides of the calculation domain, boundary conditions have to be prescribed. At solid walls, the tangency condition is imposed, and the pressure at the wall is obtained by linear extrapolation from the interior field to the wall. At outflow and farfield boundary three option exists. The state vector **w** can be extrapolated, set to the free stream conditions or found using the 1D Riemann invariant described for example in [3].

LEVELS OF PARALLELISM.

In [4] an overview is given of parallel computer projects. The architectures of these computers differ strongly, but a classification can be done in terms of instruction and memory management. In general two different types of parallel computers are distinguished, the MIMD (Multiple Instruction Multiple Data) and the SIMD (Single Instruction Multiple Data) machines. For the MIMD computers it is possible to distinguish tightly coupled machines, like the CRAY 2, and the ALLIANT FX/80 in which the memory is shared by all processors via fast memory buses, and loosely coupled machines in which each processor has its own memory. In the latter case the different processors must be connected in a network in order to exchange data. Examples of such machines are the Intel iPSC/2 and the German SUPRENUM. Many computers are hybrid machines. For example, the CRAY 2 has up to 4 processors (MIMD), but each processor is a vector processor (SIMD).

Examples of SIMD computer available on the market are the TMC Connection Machine (CM-2) with 65536 processors and the MasPar computer having 16384 processors. Writing a solver that explores the parallelism of the SIMD machines differs strongly from those of the MIMD machines due to the different hardware concepts.

Granularity of parallelism

One can distinguish several levels of parallelism. The simplest and most easy to use is so called *fine grain parallelism*, which is provided on an instruction level. An example of this approach is the ability to perform multiple floating point operations at the same time on a single processor. For example, the multiplication and the addition in the do-loop

```
do i=1,1000
   x(i) = a(i) * b(i) + c
end do
```

can be carried out simultaneously. This form of parallelism is often used on pipelined vector computers as the Cyber 205, but also on the NEC SX3 which can have up to 4 pipelines per processor. Computers from MULTIFLOW uses this VLIW (Very Long Instruction Word) technology to enhance performance. The longest instruction word on that machine can perform up to 28 operations per clock cycle. Since data suitable for vectorization in principle always can be performed using a VLIW instruction this has been launched as an alternative to vectorization. On SIMD machines, like the CM-2, fine grain parallelism represents the mapping of the arrays to the underlying hardware. Parallel data structures are spread across the data processors, typically with a single element stored in each processor. For the example shown above, it could mean that the do-loop is spread over 1000 processors.

A second level of parallelism, called *medium grain parallelism,* can be used on computers having a shared memory. On this level do-loops are spread over multiple processors. For example, if the do-loop shown in the example above would be carried out on a 10 processor machine, each processor would process 100 elements. When the splitting of a do-loop is done manually by the inclusion of compiler directives it is called micro tasking, when performed automatically by the compiler it is named auto tasking. This form of parallelism is only efficient if the startup and synchronization costs are very low. A disadvantage of shared memory computers is that the maximum number of processors that can share the same memory is probably below 32, due to limitations in memory bus speed and memory conflicts.

Fine grain and medium grain parallelism are both attractive from the user point of view, since the present generation of compilers can recognizes these forms of parallelism automatically. Hence no special programming expertise and no knowledge about the computer architecture is required

The last level of parallelism is called *coarse grain parallelism,* which can be performed on both tightly and loosely coupled MIMD computers, but is especially suited for computers having a local memory. On this level, the program is decomposed in different tasks, which are executed in parallel. On tightly coupled machines this can be done via compiler directives, spreading these task on multiple processors. More elaborate is the implementation on loosely coupled computers since code for controlling the communication between the different processors via message passing network must be designed. Programming these computers is much more difficult compared to shared memory computers since the interaction between algorithm and computer architecture is much stronger. Special attention has to be given to limit the communication between the different processors, since this can degrade the performance considerably.

STRUCTURE OF THE MULTI BLOCK SOLVER

Introduction

From a mathematical and numerical point of view, there is no difference between a single block and multi block solver, since the equations and the numerical method are exactly the same. It is on the level of the incorporation of the boundary conditions that there appears a difference since a block connectivity boundary condition must be introduced. Explicit schemes do not require special solution procedures for periodic and block connectivity boundary conditions, making these type of schemes very suitable for multi block calculations, this compared to the more complex and cumbersome implicit schemes. The step from a single block to a multi block solver is small, and mainly a matter of defining a good and efficient data structure. State variables are stored block by block, yielding the optimal storage scheme for distributed memory computers where each block will be connected to a processor with its own local memory. On shared memory machines a global array and a pointer must be added which points to the first address in the global array of the variables in each block. Before proceeding proper definitions of the computational space shown in Figure 1 should be introduced. The basic idea is that it should be possible to connect any point on the boundary of a block to any other point on any side of any block. Here this connection is done by means of patched multi blocks.

Definition of the computational domain

The following definitions in hierarchical order are used here:

1) Multiple Blocks
Several local curvilinear coordinate systems that are connected together, resulting from a decomposition of the physical domain.

2) Block
One local curvilinear coordinate system with I and J as running indices, see for example figure 1. Every block has four sides in 2D and six in 3D.

3) Side
Represents the boundary of the block. The sides are numbered as indicated in figure 1.

4) Window
On each side several boundary conditions can be imposed. The definition of the segment where a particular boundary condition can be imposed is called a window.

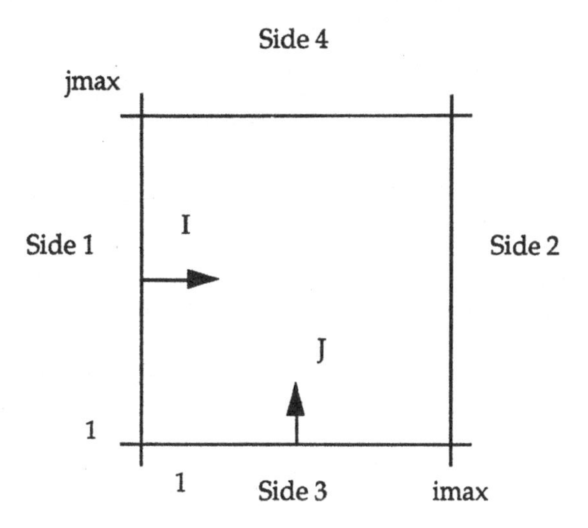

Figure 1: Single Block Computational Domain

Boundary arrays and ghost cells.

The implementation of the boundary conditions has been uncoupled from the algorithm used to update the interior points. The link between the boundary conditions and the algorithm for the interior points is made using boundary arrays and 'ghost points'. In general, finite volume flow solvers use a certain number of ghost points which are used in enforcing the boundary conditions. Here, three ghost points are used at each side of the calculation domain, and the values of the state vector w and pressure p at these points are obtained from the boundary arrays. These boundary arrays are filled in boundary condition routines in such a way that the boundary conditions are satisfied when using the ghost points. Since the difference stencils for the convective fluxes and the artificial dissipation fluxes are not the same, separate boundary arrays have been defined for each stencil (respectively `bw` and `bd`). In Table 1 the ghost cells which are used for a given boundary array are indicated with an x.

<table>
<tr><th>Array</th><th colspan="3">Ghost cells</th><th colspan="3">Interior cells</th></tr>
<tr><td></td><td>-2</td><td>-1</td><td>0</td><td>1</td><td>2</td><td>3</td></tr>
<tr><td>bw</td><td></td><td></td><td>x</td><td></td><td></td><td></td></tr>
<tr><td>bd</td><td></td><td>x</td><td>x</td><td></td><td></td><td></td></tr>
<tr><td>ba</td><td>x</td><td>x</td><td>x</td><td></td><td></td><td></td></tr>
<tr><td>bx</td><td></td><td colspan="2">x</td><td></td><td></td><td></td></tr>
</table>

Table 1: Ghost cells

In Table 1, `ba` is the boundary array for the calculation of the switch in the artificial dissipation, and `bx` contains the coordinates.

For each window, the following information is provided as input to the solver:

- Boundary condition type
- Window definition (start and end indices)
- Adjacent block number
- Side number on this block to which the window is connected
- Index direction on adjacent block
- Starting address on adjacent block
- Mapping of the coordinate systems between the two blocks (3D only)

To minimize the communication between the blocks, boundary arrays addressing adjacent blocks are updated only once per time step. Boundary conditions using only local arrays do not require communication, and are updated each Runge Kutta stage. Using this structure a time step can be made in each block independent of the other blocks, and only after completion of a time step data is exchanged between the different blocks.
A multi block calculation always implies a certain overhead compared to a single block calculation on the same calculation domain. This overhead is for a great part associated with imposing the boundary conditions. To reduce this overhead, the procedure used to fill the boundary arrays has been optimized as much as possible to obtain vectorizable do-loops everywhere [3].

RESULTS.

The ALLIANT FX/80

As a first step, the multi block approach was implemented on a shared memory computer [5]. This was done in such a way that, instead of sequentially filling up the blocks to be calculated and then calling the subroutine to perform the actual updating of the state variables, this subroutine was copied on several processors to run independently. This was achieved on the ALLIANT FX/80 using the `cvd$ cncall` compiler directive. In order to compare the performance of the medium grain with the course grain parallelism a mesh with 129x65 grid points was used. Automatic parallelization was obtained on the do-loop level by the compiler, yielding a performance of 18.1 Mflops. The domain decomposition was performed by a preprocessor dividing the single mesh in 8 blocks with 32x32 points each, one block for each processor. This is the minimum size of the blocks for performing a comparison since the vector length on the ALLIANT is 32. This resulted in a speed of 18.3 Mflops. From these results it is clear that the medium grain parallelism performed automatically by the compiler, and the coarse grain parallelism achieved by means of domain decomposition both gave approximately the same performance. There is therefore no reason to use domain decomposition on this computer since the expected gain is very small.

The CRAY 2/4-256

The 3D multi block solver was implemented on the CRAY 2/4-256 of the Swiss Federal Institute of Technology in Lausanne. Calculations were carried out for the flow over a wedge, using respectively one and eight blocks of 56x18x18 grid points. Medium grain parallelism was studied using the auto tasking facility (fpp & fmp), while coarse grain parallelism has been investigated by forcing the routine performing the time integration to run on four processors by using the compiler directive `cmic$ do parallel` [6]. The timings were obtained on a dedicated computer, using the cft77 Version 4.0 compiler, and operating system Unicos 5.0.

From Table 2, it can be seen that when using auto tasking, the code runs on 1.72 processor, independent of the number of blocks used. Auto tasking distributes the do-loops over the different

		1 block	8 blocks
cft77	CPU time	31.25	237.30
	elapse time	31.30	239.51
auto tasking	CPU time	47.77	363.61
	elapse time	27.79	212.14
	ratio CPU/elapse	1.72	1.71
macro tasking	CPU time	31.32	338.04
	elapse time	31.34	84.89
	ratio CPU/elapse	1.00	3.98

Table 2: Timing results in seconds, scaled problem on the CRAY 2.

processors. However, the solver has been written using do-loops over all grid points which were not distributed over the different processors using auto tasking on the CRAY. Tests using double or triple nested do-loops showed that in this case automatic parallelization was obtained, i.e. the vectorized inner do-loop was sent to a processor, while the outer do-loops were distributed over the four processors. Macro tasking was much more efficient, since for 99.5% of the CPU time all four processors of the CRAY were busy. This means that the cost of the exchange of information between the different blocks is in the order of 0.5% of the CPU time. Comparing the cost of the 8 block macro tasking calculation to that of the single processor calculation shows that the true speed-up is only a factor 2.78, which implies an overhead of 42%. Although the code used the four processors efficiently, startup and synchronization times were so high that some part of the gain was lost.

A 32 node Intel iPSC/2 SX

Here the 2D multi block code was implemented on a 32 node iPSC/2 SX hypercube [7] situated at the Department of Computing Science at Uppsala, Sweden. Each node is configured with 4 Mbytes of memory, a 80386 microprocessor (16 Mhz) and a Weitec 1167 floating point accelerator. As test case the supersonic flow over a wedge was used with a single block grid of 129x33 points, and 99 time steps were made. For this fixed problem in total 99x128x32x678 = 274.931.212 floating point operations were made. In Table 3 the single block grid is cut into several blocks which were send to different nodes. It can be seen that as the block size becomes smaller the block-to-block communication and the internal boundary condition overhead in the solver limits the maximum speed-up that can be obtained.

# processors	Block size	CPU seconds	Mflops	Speed-up
1	129x33	485	0.567	1.00
2	65x33	246	1.12	1.98
4	33x33	128	2.15	3.79
8	33x17	68.0	4.04	7.10
16	17x17	37.3	7.37	13.0
32	17x9	23.3	11.8	20.8

Table 3: Timing reports for the fixed problem

Another way of measuring the performance on distributed memory machines is by scaling the problem size with the number of nodes or processors. In Table 4 timings for this case shows that the cost of communication is very small and that the potential speed-up is high. Note that the CPU seconds listed in Tables 3 and 4 are the maximum timings for all processors.

# processors	# of Mflop	CPU seconds	Mflops	Speed-up
1	274.9	485.0	0.567	1.00
2	550.9	485.8	1.13	2.00
4	1100	487.1	2.26	3.99
8	2200	487.1	4.52	7.99
16	4399	487.8	9.02	15.9
32	8798	489.0	18.0	31.7

Table 4: Timing reports for the scaled problem

FUTURE DEVELOPMENTS.

Recently the 2D Euler solver has been extended to solve the 2D Navier Stokes equations. After more validation, especially in terms of turbulence modeling, the viscous terms will be included in the parallel version and in the 3D version. Models to account for the effects of equilibrium air chemistry are available, and will be implemented in the parallel versions in the near future.

A single block version of the solver has been ported to the CM-2 [8]. The aim was merely to get the code running on the Connection Machine using the same data structures and algorithms as in the original code. The conversion was made in steps checking that the results were unchanged after each changed subroutine. This is possible since scalar arrays reside on the front end machine and parallel arrays on the Connection Machine itself. Data areas can thus be doubled and copied with explicit movement between the two machines. The performance obtained with a Virtual Processor ratio of 4, was 4.2 Mflops on a 4k processor machine. Small modifications of the data structure, e.g. the boundary arrays, are expected to increase this performance to at least 30 Mflops. Many parts of the solver could be transported directly to the CM-2 machine, but, as already mentioned, the data structure for the incorporation of the boundary conditions needs special attention since it differs from the optimal data structure for this machine. For the multi block implementation the basic problem is the mapping of blocks of different sizes to the network topology of the CM-2 in such a way that the number of idling processors is minimized.

CONCLUSIONS.

Multi block Euler solvers were successfully implemented on different parallel computers. The porting was mainly focusing on MIMD machines; the conversion towards SIMD architectures has been briefly addressed. Of the MIMD machines two where tightly coupled and one was loosely coupled. The first of the former is the ALLIANT FX/80. A comparison between auto tasking and macro tasking showed that the difference in timings was very small. However, on the CRAY 2 using 4 processors, macro tasking appeared to be far more superior to auto tasking. The Intel iPSC/2 SX represented the loosely coupled machine. For a problem in which the size was scaled with the number of processors a speed-up of 31.7 on a 32 node machine was obtained. It is clear that the potential speed-up on this type of machine is high. On an SIMD type of computer, initial timings for a single block version of the code [8] indicated that the underlying data structure should be modified to obtain high performances.

REFERENCES.

[1] Jameson, A., Steady state Solutions of the Euler Equations for Transonic Flow by a Multigrid Method, Advances in Scientific Computing, pp.37-70, Academic Press 1982

[2] Jameson, A., 2nd Joint GAMNI/SMAI-IMA Conference on Computational Aeronautical Dynamics, May 17-19 1989, Antibes, France

[3] Bergman, C.M., Development of Numerical Techniques for Inviscid Hypersonic Flows around Re-entry Vehicles, Ph.D thesis, INP Toulouse, 1990

[4] McBryan, O.A., Overview of Current Developments in Parallel Architectures, Parallel Supercomputing, Wiley & Sons Inc., 1989

[5] Bergman C.M. and Rizzi A.W., Implementation of an Multi Block Euler Solver on a Shared Memory MIMD Computer, International Workshop on Supercomputing Tools for Science and Engineering, Pisa, Italy, December 4-7,1989

[6] CFT77 Compiling System: Fortran Reference Manual, SR-3071 4.0. Cray Research Inc., 1990

[7] Bergman C.M. and Vos J.B., Parallelization of CFD Codes, Second World Congress on Computational Mechanics, August 27-31, 1990, Stuttgart

[8] Bergman C.M., Wahlund P., Implementation of an Multi Block Euler Solver on the CM-2, Abstract submitted to Third Annual Conference on Parallel CFD, Stuttgart, Germany, June 10-12, 1991

Difficulties in Parallelising Efficient Sequential Codes

D.J. Gavaghan (*)(**), C.S. Gwilliam (*),
J.S. Rollett (*)
(*) Numerical Analysis Group, Oxford University Computing Laboratory, Oxford, OX1 3QD, U.K.
(**) Nuffield Department of Anaesthetics, Radcliffe Infirmary, Oxford, U.K.

Abstract

Efficient numerical algorithms for the solution of fluid flow problems on sequential machines are the major focus of research within our group. However with the recent acquisition of a MIMD machine, attempts have been made to obtain parallel implementations of these algorithms. This paper will describe the difficulties involved in two such attempts: the solution of a triple–layer diffusion problem used in modelling electrochemical sensors; and the solution of the Navier–Stokes equations in a driven cavity using multigrid.

1 Introduction

For many years, the development of efficient numerical algorithms for the solution of fluid flow problems has been the main reasearch interest in our group and as a result a large body of code (mainly in FORTRAN) has built up. With the recent acquisition of an Inmos transputer array, attempts have been made to obtain parallel versions of some of these algorithms. This paper describes two such attempts. The first problem arises from the modelling of the diffusion processes of a membrane–covered oxygen sensor, used in determining the oxygen concentration of blood. The basic equations used in modelling the problem are presented, followed by a description of the algorithm which solves the equations on a sequential machine (in FORTRAN), emphasising those points which make it particularly efficient. We then describe how the same algorithm can be adapted to run in parallel (in Occam) on any number of processors, so that it is 60–70% efficient on problem sizes of practical interest.

In the second problem, we start by taking an existing FORTRAN code to solve the Navier–Stokes equations in a driven cavity using multigrid developed by Tasneem Shah [1]. We again present the equations and sequential algorithm, and go on to describe how in adapting the original code into a parallel FORTRAN code to run on just two transputers, we can demonstrate how those features of the sequential algorithm that make it particularly effective, are precisely those features that create load imbalance and inefficiency in parallel. We then describe how we are altering the original algorithm to allow us to create an efficient parallel code to run on the whole network.

2 Electrode Problem

Fig. 1(a) shows a schematic outline of a Clark–type oxygen sensor [2] used to determine the concentration of oxygen in blood. It typically consists of a platinum cathode

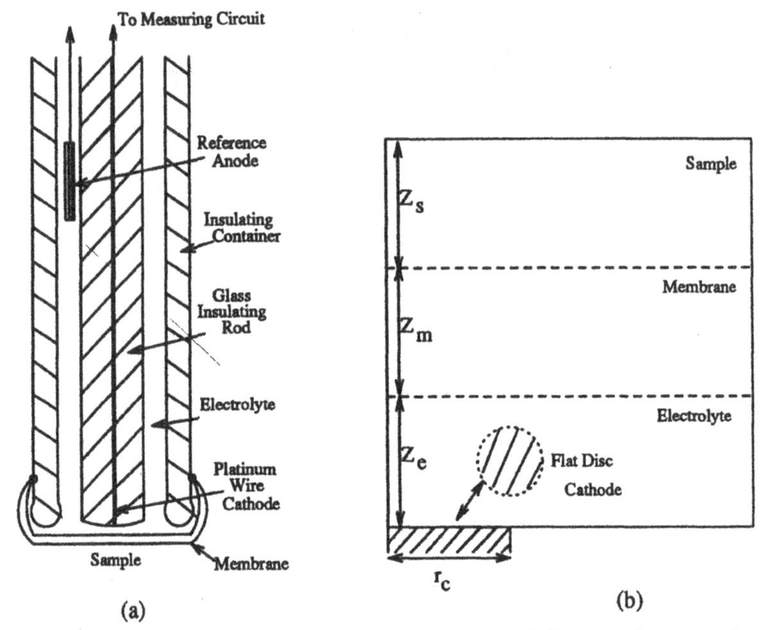

Figure 1: (a) Schematic outline of Clark oxygen sensor and (b) Two-dimensional region in which diffusion is assumed to take place.

Layer	Diffusion Coefft m^2s^{-1}	Solubility $molm^{-3}atm^{-1}$	Layer thickness (μm)	Radius (μm)
Electrolyte	$\approx 2 \times 10^{-9}$	≈ 1.0	2.0–10.0	
Membrane	1.0×10^{-10} – 1.0×10^{-12}	1.0 – 20.0	5.0 – 20.0	10.0–100.0
Sample	1.0–5.0 $\times 10^{-9}$	≈ 1.0	0.0–100.0	

Table 1: Parameter values.

sealed in a glass insulating rod immersed, together with a reference anode, in aqueous electrolyte, protected from the medium to be measured by an oxygen-permeable membrane to prevent poisoning of the electrode. When a suitable negative potential is applied to the cathode, oxygen molecules are reduced and a current flow which is proportional to the oxygen concentration of the sample. Table 1 shows the range of parameter values of interest in investigating the behaviour of the sensors. We switch the sensors on for varying times up to 20 – 30 seconds, and they can also be used in a pulsatile on/off manner. We wish to obtain numerical accuracy of around 1%.

2.1 Modelling

It is assumed that the potential applied at the cathode is sufficiently negative that all oxygen molecules reaching it will be reduced. A concentration gradient is therefore set up and transport of oxygen within the electrolyte, membrane and sample layers (thicknesses z_e, z_m, z_s) can be modelled by the diffusion equation. The cathode is assumed to be a disc, radius r_c, lying in the surface of an infinite planar insulating material, symmetry allowing two-dimensional cylindrical coordinates in (r,z), to be used (see Fig.1(b)). Since the oxygen concentration c is not continuous across material interfaces, we solve rather for the related quantity partial pressure p ($= c/\alpha_l$, where α_l is the solubility in layer l) which is continuous. Therefore within each layer, $l = e, m, s$, we assume that the partial pressure $p(r,z,t)$ obeys

$$\frac{\partial p}{\partial t} = D_l\left(\frac{1}{r}\frac{\partial p}{\partial r} + \frac{\partial^2 p}{\partial r^2} + \frac{\partial^2 p}{\partial z^2}\right) \tag{1}$$

where D_l is the diffusion coefficient in the relevant layer, starting from the initial condition

$$t < 0 \qquad p(r,z,t) = p_0 \quad \text{in } [0, r_{\max}] \times [0, z_{\max}] \tag{2}$$

and subject to the boundary conditions

$$\begin{array}{llll}
\text{On } z = 0 & p = 0 & (0 \le r \le r_c) & \\
 & \frac{\partial p}{\partial z} = 0 & (r_c \le r \le r_{\max}) & \\
\text{On } r = 0 & \frac{\partial p}{\partial r} = 0 & (0 \le z \le z_{\max}) & \textit{(symmetry)} \\
\text{On } z = z_{\max} & p = p_0 & (0 \le r \le r_{\max}) & \\
\text{On } r = r_{\max} & p = p_0 & (0 \le z \le z_{\max}) & \textit{(infinity)}
\end{array} \tag{3}$$

In addition we have internal conditions at the material interfaces that oxygen partial pressure and flux are equal

$$\begin{array}{llcl}
\text{On } z = z_e & p_{z=z_e-} & = & p_{z=z_e+} \\
 & P_e\left(\frac{\partial p}{\partial z}\right)_{z=z_e-} & = & P_m\left(\frac{\partial p}{\partial z}\right)_{z=z_e+} \\
\text{On } z = z_e + z_m & p_{z=z_e+z_m-} & = & p_{z=z_e+z_m+} \\
 & P_m\left(\frac{\partial p}{\partial z}\right)_{z=z_e+z_m-} & = & P_s\left(\frac{\partial p}{\partial z}\right)_{z=z_e+z_m+}
\end{array} \tag{4}$$

The boundary conditions on $z = 0$ ensure that there is a discontinuity in the first derivative of p at the point $(r_c, 0)$, which requires careful treatment, described later.

Once the partial pressure has been calculated the quantity of interest, the current, can be calculated using

$$i(t) = 2\pi n F P_e \int_0^{r_c} \left(\frac{\partial p}{\partial z}\right)_{z=0} r\,dr \tag{5}$$

where F is the Faraday constant, n (usually $= 4$) is the number of electrons per molecule transferred in the reduction reaction and P_e is the permeability of the membrane ($= \alpha_e D_e$).

2.2 Numerical Solution

We require a numerical solution of equation(1) in the finite region $[0, r_{\max}] \times [0, z_{\max}]$ for time $0 \le t \le T$ subject to the initial conditions (2), boundary conditions (3), and interface conditions (4). The distances $r_{\max}$, $z_{\max}$ are chosen so that the diffusion front has not reached the boundary at the time of interest, the boundary conditions are therefore time-dependent. A regular rectangular mesh is superimposed on the finite region and to avoid the problem of the singularity of the term $\frac{1}{r}\frac{\partial p}{\partial r}$ at $r = 0$, we choose the mesh lines $r_i = (i + 1/2)\,\Delta r$, $(i = 0, 1, \ldots, N)$ in the r–direction, with uniform spacing $\Delta r = r_c/n_r$. In the z–direction the mesh spacing Δz_l with $l = e, m, s$, is allowed to vary within the three layers with mesh lines at

$$\begin{array}{llll} \text{electrolyte} & z = j\Delta z_e & (0 \le j \le m_e) & \\ \text{membrane} & z = z_e + j\Delta z_m & (0 \le j \le m_m) & (6) \\ \text{sample} & z = z_e + z_m + j\Delta z_s & (0 \le j \le m_s), & \end{array}$$

where $m_l = z_l/\Delta z_l \quad l = e, m, s$.

2.3 Sequential Solution

The ADI method for solving parabolic p.d.e.'s was first suggested by Peaceman and Rachford [3] who introduced an intermediate timestep into the usual implicit Crank–Nicholson scheme, allowing the solution to be obtained by solving only tridiagonal systems in alternating directions at each half step, whilst remaining unconditionally stable and second order accurate. This property of unconditional stability is very important in the electrode problem since we wish to study the behaviour of the sensor over long time periods.

Using a timestep Δt, we use the ADI method to find an approximate solution ϕ_{ij}^n to the true solution $p((i+\frac{1}{2})\Delta r, j'\Delta z', n\Delta t)$ with $i = 0, \ldots, N$ and $j = 0, \ldots, M$, (where $M = m_e + m_m + m_s$ and $j'\Delta z'$ is a convenient notation for the non–uniform z–mesh), which, with the intermediate step at $(k + 1/2)$, $k = 0, \ldots, n-1$, can be written as the pair of equations

$$\left[1 - \frac{\nu_r^{(l)}}{2}\left(\delta_r^2 + \frac{1}{i+\frac{1}{2}}\Delta_{0r}\right)\right]\phi_{i+\frac{1}{2},j}^{k+\frac{1}{2}} = \left[1 + \frac{\nu_z^{(l)}}{2}\delta_z^2\right]\phi_{i+\frac{1}{2},j}^{k} \tag{7}$$

$$\left[1 - \frac{\nu_z^{(l)}}{2}\delta_z^2\right]\phi_{i+\frac{1}{2},j}^{k+1} = \left[1 + \frac{\nu_r^{(l)}}{2}\left(\delta_r^2 + \frac{1}{i+\frac{1}{2}}\Delta_{0r}\right)\right]\phi_{i+\frac{1}{2},j}^{k+\frac{1}{2}} \tag{8}$$

where Δ_{0r}, δ_r, δ_z are the usual central difference operators in the r– and z– directions and $\nu_r^{(l)} = D_l\frac{\Delta t}{\Delta r^2}$, $\nu_z^{(l)} = D_l\frac{\Delta t}{\Delta z'^2}$, in the three layers $l = e, m, s$.

The Dirichlet boundary conditions are imposed initially. Simple first order central difference approximations for the Neumann boundary conditions are incorporated into the tridiagonal systems, as are the approximations for the interface conditions when solving in the z–direction (cols). When solving in the r–direction (rows), the interface conditions are imposed after solving for all other rows.

At each half–step equations (7) and (8) are in the form

$$a_\nu x_{\nu-1} + b_\nu x_\nu + c_\nu x_{\nu+1} = d_\nu \text{ for } \nu = 1, \ldots, \kappa \tag{9}$$

which is a tridiagonal sytem $A\mathbf{x} = \mathbf{d}$, say, for each row in eqn.(7) where $\kappa = N - n_r$ for row 0 and N for rows 1 to $M-1$, and for each col in eqn.(8) with $\kappa = M - 1$ for cols 0 to $n_r - 1$ and M for cols n_r to $N-1$,

$$A = \begin{pmatrix} b_1 & c_1 & & 0 \\ a_2 & \ddots & \ddots & \\ & \ddots & \ddots & c_\kappa \\ 0 & & a_\kappa & b_\kappa \end{pmatrix}. \tag{10}$$

We could solve each tridiagonal system, using for example the Thomas algorithm, but since A is a constant matrix in time, it is much more efficient to decompose each tridiagonal matrix into upper (U) and lower (L) triangular matrices such that $A = LU$, just

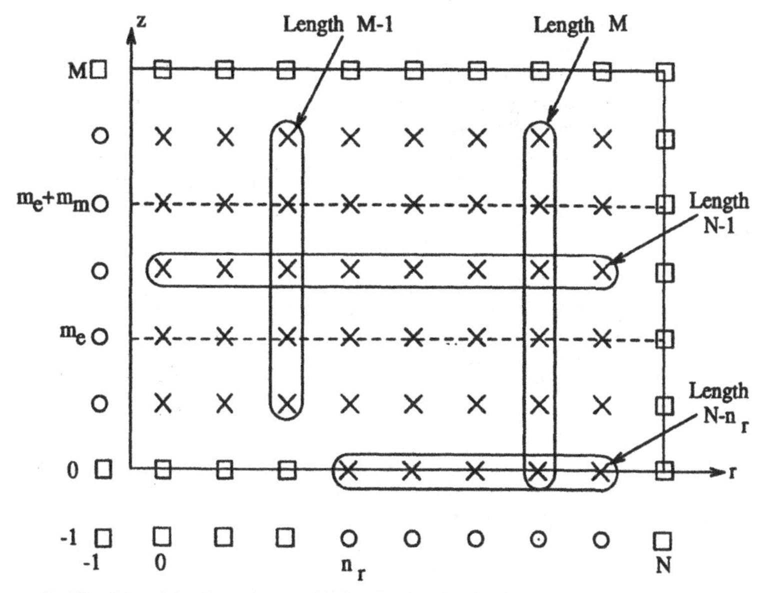

Figure 2: $N+2$ by $M+2$ mesh on which solution is obtained using ADI method, which solves at step k for vectors of length N or $N-n_r$ parallel to the r–axis, then at step $k+1/2$ for vectors of length $M-1$ or M parallel to the z–axis. Crosses represent the points at which solution is to be found, squares are Dirichlet b.c.'s, circles are the fictitious points for Neumann b.c.'s

once at the start of the calculation, store them in vector form, and use them at each half time step to obtain the solution. It can easily be verified that

$$L = \begin{pmatrix} 1 & & & 0 \\ e_2 & \ddots & & \\ & \ddots & \ddots & \\ 0 & & e_\kappa & 1 \end{pmatrix} \quad U = \begin{pmatrix} f_1 & g_1 & & 0 \\ & \ddots & \ddots & \\ & & \ddots & g_{k-1} \\ 0 & & & f_\kappa \end{pmatrix} \tag{11}$$

with

$$f_1 = b_1, \quad e_i = a_i/f_{i-1} \quad g_i = c_i \quad i = 1, \ldots, \kappa \ . \tag{12}$$

L and U can both be stored in just 3 vectors of length κ. Once in the form $LU\mathbf{x} = \mathbf{d}$, by writing $\mathbf{z} = U\mathbf{x}$ we may solve the sytem $L\mathbf{z} = \mathbf{d}$ by forward substitution using

$$z_1 = d_1, z_i = di - e_i z_{i-1}, \text{ for } i = 2, \kappa \tag{13}$$

and then obtain our solution $\mathbf{x}$ from $U\mathbf{x} = \mathbf{z}$ by backward substitution, using

$$x_\kappa = z_\kappa/f_\kappa, x_{\kappa-i} = z_{\kappa-i} - g_{\kappa-i}x_{\kappa-i+1}/f_{\kappa-i}, \text{ for } i = 1, \kappa - 1 \tag{14}$$

The forward sweep requires just 1(add)+1(mult) per mesh point, the backward sweep requires just 1(add)+1(mult)+1(div) per mesh point, and the right hand sides in equations(7) and (8) need 2(add)+4(mult), resulting in a highly efficient algorithm both in terms of calculation and storage, but by virtue of the sweeping backwards and forwards across the whole mesh in different directions at each half time step, one which seems inherently sequential.

The singularity at the electrode edge mentioned earlier, is dealt with by using the ADI solution at points distant from the singularity to calculate a locally valid truncated series solution, which has been described elsewhere [4]. The main point to note is that it requires the calculation of a matrix pseudo–inverse.

A sequential FORTRAN program was written to solve the problem on the Oxford University VAX Cluster using the algorithm given below. Without graphics routines, this code is approximately 1000 lines and uses subroutines from the NAg library to implement the singularity correction. A few test runs demonstrated that a typical case to obtain current values for a particular parameter set would take roughly 60 minutes CPU time, which is impractical given the size of the parameter space and the computing time available.

SEQUENTIAL ALGORITHM
Define an $N+2$ by $M+2$ array including fictitious mesh lines at $i=j=-1$
Initialise all variables
Calculate and store in 12 vectors the required
  LU decompositions (i)–(iv) in the r–direction.
    (i) On $j=0$
    (ii) For $j=1$ to m_e-1 (electrolyte)
    (iii) For $j=m_e+1$ to m_e+m_m-1 (membrane)
    (iv) For $j=m_e+m_m+1$ to $M-1$ (sample)
Calculate and store in 6 vectors the required
  LU decompositions (a)–(b) in the z–direction.
    (a) On $i=0$ to n_r-1 (above cathode)
    (b) For $i=n_r$ to $M-1$ (above glass rod)
Do timestepping :
    (1) Calculate rhs of equation (7)
    (2) Solve at timestep $k+1/2$ using (i)–(iv) in r-dirn for rows
        by sweeping forwards then backwards across each row.
    (3) Impose interface conditions to give rows m_e, m_e+m_m
    (4) Calculate rhs of equation(8)
    (5) Solve at timestep $k+1$ using (a)–(b) in z-dirn
        by sweeping up then down each col.
    (6) Do singularity correction (uses a NAg routine)
    (7) Calculate current
Repeat from (1) for $k+3/2$

2.4 Parallel Solution

At the time that we first attempted a parallel solution we had available to us a 40 transputer network of Inmos T800 transputers each with 1Mb of memory and the Inmos TDS, toolset and Occam compiler [5]. An early FORTRAN compiler was also available but generated floating point code which ran 3 to 6 times slower than the equivalent Occam code, and was therefore not considered further. Our first attempt at a parallel solution used the hopscotch scheme of Gourlay [6], which is an explicit algorithm and is therefore easier to implement in parallel than the implicit ADI method. Unfortunately, for our boundary conditions, hopscotch became unstable as the timestep was increased resulting in the parallel code being only marginally faster than the sequential code. A parallel version of the ADI algorithm was clearly the next step.

At first inspection the sequential ADI algorithm shows no obvious method of decomposition into a parallel algorithm. Both the forward and backward sweeps to obtain the solution require us to calculate across entire rows to solve each tridiagonal system, then to switch direction and sweep forwards and backwards across the columns. We did not wish to go to a substructuring algorithm such as that described in section 3, since the efficiency of both storage and calculation is achieved by performing the decompositions only once. Nor was an attempt to decompose the mesh onto the transputers as groups of rows or columns feasible since communication costs would be

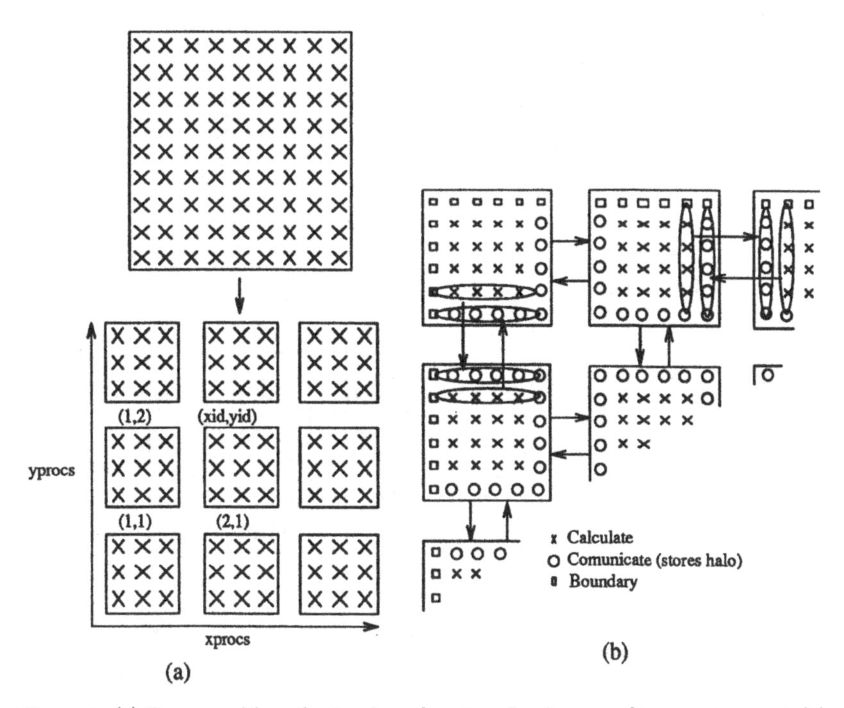

Figure 3: (a) Decomposition of a 9×9 mesh onto a 3×3 array of transputers, and (b) Storage and communication for a 4×4 submesh.

prohibitive. Instead we configure the transputers as an $xprocs \times yprocs$ rectangular array, with each processor connected to each of its neighbouring processors by up to 8 uni–directional links, northin, northout, southin,...etc. We then do the obvious spatial decomposition of our rectangular $(N+2) \times (M+2)$ mesh as rectangular submeshes of size $(N+2)/xprocs(= xnodes) \times (M+2)/yprocs(= ynodes)$ as illustrated in fig.3(a). Rather than storing just the sub mesh for which we wish to calculate the solution on a particular processor, we also store the values of the edge mesh points from all neighbouring processors in a 'halo' as shown in fig.3(b), since we require the values of immediate neighbours in calculating the right hand side of equations (7) and (8).

The problem is solved using a single Occam code which is loaded onto each processor. Each processor determines the position of its $xnodes \times ynodes$ submesh in the overall mesh from its coordinates (x_{id}, y_{id}) in the processor array, and sets up and initialises the problem accordingly. The decompositions in the r–direction are then started on the first processor in each row, sweeping forwards across both the submesh and the processor array, communicating the required last value of $\mathbf{f}$ on a processor, which becomes the first value of $\mathbf{f}$ on the adjacent processor, and similarly in the z–direction. Each processor stores only those parts of the vectors making up L and U that it requires to calculate the solution on its own submesh. The code has the same structure as that shown below which calculates $\mathbf{z}$.

The timestepping algorithm uses a similar idea, as illustrated in Fig.4. and in the Occam code below. At the first half step we begin the forward sweep to find $\mathbf{z}$ (zx in the code) for the first (sub-) mesh line in the r–direction on each processor in the first processor column. When these processors reach the end of their first r–line, they each communicate their final value of $zx[xnodes]$ to the adjacent processor in the second column, which each begin their first mesh line in the r–direction, whilst the first processors start their second and so on. The calculation proceeds across the array until

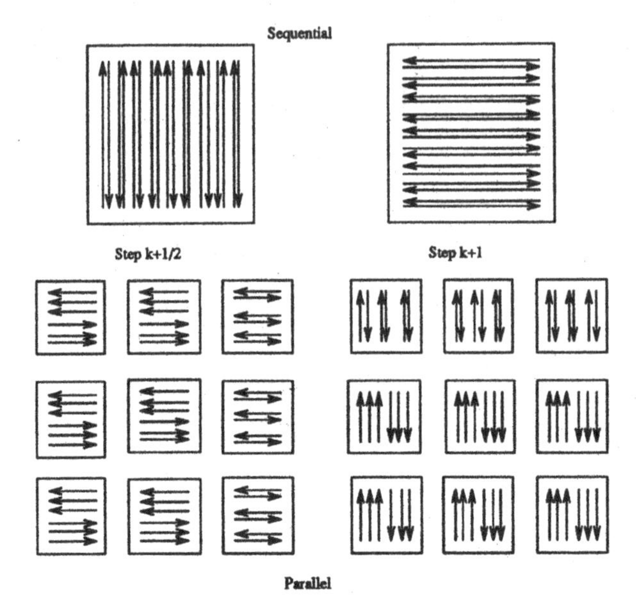

Figure 4: The sequential algorithm sweeps forwards then backwards across each r–line in turn, then repeats for z. The parallel algorithm sweeps forwards along all r–lines on the first $p-1$ processors, forwards and backwards on th p th processor, then backwards along each r–line on the first $p-1$ processors, then repeats for z.

all processor columns are working (after $(xprocs - 1)$ stages).

```
IF
  xid=1
    SEQ
      zx[1]:=dx[1]
      SEQ i=2 FOR xnodes-1
        SEQ
          zx[i]:=dx[i]-(ex[i]*zx[i-1])
      eastout ! zx[xnodes-1]
  TRUE
    SEQ
      westin ? zx[0]
      SEQ i=1 FOR xnodes
        SEQ
          zx[i]:=dx[i]-(ex[i]*zx[i-1])
      IF
        xid = xprocs
          SKIP
        TRUE
          eastout ! zx[xnodes-1]
```

The first $(xprocs - 1)$ processors in each processor row calculate the forward sweep for z for each r–mesh line first (storing it over the old solution), whilst the last processor sweeps forwards then backwards along each r–line in turn. By counting the number of completed r–lines on the last processor and buffering the information transmitted by the next to last, we can ensure that it is always ready to receive a z value or transmit

Proc.																
(1,1)	F	F	F	W	W	B	W	B	W	B	U	U	U	W	W	...
(1,2)	W	F	F	F	B	W	B	W	B	U	U	U	W	W	D	...
(1,3)	W	W	F	B	F	B	F	B	U	U	U	U	W	W	D	...
(2,1)	F	F	F	W	W	B	W	B	W	B	W	U	U	U	D	...
(2,2)	W	F	F	F	B	W	B	W	B	W	U	U	U	D	W	...
⋮																

Table 2: Summary of work of processors in a 3×3 processor array, each with a 3×3 sub mesh. F=forwards sweep, B=backwards, U=upwards, D=downwards and W=wait.

a solution value as required. Once the next to last processor has finished all of its mesh lines in the r–direction, it inputs the last element of the solution vector from the last processor, and begins the backwards sweep across each mesh line in turn, until all lines in the r–direction are completed. Each processor then communicates the vectors of edge values into the haloes (see Fig.3(b)), and repeats the process in the z–direction. Since we need only communicate edge values, on an $n \times n$ sub mesh, communication is $O(n)$ whilst calculation is $O(n^2)$, so that provided we have a large enough sub mesh, the algorithm should be efficient in parallel. Table 2 summarises the work load of each processor as the calculation progresses.

PARALLEL ALGORITHM
Determine position in processor array and initialise submesh accordingly
Calculate LU decompositions in both directions
Do timestepping
(1) Calculate rhs of equation (7)
(2) Solve at timestep $k + 1/2$ in r–direction
(3) Communicate halo
(4) If interface is part of submesh, impose conditions
(5) Calculate rhs of equation (8)
(6) Solve at timestep $k + 1$ in z–direction
(7) Communicate halo
(8) If singularity is part of submesh, do correction.
(9) If submesh contains part of electrode, calculate that part of current
Repeat from (1) for $k + 3/2$
Communicate results

2.5 Results

Table 3. shows the efficiency figures for various mesh sizes and numbers of processors. Since we use the same algorithm as the sequential case, these figures are a true comparison of the speedup we can obtain by moving to a parallel machine. Had we gone to a different algorithm, we might have been able to achieve better efficiency figures but it is likely that the code would have run more slowly in real time (for example the alternative tridiagonal solver that we considered, the Wang algorithm, takes at least twice as many operations as our method). On one processor the calculation takes approximately $5 \times 10^{-5}s$ per mesh point per complete timestep, which is about 0.5 Mflops, with about 5 Mflops on 16 processors (only 16 of the original 40 processors are currently available). On problems of practical size, we obtain efficiency of about 65% on 16 processors, which gives a factor of 10 speedup over 1 processor, and a factor of about 5 over the VAX. The code can be configured to run on any number of processors, and table 3 demonstrates that if more processors becomes available greater speedups are attainable.

Mesh	Efficiency			
No. Procs.	2	4	8	16
50 × 50	71%	63%	55%	46%
100 × 100	74%	66%	61%	55%
200 × 200	75%	68%	66%	62%
400 × 400	–	69%	68%	66%
800 × 800	–	–	–	68%

Table 3: Efficiency of the parallel code.

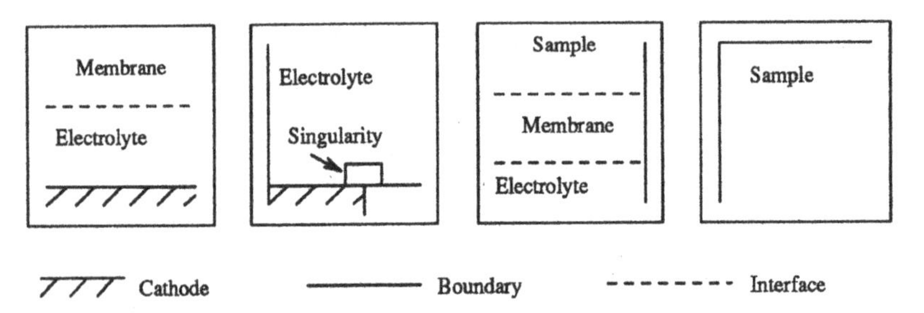

Figure 5: Four examples of the 144 different types of possible sub mesh.

2.6 Difficulties in Parallel Solution

The greatest problems in producing an Occam code to implement the above algorithm were not caused by the sequential nature of the ADI method, but rather by the need to have a single occam code which could run on any number of processors. The reason for this is the nature of the boundary conditions for this problem which result in a complicated decision structure in deciding the sub mesh, which in turn determines which parts of the code to run and the amount of communication that is necessary. Fig.5 shows a few of the 144 different possible sub meshes. In total the code is approximately 3000 lines of Occam and it took about 6 months to get it working in its final form, with the advantage of some previous parallel programming experience. The total includes about 200 lines to calculate the pseudo-inverse of a matrix (sequentially) since we do not have the advantage of a standard subroutine library. It is difficult to estimate how long the sequential FORTRAN code took to develop since it was developed in parallel with the theory and modelling. However we might estimate that it would take an experienced programmer about 6 weeks to design and code the algorithm. The debugging tools available on the transputer network were inadequate and we developed our own techniques as the code was built.

3 A Navier-Stokes code for a Driven Cavity

The second problem considered took a different approach, that of taking an existing efficient sequential FORTRAN code developed by another member of the department, and attempting to create a parallel version of it [7]. In his DPhil thesis [1], Shah considered the laminar incompressible Navier-Stokes equations in a two-dimensional Driven Cavity. The equations were used in the form

$$\frac{\partial \rho u_i}{\partial x_i} = 0, \tag{15}$$

$$\frac{\partial \rho u_i u_j}{\partial x_j} - \frac{\partial \rho u_j}{\partial x_j} u_i = -\frac{\partial p}{\partial x_i} + \frac{\partial}{\partial x_j}\left[\mu\left(\frac{\partial u_i}{\partial x_i} + \frac{\partial u_j}{\partial x_i}\right)\right]. \tag{16}$$

Let

$$S^{u_i} = \frac{\partial}{\partial x_j}\left[\mu \frac{\partial u_j}{\partial x_i}\right], \qquad i \neq j.$$

The equations were discretised using the Marker and Cell (MAC) grid [8]. The cell structure of this grid is shown in Fig. 6.

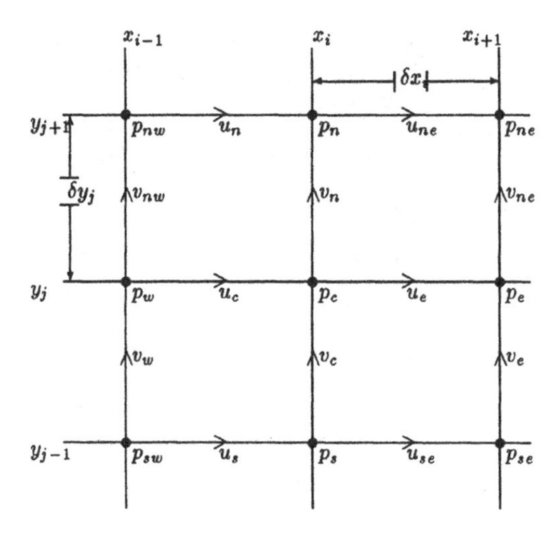

Figure 6: A typical MAC cell

The discretisation led to the following equations (using $u = u_1$, $v = u_2$)

$$A_c^u u_c = A_n^u u_n + A_s^u u_s + A_e^u u_e + A_w^u u_w + S_c^u - (p_c - p_w)/\delta x_{i-1}, \tag{17}$$

$$A_c^v v_c = A_n^v v_n + A_s^v v_s + A_e^v v_e + A_w^v v_w + S_c^v - (p_c - p_s)/\delta y_{j-1}, \tag{18}$$

$$A_e^c u_e - A_w^c u_w + A_n^c v_n - A_s^c v_s = 0. \tag{19}$$

The coefficients are

$$A_n^u = \frac{1}{\Delta y_j}\left(\frac{\mu}{\delta y_j} - \frac{(\rho v)_A}{2}\right), \qquad A_s^u = \frac{1}{\Delta y_j}\left(\frac{\mu}{\delta y_{j-1}} + \frac{(\rho v)_B}{2}\right),$$
$$A_e^u = \frac{1}{\delta x_{i-1}}\left(\frac{2\mu}{\Delta x_i} - \frac{(\rho u)_A}{2}\right), \qquad A_w^u = \frac{1}{\delta x_{i-1}}\left(\frac{2\mu}{\Delta x_{i-1}} + \frac{(\rho u)_B}{2}\right),$$
$$A_n^v = \frac{1}{\delta y_{j-1}}\left(\frac{2\mu}{\Delta y_j} - \frac{(\rho v)_a}{2}\right), \qquad A_s^v = \frac{1}{\delta y_{j-1}}\left(\frac{2\mu}{\Delta y_{j-1}} + \frac{(\rho v)_b}{2}\right),$$
$$A_e^v = \frac{1}{\Delta x_i}\left(\frac{\mu}{\delta x_i} - \frac{(\rho u)_a}{2}\right), \qquad A_w^v = \frac{1}{\Delta x_i}\left(\frac{\mu}{\delta x_{i-1}} + \frac{(\rho u)_b}{2}\right),$$
$$A_e^c = \frac{\rho}{2\Delta x_i} = A_w^c, \qquad A_n^c = \frac{\rho}{2\Delta y_j} = A_s^c,$$
$$A_c^u = A_n^u + A_s^u + A_e^u + A_w^u, \qquad A_c^v = A_n^v + A_s^v + A_e^v + A_w^v,$$

$$(\rho v)_A = \frac{\rho}{2}(v_n + v_{nw}), \qquad (\rho v)_B = \frac{\rho}{2}(v_c + v_w),$$
$$(\rho u)_A = \frac{\rho}{2}(u_c + u_e), \qquad (\rho u)_B = \frac{\rho}{2}(u_c + v_w),$$
$$(\rho v)_a = \frac{\rho}{2}(v_n + v_c), \qquad (\rho v)_b = \frac{\rho}{2}(v_c + v_s),$$
$$(\rho u)_a = \frac{\rho}{2}(u_{se} + u_e), \qquad (\rho u)_b = \frac{\rho}{2}(u_c + u_s),$$
$$\Delta x_i = \frac{1}{2}(x_{i+1} - x_{i-1}), \qquad \Delta y_j = \frac{1}{2}(y_{j+1} - y_{j-1}),$$
$$\delta x_i = x_{i+1} - x_i, \qquad \delta y_j = y_{j+1} - y_j.$$

3.1 Method of Solution

The non-linear equations (17) to (19) were solved using a multigrid method on a series of staggered grids. For non-linear problems a global linearisation process and a linear multigrid method can be used. Alternatively, a local linearisation of the non-linear equations with a smoothing method leads to a non-linear multigrid method, called the full-approximation scheme (FAS). Details can be found in [9, 10, 11].

3.2 The Multigrid Algorithm

Consider the non-linear problem on a mesh Ω_l

$$N_l u_l = f_l \tag{20}$$

where N_l is a non-linear operator. The non-linear multigrid algorithm proceeds as follows

1. Perform ν_1 pre-relaxation iterations of the smoothing method to give $\overline{u}_l$.

2. Calculate the defect $\qquad d_l = f_l - N_l\overline{u}_l$.

3. Restrict $\overline{u}_l$ and d_l to the next coarser grid.

$$\overline{u}_{l-1} = \hat{R}_l^{l-1}\overline{u}_l \qquad\qquad d_{l-1} = R_l^{l-1}d_l.$$

   $\hat{R}_l^{l-1}$ and R_l^{l-1} may be different as they operate on different spaces.

4. The defect equation becomes

$$N_{l-1}(\overline{u}_{l-1} + \overline{v}_{l-1}) - N_{l-1}\overline{u}_{l-1} = d_{l-1}$$

   or

$$N_{l-1}w_{l-1} = d_{l-1} + N_{l-1}\overline{u}_{l-1}$$

   where

$$\overline{v}_{l-1} = w_{l-1} - \overline{u}_{l-1}.$$

   Using iterations of ν_1 the smoothing method, find an approximation $\tilde{w}_{l-1}$ to w_{l-1} and hence find $\tilde{v}_{l-1}$, an approximation to $\overline{v}_{l-1}$.

5. Continue restricting and smoothing until the coarsest grid is reached on which

$$N_0 w_0 = d_0 + N_0 \overline{u}_0.$$

This is not solved exactly but an approximation $\tilde{w}_0$ to w_0 is found by performing ν_2 iterations of the smoothing method.

6. Calculate $$\overline{v}_0 = \tilde{w}_0 - \overline{u}_0.$$

7. Prolong to the next finer grid

$$\hat{v}_1 = P_0^1 \overline{v}_0.$$

8. Calculate

$$w_1 = \tilde{w}_1 + \hat{v}_1 \qquad \overline{u}_1^{new} = \overline{u}_1 + \hat{v}_1.$$

Smooth

$$N_1 w_1 = d_1 + N_1 \overline{u}_1^{new}$$

using ν_3 iterations, to give a new $\tilde{w}_1$ and hence $\tilde{v}_1$.

9. Continue prolonging ($\hat{v}_j = P_{j-1}^j \tilde{v}_{j-1}$) and smoothing ($\nu_3$ times) to find $\tilde{v}_j$, until the finest grid is reached.

10. Compute $\overline{u}_l + \hat{v}_l$ on Ω_l

$$u_l^{new} = \overline{u}_l + \hat{v}_l.$$

11. Perform ν_3 smoothing relaxation sweeps on the new solution, u_l^{new}.

This algorithm is repeated until u_l converges.

3.3 Restriction and Prolongation Operators

Consider two grids, a coarse grid Ω_H and a fine grid Ω_h. Denote the coarse grid values by U, V, P and the fine grid ones by u, v, p. Let $u_{i,j} = u_{i+\frac{1}{2},j}$, $v_{i,j} = v_{i,j+\frac{1}{2}}$. If, on the coarse grid,

$$X = IH \qquad Y = JH$$

then on the fine grid

$$x = ih \qquad y = jh$$

where

$$i = 2I - 1 \qquad j = 2J - 1.$$

Both the restriction operators are defined by

$$U_{I,J} = \tfrac{1}{2}[u_{i,j} + u_{i,j-1}], \quad V_{I,J} = \tfrac{1}{2}[v_{i,j} + v_{i-1,j}],$$

$$P_{I,J} = \frac{1}{4}[p_{i,j} + p_{i,j-1} + p_{i-1,j} + p_{i-1,j-1}].$$

Using bilinear interpolation, the prolongation relations are derived. Each coarse grid node is used to derive four fine grid values. For the u-velocity

$$u_{i,j} = \tfrac{1}{4}[3U_{I,J} + U_{I,J+1}], \quad u_{i,j+1} = \tfrac{1}{4}[U_{I,J} + 3U_{I,J+1}],$$

$$u_{i+1,j} = \frac{1}{8}[3U_{I,J} + U_{I,J+1} + 3U_{I+1,J} + U_{I+1,J+1}],$$

$$u_{i+1,j+1} = \frac{1}{8}[U_{I,J} + 3U_{I,J+1} + U_{I+1,J} + 3U_{I+1,J+1}].$$

The v-velocity relations are found by the equivalent relations obtained by rotating the co-ordinates by 90°. For example

$$v_{i,j} = \frac{1}{4}[3V_{I,J} + V_{I+1,J}].$$

As the pressure is calculated at the nodes the weighting is different and the relations are

$$p_{i,j} = \frac{1}{16}[9P_{I,J} + 3P_{I+1,J} + 3P_{I,J+1} + P_{I+1,J+1}],$$

$$p_{i+1,j} = \frac{1}{16}[3P_{I,J} + 9P_{I+1,J} + P_{I,J+1} + 3P_{I+1,J+1}],$$

$$p_{i,j+1} = \frac{1}{16}[3P_{I,J} + P_{I+1,J} + 9P_{I,J+1} + 3P_{I+1,J+1}],$$

$$p_{i+1,j+1} = \frac{1}{16}[P_{I,J} + 3P_{I+1,J} + 3P_{I,J+1} + 9P_{I+1,J+1}].$$

These relations are slightly modified near the boundaries to avoid using boundary pressure values. The above prolongation relations are used to prolong the defects.

3.4 The Smoothing Method

In [1], Shah derived a procedure called Symmetrical Coupled Gauss Seidel/Line Solver or (SCGS/LS). Equations (17), (18) and (19) are written, assuming that $\delta x_i = \delta x$ etc., as

$$\begin{aligned}(A_c^u)_{i,j}u_{i,j} &= (A_n^u)_{i,j+1}u_{i,j+1} + (A_s^u)_{i,j-1}u_{i,j-1} + (A_e^u)_{i+1,j}u_{i+1,j} \\ &\quad + (A_w^u)_{i-1,j}u_{i-1,j} + (S_c^u)_{i,j} + (p_{i,j} - p_{i+1,j})/\delta x,\end{aligned}$$

$$\begin{aligned}(A_c^v)_{i,j}v_{i,j} &= (A_n^v)_{i,j+1}v_{i,j+1} + (A_s^v)_{i,j-1}v_{i,j-1} + (A_e^v)_{i+1,j}v_{i+1,j} \\ &\quad + (A_w^v)_{i-1,j}v_{i-1,j} + (S_c^v)_{i,j} + (p_{i,j} - p_{i,j+1})/\delta y,\end{aligned}$$

$$(u_{i,j} - u_{\cdot-1,j})/\delta x + (v_{i,j} - v_{i,j-1})/\delta y = 0.$$

Consider the equations for $u_{i,j}$, $u_{i-1,j}$, $v_{i,j}$, $v_{i,j-1}$

$$\begin{aligned}&(A_c^u)_{i-1,j}u_{i-1,j} = F_{i-1,j}^u, \qquad (A_c^u)_{i,j}u_{i,j} = F_{i,j}^u,\\ &(A_c^v)_{i,j-1}v_{i,j-1} = F_{i,j-1}^v, \qquad (A_c^v)_{i,j}v_{i,j} = F_{i,j}^v,\\ &(u_{i,j} - u_{i-1,j})/\delta x + (v_{i,j} - v_{i,j-1})/\delta y = 0.\end{aligned}$$

where, for example,

$$\begin{aligned}F_{i,j}^u &= (A_n^u)_{i,j+1}u_{i,j+1} + (A_s^u)_{i,j-1}u_{i,j-1} + (A_e^u)_{i+1,j}u_{i+1,j} \\ &\quad + (A_w^u)_{i-1,j}u_{i-1,j} + \mu/\delta x\delta y(v_{i+1,j} - v_{i,j} - v_{i+1,j-1} + v_{i,j-1}) \\ &\quad + (p_{i,j} - p_{i+1,j})/\delta x.\end{aligned}$$

[$(S_c^u)_{i,j}$ is now written in expanded form.] In SCGS/LS the equations are rewritten in terms of the corrections and residuals along a paraxial line (say the y-axis) giving

$$(A_c^u)_{i-1,j}\delta u_{i-1,j} + (\delta p_{i,j} - \delta p_{i-1,j})/h = R_{i-1,j}^u \tag{21}$$

$$(A_c^u)_{i,j}\delta u_{i,j} - (\delta p_{i,j} - \delta p_{i+1,j})/h = R_{i,j}^u \tag{22}$$

$$(A_c^v)_{i,j-1}\delta v_{i,j-1} + \delta p_{i,j}/h = R_{i,j-1}^v \tag{23}$$

$$(A_c^v)_{i,j}\delta v_{i,j} - \delta p_{i,j}/h = R_{i,j}^v \tag{24}$$

$$(\delta u_{i,j} - \delta u_{i-1,j})/h + (\delta v_{i,j} - \delta v_{i,j-1})/h = R_{i,j}^c \tag{25}$$

where δ's are the corrections and $h = \delta x = \delta y$.

δu and δv are eliminated from 21–24 to give

$$\begin{aligned}
\delta u_{i-1,j} &= \left[R^u_{i-1,j} - \frac{(\delta p_{i,j} - \delta p_{i-1,j})}{h}\right] / (A^u_c)_{i-1,j}, \\
\delta u_{i,j} &= \left[R^u_{i,j} + \frac{(\delta p_{i,j} - \delta p_{i+1,j})}{h}\right] / (A^u_c)_{i,j}, \\
\delta v_{i,j-1} &= \left[R^v_{i,j-1} - \delta p_{i,j}/h\right] / (A^v_c)_{i,j-1}, \\
\delta v_{i,j} &= \left[R^v_{i,j} + \delta p_{i,j}/h\right] / (A^v_c)_{i,j}.
\end{aligned}$$

These are substituted into (25) giving the following system of equations to be solved

$$(A^p_c)_{i+1,j}\delta p_{i+1,j} + (A^p_c)_{i,j}\delta p_{i,j} + (A^p_c)_{i-1,j}\delta p_{i-1,j} = R^p_{i,j}, \qquad (26)$$

where

$$\begin{aligned}
&(A^p_c)_{i+1,j} = -\tfrac{1}{h^2 (A^u_c)_{i,j}}, \quad (A^p_c)_{i-1,j} = -\tfrac{1}{h^2 (A^u_c)_{i-1,j}}, \\
&(A^p_c)_{i,j} = \tfrac{1}{h^2}\left[\tfrac{1}{(A^u_c)_{i,j}} + \tfrac{1}{(A^u_c)_{i-1,j}} + \tfrac{1}{(A^v_c)_{i,j}} + \tfrac{1}{(A^v_c)_{i-1,j}}\right], \\
&R^p_{i,j} = R^c_{i,j} - \tfrac{1}{h}\left[\tfrac{R^u_{i,j}}{(A^u_c)_{i,j}} - \tfrac{R^u_{i-1,j}}{(A^u_c)_{i-1,j}} + \tfrac{R^v_{i,j}}{(A^v_c)_{i,j}} - \tfrac{R^v_{i,j-1}}{(A^v_c)_{i,j-1}}\right],
\end{aligned}$$

which is again a tridiagonal system in the form of equation(9). This time, however, the coefficients vary from iteration to iteration and the decompositions cannot therefore be done only once. Also, the order of solution of the rows (or columns) is important since it affects the rate of convergence of the method, and in this case we solve each row successively moving up the grid from the boundary (and across for columns). The SCGS/LS algorithm is then

1. For $i = 1, ..., N$ solve the set of tri-diagonal equations (26) to find the pressure corrections δp.

2. Substitute the δp's into the equations for δu and δv to find the velocity corrections along the j line.

3. These corrections are added to the latest calculated values of u, v, p for that line, the coefficients are updated and then the corrections for the next column are calculated, until the whole domain has been swept. A local linearisation is carried out at each stage.

4. Multiples of the corrections are added to the previous approximations to get the new approximations

$$\begin{aligned}
u^{new}_{i,j} &= u^{old}_{i,j} + \omega_u \delta u_{i,j}, \quad v^{new}_{i,j-1} = v^{old}_{i,j-1} + \omega_v \delta v_{i,j-1}, \\
v^{new}_{i,j} &= v^{old}_{i,j} + \omega_v \delta v_{i,j}, \quad p^{new}_{i,j} = p^{old}_{i,j} + \omega_p \delta p_{i,j},
\end{aligned}$$

   where ω_u, ω_v and ω_p are relaxation parameters.

5. The domain is swept again, using the same algorithm, with i fixed instead of j and slightly different equations. An alternating sweep direction (left-to-right and then right-to-left) can also be used.

The form of the Navier-Stokes equations used has ensured that the system is diagonally dominant. Consider the value $p_{i,j}$. A sweep of the smoothing algorithm in each direction also involves $p_{i-1,j}, p_{i+1,j}, p_{i,j-1}, p_{i,j+1}$. However, $p_{i,j}$ is updated twice (once for each sweep direction) and the other p values are updated once.

3.5 The Parallel Code

The algorithms described above were shown by Shah to be very effective on a sequential machine. However, to solve a very small case still takes around 17.3s CPU time on the VAX cluster. We therefore decided that it would be a useful exercise to attempt to adapt Shah's code to run on the transputer network. Since the original code is in FORTRAN, and we now have available a FORTRAN compiler which generates better floating point code, we decided to produce a parallel FORTRAN version, and make use of as much of the original code as possible. In order to get a feel for the likely difficulties involved, Shah's code was first adapted to run on just two transputers.

The domain was divided down the middle. Each transputer held half of the grid plus the first column of u, v, p points on the other side of the dividing line (those marked ○), as shown in figure 7. The restriction and prolongation operators are easily parallelised

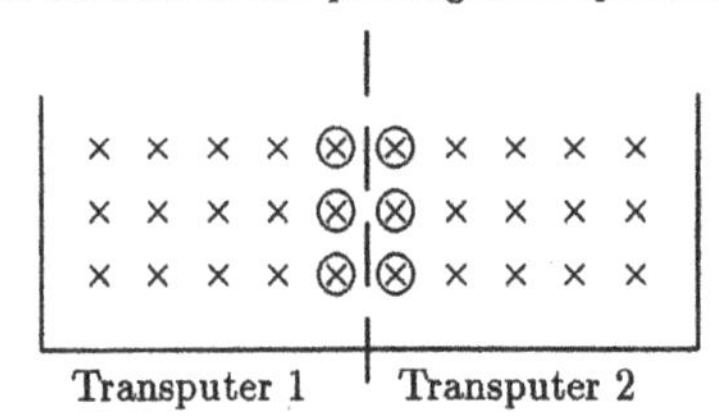

Figure 7: Division of the Grid for Two Transputers

since each value that is updated only needs to know the values of surrounding points on the previous grid. There is no need to transmit any values until the end of the operation.

However, the smoothing method is highly sequential. The local nature of the linearisation means that the coefficients for the second line require updated values from the first line. When the grid is swept column by column, from left-to-right, the coefficients of column m can not be calculated until column $m-1$ has been finished. Transputer 2 can not work on its first column until transputer 1 has finished its last column. This leads to inefficiency since it is not possible to ensure that both transputers work for most of the time. The problem increases with the number of processors. When the grid is swept row by row, from bottom-to-top, the solution of the tri-diagonal system of equations causes a problem. Transputer 1 does the forward substitution on the values of its first row. Then it has to wait for transputer 2 to do the forward and backward substitutions on its first row, before transputer 1 can do the backward substitution on the same row. Transputer 1 can not work on the next line whilst it is waiting. The tri-diagonal solver takes up little time in the two transputer case but the problem would become worse if there were more transputers in each row. This problem can be overcome by using a different tri-diagonal solver, for example the Wang Algorithm [12], which uses a substructuring of the usual Gaussian elimination method to obtain a much smaller tridiagonal system which must be solved in parallel, the subsystems can then be solved sequentially on each processor. This method can be shown to be optimally efficient on sufficiently large systems, and although it requires twice as many operations as the usual sequential algorithm, the improved efficiency and communications requirements result in a faster parallel implementation on a large processor array.

3.6 Global Linearisation

The Wang algorithm cannot, however, overcome the problems caused by the necessity to sweep each row in turn for the SCGS/LS method. This cannot be implemented efficiently in parallel since each complete row of the mesh must be completed before the next can begin, and as each mesh row is strung across several processors, only one column of processors can be working at any time. Shah's SCGS/LS must therefore be dropped and a global linearisation used instead. The equations remain the same, except that old values of u ,v and p are used instead of new ones, and the values are updated at the end of the sweep instead of at the end of each line. This method was found to require only a few more sweeps than the local linearisation (in fact, for large cases, the number of sweeps was identical), when the relaxation parameters were altered. This is similar to using the Jacobi method instead of the Gauss Seidel method to solve a system of linear equations. The smoothing method becomes

- All the coefficients are calculated,
- The tri-diagonal solver is implemented in a staggered manner,
- All the values are updated.

The use of all old values means that transputer 2 can work on its columns at the same time as transputer 1 is working on the columns it holds. The staggering of the tri-diagonal solver allows transputer 1 to work on the next row whilst transputer 2 is doing the forward and backward substitutions. If there are n^2 transputers, all of them can be working during the smoothing method and, as we now have an efficient parallel implementation on two transputers, there is every reason to hope that efficiency can be maintained for more processors.

4 Discussion

During the five years in which we have been involved in developing parallel algorithms, there has been very rapid growth in the development of parallel hardware and a wide variety systems is now available. It now seems to us to be necessary to further develop the operating systems and software tools to allow these systems to be used to their full potential. We describe here some of the changes which would help us, in particular, in the development of parallel numerical algorithms.

When we began our work on the electrode problem, the best software tools available to us were the software to support the Occam language which was designed specifically for the transputer. Since then, compilers have been developed for the transputer in the more traditional scientific languages of FORTRAN and C, for which the penalty for not using occam is only a factor of about 1.5. It now seems likely that the advantages of familiarity and portability of these languages will to outweigh the elegance and theoretical rigour of Occam, which is likely to be relegated, at best, to the role of an intermediate language.

In general, the porting of a sequential numerical code to a distributed memory system demands changes at several levels. In most cases, storage arrays must be divided up between processors and basic units of algorithms need to be redesigned or even replaced to allow an efficient parallel implementation. Some of the lower level changes may soon be handled painlessly by use of appropriate libraries, if (as?) these become available, and by (partially) parallelising compilers. Higher level changes will probably continue to demand the re–thinking of algorithm structure and research to ensure that the revised algorithms are as effective as those tuned to sequential systems.

At present, the environment for program development on MIMD machines is hostile, and it is usually advisable to test an algorithm on a sequential machine before moving it to a parallel system. This is partly due to the novelty and variety of the parallel architectures currently available, partly to the small scale on which such architectures are produced and supplied, and partly intrinsic in the complexity of parallel work. Since, particularly for numerical algorithms, the scalable nature of distributed memory systems gives them the potential to handle cases too large for the storage of any feasible sequential machine, part of the development work will need to be done on the parallel system. Currently available MIMD systems usually consist of a network of processors (configured as a rectangular array, tree, hypercube etc.) connected to a host machine. Difficulties in communication between different parts of the network and the host machine appear to be entirely due to the present state of parallel operating systems, and standardisation of communication proceedures for all systems is necessary if parallel codes are to become portable between different types of hardware, or even between different operating systems for the same hardware. Generally there is a need for recognition of the inherent and peculiar difficulties of locating errors in parallel codes and for the appearance of tools for program testing which reflect these difficulties, as adequately as do the equivalent tools for sequential systems.

It is to be hoped that as the technology matures, many of the disincentives to development on parallel systems will be removed. This will have a great effect on the willingness of programmers to commission work on such systems and on the lead times for doing so.

References

[1] Shah, T.M. Analysis of the Multigrid Method, DPhil thesis, Oxford University Computing Laboratory, Oxford, 1989.

[2] Clark, L.C. Monitor and Control of Blood and Tissue Oxygen Tensions, Trans. Am. Soc. Art. Intern. Organs, Vol.2, pp. 41–60, 1956.

[3] Peaceman, D.W. and Rachford, H.H. The Numerical Solution of Parabolic and Elliptic Differential Equations, J. Soc. Indust. Appl. Math., Vol.3, pp. 28–41, 1955.

[4] Gavaghan, D.J. and Rollett, J.S. The Correction of Boundary Singularities in Numerical Simulation of Time-dependent Diffusion Processes at Unshielded Disc Electrodes, J. Electroanal. Chem., Vol. 295, pp. 1–14, 1990.

[5] Inmos. Transputer Development System, Prentice Hall, London, 1988.

[6] Gourlay, G.R. Hopscotch: A Fast Second Order Partial Differential Equation Solver, J. Inst. Maths. Applics., Vol.6, pp. 375–390, 1970.

[7] Gwilliam, C.S. A Parallel Algorithm for the Navier-Stokes Equations in a Driven CavitY, MSc thesis, Oxford University Computing Laboratory, Oxford, 1990.

[8] Harlow, F.H. and Welch, J.E. Numerical Calculation of Time-dependent Viscous Incompressible Flow of Fluid with Free Surface, The Physics of Fluids, Vol.8(12), pp. 2182–2189, Dec 1965.

[9] Brandt, A. Multi-level Adaptive Solutions to Boundary-value Problems, Mathematics of Computation, Vol.31(138), pp. 333–390, April 1977.

[10] Sivaloganathan, S. and Shaw, G. A Multigrid Method for Recirculating Flows, International Journal for Numerical Methods in Fluids, Vol.8, pp. 417–440, 1988.

[11] Stüben, K. and Trottenberg, U. Multigrid Methods: Fundamental Algorithms, Model Problem Analysis and Applications, Springer Verlag Lecture Notes in Mathematics, Vol.960, pp. 1–176, 1981.

[12] Johnsson, S.L. Saad, Y. and Schultz, M.H. Research Report, YALEU/DCS/RR-382, Oct. 1985.

Mixed Parallelism for Single Alternating Group Explicit Method

M. Carmignani, A. Genco, G. Pecorella

Dipartimento di Tecnologia e Produzione Meccanica, Università degli Studi di Palermo, Viale delle Scienze, 90128 Palermo, Italy

ABSTRACT

The paper deals with parallel implementation policies for D.J. Evans S_AGE (Single Alternating Group Explicit) unconditionally stable methods concerning the solution of parabolic problems in two space dimensions.
Space and time mixed parallelisms are investigated with reference to distributed memory computing systems in order to achieve the best relative speedup by means of the best processing elements distribution.
Particular care has been given to startup period minimization together with suitable hypotheses of domain partitioning and of calculation procedural flow.
Some communication strategies are also proposed in terms of data transfer logic, linking resources and their concurrency capabilities.
On these bases, taking into account different parallel system features and different computation policies, a general analytical expression is built allowing to pre-evaluate the performances of different implementations of the S_AGE method.
The speedup evaluations are presented with reference to the parameters determining the actual working environment of the system. They are: the number of time steps, the sampling degree of the domain, the total number of available processors, the distribution of the processors over space and time, the communication capabilities of the processors, the ratio between computation and communication times.

INTRODUCTION

Partial differential equations are much more often employed for the analysis of engineering problems which can be numerically faced by means of propagation or balancing models.
In particular our study was determined by a technological issue concerning heat transfer during welding processes.
D.J.Evans et al. [6,7,8,9], on the basis of Saul'yev's [13] finite difference schemes, obtained explicit methods to approximate the solution of multi-dimensional parabolic partial differential equations.
Such methods work on groups of adjacent points in the sampling grid and are identified according to the position of the groups in the grid (Fig.1a).

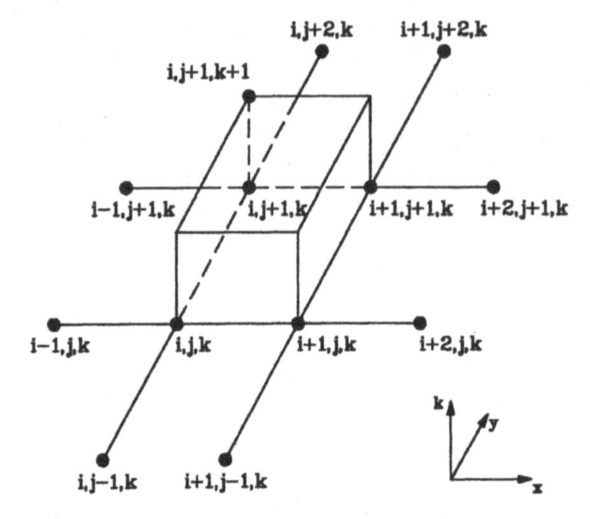

Fig.1a Local computation scheme

The four main adopted schemes (GEC: Group Explicit Complete, GEU: Group Explicit with Ungrouped ends, GEL: Group Explicit with Left ungrouped point and GER: Group Explicit with Right ungrouped point) can be also coupled by means of Single (S_AGE) or Double (D_AGE) Alternate rotation (Fig. 1b).

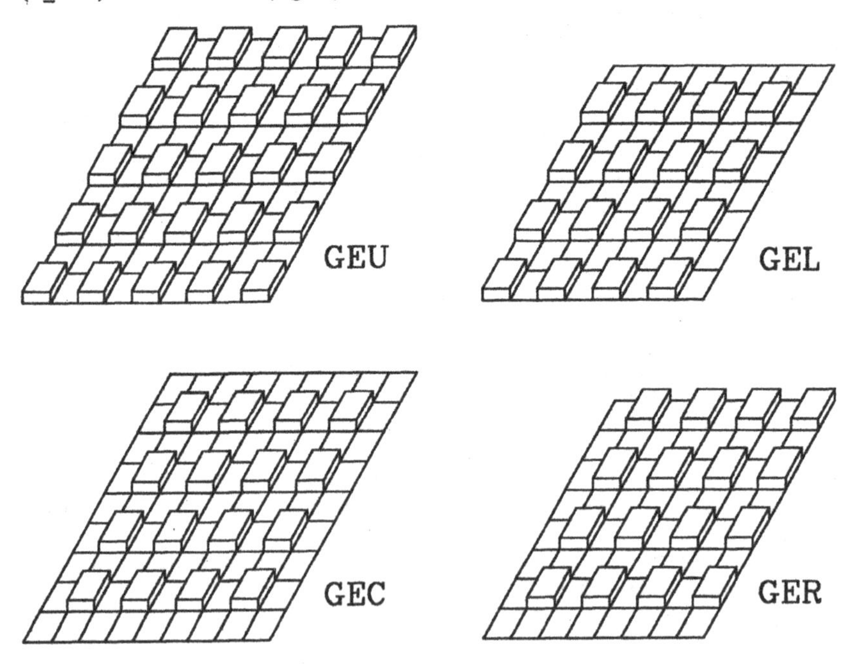

Fig.1b Group explicit methods

The computation methods and related performance evaluations proposed by Evans [1],

concerning Burgers [3] one-dimensional non linear parabolic p.d.e., were carried out by means of a parallel approach based on the strictly local characteristic of computations.
In two recent works [4,5] the Authors dealt with implementative solutions for finite difference explicit methods applied to parabolic equations in two space dimensions, together with different GEC scheme implementations in the hypothesis of space and time parallelism.
Now it is proposed a parallel implementation policy for the S_AGE method based on the periodic rotation of GER and GEL at alternate time levels.
Such scheme is unconditionally stable and thus it allows to analyse suitable time steps only depending on application requirements, with no constraint about the sampling degree of the space domain.

MIXED PARALLELISM

In the following a two-dimensional domain is assumed and, under certain conditions, a trial is carried out to increase the global parallelism degree considering the possibility to concurrently perform the computations related to the consecutive time levels along which the whole process evolves.
In space parallelism the domain is partitioned in a number of subdomains equal to the number of processors employed to carry out the computations related to the same intermediate solution. In particular, they determine square or rectangular matrices defined by subsets of grouped points of the grid.
Any Evans's scheme, other than GEC, cannot be subdivided in a number of equal sub-schemes reflecting the original. That because the interior subdomains have unproper boundary elements in the sense that they are boundary elements only for computation partitioning purpose and cannot be considered as ungrouped points.
In practice, these interior subdomains must be computed by means of a GEC scheme and, for this reason, they need to overlap eachother along two common rows or columns.

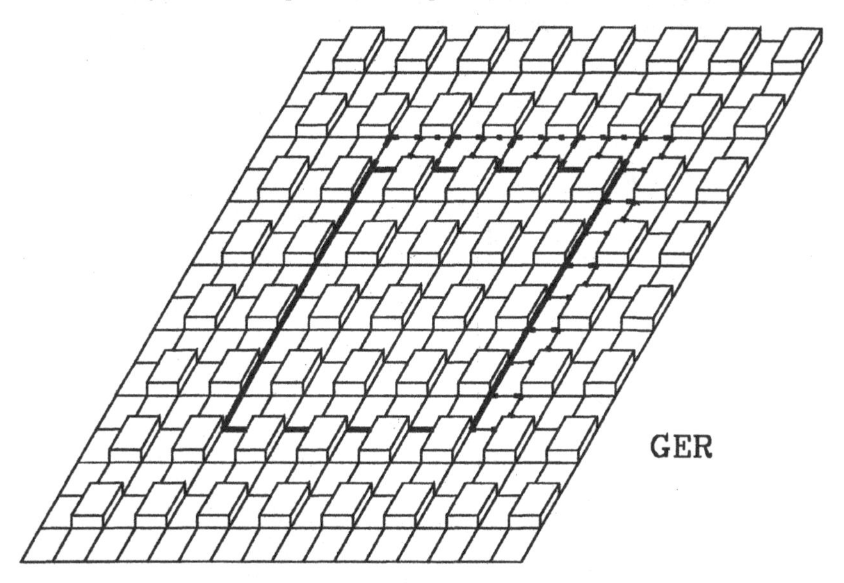

Fig.2a GER expanded subdomain

In particular a GER scheme partitioning determines each interior subdomain to overlap the neighbours on the right and top side while a GEL scheme partitioning determines subdomain expansion on left and bottom side (Fig. 2a, 2b).

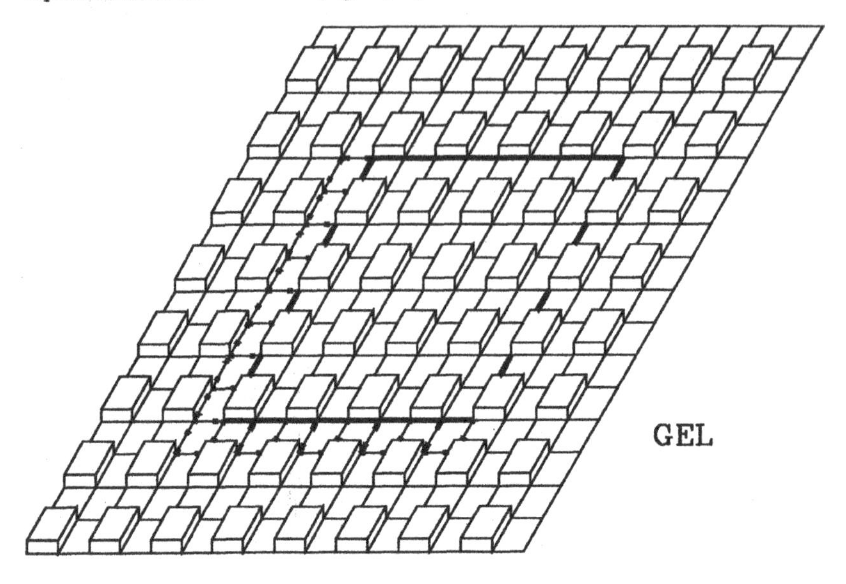

Fig.2b GEL expanded subdomain

On the basis of such considerations, each PE working on a partition must hold the matrix of elements covering the expanded subdomain. Moreover it must reference its nodes by means of an absolute addressing (referred to the whole integration domain) because it must differently consider the actual boundary elements if it works on a boundary partition.

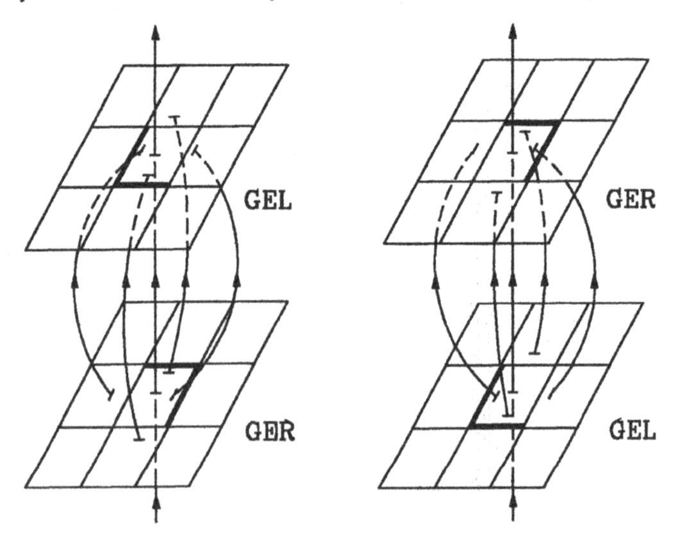

Fig.3 Six link communication scheme

It is also to be taken into account that at each alternate time level a different scheme (GER, GEL) must be employed.

This induces different forward communication schemes between processing elements working on different consecutive time levels depending on data flow from GER to GEL or from GEL to GER.
Each PE must transfer all its calculated values to its correspondent PE at the next time level. Moreover, for what concerns the boundary groups, each PE must transfer the calculated values also to the neighbours of its correspondent at the next level, sited at its expanded sides (Fig. 3).
We can also consider a striped partitioning of the domain (Fig. 4). In such a case the interior subdomains have only one expanded boundary side, right or left according to GER or GEL scheme.

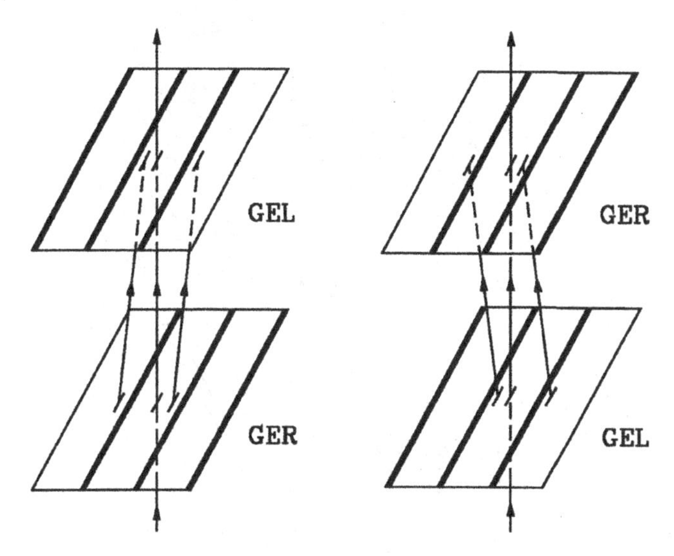

Fig.4 Four link communication scheme

This hypothesis induces a reduced communication scheme requiring only two forward links for each partition (Fig. 4,5).

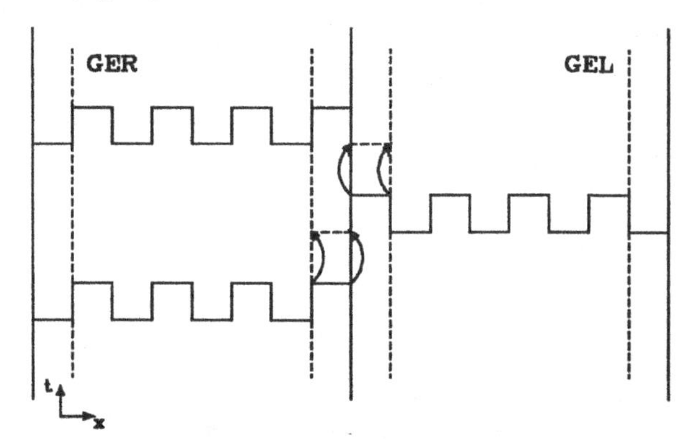

Fig.5 Data flow across overlapping boundaries

In time parallelism each PE working for a particular space partition can start a time step only

if it can access the data evaluated by the correspondent and neighbouring PEs working at the previous time step. For this reason each PE must first elaborate those points at the subgrid boundary which are necessary to the neighbouring PE of the next step.
As shown in Fig. 6, this points single out the groups sited at the upper row of the subgrid plus the first two at the lower row. They are m + 2 at all, being m > 2 the number of groups in a row of a space squared sub-domain.

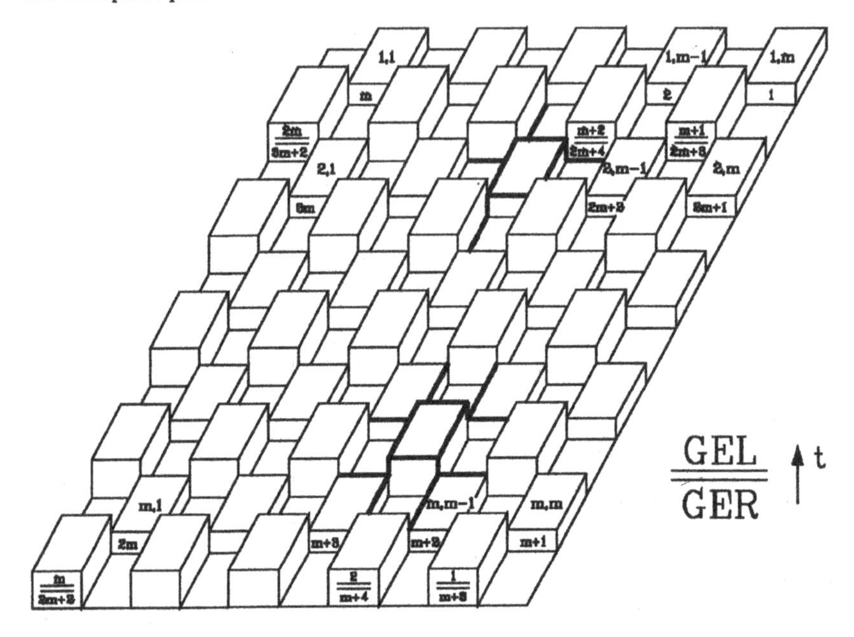

Fig.6 Initial computation path

RELATIVE SPEEDUP

Let us consider the general formula of the relative speedup of a P-processor computation over N data as the ratio between the sequential execution time and the parallel execution time, e.g. Brochard [2]. So we have:

$$Sp(N,P) = \frac{T(N,1)}{T(N,P)} \tag{0}$$

where:
T(N,1) is the execution time of the sequential algorithm,
T(N,P) the execution time using P processors, evaluated as:

$$T(N,P) = T_{calc}(N,P) \# T_{comm}(N,P) \# T_{cont}(N,P) \tag{1}$$

and
$T_{calc}(N,P)$ is the execution time of calculation,
$T_{comm}(N,P)$ the communication time,
$T_{cont}(N,P)$ the control part of the parallel process,
(instead of +) a generic combine operator taking into account overlapping conditions.

In order to evaluate such T(N,P) components, we can consider the mixed parallelism as the product of two factors which are time and space functions of the parameters determining the working environment. So, in general, we can also write:

$$T(N,P) = Ft \cdot Fs.$$

In order to achieve effective time parallelism over computations related to the same instance of a problem, we have also to consider that the starting period of the whole system must be reduced to a quantity smaller than the execution time of an entire step times the number of elements in the pipeline.

Hence, we must separately consider the time needed to start up the pipe and the above formula will be written as:

$$T(N,P) = T(N,P)_{start} + T(N,P)_{par} = \\ = Ft_{start}(...) \cdot Fs_{start}(...) + Ft_{par}(...) \cdot Fs_{par}(...) \qquad (2a)$$

where (...) represents a list of parameters.

Let K be the total number of intermediate approximated solutions of the process being studied and L the length of the pipeline. We can assign the value L-1 to Ft_{start} and the value K/L to Ft_{par}. So we can rewrite (2a) as follows:

$$T(N,P) = (L-1) \cdot Fs_{start}(...) + \frac{K}{L} \cdot Fs_{par}(...). \qquad (2b)$$

According to (1), in the following, we shall examine Fs_{start} and Fs_{par} as expressions involving three components due to computation, communication and control execution time.

CALCULATION

As mentioned in the mixed parallelism paragraph, m+2 groups must be computed by each PE in a pipe element during the starting period and thus, the Fs_{start} component due to computation time can be evaluated as:

$$T_{calc}(N,P)_{start} = (m+2)\tau_g. \qquad (3a)$$

The same component of Fs_{par} can be evaluated as:

$$T_{calc}(N,P)_{par} = m^2 \tau_g, \qquad (3b)$$

being τ_g the computation time of a group of four points in the sampling grid.

SEMAPHORE HANDLING

Each time a processor is going to perform the computations related to a group, it must verify if the values of the twelve points intervening in the explicit formulas have been updated in order to represent the surface status at the end of the previous time level.

To this aim a semaphore based control algorithm is provided allowing the synchronization between time steps.

Considered that each interior point of the grid is employed for calculations related to three different groups, each time a processor receives a calculated value from the previous level, it must initialize the related semaphore to a value of three.

Each time it is going to use a datum, it must control that the value of the related semaphore is

greater than zero and then it must decrease this value by one .
For performance evaluation purposes we can consider that each access to a semaphore involves one test plus one arithmetical operation and so, if we represent the related execution time by means of the generic quantity τ_h, the Fs_{start} component due to semaphore handling can be expressed as:

$$T_{cont}(N,P)_{start} = 16\,(m+2)\,\tau_h, \tag{4a}$$

while, for the same component of Fs_{par} we have:

$$T_{cont}(N,P)_{par} = 16\,m^2\,\tau_h. \tag{4b}$$

DATA COMMUNICATION

Each PE working for an interior subdomain, according to square partitioning, has four neighbours, but, due to scheme alternance, it has only two expanded sides where it needs values of points calculated by the neighbours working at the previous time level and where it must transfer its calculated values to the neighbours at the next time level.
So, as shown in Fig. 3, the best condition to carry out all the communication flows at the same time will consider six links each PE.
By this way, each PE receives $4m^2$ data from its correspondent at the preceding level and 4m from each of the two neighbours.
It also sends the same data amount to the PE at the subsequent level.
Hence, the Fs_{start} component due to data communication can be evaluated as:

$$T_{comm}(N,P)_{start} = \{1,2\}\ ((m+2)\ \#_1\ m\ \#_2\ 2)\ 4\,\tau_d, \tag{5a}$$

and the same component of Fs_{par} can be evaluated as:

$$T_{comm}(N,P)_{par} = \{1,2\}\ (m^2\ \#_1\ m\ \#_2\ m)\ 4\,\tau_d. \tag{5b}$$

where:
{1,2} represents a real value belonging to the closed interval. This quantity depends on the ability of a communication link to carry out data reception and transmission at the same time.
$\#_1$ and $\#_2$ are operators returning a value belonging to the closed interval $[\max(op_1,op_2),(op_1+op_2)]$ depending on parallelism between different communication links,
τ_d is the communication time of each datum.

EXTENDED FORMULA COMPOSITION

On the basis of the above formulas the final expression of Fs_{start} and Fs_{par} can be written as follows:

$$Fs_{start} = [(m+2)\tau_g + 16(m+2)\tau_h]\ \#\ [\{1,2\}4((m+2)\ \#_1\ m\ \#_2\ 2)\tau_d],$$

$$Fs_{par} = [m^2\tau_g + 16m^2\tau_h]\ \#\ [\{1,2\}4(m^2\ \#_1\ m\ \#_2\ m)\tau_d]$$

where:
τ_g is the calculation time of a group,
τ_d the communication time of each datum,
τ_h the time spent on accessing a semaphore.

Finally, in order to complete (0), we have only to define the term T(N,1) which can be easily expressed as $K \cdot M^2 \cdot \tau_g$, being M the number of groups in a row of the whole domain.

FOUR LINK HYPOTHESIS

Let us consider the case of four communication links for each PE, that, in general, is the minimum requirement to carry out mixed parallism where two links are employed for space communication and the other two for time communication.
In such a condition, as mentioned above, we can perform a space parallelism only by means of a striped partitioning of the domain (Fig. 4,5).

To this aim, we shall assume the whole domain to be rectangular with (M·M') groups. The m term will assume a value equal to the ratio between M' and n, being n the number of space partitions.
Each PE running for an interior subdomain has two neighbours on the same plane and two adjacent time levels. So, similarly to the case of six links, it must receive and send all the grouped points values from and to its corresponding ones at the adjacent levels.
With a stripe partitioning each subdomain has only one expanded side. So, in order to evaluate the relative speedup, we must modify the (3a,3b,4a,4b,5a,5b) as follows:

$$T_{calc}(N,P)_{start} = (m + 2)\tau_g, \qquad (3a)'$$

$$T_{calc}(N,P)_{par} = (M \cdot m)\,\tau_g, \qquad (3b)'$$

$$T_{cont}(N,P)_{start} = 16\,(m + 2)\,\tau_h, \qquad (4a)'$$

$$T_{cont}(N,P)_{par} = 16\,(M \cdot m)\,\tau_h, \qquad (4b)'$$

$$T_{comm}(N,P)start = \{1,2\}\; 4\,((m + 2)\;\#_2\; 2)\,\tau_d, \qquad (5a)'$$

$$T_{comm}(N,P)_{par} = \{1,2\}\; 4\,((M \cdot m)\;\#_2\; M)\,\tau_d, \qquad (5b)'$$

$$T(N,1) = K \cdot M \cdot M' \cdot \tau_g.$$

IMPLEMENTATIVE SPECIFICATIONS AND RESULTS

In our study we tried to give an enough detailed evaluation of the computation, control and communication components with reference to the technical specifications of the transputer internal architecture, as shown in [10,11] and, more precisely, to the IMS T800-20 model.
The arithmetical complexity [8,12] of the S_AGE scheme, according to Saul'yev's stable asymmetric equations [13], involves 11 additions, 6 multiplications and 1 division, requiring 49.4 µs in order to fetch instructions, to transfer data from external to the internal memory and to carry out the execution with 32 bit real operands. This evaluation applies to each group of four neighbouring points on the two-dimensional sampling grid, that is the τ_g term in the above formulas.
For what concerns the τ_d term, that is the time spent to communicate a 32 bit real value, we consider a timing of 2.5 µs with refence to the same architecture.
The # operand may have different meanings depending on the possibility of executing communications and calculations at the same or different time. This depends on hardware and/or software features and it is not always generalizable. In the worst case the communication overhead must be added to the computation time while, in the best case, the biggest quantity between computation and communication time should be considered. We adopted a mean value in order to carry out the charts included in this paper.
Figs. 7a-b, and 8a-b show different curves each representing the trend of the speedup referred to a particular number of available PEs.

The trend depends on time parallelism evaluated as the ratio between the total number of available PEs and those ones employed for carrying out the processes belonging to one time step.

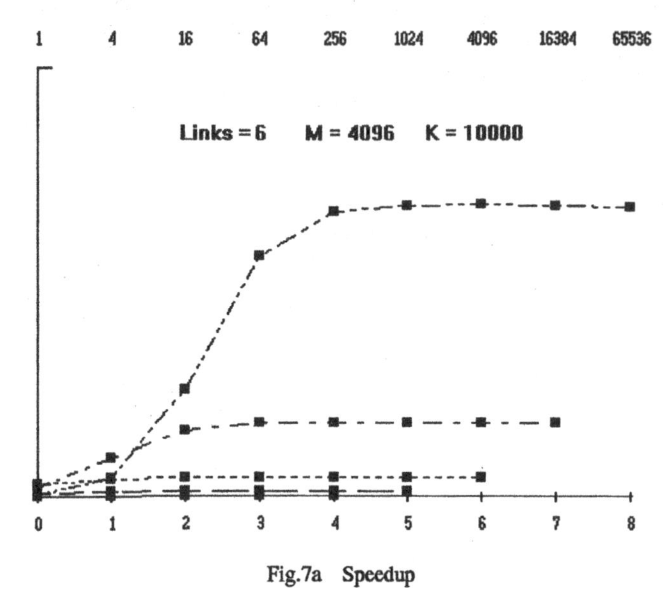

Fig.7a Speedup

The Y axis of Figs. 7a and 8a gives the percentage values referred to the maximum speedup, respectively with six and four communication links.

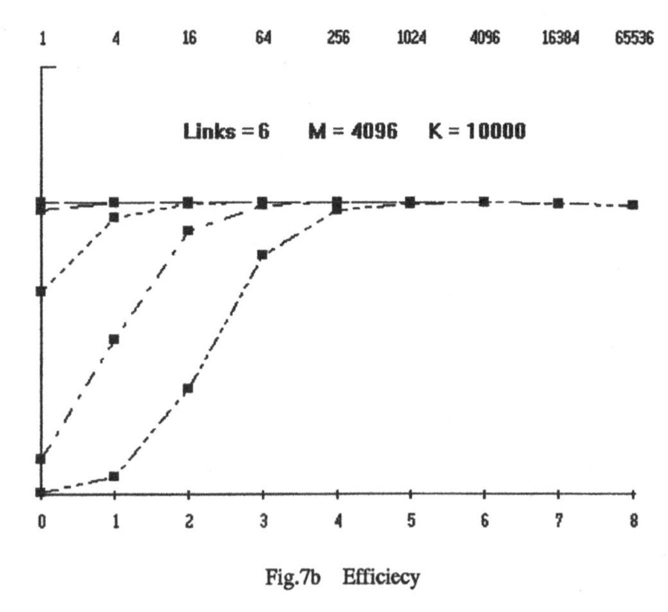

Fig.7b Efficiecy

The Y axis of Figs. 7b and 8b gives the efficiency rates, respectively with six and four communication links.

The X axis reports the different exponents to be given to the base 4 in order to indicate the number of PEs employed in the space parallelism. Moreover, at the top of the charts, they indicate the maxima linear speedups which could be reach by a system with that total number of PEs.

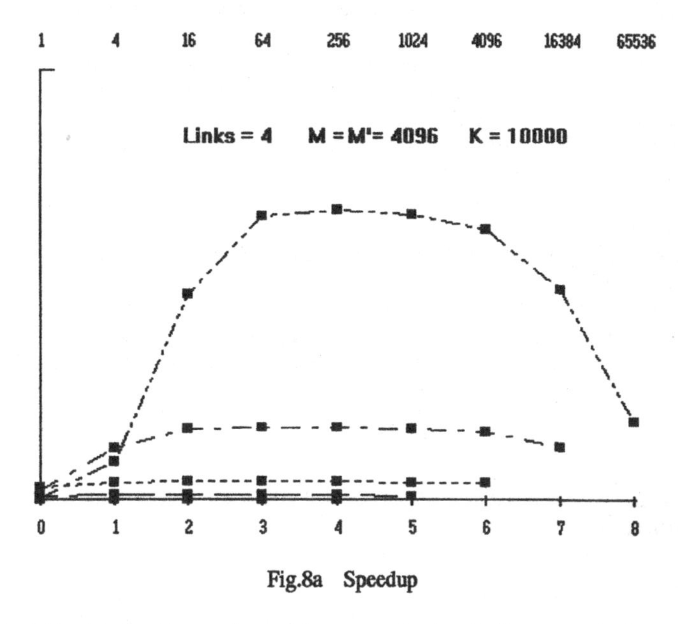

Fig.8a Speedup

The virtual line binding the maxima of the curves in fig. (7-8)b represents the loosing of efficiency at increasing the total number of available PEs.

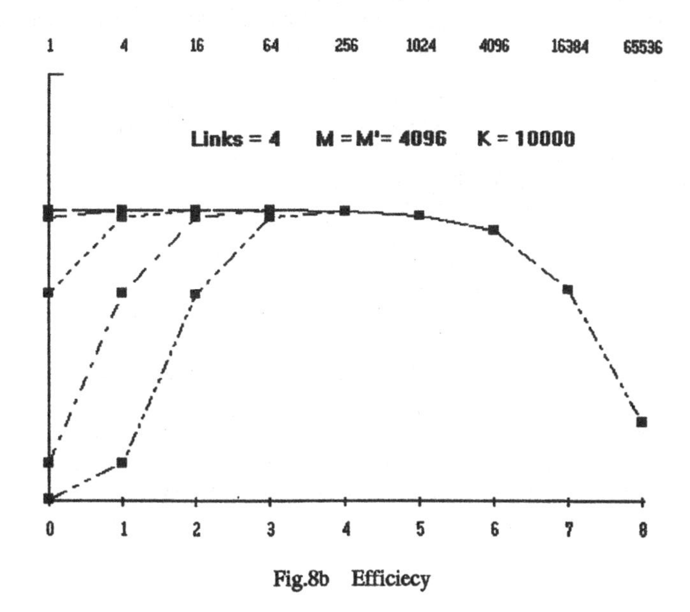

Fig.8b Efficiecy

CONCLUSIONS

An Implementative policy for the S_AGE method based on the alternate rotation of GEL and GER schemes has been proposed.
Both space and time parallelism have been investigated in order to obtain the highest relative speedup.
The adopted model, allowing to pre-evaluate the performances of different parallel implementative solutions, gives the possibility of better tuning the application parameters in terms of problem resizing and partitioning, both in time and space domains.
The applicative instance which gave rise to this study was a numerical model for simulating transient conduction in tungsten inert gas welding.
The initial and boundary conditions, deriving from the technological problem, are extensively analysed in [14] and [15].
To the aim of this paper, we considered initial and boundary values as known data not intervening in the performance evaluations.
The charts in Figures 7 and 8 was drawn by means of a software developed by the Authors, giving the possibility to grafically analyse different implementative solutions.
Such graphical package was carried out by means of the Microsoft Software Development Kit for the MS/DOS Windows environment.

ACKNOWLEDGEMENT

This work has been supported by the Ministero della Università e della Ricerca Scientifica e Tecnologica of Italy.

REFERENCES

1. Bekakos, M.P. and Evans, D.J. Relative Performance Comparisons for the Group Explicit Class of Methods on MIMD, SIMD, and Pipelined Vector Computers, Parallel Computing, Vol.10, pp. 357-364,1989.
2. Brochard, L. Efficiency of Some Parallel Numerical Algorithms on Distributed Systems, Parallel Computing, Vol. 12, pp. 21-44, 1989.
3. Burgers, J.M. Adv. Appl. Mach - Vol.1, pp. 171, 1948.
4. Carmignani, M. and Genco, A. Implementative Solutions for D.J.Evans Parallel Algorithms (Ed. Laforenza, D. and Perego, R.), pp. 321 to 326, Proceedings of the International Workshop on Supercomputing Tools for Science and Engineering, Pisa, Italy, 1989, Franco Angeli, Milano, Italy, 1990.
5. Carmignani, M., Genco, A. and Pecorella, G. Speedup Evaluations for the GEC Method using Space and Time Single or Mixed Parallelism (Ed. Messina, P. and Murli, A.), Proceedings of the Int. Conf. Parallel Computing: Achievements, Problems and Prospects, Capri, Italy, 1990, Elsevier, Amsterdam, 1991, in press.
6. Evans, D.J. and Abdullah, A.R.B. Group Explicit Methods for Parabolic Equations, Int. Jour. Comp. Math., Vol.14, pp.73-105, 1983.
7. Evans, D.J. and Abdullah, A.R.B. A New Explicit Method for the Solution of $\delta u/\delta t = \delta^2u/\delta x^2 + \delta^2u/\delta y^2$ - Int. Jour. Comp. Math., Vol. 14, pp. 325 -354, 1983.
8. Evans, D.J. New Parallel Algorithms for Partial Differential Equations, (Ed. Feilmeier, M., Jöubert, J. and Scendel, U.), pp. 1 to 56, Proceedings of the International Conference on Parallel Computing 83, Berlin, Germany, 1983, Elsevier Science Publishers B.V., North Holland, 1984.

9. Evans, D.J. Parallel Algorithm Design (Ed. Evans, D.J. and Sutti, C), pp. 1 to 24, Proceedings of the Int. Meet. on Parallel Computing, Verona, Italy, 1988, Adam Hilger, Bristol and Philadelfia, 1989.
10. Carling, A. Parallel Processing,the Transputer and Occam, Sigma Press, Wilmslow and Cheshire, U.K., 1988.
11. INMOS Transputer Reference Manual, Prentice Hall, New York, London, Toronto, Sydney and Tokyo, 1988.
12. Kronsjö, L. Computational Complexity of Sequential and Parallel Algoritms , John Wiley & Sons, Chichester and New York, 1987.
13. Saul'yev, V.K. Integration of Equations of Parabolic Type by the Method of Nets , Macmillan, New York,1964.
14. Carmignani, M., Masnata, A., Ruisi, V. and Tortorici A. A Mathematical Model of Thermal Cycles in Welding (Ed. Lewis, R.W. and Morgan, K.) pp. 1-154 to 1-163, Proceedings of the 5th Int. Conf. on Numerical Methods in Thermal Problems, Montreal, Canada, 1987, Pineridge Press, Swansea, U.K., 1987.
15. Masnata, A. and Micari, F. In Process Control of Weld Penetration by a Numerical Method (Ed. Lewis, R.W. and Morgan, K.), pp. 2-1412 to 2-1421, Proceedings of the 6th Int. Conf. on Numerical Methods in Thermal Problems, Swansea, U.K, 1989, Pineridge Press, Swansea, U.K., 1989.

The Parallel Processing Concept on the Siemens Nixdorf S Series and its Use in Engineering Applications

E. Schnepf

Siemens Nixdorf Informationssysteme AG, Division Supercomputer, Otto-Hahn-Ring 6, D-W-8000 Munich 83, Germany

ABSTRACT

The architecture of the multiprocessor models within the Siemens Nixdorf S Series is described. With the unique dual scalar architecture, these models include four scalar units and two vector units, where each vector unit is shared by two scalar units. Several multiprocessing requirements can be fulfilled. A single job without parallelization reaches a high performance thanks to the powerful scalar and vector units. The throughput of such a supercomputer is optimized for the needs of big computer centers. The elapsed time of large scale applications can be reduced by parallelizing the source code with the help of the new compiler FORTRAN77 EX/PP. This compiler has been designed with the target to allow engineers parallelizing their programs easily with a high degree of automatism. On the other hand, the parallelization concept is guided by the aim to maintain the portability of source programs as much as possible. Some examples show the application of the parallelization techniques.

INTRODUCTION

To provide the computational power which is requested by big research centers in science and industry, it is efficient to build multiprocessor systems which combine high-speed vector units with a number of powerful scalar units.

To meet these requirements, Siemens announced in 1989 the S Series, the successor of the successful VP Series of which about 90 systems have been installed worldwide. The S Series consists of 10 models which cover a peak performance range from 500 MFLOPS to 5 GFLOPS (billion floating–point operations per second). There are various configurations offered, including monoprocessors, dual scalar models, where two scalar units share one vector unit, and multiprocessor models with four scalar units and two vector units.

SYSTEM ARCHITECTURE

The Siemens Nixdorf S Series, which is equivalent to the Fujitsu VP2000 Series, is based on four models (S100, S200, S400, S600) with different peak performance of the vector unit between 500 MFLOPS and 5 GFLOPS. Every model can be equipped either with one scalar unit and one vector unit (model /10) or, using the unique dual scalar architecture, with two scalar units sharing one vector unit (model /20). Additionally, there are two multiprocessor models S200/40 and S400/40 available with four scalar units that share two vector units. The three different configurations are shown in figure 1. The connection of the scalar units (SU) and vector units (VU) with the main memory (MSU) can also be seen in this figure.

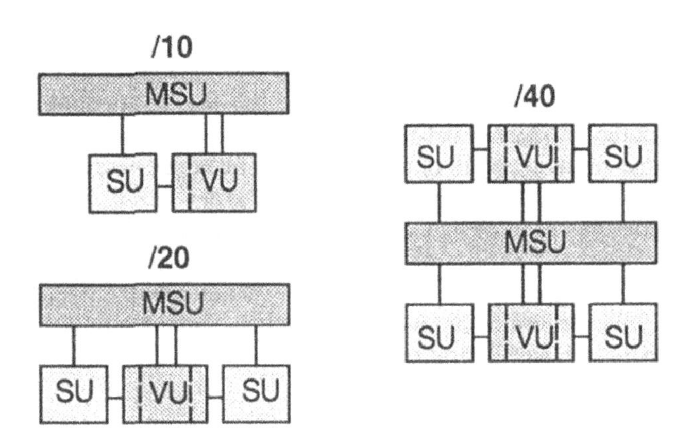

Figure 1. S Series models overview

The whole S Series is field upgradable from the low–end model S100 to the high–end models S400/40 and S600/20. In the case of an upgrade, no replacement of hardware has to be done, just frames and units are added. An overview on the various upgrade paths and the corresponding values for the sustained scalar and peak vector performance is shown in figure 2.

The architecture of the monoprocessor models has been described in [1]. In that paper the vector unit and the usage of the vector registers, together with the vector register optimization ,are described in detail.

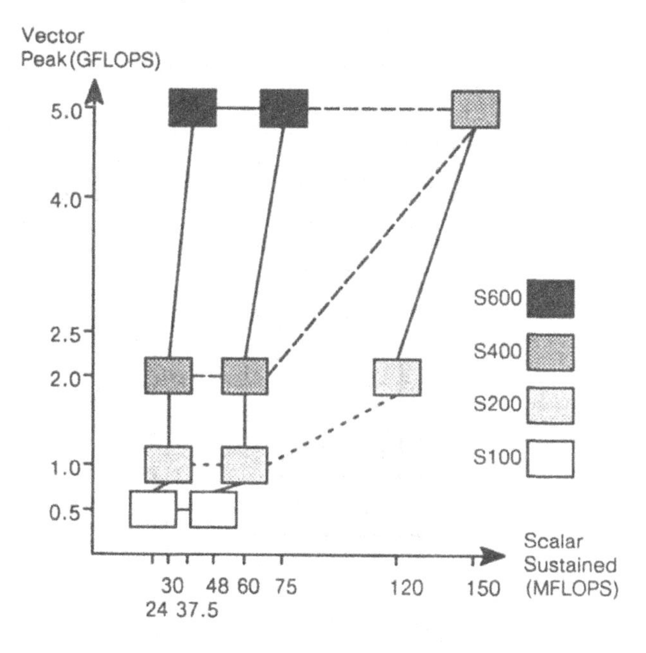

Figure 2. S Series upgradability and performance

DUAL SCALAR ARCHITECTURE

The dual scalar architecture is the key for the efficiency of the multiprocessor models. Experiences with representative applications in industry and science have shown that the vector unit is not fully utilized, as most applications include a significant scalar part, in spite of all sophisticated vectorization techniques.

The only solution to improve the vector unit utilization rate is to feed the vector pipelines with more vector instructions, coming from a job which executes on a second scalar unit. In the models /20 a complete second set of registers is assigned to the second scalar unit (see figure 3). This enables an efficient switching between tasks that are running in parallel on the two scalar units. The whole control of the shared vector unit including the switching method is performed by the hardware control program, not by the operating system.

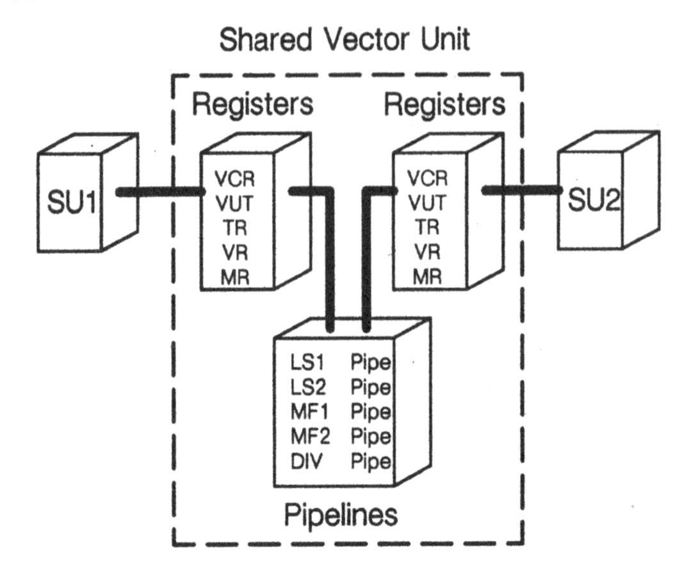

Figure 3. Shared Vector Unit

MULTIPROCESSOR ARCHITECTURE

In the multiprocessor models S200/40 and S400/40 the dual scalar architecture has been doubled, resulting in four scalar units sharing two vector units. Several multiprocessing aspects can be fulfilled with this architecture.

- A single job reaches a high performance on one processor thanks to the powerful scalar and vector units.
- The throughput and the resource utilization of such a supercomputer is optimized for the needs of big computer centers.
- Large scale applications can be accelerated by parallelization.

Due to these characteristics, the models S200/40 and S400/40 are the optimal solution for computer centers in science and industry.

The parallel execution of two parallelized jobs is shown in figure 4. Here each job runs on two scalar units and submits vector instructions to the shared vector units. This results in an optimal use of the resources.

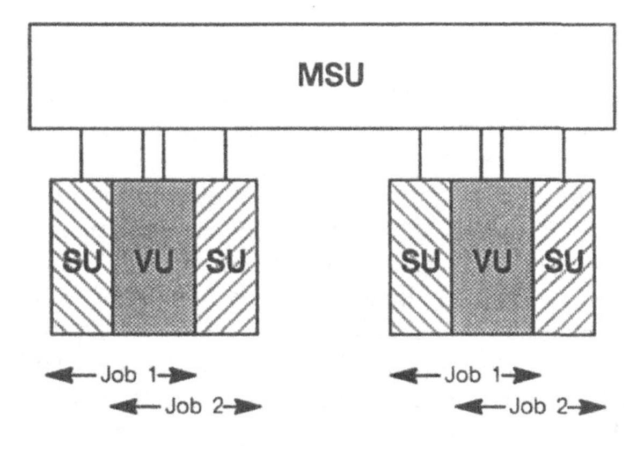

Figure 4. Parallel execution on a /40 model

PARALLELIZATION

The new compiler FORTRAN77EX has been designed, guided by two aims: the capabilities of the supercomputer can easily be used, and the portability of source programs is kept whenever possible. The scalar compiler has been enhanced with interprocedural optimization like the automatic inlining of subprograms. The sophisticated vectorization techniques of FORTRAN77EX/VP have been enhanced to utilize the multifunctional pipelines in the vector units efficiently. This is achieved by the global vector register optimization.

The new parallelizing compiler FORTRAN77EX/PP performs the parallelization of nested loops automatically if the data dependencies allow for it. If the underlying algorithm of a source program results in parallelizable loops, the effort for the programmer is reduced to a minimum amount of work for introducing, probably, some optimization control lines (OCLs) to clarify data dependencies for vectorization and parallelization. OCLs are also used to enable the parallelization of subprograms.

The use of OCLs has several advantages. The portability of the source code is maintained, as these OCLs are treated like comment lines by other compilers. For debugging purposes, OCLs can easily be ignored by the compiler. A syntax error in an OCL does not affect the semantics. Parallelized programs including OCLs can easily be resequentialized and ported to other sequential or parallel systems.

For the realization of some synchronization means, e.g. the POST/WAIT–mechanism and barrier constructs, the call of service subroutines is better suited than the use of OCLs. It is important in these cases to code carefully the synchronization mechanism. The erroneous use of OCLs would cause wrong results. Therefore, it has been decided to implement service subroutines for the mentioned synchronization constructs in the parallel processing library of the compiler.

APPLICATION EXAMPLES

The first example is a kernel taken from the program package FIDISOL, a subroutine library for the numerical solution of elliptic and parabolic differential equations, see [2]. In the solution process of FIDISOL the matrix of the resulting linear system is stored in diagonal form. The sparse diagonals are stored in packed form to reduce the main memory requirement. The linear system is solved by generalized CG (conjugate gradient)-type methods. The most time consuming part in this solution is the matrix-vector-multiplication (MVM), which has to be performed many times with varying right-hand sides. This kernel operation is perfectly vectorizable, see [3]. To reduce the elapsed time of this application, parallelization has to be applied to the MVM. Naturally, the diagonals of the matrix can be processed on different processors, in the case of the S Series multiprocessors on two vector units.

The original source of the kernel is shown in figure 5. For simplicity some index calculations are omitted.

```
         DO 20 ID=1,ND
           LD = ... (ID)
           ... (index calculations)
           DO 10 I=1,LD
             X(IP(I+IP0)+IX0)
        &     =X(IP(I+IP0)+IX0)
        &     +A(I+IA0)*B(IP(I+IP0)+IB0)
10         CONTINUE
           ... (index calculations)
20       CONTINUE
```

Figure 5. Kernel of the matrix-vector-multiplication

It is important, that the vector processor is equipped with a load/store pipeline which is able to access data indirectly with high speed. Only the non-zero matrix elements are stored and processed. Therefore, the arrays for the solution and the right-hand-side are accessed by means of index vectors.

The parallelization of this kernel shows, that the numerical algorithm has to be prepared in a suitable way to enable automatic parallelization. The array of the solution X is accessed for each diagonal. To allow the parallel processing of the diagonals, as many arrays for the solution have to be allocated as tasks, respectively pieces, should run in parallel. After the parallel processing of the diagonals, a final sum up of the solution arrays has to be performed to build the final solution.

This additional programming work has to be done by the programmer in order to prepare the source code for parallelization. This reformulation of the MVM kernel is done totally in standard FORTRAN77. The rewritten kernel can also be processed sequentially for debugging purposes. The parallelized kernel is shown in figure 6. Two OCLs are included to indicate that there are no recurrencies. The compiler parallelizes automatically the outermost loop and realizes the necessary FORK/JOIN-mechanisms.

```
!OCL LOOP,NOPREC
      DO 30 IPROC=1,NPROC
         DO 20 ID=1,ND,NPROC
            ... (index calculations)
*VOCL LOOP,NOVREC
            DO 10 I=1,LD
              X(IP(I+IP0)+IX0,IPROC)
     &       =X(IP(I+IP0)+IX0,IPROC)
     &       +A(I+IA0)*B(IP(I+IP0)+IB0)
10          CONTINUE
            ... (index calculations)
20       CONTINUE
30    CONTINUE
      ... (process remaining diagonals
           if ND is not a multiple of
           NPROC)
      ... (sum up the solution arrays
           X(.,IPROC))
```

Figure 6. Parallelized kernel MVM

A second example is taken from a CFD code to simulate the Euler equations. Here two types of parallelization are applicable, the parallelization of subprograms and the parallelization of nested loops. At first, three subroutines are called to evaluate the right-hand-side of the resulting system of equations:

```
CALL XIFLUX
CALL ETFLUX
CALL ZEFLUX
```

The array for the right-hand-side is updated in each of the three subprograms. For the parallelization it is necessary to use three different arrays for these subprograms and, finally, to sum up these arrays to build the right-hand-side. The parallelization itself is enabled by OCLs:

```
!OCL PARCALL
      CALL XIFLUX
      CALL ETFLUX
      CALL ZEFLUX
!OCL END PARCALL
      CALL SUMUP
```

After the evaluation of the right-hand-side, the solution of the resulting system of equations is done by a Gauss-Seidel solver. With the help of a red-black ordering method for the innermost loop the usual recursive algorithm is vectorizable. The same technique has now to be applied to the outer loop to parallelize the nested loops. The loop structure of the resulting solver is shown in figure 7.

```
      DO 100 IT=1,NIT
        ...
        DO 31 KK=2,3
!OCL LOOP,NOPREC
          DO 30 K=KK,NK,2
            DO 20 J=2,NJ
              DO 11 II=2,3
*VOCL LOOP,NOVREC
                DO 10 I=II,NI,2
                ...
10              CONTINUE
11            CONTINUE
20          CONTINUE
30        CONTINUE
31      CONTINUE
        ...
100   CONTINUE
```

Figure 7. Loop structure of the parallelized solver

CONCLUSION

The architecture of the multiprocessor models of the S Series fulfills the requirement from the engineer (to obtain the results of his numerical application in the shortest possible time) as well as the requirement from the supercomputer center (to obtain highest throughput and best utilization of the computer resources) without any conflict. The corresponding parallelization concept is easy and powerful and has been oriented at the standard FORTRAN syntax. The design of the new compiler takes the burden of hand parallelization as far as possible from the programmer.

REFERENCES

1. Schnepf, E. Highly Sophisticated Vector Register Usage in the Siemens Vector Processor System and its Impact on Engineering Applications, in Proceedings of the 1st Int. Conf. on Applications of Supercomputing in Engineering, Southampton, UK, 1989. Elsevier, Amsterdam, 1989.
2. Schönauer, W. Scientific Computing on Vector Computers, Elsevier, North Holland, Amsterdam, New York, Oxford and Tokyo, 1987.
3. Weiss, R., Häfner, H. and Schönauer, W. Tuning the Matrix-Vector-Multiplication in Diagonal Form, in Parallel Computing 89 (Ed. Evans, D.J., Joubert, G.R. and Peters, F.J.), Elsevier, North Holland, Amsterdam, 1989.

Parallelism in the Boundary Element Method: fine grain and course grain

A.J. Davies

Division of Mathematics, Hatfield Polytechnic, Hatfield, AL10 9AB, U.K.

ABSTRACT

The boundary element method has been a well-established procedure for the solution of a wide variety of engineering problems for over two decades. Many large commercial codes are available besides the many thousands of programs written by individual research workers. Predominantly these codes have been produced to be run on convential sequential computers.

During the past ten years there have been important advancements in computer architectures, in particular we have seen the development of parallel computers. These machines have been found to be invaluable for reducing solution time or for producing simpler code when used with finite difference[1] and finite element[2] methods applied to large engineering problems. Surprisingly, very little work has been done on the use of such architectures for boundary element problems.

Whenever a parallel computer is to be used for the solution of a problem it is essential that the chosen algorithm maps onto the architecture in a suitable manner. The parallelism inherent in the algorithm must have a natural representation in terms of the parallelism of the chosen machine.

The numerical aspect of the boundary element method has, essentially, three distinct phases: the equation set-up phase, the solution phase and the recovery of field variables phase. Most of the published work has been associated with exploiting the parallel nature of the solution phase using a vector processor[3].

All three phases exhibit both fine-grain and coarse-grain parallelism, for example in the set-up phase the calculation of the coefficients may be accomplished by treating the elements in parallel, fine-grain, and developing a loop over the Gauss quadrature points[4]. Alternatively the Gauss quadrature can be performed in parallel, coarse-grain, with loops over the elements[5]. The fine-grained parallelism is suitable for an array processor consisting of a large number of simple processing elements and the coarse-grained parallelism is suitable for a transputer array consisting of a small number of far more sophisticated processors. In this paper we shall discuss the details of such parallel features and suitable strategies for implementing them.

KEYWORDS

Boundary element method, parallel processing.

GRANULARITY AND PARALLEL COMPUTER SYSTEMS.

Parallel computer systems comprise a set of sequential processors interlinked in some way. The grain size associated with a parallel machine is a measure of the number and the complexity of the basic operations that are performed on each processor. In a fine-grained system a relatively small number of operations is performed on each processor; conversely on a coarse-grained system a relatively larger number of operations is performed on each processor.

In a similar manner, problems themselves can also be classified as fine-grained or coarse-grained with reference to the subproblems into which the overall problem may be divided. The parallel implementation of any particular problem requires that a suitable mapping is found from the problem to the architecture. Coarse-grained parallel problems are mapped to coarse-grained architectures and fine-grained problems to fine-grained architectures.

According to the taxonomy of Flynn[6] all computing machines fall into one of four catagories. From the point of view of the boundary element method the overwhelming majority of implementations have been on the single-instruction single-data, SISD, type of architecture. In such a computer a single processor works on a single data item and progresses through the algorithm in a straightforward beginning-to-end process. This is the basis of the conventional, von Neumann computing architecture, usually referred to as a sequential computer.

The second category is the multiple-instruction single-data, MISD, machine, in which a single data item is processed by more than one processing unit. Currently there are no boundary element applications of such architectures. The other two catagories are the single-instruction multiple-data, SIMD, and the multiple-instruction multiple-data, MIMD, machines. It is amongst these two catagories of architecture that we shall consider the application of the boundary element method. In general SIMD machines are associated with fine-grained parallelism and MIMD machines are associated with coarse-grained parallelism. The major difference is that a computer with an SIMD architecture behaves synchronously whereas a computer with an MIMD architecture behaves asynchronously. This means that the same instruction is performed on all processors of an SIMD machine whereas the processors in an MIMD machine can perform different instructions at the same time. Amongst the SIMD machines are the so-called array processors and the vector pipeline machines. Array processors usually comprise a large number, often thousands, of relatively simple processing elements which are interconnected in a fashion depending on the machine. This interconnection is usually very tight so that there is the facility of rapid interchange of data between neighbouring processors. A master control unit broadcasts the same instruction to each of the elements which has its own memory. Consequently this instruction is effected simultaneously on a multiplicity of data items.

Vector pipeline machines usually comprise only one processing unit. The parallelism is incorporated at the level of the arithmetic operations in the CPU.

The MIMD machines usually comprise a relatively small number of processors each with its own memory. These processors are considerably more sophisticated than those of an array processor and interconnection between the processing elements is usually much looser than in an array processor. Each processor can perform a different task on the data in its memory. Consequently in such a machine, we have an independent set of instructions acting on a multitude of data items.

Boundary element implementations have been developed on both SIMD and MIMD architectures. Bozek et al[3] have implemented the method on a vector pipeline machine and Davies[4,7] has implemented the method on an array processor. Davies[5] has implemented the method on a transputer array.

THE BOUNDARY ELEMENT METHOD

Consider the boundary-value-problem

$$L\phi = f \qquad \text{in } V$$

subject to (1)

$$B\phi = g \qquad \text{on } S,$$

where L is a differential operator defined in the region V and B is a differential operator defined on the closed boundary, S, of V.

This problem can be re-cast in the form of an integral equation given by

$$\alpha(r)\phi(r) + \int_S \phi(r) \frac{\partial}{\partial n'} G(r,r')dS' - \int_S G(r,r')q(r')dS' = \int_V f(r')G(r,r')dV', \tag{2}$$

where $q = \partial\phi/\partial n$.

α is a parameter which depends on the surface geometry, $G(r,r')$ is the so-called fundamental solution associated with the differential operator, L, and r is the position vector of the field point P on the boundary and r′ is the position vector of the source point P′ on the boundary.

For homogenous problems $f = 0$ and the equation (2) reduces to a boundary integral equation. In the direct boundary element method approach to the approximate solution of equation (2) the boundary S is divided into a set of elements and the potential, ϕ, and flux, q, are approximated in a piecewise manner over the elements. This yields a system of linear equations for the unknown nodal values of ϕ or q.

In the indirect boundary element method a suitable source or dipole distribution is sought on the boundary which yields the required potentials and fluxes on the boundary. The indirect method exhibits the same general parallel features as does the direct method.

To illustrate the parallelism inherent in the boundary element method consider the following potential problem.

$$\nabla^2 \phi = 0 \text{ in } D$$

subject to (3)

$$\phi = \phi_0 \text{ on } C_0$$

$$q = q_1 \text{ on } C_1,$$

where D is a two-dimensional region bounded by the closed curve $C = C_0 + C_1$ shown in figure 1.

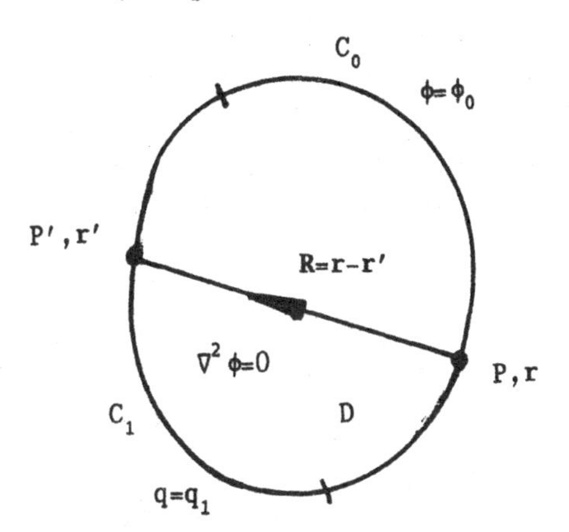

Figure 1. The two-dimensional region D for the potential problem (3).

We shall concentrate on two-dimensional problems with Dirichlet and Neumann boundary conditions. The concepts carry over to problems in three dimensions and it is not difficult to incorporate a Robin boundary condition in the form $q + \sigma\phi = h$ on some part, C_2, of the boundary[7].

The fundamental solution for the two-dimensional Laplacian operator is given by

$$G(r,r') = -\frac{1}{2\pi} \ln|r-r'|$$

$$= -\frac{1}{2\pi} \ln R \ .$$

With this fundamental solution the potential problem (3) may be replaced by the equivalent boundary integral equation in the form[4,8]

$$\alpha(r)\phi(r) = \oint_C (\phi \frac{\partial}{\partial n} (\ln R) - q\ln R)ds. \qquad (4)$$

Using the direct boundary element method with the boundary, C, divided into N elements, see figure 2, and we assume that the boundary potential and flux are approximated by $\tilde{\phi}$ and $\tilde{q}$ respectively. In the case of linear elements of length 2l, we have N nodes and the boundary is approximated in a piecewise linear manner. The approximations $\phi(s)$ and $q(s)$ are obtained as linear interpolants of the nodal values within each element.

$$\text{i.e. } \tilde{\phi}(s) = \sum_{j=1}^{N} u_j(s)\phi_j \text{ and } \tilde{q}(s) = \sum_{j=1}^{N} u_j(s)q_j, \qquad (5)$$

where $\{u_j(s);\ j = 1, 2, \ldots, N\}$ is the usual set of piecewise linear hat functions and ϕ_j and q_j are the nodal values of ϕ and q respectively.

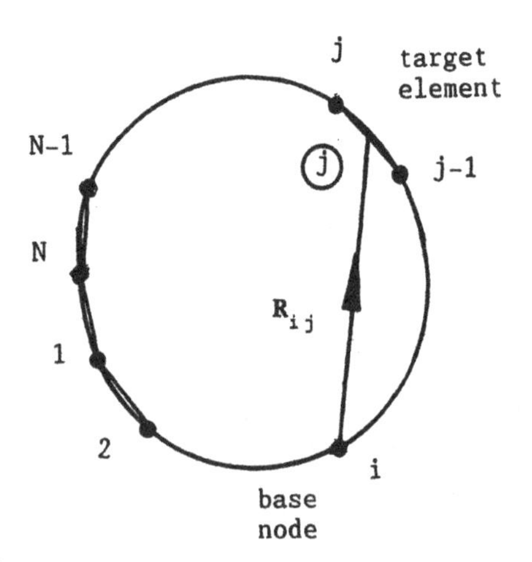

Figure 2. Linear boundary elements for the potential problem (3).

The boundary element approximation then leads to a set of algebraic equations [4,8] in the form

$$\mathbf{Ax} = \mathbf{b}, \qquad (6)$$

where the terms in the matrix **A** and the vector **b** are obtained from integrals over the boundary elements and **x** is a vector of unknown boundary potentials or fluxes.

Typically, the terms in the coefficient matrix contain integrals of the form

$$h_{ij} = l_j \int_{-1}^{1} L(\xi) \frac{R_{ij}(\xi)\cdot\hat{n}}{|R_{ij}(\xi)|^2} d\xi$$

or (7)

$$g_{ij} = l_j \int_{-1}^{1} L(\xi) \ln |R_{ij}(\xi)| \, d\xi ,$$

where $L(\xi)$ is the usual Lagrange linear interpolation polynomial.

Then either $a_{ij} = H_{ij} = h_{ij} + h_{ij-1} - \alpha_i \delta_{ij}$

or $a_{ij} = G_{ij} = g_{ij} + g_{ij-1}$.

If the base node, i, is not in the target element, j, then these integrals are non-singular and may be evaluated using the usual Gauss quadrature procedure. If, however, the base node, i, is in the target element, j, then these integrals contain a singularity. The singular integral for g_{ij} may be performed either analytically or using a special logarithmic quadrature rule. The singular integrals for the terms h_{ij} do not need to be evaluated since the diagonal terms satisfy the equation

$$H_{ii} = -\sum_{j=1}^{N}{}' H_{ij} .$$

Finally once the boundary values of $\tilde{\phi}$ and $\tilde{q}$ are known the interior potentials may be found from the integral

$$\phi(r) = \frac{1}{2\pi} \oint_C (\tilde{\phi} \frac{\partial}{\partial n} (\ln R) - \tilde{q} \ln R) ds. \quad (8)$$

PARALLEL IMPLEMENTATION

In the previous section we have seen that there are three phases in the boundary element method: the equation set-up phase, the solution phase and the field recovery phase. All three phases show both fine-grained and coarse-grained parallelism.

As has already been mentioned there has been very little published work concerning the parallel implementation of the boundary element method.

Most workers have concentrated on the equation solution phase. For example, in the application of the boundary element method to elastostatic problems, the method of substructures provides a coarse-grained parallelism that has been exploited using a vector processor[3,9,10].

A fine-grained parallel solution of the equations has been used for potential problems on the DAP[4,7]. Standard library solution routines incorporating Gauss-Jordan elimination which is particularly suitable for SIMD machines, were used.

Coarse-grained parallelism in the set-up phase has been exploited on a vector processor for time-dependent problems[11] in which the granularity is associated with the time-stepping procedure.

The equation set-up phase for potential problems illustrates very well how the boundary element method exhibits both fine-grained[4] and coarse-grained[5] parallelism and we illustrate the two types of parallelism as follows:

Consider the development of the terms a_{ij} as given by equation (7) using an n-point Gauss quadrature rule.

On a conventional sequential machine there are three loops: an inner loop over the quadrature points, an intermediate loop over the target elements and an outer loop over the base nodes.

In the fine-grained parallel approach, implemented on the DAP[4], each processor corresponds to a boundary node and the only loop performed is that associated with the quadrature i.e. the integrals are performed in parallel and the elements in the coefficient matrix **A** are produced simultaneously.

In the coarse-grained parallel approach, implemented on a transputer array[5], the quadrature loop is considered to be the outer loop and each processor is considered to be associated with a Gauss point i.e. the values of the integrands are evaluated in parallel and then accumulated to produce the coefficient matrix **A**.

The fine-grained and coarse-grained parallelisms for the calculation of h_{ij} are illustrated schematically in figures 3 and 4.

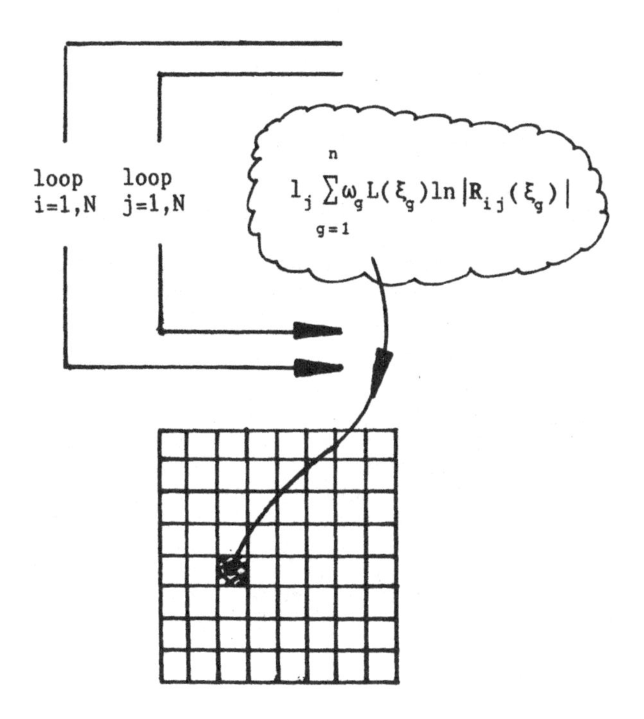

Figure 3. Calculation of h_{ij}, fine-grained mapping to an SIMD machine.

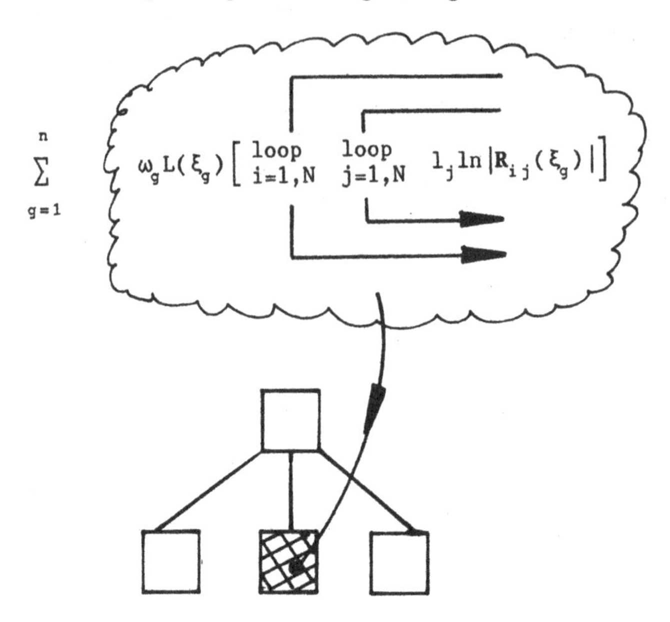

Figure 4. Calculation of h_{ij}, coarse-grained mapping to an MIMD machine.

Since the field recovery phase is very similar in nature to the set-up phase, directly analagous fine-grained and coarse-grained procedures may be developed.

For integral formulations of non-homogeneous boundary-value-problems, which involve domain integrals[8], a clear coarse-grained parallelism exists in the set-up phase. Similarly if the dual reciproity method[12] is used extra boundary integrals are introduced yielding a fine-grained parallelism in the set-up phase. In each case the corresponding contribution to the right-hand side vector, b, may be calculated on one or more processors, in parallel with the calculation of the coefficients a_{ij}.

CONCLUDING REMARKS

The advantages of exploiting a parallel computing environment are two-fold: either to reduce computing time or to introduce simpler code. Both advantages have been reported in the context of the boundary element method. The parallel implementation of linear elements on the DAP has been shown to yield a speed-up with a factor of the order of one hundred compared with an equivalent code running on a sequential computer[4]. Most parallel computers may be programmed using high level languages. Indeed extensions of FORTRAN are available to handle vector data structures on vector processors and the DAP uses an extension which handles matrix data structures. The parallel code written for quadratic elements on the DAP[7] is considerably more transparent than the equivalent sequential code.

The boundary element method has inherent parallelisms which may be exploited on either SIMD or MIMD architectures and there is every reason to expect both types of implementation to be developed further. As parallel computing environments become cheaper and more widely available, in particular when such architectures are available on desk-top computers, so the boundary element method will undoubtedly be seen as a very attractive modelling process. It is already well-established as an important technique in CAD using sequential codes[13] and parallel codes will certainly be developed in the near future. In fact a parallel implementation, using an integrated array processor, has been developed as part of a CAD package[14] for electric devices.

In electromagnetic field calculations the hybrid boundary element/finite element approach has been found to be particularly appropriate. The finite element method is used to analyse the finite regions containing magnetic material and the boundary element method is used for the infinite free space region. Since both methods have been successfully implemented in parallel environments it would be expected that the hybrid method would also be well-suited to parallel computation on both fine-grained and coarse-grained architectures.

REFERENCES

1. Ortega, J. M. and Voight, R. G. Solution of partial differential equations on vector and parallel computers. SIAM Review, 27, 1-96, 1985.

2. Lai, C. H. and Liddell, H. M. A review of parallel finite element methods on the DAP. Applied Mathematical Modelling, 11, 330-340, 1987.

3. Bozek, D. G., Ciarelli, D. M., Ciarelli, K. J., Hobous, M. F., Katrick, R. B. and Kline, K. A. Vector processing applied to boundary element algorithms on the CDC Cyber-205. EDF Bulletin de la Direction des Etudes et Recherches, Série C-Mathématique, 1, 87-94, 1983.

4. Davies, A. J. The boundary element method on the ICL DAP, Parallel Computing, 8, 348-353, 1988.

5. Davies, A. J. The boundary element method on a transputer array. Paper submitted for inclusion at the 13th International Conference on Boundary Element Methods, BEM XIII, to be held in Tulsa, Oklahoma, USA, August 1991.

6. Flynn, M. Some computer organizations and their effectiveness. IEE Transactions on Computing, C-21, 948-960, 1972.

7. Davies, A. J. Quadratic isoparametric boundary elements: an implementation on The ICL DAP. In Boundary Elements X (Ed. Brebbia, C. A.), 3, 657-666. Proceedings of the 10th International Conference on Boundary Element Methods, Southampton, UK, C.M.P., 1988.

8. Brebbia, C. A. and Dominguez, J., Boundary Elements an introductory course, C.M.P., 1989.

9. Kane, J. H., Kashava Kumar, B. L. and Saigal, S., An arbitrary condensing, noncondensing solution strategy for large scale, multi-zone boundary element analysis. Computational Methods in Applied Mechanics and Engineering, 79, 219-244, 1990.

10. Kline, K.A., Tsao, N. K. and Friedlander, C. B. Parallel processing and the solution of boundary element equations. In Advanced Topics in Boundary Element Analysis (Eds. Cruse, T. A., Pifko, A. B. and Armen, H.), ASME-AMD, 1985.

11. Dohner, J. L., Shoureshi, R. and Bernhard, R. J. Dynamic modelling of three-dimensional sound propagation. ASME paper 86-WA/NCA-18, presented at the winter annual meeting, Anaheim, California, 1986.

12. Partridge, P. W., Brebbia, C. A. and Wrobel, L. C. The Dual Reciprocity Boundary Element Method. C.M.P. 1990.

13. Danson, D., Brebbia, C. A. and Adey, R. A. BEASY System. Advanced Engineering Software, 4, 68-74, 1982.

14. Calitz, M. F. and du Toit, A. G. CAD system for cylindrically symmetric electric devices. IEEE Transactions on Magnetics, MAG-24, 1, 427-430, 1988.

Tds2gks: Breaking the TDS Text Barrier

G.J. Porter, S.N. Lonsdale, B.M. Hutson

Department of Electrical Engineering, University of Bradford, Richmond Road, Bradford, West Yorkshire, U.K.

ABSTRACT

This paper will describe a new graphics interface (GI) for the TDS, which provides the application programmer (AP) with access to the GKS library available on a SUN workstation. The focus of this contribution will describe how this GI was implemented. In addition, a method for its future expansion to support other graphics libraries such as Ghost and Simpleplot will be outlined.

1. INTRODUCTION

Recent years have seen the continued expansion and exploitation of parallel systems in both education and industry. The Inmos Transputer has aided this expansion by providing a low cost, flexible processor which is both easy to program and can be utilised in a straightforward manner in a user configurable network. An integral part of the available package is the Transputer Development System[4]. This provides the user with a complete development environment, enabling the rapid prototype and testing of parallel algorithms. However, the TDS is limited by its available Human Computer Interface (HCI) and graphics display facilities. As standard, the TDS will only provide a text orientated HCI, thus precluding any graphics display by the application programmer.

This paper will describe the features of a special server program, which allows the transputer based, Occam programmer access to the library of GKS graphics routines.

The server runs on a sun workstation and communicates with a special Occam harness program running on the root transputer. The Occam programmer has a simple predefined soft-channel interface to the harness, which translates the commands into the communication protocol format before transmitting them to the sun based server. In this way the application programmer sees only a simple channel interface, down which he sends GKS commands and parameters.

1.1 The Transputer and the TDS

The Transputer[1] is a single chip microcomputer, with internal memory and four asynchronous, serial communication links. It is derived around a flexible architecture that allows for easier generation of multiprocessor systems and the simplified distribution of programs.

The Transputer Development System (TDS) is a programmers shell environment, developed by Inmos to aid program development. It includes an integral editor, compiler, linker, configurer and network loader. The system runs on a transputer and communicates with the host computer via a server program. This allows the transputer to have simple access to the hosts keyboard, screen and file system.

The Occam[2] programming language was developed by Inmos concurrently with the original transputers. It is designed to represent the lowest level of programming language that the programmer will need, whilst still having all the benefits of a high level language. The transputer has microcoded support for some Occam constructs and as such executes Occam programs efficiently. Occam is still the chosen language for experienced transputer programmers, due to its flexible structure and the ease with which parallel programs can be developed.

1.2 The Graphics Kernel System : GKS

The Graphics Kernel System (GKS) is a set of language and device independent operations that facilitate the programming of two dimensional colour graphics applications. The GKS standard is defined by the British Standard Institute[3] (BS9390), the International Standard Organisation (ISO 7942) and the American National Standards Institute (ANSI X3-124 1985), making it a world standard set of graphics routines.

The available version of GKS on the SUN system was a Fortran binding, UNIRAS's UNIGKS[5]. All references to GKS in this paper will refer to this binding.

1.3 TransLogic

The tds2gks server was written as part of the TransLogic[7,8] parallel digital logic simulator project. Soon after starting this project it became evident that the available interface between the closed environment of the TDS and the Sun workstation or the IBM PC would not be sufficient. It was necessary to extend this interface to include a graphics capability. Figure 1.1 and 1.2 illustrates how the tds2gks server interfaces with the TransLogic simulator.

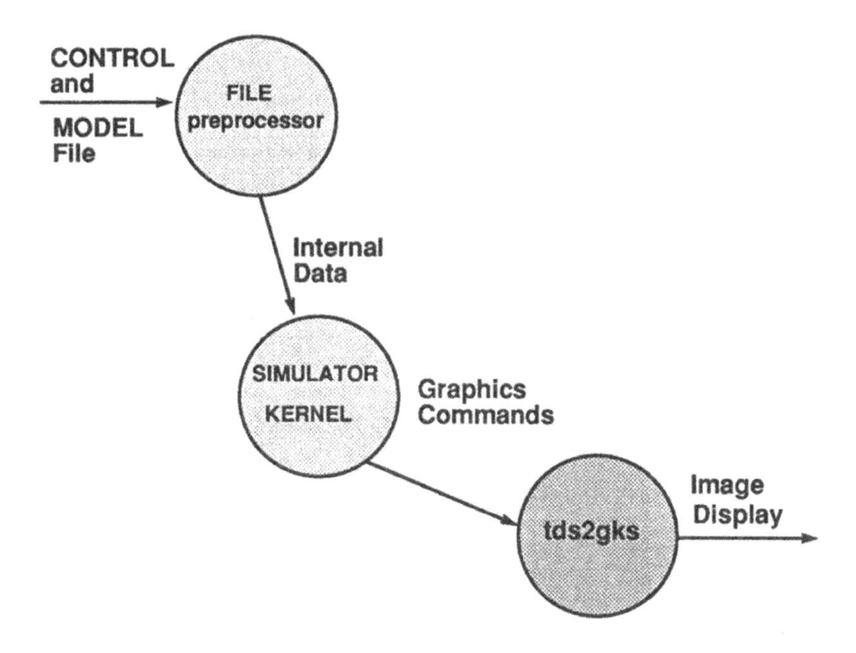

Figure 1.1 TransLogic Structure

A decision was quickly reached, that the interface would be designed to allow access to GKS as its primary aim, but would be designed in such a way that support for other standards could be simply achieved. A second criteria was that there should be no visible change in the actions of the TDS or in the way that it interfaces with the user. Thus careful consideration of the interaction of the GKS system and the TDS was required.

1.4 Overview

The server was split into two parts, one residing on the actual transputer and the other on the Sun workstation. This is illustrated in figure 2.1. The transputer harness provided a standard channel interface for all communication with the application program. The Sun based server was responsible for the control of the TDS and the GKS servers and the instigation of the received GKS commands.

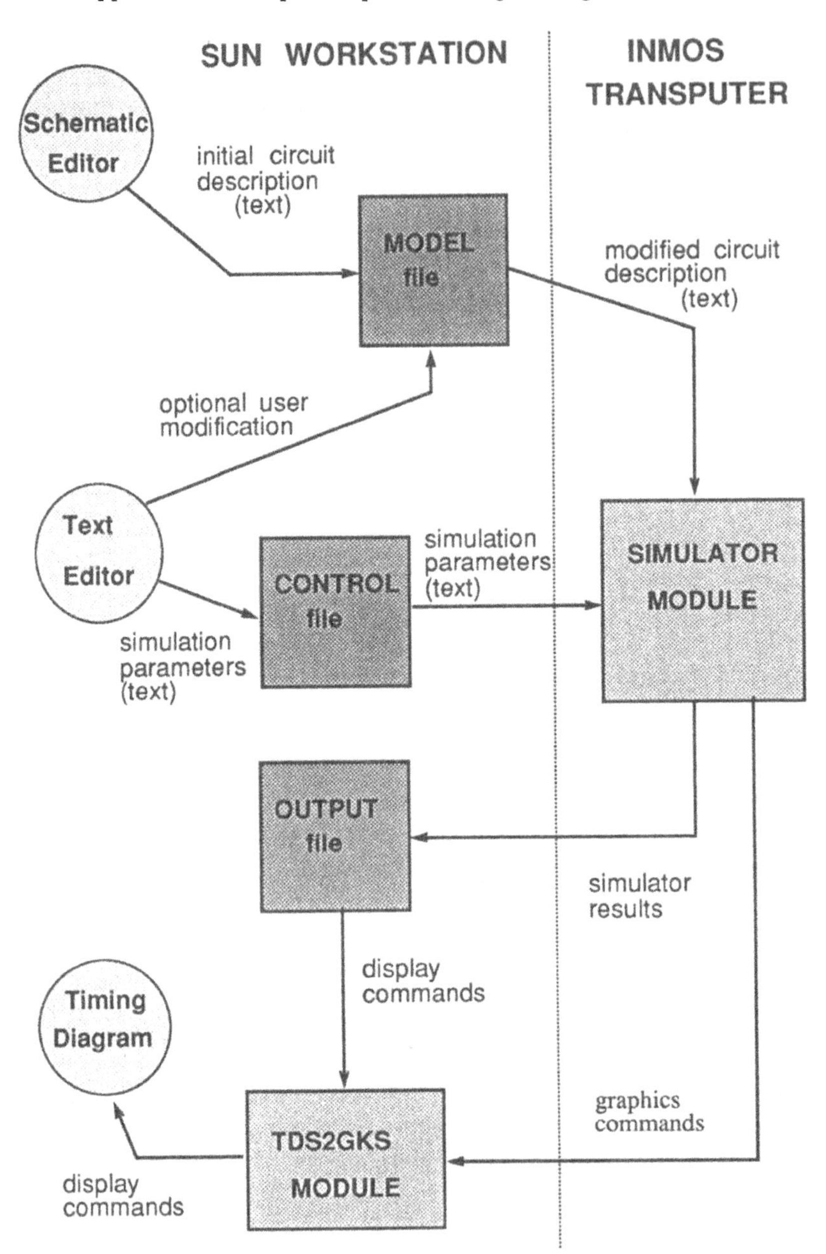

Figure 1.2 TransLogic Structure

2. COMMUNICATION

To communicate effectively between the SUN and the transputer, a special communication method and protocol had to be established. This allowed the transfer of commands and parameters to the SUN server and the return of status messages and results to the Transputer Server.

2.1 Communication Methodology

As it was not possible to directly alter the TDS server, the only available method of passing data from and to the transputer was via files. Thus two files, *tmp.pipe.in* and *tmp.pipe.out* were used to transfer data to and from the transputer based server. Each file acting as a pipe in a single direction, from the Sun to the Transputer and from the Transputer to the Sun. The transputer based GKS.server taking commands from the application program and placing them in the output file, the Sun server reading the file, acting on the commands and placing any results in the input file. This is shown in figure 2.1.

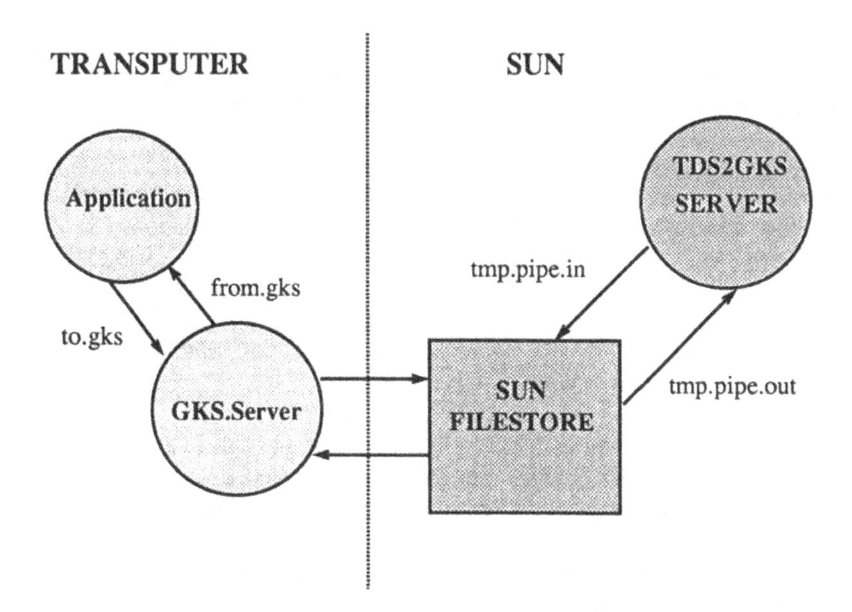

Figure 2.1 Communication Methodology

The Occam channels *to.gks* and *from.gks* are used by the application program to send data to the transputer GKS.server and to receive results from it. Thus the programmer sees only a pair of channels as his interface, effectively seeing the GKS server as a parallel process executing on another processor. If at a latter date, GKS becomes available for the transputer the application programs will not need altering. A simple change of the GKS.server program will enable the use of the transputer version rather than the Sun version.

2.2 Communication Protocol

To allow a strict control of the data transfer between the transputer and the Sun servers, a fixed format or protocol was used for the data. The data was transferred as :

GKS Routine Code	INTEGER
Number of Integers	INTEGER
Number of Reals	INTEGER
Number of Character Arrays	INTEGER
All integer values	INTEGER(S)
All real values	REAL(S)
All characters	CHAR(S)
ASCII '0'	NULL

The NULL character was used to mark the end of the list, telling the reading server that the transfer was complete and that the data block could be used. It also acted as a partial confidence check.

2.3 Communication Synchronisation

Normally, under Occam's rules all communication is synchronised by the channel mechanism[9], however it was not considered to be efficient for the Transputer to wait while the Sun workstation completes the communication. Thus, the application program considers its communication to be complete when it has transferred its data packet to the GKS.server. No consideration was taken of the effects this could have within the program.

This decision was justified by the fact that all of the GKS graphics operations would be carried out within a strict first come first served order. The GKS.server program not accepting another command block from the application program until it had completed the current transfer and received and transmitted to the application program any results. Also, if the application programmer required a more strict form of synchronisation within his program, he could simple follow each command with a GKS Enquire function. This would force the current process to wait for the results of the enquire function before continuing, thus introducing synchronisation.

Figure 2.2 illustrates the sequence of events for non-synchronous communication, were no synchronisation message is returned to the transputer. Figure 2.3 illustrates a synchronous communication, with the Sun server returning a message via the input file to the server.

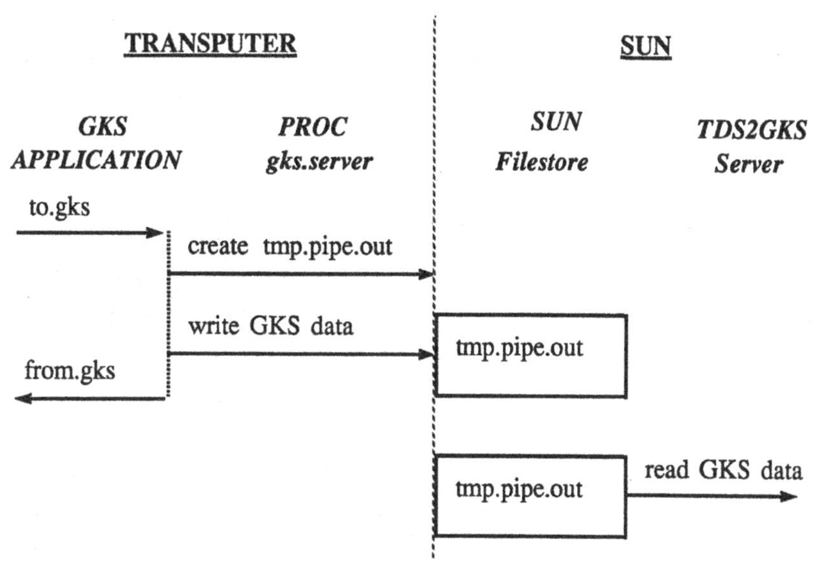

Figure 2.2 Non-synchronous Communication

3. SUN SERVER

This portion of the server was responsible for receiving commands from the transputer based server, executing the corresponding GKS function and returning any results. It was also responsible for ensuring that the normal TDS environment was available to the programmer.

3.1 Server Structure

As it was necessary for the server to allow access to the TDS and GKS routines in parallel, the server was constructed from three concurrent threads. The Unix Fork, Wait and Exec system calls were used to achieve this structure, as shown in figure 3.1. This illustrates how the server spawns two child processes, one the TDS and the other the actual GKS server. In this way the parent process can monitor the execution of the TDS and kill off the GKS server when the TDS is terminated.

3.2 GKS Child Process

This is the actual worker process that executes the GKS commands on the SUN workstation. It accepts commands from the input file, calls the corresponding GKS library routine and sends any results back via the output file.

As it will execute in parallel with the TDS server, and to reduce the amount of slow file checks it must make, whether the input file is found to be empty the child process will sleep for one second. This overcomes the problem of it continuously reading an empty file and utilising processor time.

3.3 Server Termination

The server is required to terminate correctly whenever the TDS is terminated. It must ensure that the GKS server has killed all open window and that it has terminated correctly before allowing itself to complete. This is controlled by a single Unix Signal from the TDS server when it is about to terminate, as shown in figure 3.1.

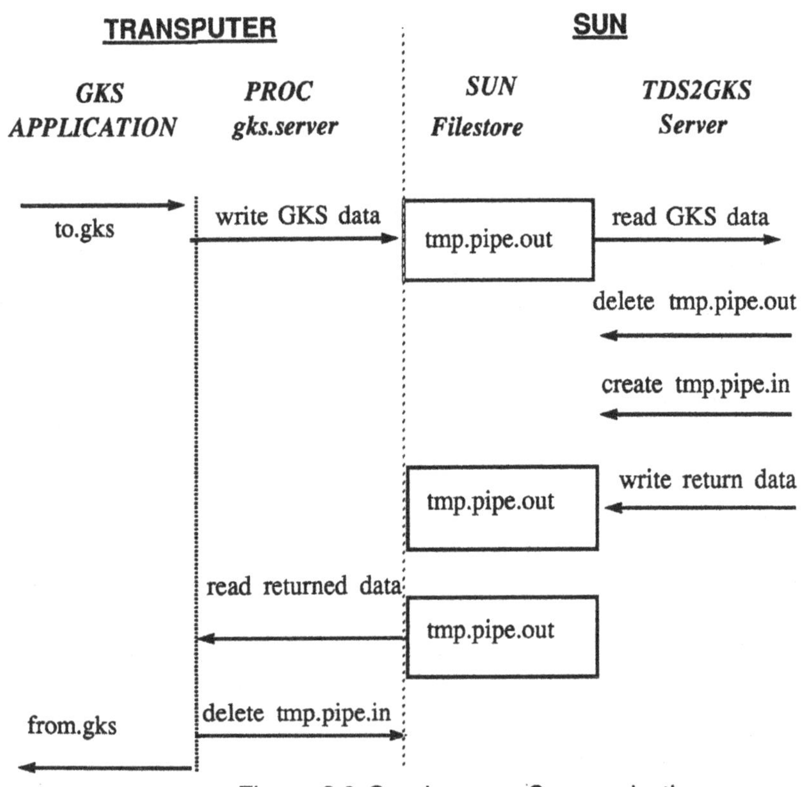

Figure 2.3 Synchronous Communication

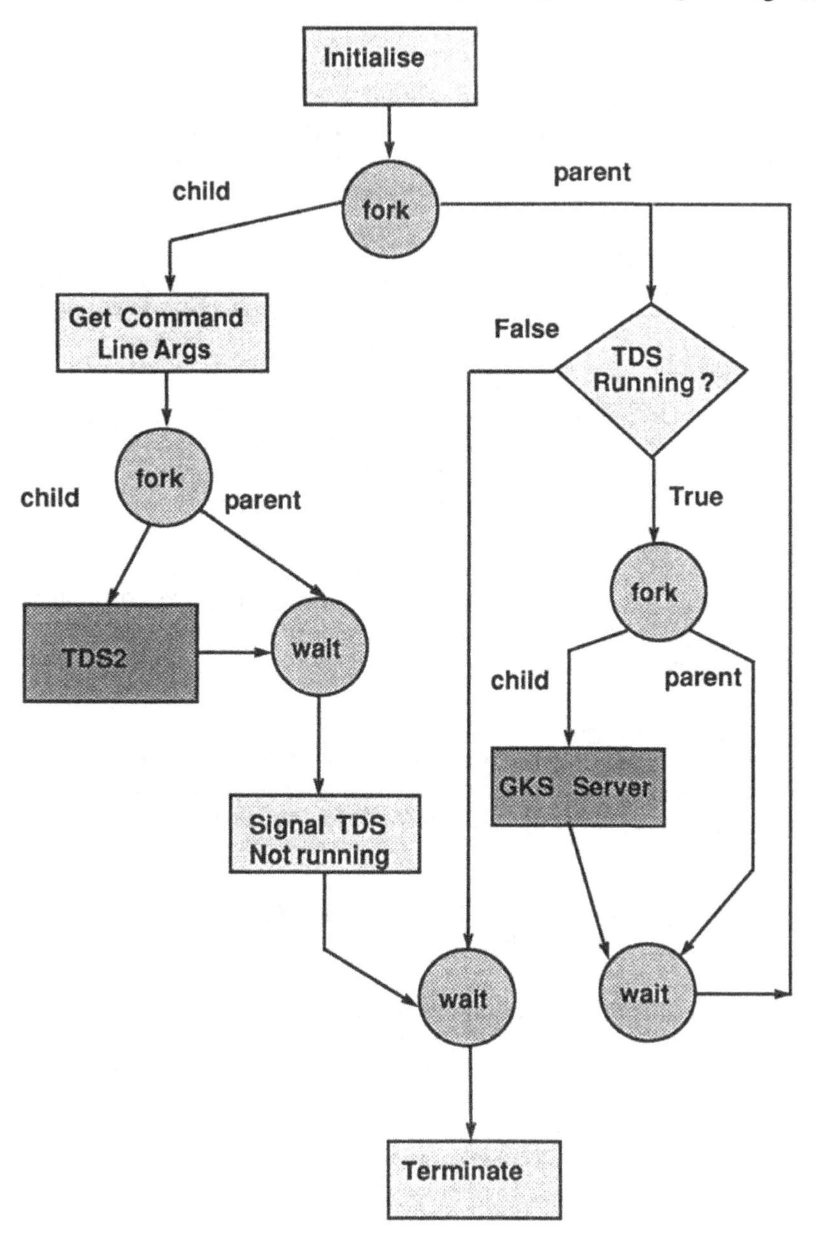

Figure 3.1 SUN Server Structure

4. TRANSPUTER SERVER

The Transputer Server or Harness program was responsible for accepting commands and data from the application program, converting them to the communication protocol format and transmitting them to the GKS child process on the Sun via the output file.

4.1 Occam Channel Protocol

To communicate with the application program via the *to.gks* and *from.gks* channels the harness has a predefined Occam protocol. This is:

```
PROTOCOL GKS.RECORD IS
    INT;                  -- GKS Command Code
    INT : : []INT ;       -- No and array of integer params
    INT : : []REAL32 ;    -- No  and array of real params
    INT : : []BYTE :      -- No  and array of characters
```

The use of this protocol allows the Occam compiler to do runtime checks on the channel communication, aiding program development and debugging.

4.2 Harness Application

The harness program and the application program are run as two parallel processes on the transputer, and as such must be declared at compile time. A construct such as the following is recommended:

```
PAR
  gks.server(from.filer,to.filer,from.gks,to.gks)
  application(.......,from.gks,to.gks)
```

4.3 Harness Initialisation and Termination

According to the definition of a PAR statement in the syntax of Occam, it will never complete until all its processes have complete. Thus as the application program and the server run under a PAR construct, they must both complete before the PAR and hence the program can complete. To achieve this it is necessary to ensure that the harness program will terminate when instructed to do so by the application program.

Before it will utilise the filer channels, the harness process will expect to receive the 'gks.initialise' command. This must be the first command it receive, otherwise any commands sent will be ignored. The harness will then enter a loop, passing commands and data as described above.

The harness process will be terminated by the receipt of a 'gclks' or a 'geclks' command from the application process. It will then pass on the command to the Sun based server and terminate without error.

5. SUPPORT OF OTHER GRAPHICS STANDARDS

As mentioned in section 1.3, the initial design of tds2gks took into account the need to support graphics standards other than GKS. To this end the number of GKS specific sections was kept to a minimum, these being: the GKS child process and the transputer server initialisation and termination sequences.

To alter tds2gks to support another standard such as PHIGS would require that a child process be written to call the correct PHIG routines and that the initialisation and termination controls of the transputer server be changed to those of PHIGS. The source code or a working version of tds2gks can be made available by contacting the author at his above address.

5.1 PC Based Graphics Support

A graphics support package for the TDS running under DOS has just been completed. This allows the application programmer access to all of the available Turbo C graphics library functions. It utilises a TSR to overcome the problems of DOS not being multi-tasking and provides the user with full control of the graphics server via a menu routine called up with a Hot-Key.

It is hoped to expand this server to allow the user access to other libraries and to emulate GKS on a PC. Again source code or a working version of this program can be obtained from the author at the above address.

6. SUMMARY

This paper has outlined the structure and function of tds2gks, a Sun based server program that allows an application programmer to access GKS routines via a simple soft channel interface. It has also briefly introduced a PC based program that will allow the user access to any of the Turbo C graphics commands via a simple channel interface.

An example of the output from tds2gks via the TransLogic simulator is shown in figure 6.1.

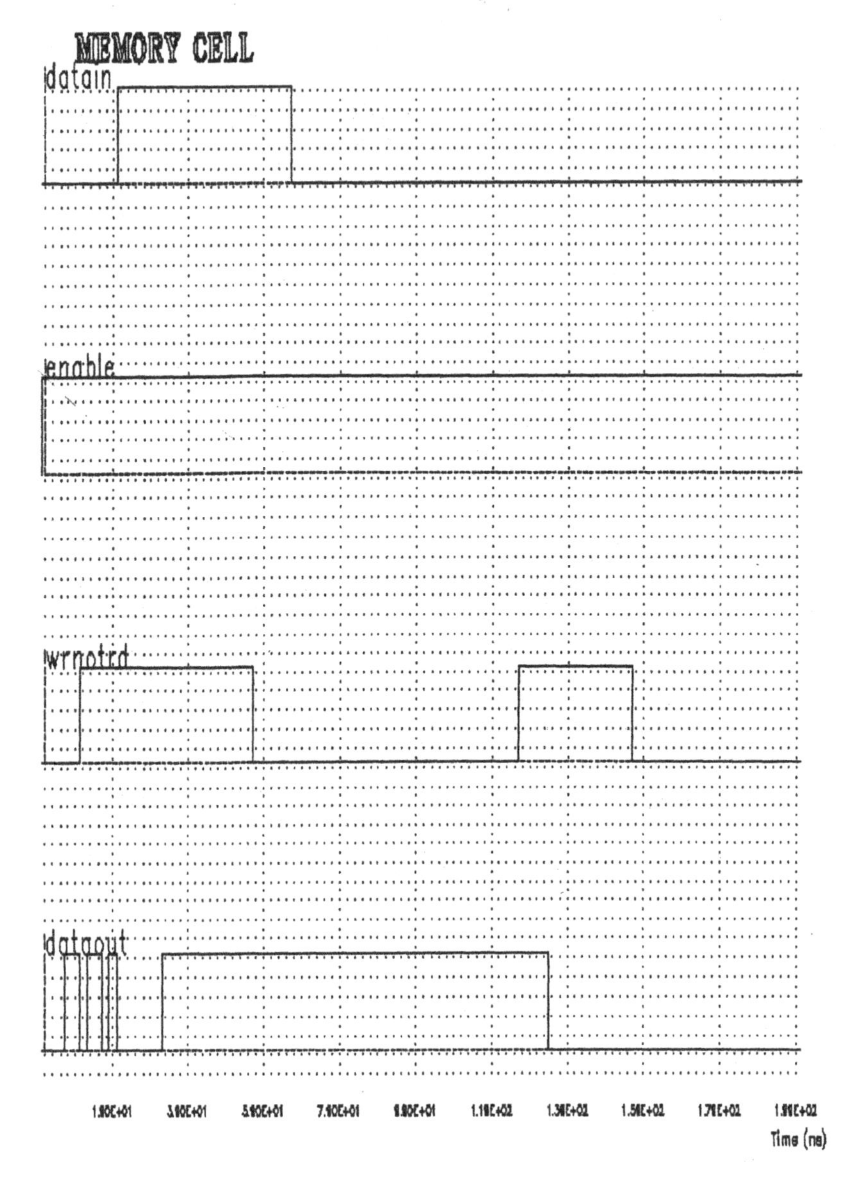

Figure 6.1 Example Output from TransLogic using tds2gks

REFERENCES

1. Inmos Ltd The Transputer Reference Manual, Prentice-Hall International, 1988

2. Inmos Ltd, Occam 2 Reference Manual Prentice Hall International, 1988

3. A set of functions for computer graphics programming, the Graphics Kernel System (GKS) BS 6390: 1985 (ISO 7942 - 1985)

4. Inmos Ltd, The Transputer Development System Prentice Hall International, 1988

5. Computer Graphics: Graphics Kernel System (GKS) language binding, Part 1, Fortran. Draft for Development. DD 114: Part 1: 1986 (ISO/DIS 8651/1)

6. Porter, G.J. and Lonsdale, S.N., TDS2GKS : A Transputer - GKS Graphics Server, Research Report No 466, University of Bradford, Postgraduate School of Electrical and Electronic Engineering and Information Systems, June 1990

7. Porter G.J. and Hutson, B.M., A Transputer Based Digital Logic Simulator, Research Report No 467, University of Bradford, Postgraduate School of Electrical and Electronic Engineering and Information Systems, June 1990

8. Porter, G.J., Hutson, B.M. and Lonsdale, S.N., TransLogic - a Brute Force Digital Logic Simulator, (Ed. Fielding D.L.),pp. 66-76,Proceedings of the Fourth International Conference of the North American Transputer User Group, Ithaca, New York, 1990.

9. Porter, G.J. and Rhodes, R.L. The transputer at Machine Level, Research Report No 440, University of Bradford, Postgraduate School of Electrical and Electronic Engineering and Information Systems, December 1989.

Operating System Support for a Parallel Cellular System

F.F. Cai, M.E.C. Hull

Department of Computing Science, University of Ulster at Jordanstown, Co. Antrim, BT37 OQB, U.K.

Abstract

This paper reports the design and evaluation of operating system policies for a parallel access machine based on the loosely-coupled cellular architecture. A hierarchical model is presented for the implementation of the distributed operating system, which disperses the overall control to many user abstract machines executing in parallel. Various cell management strategies are then put forward which attempt to maximise the utilisation of the system resources, while minimising communication overheads. A comprehensive simulation study has been conducted whose outcome provides a useful insight into the effectiveness of the proposed policies, which may be tailored to the needs of various other multi-user parallel processing systems.

1. Introduction

Whilst the recent advances in architecture of various parallel systems have offered the necessary hardware base for parallel processing, it is the support of an operating system which realizes that parallelism by utilizing the system resources effectively [4,6].

This paper reports the design and evaluation of operating system strategies for a parallel machine based on cellular architecture. A cellular system may essentially be visualised as a homogeneous array of intelligent memory cells, which offers a high level of availability and flexibility while maintaining a simple yet extensible structure. A large cellular system can easily be built up of several sub-systems through the use of a reconfigurable interconnection network. As a general purpose parallel machine, it is an extension in computer power over and above that provided by conventional architectures.

To exploit the parallelism from the hardware, the operating system must create a parallel execution environment in which overall tasks are distributed onto available resources. A hierarchical model of a distributed operating system for the cellular machine is proposed, which disperses the overall control to many parallel executing user abstract machines in both single user and multi-user configurations.

One of the most important resources in the cellular system is cells of which the effective management amongst users is crucial to the overall system performance. It is also an important requirement to facilitate fast communications between parallel processes [3] residing in cells within the user's execution environment, given the fact that a high degree of system parallelism realised by a large collection of co-operating cells may also result in frequent interaction between them, potentially causing heavy traffic on various system buses.

To this end, various cell allocation policies have been proposed to capture the modular feature of the cellular architecture. These strategies attempt to maximise the utilisation of system cell resources, while minimising the communication overhead. These requirements are also of paramount importance to other truly parallel systems.

A software simulation has been carried out to evaluate the performance factors of these strategies within a cellular-like Group Processor System (GPS) environment. Different types of benchmark programs were executed ranging from sequential to parallel in nature. The outcome of this study has provided a useful insight into the effectiveness of the proposed policies. These operating system policies, considered together with various other aspects of database management support (for example, buffer management [2,5]), can be tailored to the needs of parallel database machines.

The remainder of the paper is organised as follows. After a brief outline of the cellular architecture in the next section, a hierarchical operating system model is presented in Section 3. The cell management strategies are then discussed in details in Section 4 and 5, followed by a presentation of the outcome from the simulation experiment. Finally, a summary is given in Section 7.

2. Loosely-Coupled Cellular Architecture

A cellular system may be visualised as a homogeneous array of intelligent memory cells, with direct coupling to the outside world, i.e., the users. This section describes briefly the main features of the loosely-coupled Group Processor System [7,8] based on the cellular architecture.

The realisation of the GPS is based on the principle that a complete system is built up of sub-systems of common elements which are cells, modules and bus structure. A cell, in its simplest concept, can be viewed as a typical uniprocessor system. It provides the processing ability through its processing unit, with the available local memory accommodating the storage needs. Moreover, the input/output communication capability supported by the structure allows each cell to interact with others. This offers a hardware base for the efficient parallel execution of processes mapped to the different cells throughout the system.

A module has an array of cells and is coupled with other modules in the system via a flexible interconnection network. Co-operation between modules is achieved by allowing communication through a common bus structure. This powerful structure consists of a number of functionally dedicated buses (including global, inter-module, intra-module and input/output buses) that are available for use by any cell within any module.

With many of these intelligent cells per module, the system has the resources needed for a fully distributed machine. A multi-user image of the system is illustrated in Figure 1, where different users are working in their own environment within a *reconfigurable* user boundary.

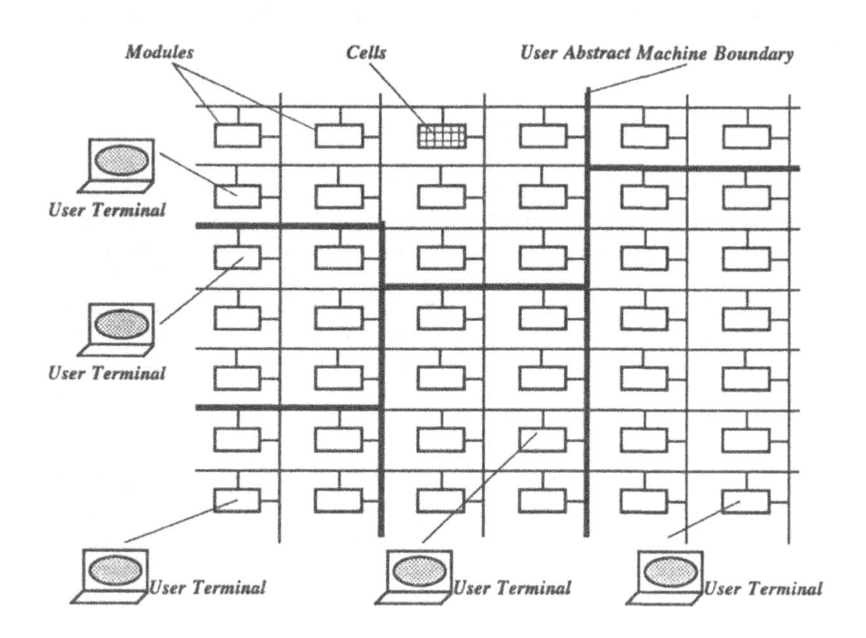

Figure 1. Multi-user Group Processor System

In order to appreciate the cellular hardware/software features the concept of an *abstract processor* is introduced. An abstract processor is a group of cells configured so as to execute a given process. These cells are interconnected by physical, as well as logical, communication paths. In other words, it is the equivalent of a multiprocessor/local memory architecture, with a wide range of possible interconnection strategies.

The concept of an abstract processor is not too far distant from that of a virtual processor. Whereas a virtual processor eventually maps to the physical processor when the process is being executed, the abstract processor may map to one or more cells in the cellular system. With many such abstract processors within a machine, the Group Processor structure can be seen as interconnecting abstract processors.

Referring to Figure 1 again for a conceptual relationship between various users and their *abstract machines*. These users are working under their own abstract machine boundaries created by the operating system. The execution of processes is performed by cooperation among those logically-threaded executing abstract processors within the abstract machine. Each user's abstract machine may grow during processing, simply by expanding its logical boundary. As a general purpose computer, a cellular GPS is an extension in computation power over and above that provided by more conventional architectures.

3. A Hierarchical Operating System Model

The realisation of multi-user operations in the GPS is supported by its parallel access hardware structure, which provides the potential base for a high degree of distributed processing. However, it is this loosely coupled architecture that imposes inherent complexity in overall system management. Clearly, the control of this complexity is crucial to the operating system (OS) design. A good solution to this problem is to construct a well organised control hierarchy of multi-level abstraction [1], which provides a simple image in each layer's implementation whilst reflecting the complex nature of the system at the global level.

To match the hardware features of the GPS, a distributed operating system structure is proposed, as shown in Figure 2, which partitions the overall control into multi-level management by providing a hierarchy of different abstractions.

From the logical structure of the highest level, a global OS works as a system oriented manager which resides in a central command processor to create a multi-user environment. It is important to note that this global OS does not provide centralised control, but acts as an arbitor to dispense the control of resources following the system start-up, to software created abstract machines. A set of local OSs reside in each user's abstract machine. They are created by the global OS when their users log in. The local OS is responsible for the control over the execution of its user programs, and communicating with the global OS when necessary.

This layer of control can be further distributed to lower levels, when the program is divided into smaller parallel processes which are mapped onto abstract

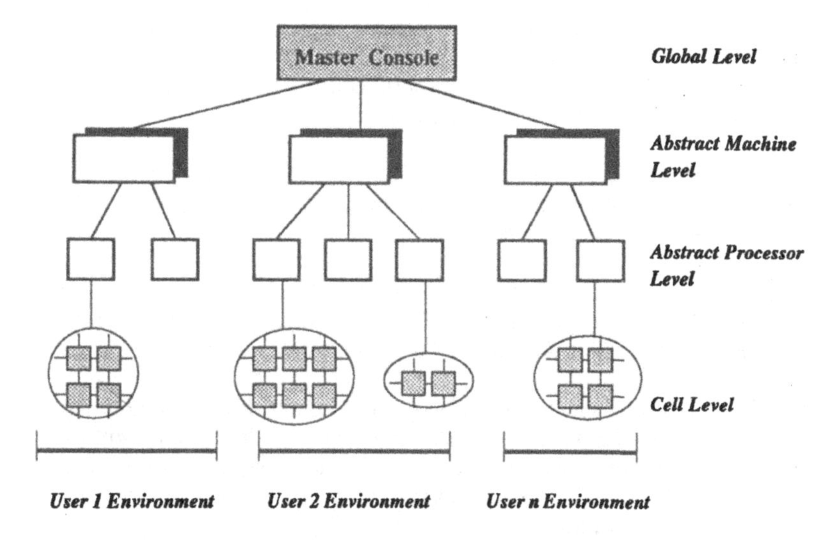

Figure 2. Multi-Level Distributed Control

processors and physical cells available in the user abstract machine. Each abstract processor or cell is a simple but complete uniprocessor system, in which resides a copy of simple OS "primitives" which schedules and executes processes within the processor, provides system calls where needed, and handles queueing and routing of messages for the processes. It is therefore capable of facilitating the necessary self-management for its process execution.

This logical structure of this control hierarchy creates a fully distributed environment where the various levels are implemented in parallel. Furthermore, the complexity of the overall system management is minimised by dispersing the control from global level down to users's abstract machine level, the abstract processor level, and finally the cell level.

This structure also provides a more flexible communication mechanism than a traditional tree organisation, where nodes at the same level can only pass messages through their common parent node. In contrast, the proposed structure facilitates direct interaction where necessary between abstract processors, or cells, at the same level within the user boundary. This realisation is supported by the system interconnection network.

4. A Cell Management Strategy

The GPS machine in which overall tasks are performed by cells executing in parallel is faced with a challenging problem of managing a large number of cells in an effective manner. This problem is made even more complex by the frequent interaction between executing cells, given the fact that each cell is relatively small and application programs are generally mapped onto a collection of co-operating cells. An efficient cell management scheme is therefore needed to utilize the system cell resources effectively, while minimising the communication overhead.

To tackle this problem, a multi-level cell management strategy is proposed, as shown in Figure 3. This strategy exploits the hardware feature of the GPS, and matches the OS model described earlier.

Under this scheme, a global allocator residing in the global OS performs the cell allocation function on the system level. It receives requests from all the systems users, and process their demands on a first-come-first-serve basis.

The global allocator is related to a global pool of available cell resources, termed the *free-cell pool*. These cells are organised in a table list with information on the location of each cell. When a new user enters the system the minimum number of cells is allocated from this pool by the global allocator, enabling the user to start its operation.

A user, when requiring more cells, interacts with the global allocator by sending a message to the global OS. The message usually contains the following information: the logical number of the user, the number of cells requested by the user, and the location of the cell making the request. Depending on which cell allocation policy is implemented, some parts of the imformation may not be required by the global allocator. In such cases they are simply ignored.

Upon receiving a user request for additional cells the global allocator chooses cells from the global free-cell pool and assigns them to the user. The user is informed if the global pool is exhausted.

The global allocator also receives deallocation messages from users after the release of previous executing cells, or the termination of user programs. These freed cells will then be collected by the global allocator and returned to the global free-cell pool.

On the user abstract machine level, a local allocator resides in the local OS of each user machine. It is also associated with a pool of locally available cells, named the *ready-cell pool*, They are given by the global allocator under certain allocation policy, but not currently being used by the user program. This pool is implemented as a linked queue data structure, which is organised in the order of the physical locations of its cell entries.

The local allocator handles requests on a demand basis from executing cells within its own machine environment. When receiving a new cell request the local allocator checks its ready-cell pool first and , if it is not empty, satisfies the demand by assigning a cell from this pool. Only when there is no available cell in the pool will the local allocator send a message to the global OS.

A user machine environment is composed of a collection of executing cells.

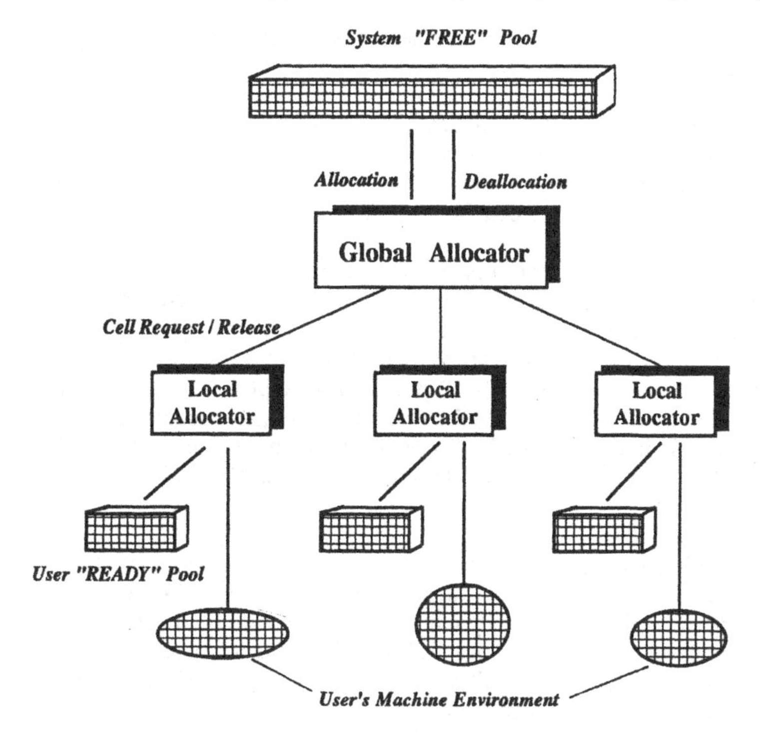

Figure 3. Multi-Level Structure for Cell Management

When the process or data size is beyond the limitation of the cell size, a new cell request is despatched to the local allocator.

The advantage offered by this structure is that the overall cell management of the system is distributed to many parallel executing user machines. Every local OS has the power to allocate and deallocate cells within its domain, provided its local ready-cell pool is available. With its decentralised structure, the scheme reduces the access to the global OS shared by all users, and consequently minimises the potential bottleneck on the system global bus. This is desirable to not only the cellular system, but also other multi-user parallel machines in general.

5. Cell Allocation Policies

Various cell allocation policies are suggested in this section, which are concerned with *when* to allocate a cell, *which* cell should be chosen for allocation, and *how many* cells should be allocated. These policies have been evaluated in a simulation study whose outcome will be reported in Section 6.

5.1 Demand Allocation (DA)

Under this policy a cell is allocated to a user only when it is actually needed. Since the global allocator assigns cells only on a demand basis, the local ready-cell pool in a user machine does not exist.

In this scheme every demand for additional cells from any executing cells will first be sent to its local allocator, who then simply passes the message to the global OS through the global bus. Upon receiving the request the global allocator assigns an arbitrary cell from the free-cell pool to the user.

Obviously, one of the merits of this policy is its simplicity in implementation; no action other than merely transfering the message is required for the local allocator, and only minimum information is needed by the global allocator. There is an added advantage that, by allocating only those cells which are currently needed, the global free-cell pool can be kept as big as possible to accommodate more potential user demands. However, this scheme can result in frequent calls to the global allocator, potential global bus contentions, and consequently, delays in the global OS's response to user requests.

5.2 Predictive Allocation (PA)

Anticipating future needs, this policy allocates a number of extra cells to the user program when it starts execution. Since these pre-allocated cells are available to the user before they are actually required, they are kept in the user's ready-cell pool managed by the local allocator. When the execution proceeds, requests for additional cells are first dealt with by this local pool. Only when it is exhausted will a global OS call be delivered.

This policy potentially offers better performance than DA in terms of less global bus accesses and a faster allocation time. With these pre-allocated cells, the local allocator may accommodate possible cell demands in the near future. The burden placed on the global allocator is substantially lightened. Furthermore, when global allocator allocates a block of physically contiguous cells from the free-cell pool, this scheme makes it possible that a group of neighbouring cells are availble to the user program. This may diminish the communication overhead on the system inter-module buses, for many of the co-operating cells may reside in the same module.

The disadvantage of this strategy, however, is the number of possibly "wasted" cells which have been assigned to the local allocator but are not currently being used. Worse still, some of them may never be used by the program. This in fact reduces the size of the global free-cell pool, from which other users may demand additional resources.

5.3 Nearest Possible Allocation (NPA)

It is expected that, by keeping the executing cells in a user environment as close in location as possible, the communication overhead may be minimised. The NPA policy attempts to achieve this objective by intentionally allocating the "nearest" possible cell to the user request.

When a local allocator receives a request from one of its executing cells, it passes the message to the global allocator, together with the information concerning the physical location of the requesting cell. Accordingly, the global allocator searches through its free-cell list for the nearest possible cell entry. The resulting optimal match is then made available to the user in question.

By allocating cells within close proximity to each other, this policy is likely to provide a linear space for the user program execution. This is realised, however, at the cost of more OS overhead in terms of list-searching. If a long list is involved the allocation process could be slowed down to some degree.

5.4 Exact Fit Allocation (EFA)

For some programming languages using static compilation techniques, the exact number of cells needed for processing is known prior to program execution. In these cases the EFA policy offers another choice.

Under this policy the total number of cells required for the execution is allocated once at the start of the processing by the global allocator, and they are kept in the user's ready-cell pool. Any subsequent cell requests will be promptly satisfied by the local allocator. During the lifetime of the program execution all the cells in the pool will have been used at least once.

There are apparently several positive reasons to favour this policy. Firstly, there is no more than one global OS call from each user program so that the global bus traffic is greatly reduced. Secondly, a fast allocation is achievable by keeping all the required cells in the user environment, which may lead to an improvement in the program elapsed time. Thirdly, the module structure and linear work space for the program execution can be secured by allocating a block of contiguous cells. Finally, the user boundary is set at the start of the processing and remains static, so the OS management is simplified.

However, a number of drawbacks also exist. One is the number of temporarily "idle" (hence "wasted") cells which are allocated to the user well before they are required. Another limitation is the suitability to the type of programming languages. If the exact number of cells needed is unknown in advance this policy may not be feasible (whereas the PA policy offers a practical alternative). Nevertheless, the EFA policy may be used as a performance measurement for the upper bound in evaluating different allocation policies.

6. Performance Evaluation

To investigate the potential effectiveness of the above cell allocation policies, a software simulation study has been carried out. This section reports some of the outcome from this study.

6.1 Set-up of Simulation Environment

In order to establish a simulation model, it is first necessary to specify the simulated GPS environment. This concerns both the system hardware configuration and software parameters.

6.1.1 System Hardware Configuration
The basic machine architecture of the GPS simulated in this study consists of three main components as described in Section 2. They are cells, modules and various system buses. The system configuration is flexible in the simulation, allowing a number of alternatives available by varing the following parameters:

1) The number of mudules in the system;
2) The number of cells per module;
3) The number of global buses;
4) The number of inter-module buses;
5) The number of intra-module buses;
6) The type of input/output bus configurations;
7) The type of interconnection structures.

6.1.2 System Software Parameters
The simulator consists of a class of software variables which tune the system performance. Some assumptions were made regarding various timing parameters for different OS operations.

The execution time of the system is measured by time units of equal length. The time needed for the global OS to allocate a cell is assumed to be 1 time unit (including the list-search overhead), so is the time for a local OS to process a cell request. The total length of time incurred by communication between two cells is the sum of the message length specified by the real programs and communication delays caused by bus conflicts. The intra-module delays (due to interaction between cells residing in the same module) are assumed to be 1 time unit, whereas the inter-module delays (interaction between cells in different modules) depend on the communication path length between two interacting entities. Also, the system speedup ratio is defined as the division of the sequential execution time by the parallel execution time.

6.1.3 Selection of Representative Benchmark Programs
A set of real programs were executed as benchmarks in the simulation to produce a more realistic and accurate output than those generated by using only random numbers as their input data.

Three benchmark programs which have been selected, range from sequential to parallel in nature and vary from numerical to non-numerical computations. The Matrix Multiplication program is chosen for numerical operations, the Quicksort program for non-numerical computations, and the Tower of Hanoi [9] program for purely sequential applications. These programs, while being small in quantity, were designed to illustrate some problems common to parallel processing.

6.2 Examples of Evaluation Results
In order to examine the effect of the cell allocation policies on the system performance, the benchmark programs were executed in both single user environment (where only one of them was running at a time), and multi-user environment (in which they were running in parallel).

Figure 4 lists these representative programs, together with the total number of cells required for each during processing. The Sequential Execution Time was obtained by running a program in a simulated sequential machine environment, whilst the Ideal Parallel Execution Time was acquired in a presumably "ideal" situation by ignoring any overhead in the OS management.

Limited paper size makes it impossible to present a whole range of comprehensive results obtained from the simulation study. Consequently, the emphasis of this section is only on two factors, namely, system speedup and communication overhead.

6.2.1 System Speedup

Figure 5 summarizes the program elapsed time results, which enables the reader to quickly compare the speedup factors achieved under different cell allocation policies (corresponding to 4 entries) in both single user and multi-user environment. The table has been broken down under four main headings, namely, overall system execution time, overall system speedup ratio, separate user program execution time, and separate program speedup. The latter two items are each further divided into three sub-headings associated with each user program. The numbers shaded denote the results for the single user configuration. These results can be compared with their counterparts for multi-user environment which are given in their immediate left column under the same sub-heading.

In the first entry the bus contention in the multi-user environment under the DA policy contributes to a degradation of the program speedup (compared with single user results). The execution time for each user program is prolonged considerably. For example, the speedup achieved for user 1 (matrix multiplication) is nearly halved from the single user result of 11.3 to 7.

Since inter-module communications are more expensive than intra-module ones, it is expected that, by mapping different user programs into separate modules, the communication overhead within a program should be minimised, and the execution time for each program generally maintained at a similar level to that of the single user results. This argument has been vindicated by the other three entries 2, 3 and 4 in the table, which demonstrate the effect of the PA, EFA and NPA allocation policies,

User Number	Benchmark Program	Number of Cells Used	Sequential Execution Time	Ideal Parallel Execution Time
User 1	Matrix Multiplication(10*10)	181	21192	849
User 2	Quicksort (200 Data)	65	76402	18590
User 3	Tower of Hanoi (7 disks)	122	2624	1221
Total		368	100236	20660

Figure 4. Benchmark Programs

Policy	System Execution Time	System Speedup	User Program Execution Time						User Program Speedup					
			User 1		User 2		User 3		User 1		User 2		User 3	
Demand Allocation	21737	4.61	2994	1864	20737	19247	2579	2120	7.1	11.3	3.7	4	1.0	1.2
Predictive Allocation	20361	4.92	1605	1679	19361	18923	1894	1698	13.2	13.4	3.95	4.1	1.4	1.5
Exact Fit Allocation	20173	4.97	1240	1239	19173	18923	1692	1614	17.1	17.1	3.98	4.1	1.56	1.6
Nearest P. Allocation	20350	4.92	1304	1303	19350	18923	1679	1623	16.2	16.3	3.95	4.1	1.57	1.6

Figure 5. Speedup Results

respectively. These results are very favourable. The execution time and speedup results for each program in single and multi-user environment are very close indeed.

This observation effectively indicates that increasing the number of users in the system does not necessarily lead to a marked slowdown of execution for each single job, as long as they are separated from each other in their processing environment.

It is important to note the overall system speedup results, which are calculated by the formula in Section 6.1.2. The sequential time used here is the sum of the execution times of all three programs. Since the overall execution time is dependant on the completion of the longest program in the system, the overall speedup factor is also limited by that program. In this case the long execution time of user 2's program (Quicksort) prevented the system from achieving a higher speedup ratio.

Another interesting finding in this table is that the speedup ratio achieved by user 3 (Tower of Hanoi program) was not as significant as the other two application, reflecting the sequential nature of the program. Indeed, for programs where little parallelism exists, variations in cell allocation policies do not have significant impact on the improvement of program execution time. This is because parallel architectures are of little benefit to sequential computations.

6.2.2 Communication Overhead

Bus contention is another crucial factor affecting overall system performance, especially in a multi-user configuration [10]. Figure 6 outlines the simulation results regarding various system bus load under different allocation policies.

Under the DA policy (first entry) any request for additional cells is passed by the local allocator to the global OS via the global bus. This resulted in a high number of global calls (being 344). In contrast, the PA policy produced a considerable reduction in the number of global OS access, thanks to the provision of pre-

Policy	Global Calls	Inter-Module Communication	Intra-Module Communication
Demand Allocation	344	3340	17
Predictive Allocation	125	1890	1686
Exact Fit Allocation	1	1110	2590
Nearest Possible Allocation	344	1695	1662

Figure 6. Communication Overhead (Multi-User Environment)

allocating cells to the user. It is also self-evident that the EFA policy generated only one global call, as all the cells required for program execution were allocated at its start-up.

Another drawback of the DA policy is demonstrated by its poor inter-module communication results. As this policy allocates a "random" cell to the requesting user without any consideration of its physical location, an undesirable situation is created where each user's executing cells are scattered in several modules and every module is possibly shared by several users. This only seems to contribute to an unnecessarily heavy traffic on inter-module buses.

Remarkable improvements have been witnessed by implementing other allocation policies, as shown by entries 2,3 and 4 in the table. By allocating either a group of physically contiguous cells or a "nearest" cell to the user request, these policies exploit the communication locality in the user programs, hence minimising the bus contention and system communication overhead.

Although this section has presented only a limited set of simulation results due to the paper size, it has nevertheless provided some positive evidence of the usefulness of the GPS architecture. It has been shown that its overall performance in multi-user environment is, in most cases reported so far, close to that of the single user configuration. This indicates that, whilst the performance of conventional time-sharing computer systems usually degrade when several users share the system, the multi-user parallel GPS may maintain its performance, if suitable cell allocation and other operating system strategies are implemented.

7 Summary

This paper has investigated operating system policies for a cellular Group Processor System. A logical control hierarchy has been suggested as a possible model for the operating system implementation. The design aim of this scheme is to provide an environment in which distributed and parallel processing can be performed in a structured fashion.

The efficient management of cells, being one of the most important resources in the cellular system, presents a serious challege to the operating system designers. Various cell allocation policies have been proposed to tackle this problem. The performance factors of these policies have been evaluated in a simulation study. A set of benchmark programs were executed, varying in their computational nature. The outcome of this simulation provides an insight into performance factors which affect the operating system implementation in a cellular machine.

The conclusions drawn from this study should be useful not only to the understanding of the effectiveness of the proposed cell allocation policies in particular, but also to the realisation of the importance of resource management for parallel systems in general. These operating system policies, if considered together with various other aspects of database management support, can be tailored to the needs of parallel database machines, where efficient memory utilisation and fast communication capability are central to the overall performance. Built upon the general ideas described in this paper, a flexible and effective operating system buffer management scheme has been proposed [2,5] and evaluated. It is hoped that such a study will contribute to the existing knowledge of designing efficient operating systems for various parallel processing machines.

Acknowledgement

The main basis of the work presented in this paper was undertaken by the first-named author, directed by Dr G.E. Quick, while at West Glamorgan Institute of Higher Education, Swansea,UK.

References

1 Alia,V. & Gysin,G. "Evolution of Future Microcomputer Operating Systems", Computer Design (July 1984).
2 Cai,F.F., Hull,M.E.C. & Bell,D.A. "Predictive Buffer Management Based on the Resident Set Model", in Computing and Information, Elsevier Science Publishers B.V. (1989).
3 Dubois,M & Briggs,F.A. "Efficient Interprocessor Communication for MIMD Multiprocessor System", ACM Conf. Proc. 8th Annual Symposim on Computer Arch. (1981).
4 Deitel,H.M. "An Introduction to Operating Systems", Addison-Wesley Publishing Company (1984).
5 Hull,M.E.C., Cai,F.F. & Bell,D.A. "Buffer Management Algorithms for Relational Database Management Systems", Information & Software Technology, Vol 30, No 2 (1988).
6 Joseph,M., Prased,V.R. & Natarajan,N. "A Multiprocessor Operating System", Prentice-Hall International, Inc. (1984).
7 Quick,G.E. "Intelligent Memory:- A Parallel Processing Concept", ACM SIGARCH, Vol7, No 8 (1979).

8 Quick,G.E. "Intelligent Cellular System:- A New Direction for Total System Design", NATO-ASI on Relational Database Machine Architecture (1985).
9 Sedgewick,R. "Algorithms", Addison-Weslaey Publishing Company (1983).
10 Treleaven,P.C. "The New Generation of Computer Architecture", ACM Conf. Proc. 10th Annual Symposium on Computer Arch. (1983).

SECTION 2: VISUALIZATION AND GRAPHICS

Opening a Window on the Simulation of a Total Artificial Heart

M. Arabia (*), L. Moltedo (**), F. Pistella (**)

(*) ENEA - TIB, Roma, Italy

(**) Istituto per le Applicazioni del Calcolo - IAC-CNR, Roma, Italy

ABSTRACT

In the design of an artificial heart the simulation is an essential tool in evaluating its possible structural and functional characteristics, as well as in selecting optimal control strategies. Many test runs, each needing heavy computations on supercomputer, are necessary to analyse the performances of the device in different physiological and pathological states.
The visualization of cardiac functions during the simulation is very important in the study of the interactions between the circulatory system and the structural and functional parameters of an artificial heart.

1. INTRODUCTION

The design of a total artificial heart (TAH) requires the analysis and the development of technical solutions which should be reliable, simple, of minimal volume and at the same time reproduce the functional characteristics of the natural heart. They are essentially proper balancing of input and output flows for each ventricle, balancing of the outputs of the two ventricles, physiological distribution of blood in the circulatory system, regulation of the cardiac output according to the needs of the organism.
An overall evaluation of the performances of a TAH is normally given by means of the reconstruction of the cardiac surfaces for each ventricle,which represent the outflow q as a function of the input and output pressures. All the variables are averaged on the

cardiac cycle after an equilibrium state has been reached.

A wide numerical simulation is an essential tool in testing the possible technical solutions and evaluating the related performances of a TAH before constructing the device: in this way for each technical choice it is possible to generate cardiac surfaces for both ventricles and to get a 'measure' of their fitness to the natural corresponding ones in many physio-pathological reference conditions. The partial derivatives of q, also called sensitivities to preload (the input pressure) and to afterload (the output pressure) respectively, are usually assumed as typical indexes of the efficiency of the system.
It is worthwhile to mention that not only the driving system, but also the other components (ventricles, valves, tubes) must be tested, so that an interactive visualization of the cardiac surfaces is perhaps the only way to get an on-line evaluation of the effects of e.g. a change in the mechanical properties of the ventricles or of the valves.

In order to analyse data during the simulation process we use the system named VIDA -Visualization Interface for Distributed Applications- [1], that we designed at IAC. It is a portable interface which allows the researcher to use a distributed environment for computation and visualization in a transparent way. VIDA sends data that we need to visualize to VESCA3 [1], a software we realized to represent scalar and vector fields.

In section 2 the mathematical model of the system TAH-circulatory network is presented. Section 3 describes the scientific visualization software tools we have used. Finally, in section 4, two tests to evaluate the sensitivity of the system to some parameters of the TAH are discussed by means of interactive graphical visualizations.

2. THE MATHEMATICAL MODEL

For the simulation a lumped parameters mathematical model of a TAH is adopted, where:

i) both arterial (systemic and pulmonary) networks are represented by the five components impedance developed by Noordergraaf [2];
ii) the venous networks are modelled according to Guyton [3];
iii) the four cardiac valves are represented by a real diode function (low/high resistance when the valve is open/closed);

iv) the driving system works as a position generator for each ventricle only during systole (ejection), while the diastolic phases (filling) are controlled by the venous pressures and the mechanical properties of the ventricles [4].

The result is that at each time iteration we have to integrate a set of fourteen ordinary differential equations and twelve linear algebric equations and moreover to determine the state (open/closed) of the four valves. The stiffness of the system requires a sampling frequency of 10KHz.
Since each point of the surfaces is the result of the simulation of many cardiac cycles following a step variation of the inputs, and since many points are needed to construct the two cardiac surfaces, computer costs may represent a heavy limit to the simulation, unless a parallel processing of the test cases is performed.

3. SCIENTIFIC VISUALIZATION SOFTWARE TOOLS

This section briefly describes the scientific visualization software tools which have been used. After an introduction concerning the environment and the conceptual framework we deal with, VIDA design characteristics and VESCA3 functionalities are presented.

The development of large-scale numerical simulations in scientific visualization environments can be highly simplified by the use of architectures where workstations connected by means of a Local Area Network are able to interact with one another as well as, if connected by means of a high-speed line, they are able to interact with a remote supercomputer [5].
If such kind of architecture is available, a computing code which has been realized and debugged on a workstation primarily dedicated to computation is sent to a remote supercomputer where the run is started on.
At run-time the simulation flow can be followed in an interactive graphic way on a local workstation which is primarily dedicated to visualization: in other words we open a window on the computation.
The continuation of the whole simulation process is decided on the basis of the visual analysis of data produced by a part of the simulation or by a complete simulation under some conditions. These are the so-called tracking and steering techniques of a numerical simulation.

In order to get these functionalities we need an interface software between the computation and the visualization.
Among the research activities carried out within the Finalized Project "Information Systems and Parallel Computation" -sponsored by the Italian National Research Council-we designed the VIDA system (Visualization Interface for Distributed Applications) [1] which allows the integrated development of simulation programs and the corresponding interfaces for visualization in a distributed environment.
The most important functionalities VIDA is endowed with are:
- to allow the simulation steering;
- to solve the problems concerning the management of a distributed environment in a friendly to use way and, in particular, the synchronization and communication of simulation and visualization processes;
- to make the typical operations coming out from the researcher-simulation interaction easier, that is to provide a support to the management of the input and output files, to allow the user to tailor his own working environment as well as to preserve it, to make the interfacing of the data produced by simulation easier towards the available visualization packages.

To the purpose of focusing the VIDA development environment we chose:
- UNIX as a support operating system;
- X and Xtoolkits for the development of the user interface;
- RPC (Remote Procedure Call) as communication paradigm among remote processes;
- NFS (Network File System) for the file system management.

As far as our application is concerned the visualization has been realized by VESCA3 software [1] oriented to the 3D representation of scalar and vectorial 2D-3D fields. It accepts sets of multidimensional data describing stationary as well as unstationary phenomena.
Since the numerical simulations return irregularly distributed data, the scattered data are interpolated using the Interpolation Library included in UNIRAS software [6].
VESCA3 software includes two classes of functionalities oriented to the image creation and

the image manipulation. Image creation functionalities allow the visualization of surfaces as wire frame or shaded images and vector plots by means of icones whose fundamental parameters can be mapped to different variables of the problem. Surfaces and vector plots can be also superimposed and visualized as time-dependent image sequences. Image manipulation functionalities belonging both to scalar and to vectorial representation consist of point of view transformations, data sampling on the computational domain, scaling, rotation and translation of axes, zoom, lighting.
To the purpose of portability VESCA3 has been realized using the FIGARO-Template basic graphic library which is an implementation of PHIGS+ [7].

4. THE EXPERIMENTS

The window on the computation for the application we are interested in shows the representation of a scalar field by the visualization of a wire frame surface.
Tangent lines to the surface y=f(x,y) in a point (x_c, y_o), which may be chosen in an interactive way, on the x= x_c and y= y_o planes are also displayed.
In this section some results concerning two different tests of the sensitivities of the TAH-circulatory network system are presented and discussed.

In the first test(figures 1-6) the influence on the system of the values of the direct (low) and inverse (high) resistances which characterize the input (atrium- ventricle) and output (ventricle-artery) valves is evaluated. It is possible to observe that the sensitivity to preload pa is a decreasing function of the direct resistances of the input valves for both ventricles. As a consequence, a proper choice of the input valves is very important in determining the quality of the ventricular behaviour. More specifically, the biological valves, characterized by lower direct resistances rather than the mechanical ones, are preferable. Moreover, while the sensitivities to preload obtained for the left ventricle are acceptable if compared to the physiological range, the corresponding right values are not good enough, as confirmed by in vivo TAH testing. Since the simple control designed for our

TAH model does not provide any kind of sensitivity to afterload ps, the values obtained must be referred to a backflow caused by too low inverse resistances of the output valves, which is a technical challenge to improve.

In the second experiment the influence of the ratio between ejection and filling times on the performance of the TAH is tested. If this rate for the left ventricle is decreased from 0.9 (figures 1-2) to .8 (figures 7-8), the left filling becomes easier and as a consequence the left sensitivity to preload improves, at the expense of the right one.

CONCLUSIONS

A graphical window receiving data from VIDA, a portable interface which allows the researcher to use a distributed environment for visualization and computation in a transparent way, has been presented. This graphical facility has been applied to the simulation of the TAH-circulatory network interaction.
By means of the on-line reconstruction of the cardiac surfaces we are able to evaluate how some parameters which characterize the device influence the system .
As a final remark we should oserve that the visualization shows how in the physiological range of the input and output pressures the shape and the smoothness of the cardiac surfaces are in good agreement with the corresponding ones for the natural heart.

REFERENCES

1. Moltedo, L.and Palamidese, P. Supercomputing graphics: research and applications, Proceedings of the Workshop of Finalized Project" Information Systems and Parallel Computation", Rome, 1991.
2. Stephanie, M.T., Melbin, J. and Noordergraaf, A. Reduced Models of Arterial Systems, IEEE Trans. on Biomed. Eng. vol. 32, pp.174-176, 1985.
3. Guyton, A.C., Jones, C.E. and Coleman, T.G. Circulatory Physiology: Cardiac Output and its Regulation, W.B. Saunders Company, Philadelphia, 1973.
4. Pierce, W.S. et al. Automatic Control of the Artificial Heart, Artificial Organs,1, pp.137-142, 1977.
5. Fabiani, G., Lanzarini, M. and Moltedo L. Scientific visualization in supercomputing environment, pp.213-222, Proceedings of the International Workshop on Workstations for

Experiments, Boston 1989, Springer Verlag, 1991.
6. UNIRAS AGL-Interpolations , User Guide and Reference Manual, 1988.
7. Information Processing System - Computer Graphics - Programmer's Hierarchical Interactive Graphics System (PHIGS Plus), ISO SC24 Rev.3, 1989.

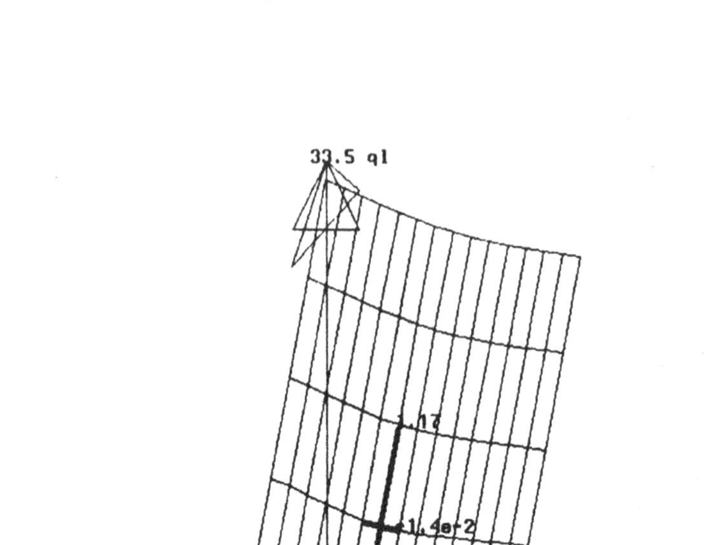

Figure 1. Outflow of the left ventricle.
Direct resistances: .0015 mm Hg/l/min.
Inverse resistances: .3 mm Hg/l/min.
Ejection time/filling time: .9

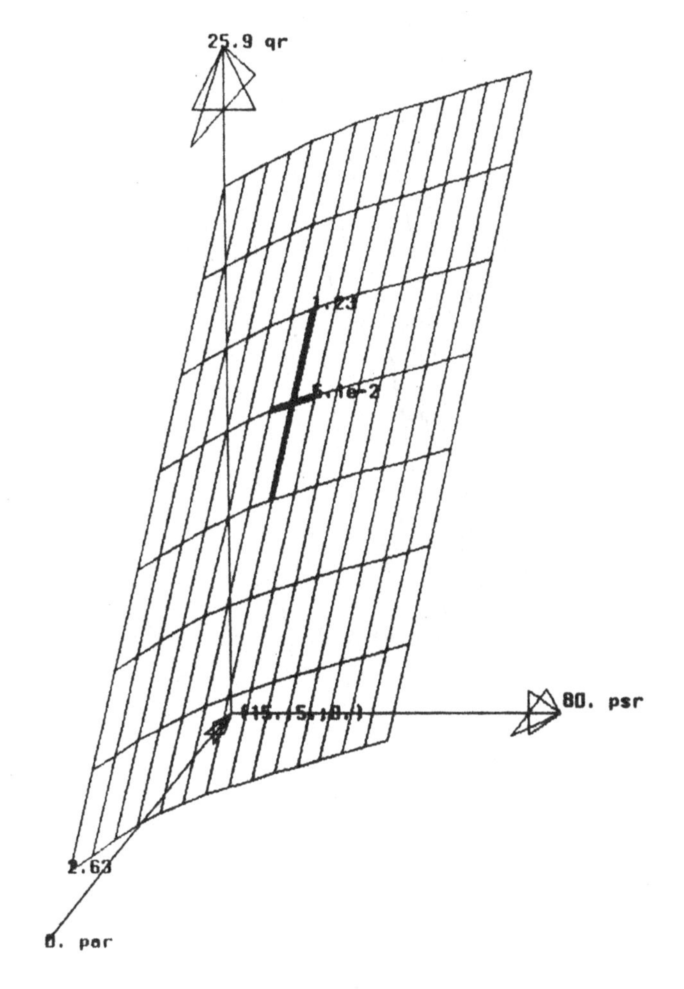

Figure 2. Outflow of the right ventricle.
Direct resistances: .0015 mm Hg/l/min.
Inverse resistances: .3 mm Hg/l/min.
Ejection time/filling time: .9

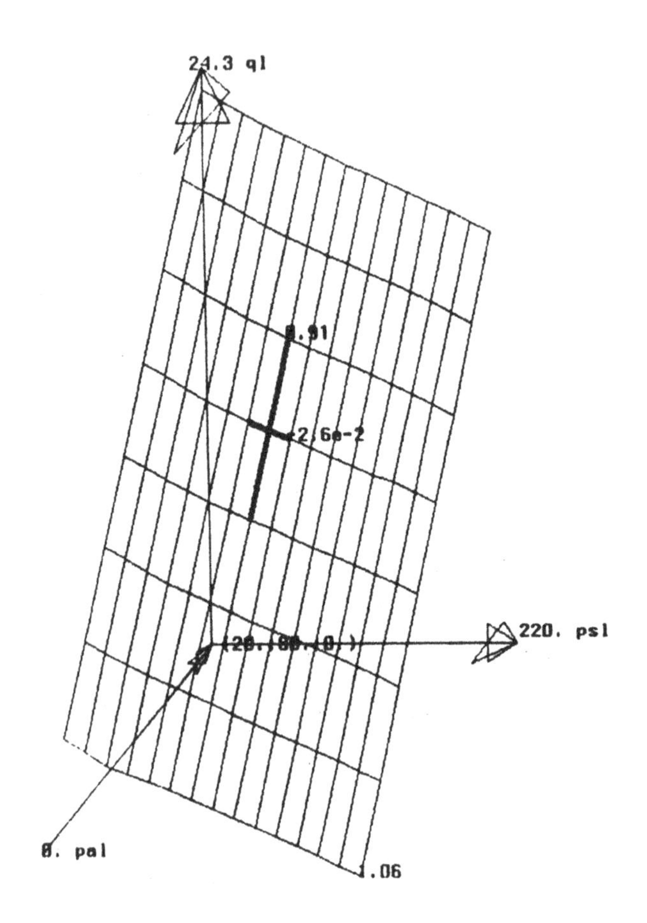

Figure 3. Outflow of the left ventricle.
Direct resistances: .0018 mm Hg/l/min.
Inverse resistances: .06 mm Hg/l/min.
Ejection time/filling time: .9

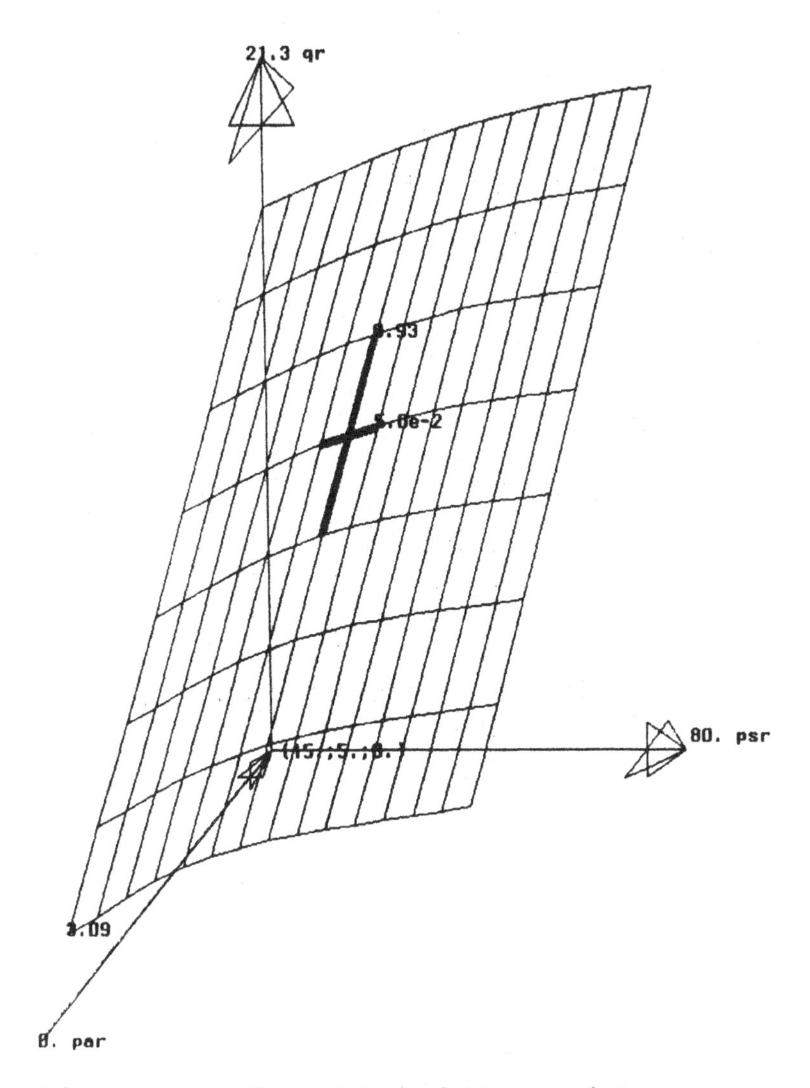

Figure 4. Outflow of the right ventricle.
Direct resistances: .0018 mm Hg/l/min.
Inverse resistances: .06 mm Hg/l/min.
Ejection time/filling time: .9

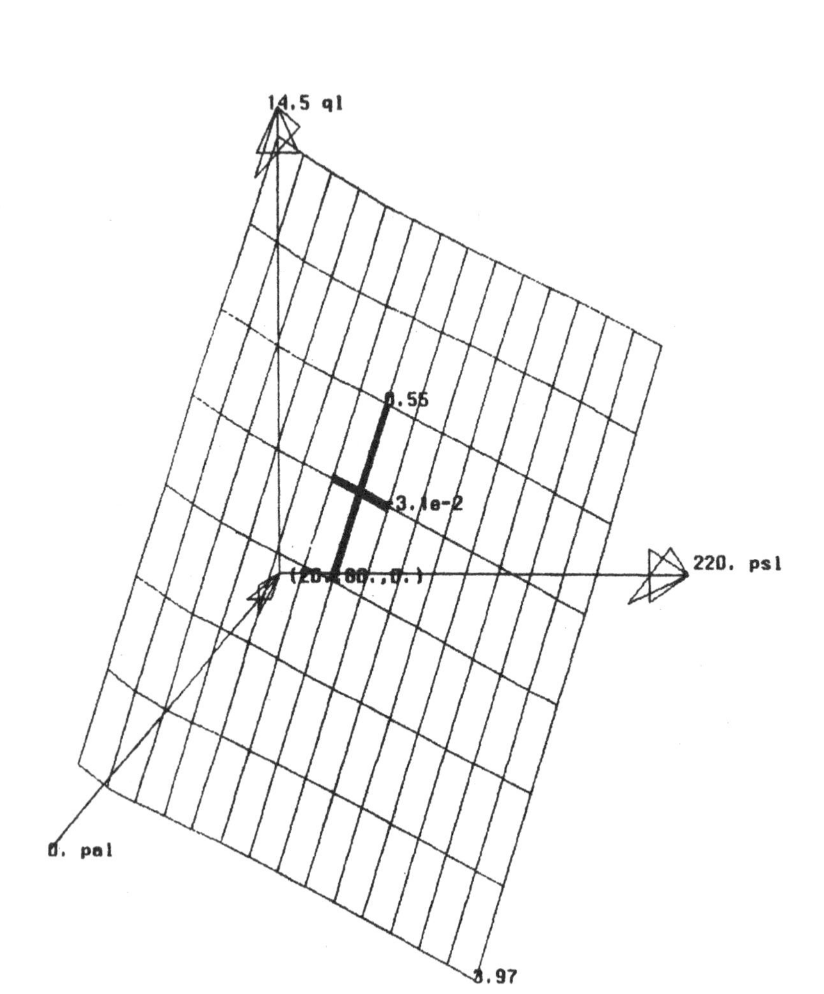

Figure 5. Outflow of the left ventricle.
Direct resistances: .003 mm Hg/l/min.
Inverse resistances: .3 mm Hg/l/min.
Ejection time/filling time: .9

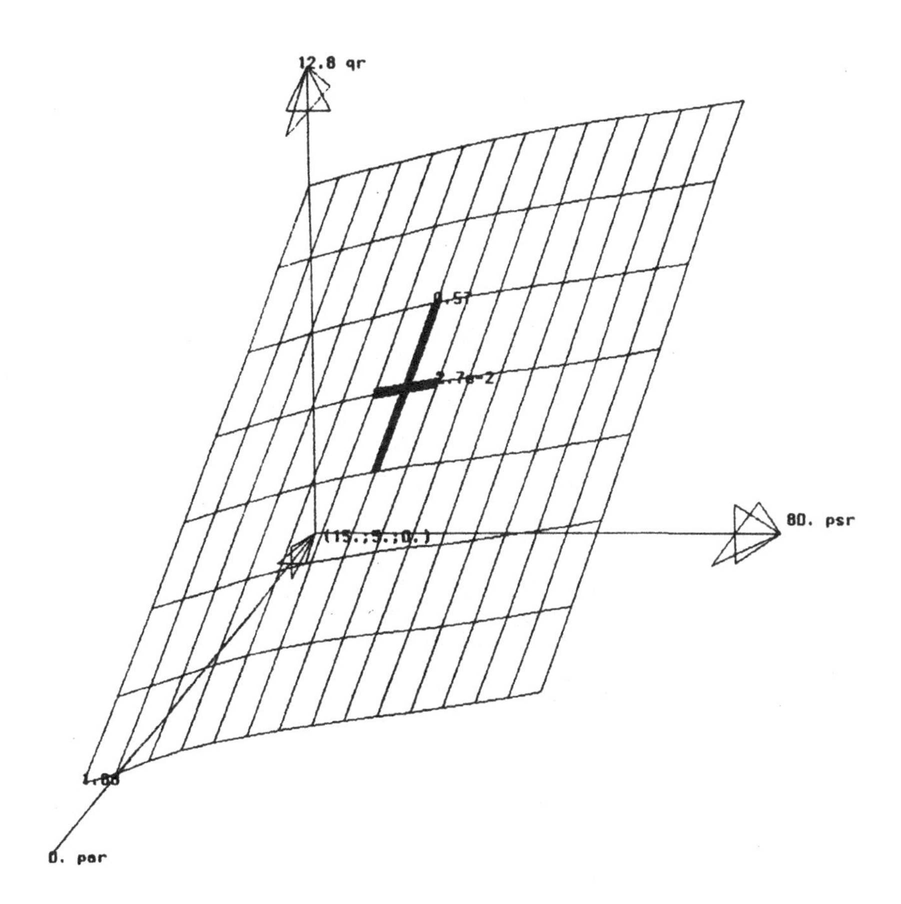


```
Figure 6. Outflow of the right ventricle.
          Direct resistances: .003 mm Hg/l/min.
          Inverse resistances: .3 mm Hg/l/min.
          Ejection time/filling time: .9
```

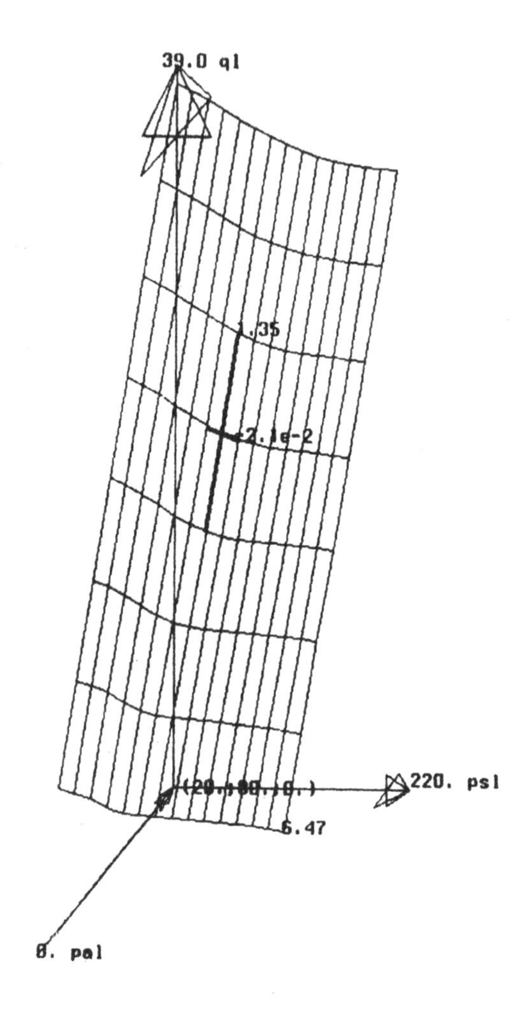

Figure 7. Outflow of the left ventricle.
Direct resistances: .0015 mm Hg/l/min.
Inverse resistances: .3 mm Hg/l/min.
Ejection time/filling time: .8

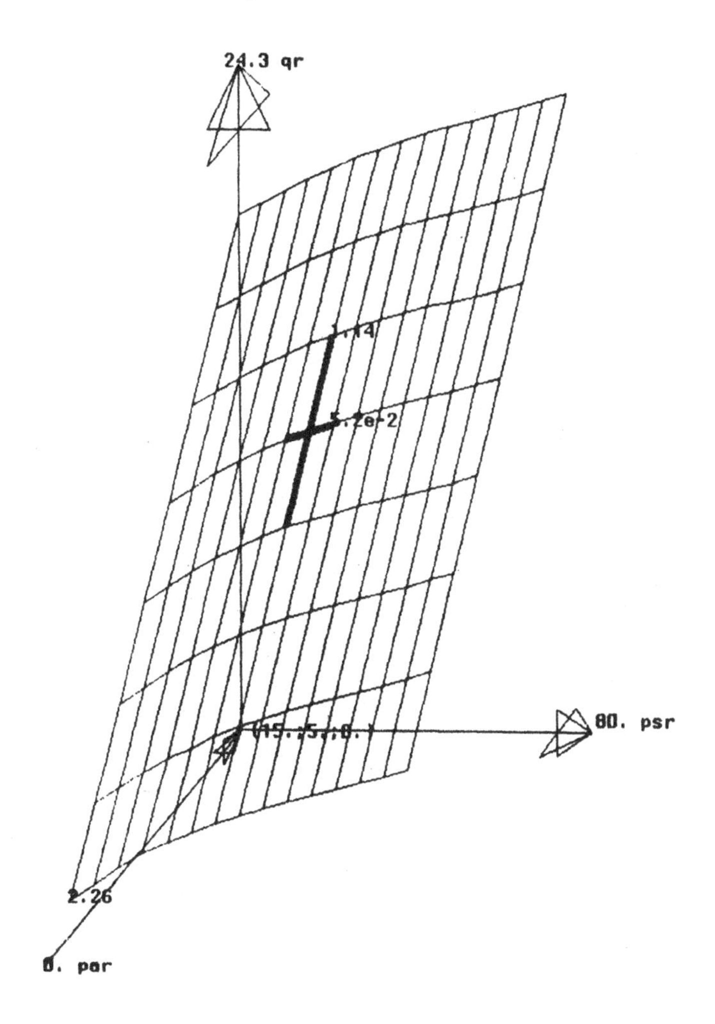

Figure 8. Outflow of the right ventricle.
Direct resistances: .0015 mm Hg/l/min.
Inverse resistances: .3 mm Hg/l/min.
Ejection time/filling time: .8

Modelling and Visualization of Impact Dynamic Processes

W. Cheng (*), I. Dilber (**)

(*) Corp. Tech. Center, FMC Corp., Santa Clara, CA, U.S.A.

(**) Corporate Office, FMC Corp., Chicago, IL., U.S.A.

ABSTRACT

Complicated dynamic processes as occurred in high velocity impact events were studied using numerical approach. A nonlinear transient finite element program was used to model the impact events. The use of supercomputers was required in these simulations due to the size of the problems and the duration of the events. Techniques designed to visualize 3D geometry as well as scientific data were developed to help engineers make sense out of the tremendous amount of data generated by the supercomputers. Resulting animation sequences provided a means for the researchers to study the dynamic failure processes and to communicate with others.

1.0 INTRODUCTION

The solid body impact has been an area of research interest for scientists and engineers for the last several decades. Its importance is evident when one considers the actual examples that may be found in the fields of tool design, foundry and machine shop operations, protective ordnance, explosions, vehicle accidents, and in many other areas. In the past, analysis of impact phenomena is only feasible to problems with simple geometry. Unfortunately, most industrial applications today are complex in geometry and are in high velocity regime. This is especially true in the design of protective and armor systems.

The advent of supercomputers and numerical tools allows some of these practical problems to be simulated. Numerical simulation of complex 3-D dynamic impact processes is performed routinely on supercomputers. Advanced state-of-the-art software is required to account for complicated dynamic phenomena such as wave propagation, large deformation, rate dependent material response including plasticity and failure. A public-domain explicit transient dynamic finite element code with some enhancements was employed for all the simulations considered in this paper. The CRAY-2 supercomputer at the National Center for Supercomputing Applications at the University of Illinois - Champaign, was used for the computations.

Being able to solve more complex problems more readily, scientists have begun to rely on scientific visualization tools to understand the results of such analyses.

However, the more powerful the supercomputers are the more data they generate, hence the need for more powerful visualization tools. The large amounts of data generated by the supercomputers, and the large amounts of data supplied by the scientists, create a problem in data handling and interpretation. To make sense out of billions of numbers in the datasets, scientific visualization techniques were developed to display them. In this work, a visualization capability was developed for the display and animation of the dynamic impact processes which could then be captured on video for future display. With the scheme developed here, one can do most of the data manipulation on a fast supercomputer, and the rendering on a 3D graphics workstation.

This paper gives a brief account of the numerical modeling of impact dynamic problems, and the importance and role of scientific visualization to these problems. The second section provides a description of the approach, material models, contact and erosion algorithms used in the modeling effort. Four numerical examples were selected to demonstrate the modelling and the computer simulation. The first example simulates an impact of a slug on a thick aluminum plate. The second example deals with an impact of a long rod on a semi-infinite target. The third example represents an impact of a thick fiber-reinforced composite plate. The fourth example involves an impact of a sharp projectile on a ceramic block. Section 3 describes the data extraction process and the animation method.

2.0 COMPUTER MODELING

The approach selected for the current study of impact dynamic processes is numerical modeling. Finite element method appeared the most appropriate choice since the effects of complex 3D geometry and boundary condition, and of material nonlinearities, damage and failure can be included. All the problems considered in this study were in the high velocity impact regime (between 1-2 km/s). An explicit Lagrangian finite element code, DYNA3D [1], was chosen for the analyses because of its capability, efficiency, and graphic support. The finite element discretization was performed using INGRID [2], the pre-processor for DYNA3D. Finite element models were developed for the following impact dynamic problems:

(i) impact of a slug on a thick aluminum plate
(ii) impact of a long rod on a semi-infinite metallic target
(iii) impact of a fragment on a thick fiber-reinforced composite plate
(iv) impact of a sharp projectile on a ceramic/metal system

Material input, initial condition such as velocity, and boundary condition such as the sliding contact interface were also defined inside the pre-processor. The version of the DYNA3D used is a vectorized version of the original DYNA3D developed by Dr. J. O. Hallquist, previously of Lawrence Livermore National Laboratory. This code was enhanced to handle some of these impact problems. The enhancements include new material models and contact erosion algorithms. The code was ported to both the CRAY-2 and the CRAY-YMP at the National Center for Supercomputing Applications (NCSA) at the University of Illinois, Champaign/Urbana, where all the calculations were performed.

2.1 MATERIAL MODELS

The materials involved in all the impact problems are metals, fiber-reinforced composites and ceramics. Different material models are required to describe the constitutive behavior and to predict the damage and failure of each material. In the case of advanced materials such as composites and ceramics, the models used were still in their infancy stage and only major failure modes were included. The use of these models was intended for capturing dominant failure mechanisms and accounting for their contribution to energy absorption of the total system, but not for modelling all the fine details of every damage mechanism over the system. In the case of metals, the behavior and failure modes at high velocity impact events are more understood and therefore, previously established material models were used. Cumulative damage of the material could be accounted for based on its strain history. Details of the material models used in this study for the three classes of materials are discussed in the following sections.

2.1.1 Metals

In the case of metals which include aluminum, titanium, tungsten and various grades of steel, elastic-plastic models were used. In most cases, the Johnson/Cook plasticity model [3] was selected to represent the material behavior. The model provides an adequate description of the flow behavior of the metals in terms of strains, strain rates and temperatures:

$$\sigma_y = (A + B\bar{\varepsilon}_p{}^n)(1 + c\ln\dot{\varepsilon}^*)(1 - T^{*m}) \tag{1}$$

where

σ_y is the flow stress
A,B,C,n, and m are input constants
$\bar{\varepsilon}_p$ = effective plastic strain

$$\dot{\varepsilon}^* = \frac{\dot{\bar{\varepsilon}}_p}{\dot{\varepsilon}_o} = \text{ effective plastic strain rate for which } \dot{\varepsilon}_o = 1s^{-1}$$

T^*= Homologous temperature

Fracture occurs when the damage parameter

$$D = \sum\frac{\Delta\bar{\varepsilon}_p}{\varepsilon_f} \tag{2}$$

reaches the value of 1.0 as ε_f, the strain at fracture, is given by (3):

$$\varepsilon_f = (D_1 + D_2\exp D_3\sigma^*)(1 + D_4\ln\dot{\varepsilon}^*)(1 + D_5T^{*m}) \tag{3}$$

where

$$\sigma^* = \frac{p}{\sigma_{eff}} = \text{ratio of pressure divided by effective stress}$$

D_1, D_2, D_3, D_4 and D_5 are materials parameters determined by tests

Typical input values to the Johnson/Cook model for various metals can be found in [3]. In other cases where little strain-rate sensitivity occurs, classical elastic-plastic model was used.

The shock behavior of the materials was described by the Gruneisen-type equation of state which is represented by (4) for compressed materials:

$$p = \frac{\rho_o C^2 \mu \left[1 + (1 - \frac{\gamma_o}{2})\mu - \frac{a}{2}\mu^2\right]}{\left[1 - (s_1 - 1)\mu - S_2 \frac{\mu^2}{\mu + 1} S_3 \frac{\mu^3}{(\mu + 1)^2}\right]^2} + (\gamma_o + a\mu)E \qquad (4)$$

and (5) for expanded materials:

$$p = \rho_o C^2 \mu + (\gamma_o + a\mu)E \qquad (5)$$

where

C is the intercept of the u_s-u_p curve
S_1, S_2, and S_3 are the coefficients of the slope of the u_s-u_p curve
γ_o is the Gruneisen Gamma
a is the first order volume correction to γ_o; and

$$\mu = \frac{\rho}{\rho_o} - 1$$

Typical values to the equation of state can be found in [3].

2.1.2 Fiber-Reinforced Composites

Existing capability for modeling failure of fiber-reinforced laminated composites due to high velocity impact is very limited. The authors developed a phenomenological model [5] to predict the impact and penetration of thick fiber-reinforced composites at high-velocity regime and has shown some success in correlating the numerical results with several ballistic tests [6,7]. The model encompasses a basic orthotropic constitutive behavior (6):

$$\begin{bmatrix} \frac{1}{E_a} & -\frac{\nu_{ab}}{E_b} & -\frac{\nu_{ac}}{E_c} & 0. & 0. & 0. \\ -\frac{\nu_{ba}}{E_a} & \frac{1}{E_b} & -\frac{\nu_{bc}}{E_c} & 0. & 0. & 0. \\ -\frac{\nu_{ca}}{E_a} & -\frac{\nu_{cb}}{E_b} & \frac{1}{E_c} & 0. & 0. & 0. \\ 0. & 0. & 0. & \frac{1}{G_{bc}} & 0. & 0. \\ 0. & 0. & 0. & 0. & \frac{1}{G_{ca}} & 0. \\ 0. & 0. & 0. & 0. & 0. & \frac{1}{G_{ab}} \end{bmatrix} \begin{Bmatrix} \sigma_{aa} \\ \sigma_{bb} \\ \sigma_{cc} \\ \sigma_{bc} \\ \sigma_{ca} \\ \sigma_{ab} \end{Bmatrix} = \begin{Bmatrix} \varepsilon_{aa} \\ \varepsilon_{bb} \\ \varepsilon_{cc} \\ \gamma_{bc} \\ \gamma_{ca} \\ \gamma_{ab} \end{Bmatrix} \qquad (6)$$

The shock property of failured composites was modeled by a linear relationship between the pressure and the natural logarithm of the relative volume given by (7)

$$p = K * \mathrm{Log}\left(\frac{V}{V_o}\right) \qquad (7)$$

where p is the pressure, K is the bulk modulus, and V and V_o are the instantaneous and initial volume. The bulk modulus K can be approximated using the rule of mixture (8):

$$K = \frac{\frac{f_f}{\rho_f} + \frac{f_m}{\rho_m}}{\frac{f_f}{\rho_f * K_f} + \frac{f_m}{\rho_m * K_m}} \qquad (8)$$

where f_f and f_m are the mass fractions for fiber and matrix, ρ_f and ρ_m are the density of the fiber and matrix, and K_f and K_m are the bulk moduli for the fiber and matrix in the two-phase system.

The onset of major failure modes as observed in numerous ballistic tests [8] was predicted by the following failure criteria:

i) Punching Failure occurs when (9) is met:

$$\left(\frac{\sigma_{ii}}{S_{ii,t}}\right)^2 + \left(\frac{\sigma_{ic}}{S_{ic}}\right)^2 \geq 1., \qquad i = a \text{ or } b \qquad (9)$$

for $\sigma_{ii} \geq 0.$

where σ_{ii} is the stress along the longitudinal and transverse material direction, σ_{ic} is the transverse shear stress, and $S_{ii,t}$, S_{ic} are the fiber breaking strength and the transverse shear strength respectively.

ii) Tensile fiber breakage occurs when (10) is met:

$$\left(\frac{\sigma_{aa}}{S_{aa,t}}\right)^2 + \left(\frac{\sigma_{ab}}{S_{ab}}\right)^2 + \left(\frac{\sigma_{ac}}{S_{ac}}\right)^2 \geq 1. \tag{10}$$

for $\sigma_{aa} \geq 0$.

where σ_{aa} is the stress along the longitudinal direction, σ_{ab} , is the inplane shear stress, σ_{ac} is the transverse shear stress, and $S_{aa,t}$, S_{ab}, S_{ac} are the fiber breaking strength, the inplane shear strength and the transverse shear strength respectively.

iii) Delamination occurs when (11) is met:

$$\left(\frac{\sigma_{cc}}{Z_t}\right)^2 + \left(\frac{\sigma_{ac}}{S_{ac}}\right)^2 + \left(\frac{\sigma_{bc}}{S_{bc}}\right)^2 \geq 1. \tag{11}$$

for $\sigma_{cc} \geq 0$.

where σ_{cc} is the transverse normal stress, σ_{ac} and σ_{bc} are the transverse shear stresses, and Z_t, S_{ac}, S_{bc} are the transverse normal tensile strength and the transverse shear strengths respectively.

Upon detection of each failure mode, a selected set of moduli are relaxed over a short duration and so are certain stress components. This model has being used to predict the ballistic performance of composite or hybrid armors.

2.1.3 Ceramics

Because of its high compressive yield strength, ceramics has been used in various armor applications. Industrial ceramics used in this application are mainly aluminum, silicon, beryllium and titanium-based. Research in the past three decades has been focused on understanding the behavior of ceramics under impact loading, and on developing new compounds and manufacturing processes. The impact penetration phenomena were studied in details by Wilkins [9] and Laible [10]. One of the first modeling efforts was due to Wilkins [9,11], in which a simplified constitutive model was used. The model assumed fracture initiated on the surface based on principal stress criterion. Fracture propagation was accounted for by considering the crack speed and the zone size. Once fracture occurred, the material stiffness and then the strength were relaxed to take into consideration of degradation. Using this model, ballistic performance of light ceramic armor systems (mainly thin composite ceramics/metal armors) was studied.

The above model is not available in DYNA3D, an elastic-plastic hydrodynamic material model [1] was used instead. In this model, the yield strength, σ_y, is determined by (12):

$$\sigma_y = \sigma_o + E_h \overline{\varepsilon_p} \tag{12}$$

where σ_O is the initial yield strength,

E_h is the plastic hardening modulus given by $\frac{E_t E}{E_t - E}$ with E being the Young's Modulus and E_t, the Tangent Modulus.

$\overline{\varepsilon_p}$ is the effective plastic strain given by $\int_0^t (\frac{2}{3} D_{ij}^p D_{ij}^p)^{\frac{1}{2}} dt$, with t denoting time, D_{ij}^p being the plastic component of the rate of deformation tensor.

The shock behavior of the ceramics was modeled by the Gruneisen equations-of-state (4) and (5). Some success was reported in a recent study [12] in which numerous ballistic tests [13] were modeled.

The impact velocities considered in the study were in the range of 1.5 km/s and the ceramics were aluminum oxide (Al_2O_3) and titanium diboride (TiB_2) sandwiched by steel plates. The ceramic material properties used in the modelling effort were obtained from [14].

2.2 CONTACT/EROSION

Penetration or/and perforation of projectiles may occur during impacts. Newly developed erosion algorithms which allow Lagrangian calculations of the processes without resorting to tedious rezoning on 3D geometry were used to facilitate the calculations. Surfaces of the projectile and target, which will be interacting with each other, are defined as the 'master' and 'slave' surfaces. Such definition allows detection and adjustment of the nodes on the master and slave surfaces so that no interpenetration occurs. This is accomplished by a symmetric treatment of the erosion algorithms [15].

2.3 NUMERICAL EXAMPLES

Example 1 involves an impact of a cylindrical slug on a thick plate. The slug is made of high strength steel and has a normal impact velocity of 1.25 km/s. The plate is 6.35-cm thick and is made of aluminum. Johnson/Cook material model was used for the high strength steel while conventional elastic-plastic material model for the aluminum. The finite element model is shown in Figure 1. Figure 2 shows the deformed geometry of the plate after the projectile was completely stopped. Extensive plastic flow followed by high speed erosion in the target material is the basic failure process. Erosion also occurred in the slug and therefore, only a fraction of the original mass of the projectile remained. A crater with diameter larger than the original diameter of the projectile was produced.

Example 2 represents an impact of a long rod on a semi-infinite target. The rod is made of tungsten and has an initial impact velocity of ~ 1 km/s. The semi-infinite target is made of 4130 hard steel. Johnson/Cook material model was used for both material. Figure 3 shows the finite element model. Again, extensive plastic flow occurs underneath the projectile and erosion is the dominant mode of material failure for the penetration to progress. Figure 4 shows a plot of the plastic strain contours 3.2 microseconds after the impact.

Example 3 models an impact of a Fragment Simulation Projectile (FSP) on a thick fiber-reinforced layered composite panel and then a witness steel plate. The FSP was made of 4340H Rc 30 steel and the witness steel plate, RHA armor steel. The composite panel was made up of about seventy layers of woven roven Kevlar-49 fiber reinforced plastics with polyester resin. The FSP has an initial impact velocity of ~2 km/s. Figure 5 which shows the pressure contours of the composites during the impact/penetration event provides a means of studying if delamination is resulted from shock-induced tensile stress waves. Failure of thick composites during high velocity impact was dominated by the punching phenomenon. Tensile fiber breakage may occur at the intermediate as well as in the final stage. The intermediate stage was a result of the distributed tensile loads resulting from the failed materials underneath the FSP before they can erode away. The tensile failure may occur at the final stage only if the fragment has slowed down to low speed so that penetration will not continue, and bulging of remaining thickness occurs resulting from delamination. The tensile failure may not occur if delamination velocity is faster than the bending wave speed and tensile stress will not build up to the fiber tensile strength, this results in a complete stopage of the fragment. Some of these failure processes can be observed in Figure 5. The failure process during impact of the fragment on the steel plate is similar to that occurs in Examples 1 and 2.

Example 4 encompasses an impact of a projectile on a ceramic/metal composite system. The projectile has a short sharp nose and a cylindrical body, and is made of tungsten. It has an initial impact velocity of ~1 km/s. The composite system consists of a 4.57-cm thick Al_2O_3 ceramics tile and a 2.54-cm thick backing plate made of RHA steel. Both tungsten and RHA steel were modeled with Johnson/Cook material model. The Al_2O_3 ceramics was modeled as elastic-plastic hydrodynamic material model with Gruneisen equation-of-state. Due to the high compressive strength of the ceramics, the sharp nose of the projectile was eroded at the early stage due to high localized contact force. This resulted in a larger contact surface. As the forces built up, the ceramic material underneath the projectile began to flow and fracture conoids started to form just outside the contact boundary due to the low tensile strength of ceramics. The high impact velocity, the geometry of the ceramics and the stiffness of the backing plate provides confinement of the ceramics inside the fracture conoids, which can still sustain high compressive loading. Penetration of the projectile is promoted by plastic flow and erosion of ceramics. It is because of these phenomena occurring in this class of impact events that elastic-plastic hydrodynamic material model might be used as the first approximation, to predict the penetration of the projectile in ceramic target. Figure 6 shows the plastic strain contours of the system during the penetration process.

3.0 DATA EXTRACTION AND SCIENTIFIC VISUALIZATION

The DYNA3D program was modified to write out all the necessary data from the simulation to a binary file as NCSA's Hierarchical Data Format (HDF) Vsets [16,17]. The HDF was developed at NCSA as a standard and portable binary file format that contains different types of data, such as numbers, images, characters, etc., in a single binary file. The file is portable between different computer architectures where the HDF libraries are installed. The Vset is an extension to the HDF idea, supporting unstructured grids such as found in finite element calculations. Since it is binary, it is more efficient and less space consuming than the ASCII format files. Especially for large data sets, this has been found to be very valuable. Furthermore, since there is no need for any conversion, the data can be transferred from the Crays to the workstations for further processing and visualization.

The Vset files are organized in groups of data, called vgroups. Each vgroup may contain multiple data sets, called vdatas. Vgroups may contain other vgroups as well as vdatas. Furthermore, each vdata can be a multiply dimensioned array, e.g. x, y, and z components for N nodes can all be defined as one vdata.

The module modified in DYNA3D creates a Vset file with the following structure. Each time step is a vgroup. Under this vgroup, there are various vdatas. The first vgroup contains the original nodal coordinate and connectivity information. Subsequent vgroups, or time steps, contain vdatas for various parameters, such as displacements, velocities, pressures, effective plastic strains, etc... Also, to detect the erosion mechanism, the deleted elements, if any, are also recorded in the vgroups. The resultant HDF file forms a database for the particular run, containing all the relevant information.

The second stage in the visualization process involves conversion of the information stored in this database to a visualization software. For our 3D visualization, we have chosen a Silicon Graphics™ 4D series workstation and NCSA's Polyview software [18]. A software is developed to read the database file and convert the information into polygonal information and written into another Vset file which can then be displayed and animated by Polyview. Since Polyview is a polygon rendering software, the finite element connectivity information has to be converted to surface (or polygon) connectivity. Multiple surface definitions are eliminated during the conversion. Also the deleted elements are eliminated and the parametric data and the connectivities are modified accordingly. The resultant Vset file contains vgroups for each time step, each vgroup containing the displaced (if requested) geometry and the scalar values. The frames are then loaded to Polyview. Once loaded, the object can be rotated, translated, zoomed in and out, rendered in 5 different ways, and once a desired view is selected, can be saved into a NCSA HDF file as 256 color images. The image files can be displayed on any hardware, either a Cray, or a Macintosh, supporting the HDF libraries. They can also be recorded to a VCR for "live" animation, and also for presentations and sharing the results of the research with colleagues. Such an animation is shown as part of the presentation of this paper. Most of the figures contained in this article are clips from such images.

The use of the HDF Vset file enables us to make use of the speed of the Cray supercomputer for the number crunching part, and use the graphics workstation for the image rendering and animation without the disadvantages of large ASCII files that needs to be transferred between the computers. The data files can also be shared with colleagues having access to HDF libraries.

The 2D animations were obtained using a similar approach, but using the post-processor to DYNA to extract the data. The procedure was developed in an earlier work and more information can be found in [7].

4.0 CONCLUSIONS

A combination of state-of-the-art computer tools and supercomputers have proven very valuable for the modeling and understanding of dynamic processes during high velocity impact events. While some practical impact problems can be modelled with the existing analytical and numerical capability, and the current computer technology, further research remains necessary to refine the current approach and to tackle other problems such as those in the hypervelocity regime. Research areas may include obtaining a better understanding of material behavior at high loading rates and pressures, designing methods of characterizing material behaviors at these regimes, developing efficient numerical techniques and algorithms to treat large deformation and high shock pressures, and producing faster computer systems to make these calculations possible in production environment.

Scientific visualization techniques developed here have enhanced our ability to understand the failure processes during high velocity impact events and predict system performance. New techniques should be developed to further enhance the human ability to absorb large quantity of data and provide means to communicate to others. Today, state-of-the-art scientific computation requires multiple computer architectures, the supercomputers for solving larger models faster, and fast graphics workstation to visualize the results. The data transfer between the architectures should be as efficient as possible since we are dealing with increasingly large amount of data. As the simulation models become more complex, so should the visualization software. In this work, we have created a portable binary DYNA3D database file and linked it to a powerful visualization/animation tool on a fast graphics workstation. However, we still need new techniques for the display and presentation of complex engineering results. Multi-media computation and multi-media post-processing of the data is necessary. Data sonification has demonstrated to be a great potential method of displaying scientific data. While these tools are proven beneficial, they are, however, far from being production tools. Efforts should be devoted to integrating these tools into a general-purpose data analysis package.

REFERENCES

[1] Hallquist, J. O., "DYNA3D User's Manual," UCID-19592, Rev. 4, April, 1988.

[2] Hallquist, J. O.; Stillman, D. W., "INGRID: A Three-Dimensional Mesh Generator for Modeling Nonlinear System," UCID-20506, Rev. 1, July, 1985.

[3] Cook, W. H.; Johnson, G. R., "A Constitutive Model and Data for Metals Subjected to Large Strains, High Strain Rates and High Temperatures," presented at the Seventh International Symposium on Ballistics, The Hague, The Netherlands, April, 1983.

[4] "LASL Shock Hugoniot Data," S. P. Marsh edited, University of California Press, Berkeley, California, 1980.

[5] Cheng, W. L.; Langlie, S. L., "A High Velocity Impact Penetration Model for Thick Fiber-Reinforced Composites," Composites and Other New Materials for PVP: Design and Analysis Considerations, PVP - Vol. 174, ASME, 1989.

[6] Cheng, W. L.; Langlie, S. L., "Numerical Simulation of High Velocity Impact on Fiber-Reinforced Composites," Shock and Wave Propagation, Fluid-Structure Interaction, and Structural Response, PVP - Vol. 159, ASME, 1989.
[7] Cheng, W. L.; Dilber, I.; Langlie, S. L., "Computer Simulation of High-Speed Impact Response of Composites," Proceedings of Supercomputing '90, IEEE Computer Society Press, 1990.
[8] Mehlman, M.; Vasudev, A., "A Comparative Study of the Ballistic Performance of Glass Reinforced Plastic Materials," SAMPE Quarterly, Vol. 18, No. 4, July 1987.
[9] Wilkins, M.L.; Fline, C.F.; & Honodel, C.A., "Fourth Progress Report on Light Armor Program", Report UCRL-50694, Livermore Lawrence Radiation Laboratory, University of California, 1969.
[10] Laible, R.C., "Ceramic Composite Armor" in Laible, R.C. (Ed.), Ballistic Materials and Penetration Mechanics, Elsevier, 1980.
[11] Wilkins, M.L., "Computer Simulation of Penetration Phenomena" in Laible, R.C. (Ed.), Ballistic Materials and Penetration Mechanics, Elsevier, 1980.
[12] Hallquist, J.O., Private Communications, 1991.
[13] Woolsey, P.; Dokidko, D.; & Mariano, S.A., "Alternative Test Methodology for Ballistic Performance Ranking of Armor Ceramics" U.S. Army Materials Technology Laboratory, Report MTL TR 89-43, 1989.
[14] Ravid, M.; Bodner, S. R. & Holcman, I., "Analytical Investigation of the Initial Stage of Impact of Rods on Metallic and Ceramic Targets at Velocities of 1 to 9 km/sec", Proceedings of the Twelve International Symposium on Ballistics, San Antonio, Texas, 1990.
[15] Hallquist, J. O.; Ong, A. C. J.; Sewell, D. A., "Penetration Calculations Using an Erosion Algorithm in DYNA," presented at the 1990 International Symposium on Ballistics, San Antonio, Texas, U.S.A., October, 1990.
[16] "NCSA HDF Calling Interfaces & Utility," Version 3.1, National Center for Supercomputing Applications, University of Illinois at Urbana-Champaign, UD-254, July, 1990.
[17] "NCSA HDF VSET," Version 2.0, National Center for Supercomputing Applications, University of Illinois at Urbana-Champaign, UD-259, November, 1990.
[18] "NCSA Polyview for the Silicon Graphics™ 4D Series Workstation," Version 1.0, National Center for Supercomputing Applications, University of Illinois at Urbana-Champaign, UD-256, August 1990.

ACKNOWLEDGEMENTS

Scientific research was funded by FMC, Corporate Technology Center, under the Corporate Research & Development Project. Scientific visualization development was supported by FMC, Corporate High Performance Computing. The authors would like to thank FMC management for their support in the areas of scientific research and visualization, and the National Center for Supercomputing Applications, University of Illinois at Urbana-Champaign for their collaboration efforts.

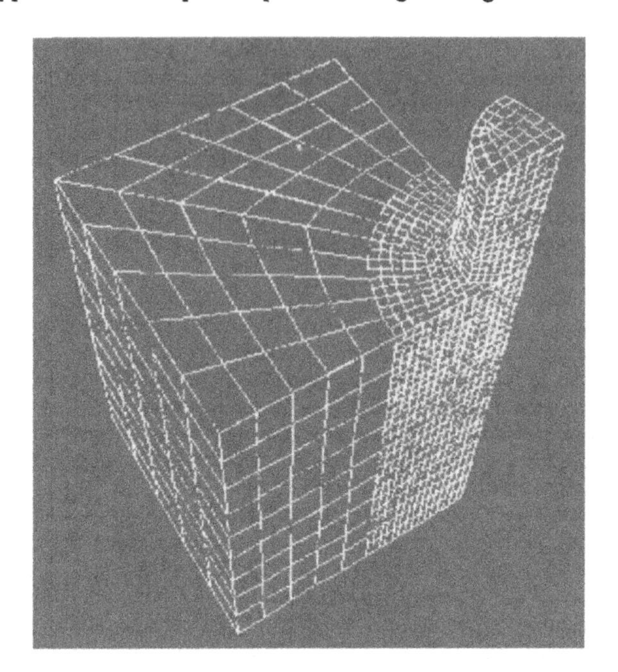

Figure 1. Example 1 Finite Element Mesh.

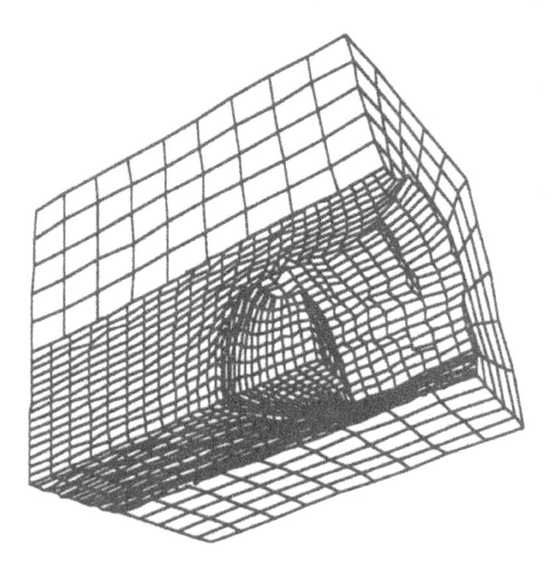

Figure 2. Example 1 Deformed Geometry at 39 μsec.

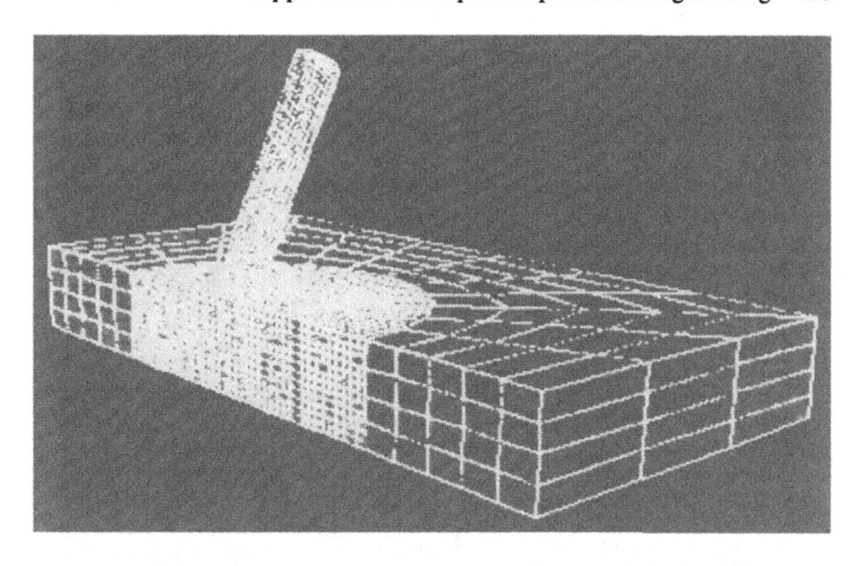

Figure 3. Example 2 Finite Element Mesh

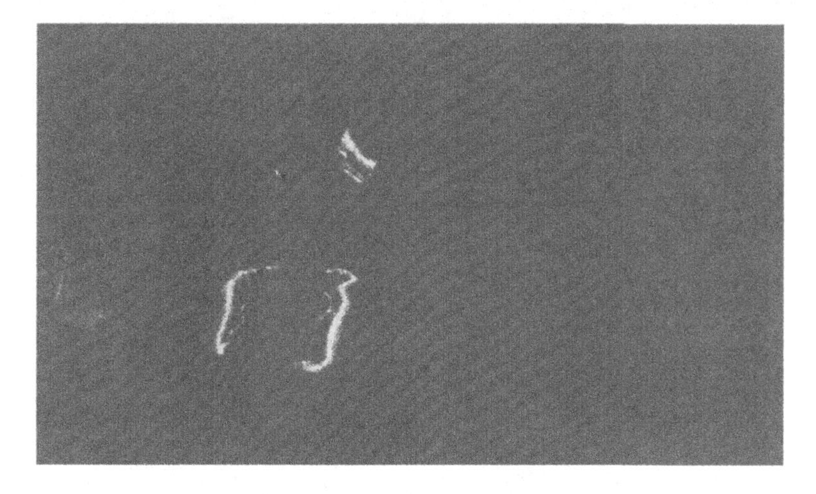

Figure 4. Example 2 Plastic Strain Contours on Undeformed Geometry at 3.2 μsec.

Figure 5. Example 3 Mesh, Failure Modes and Pressure Contours.

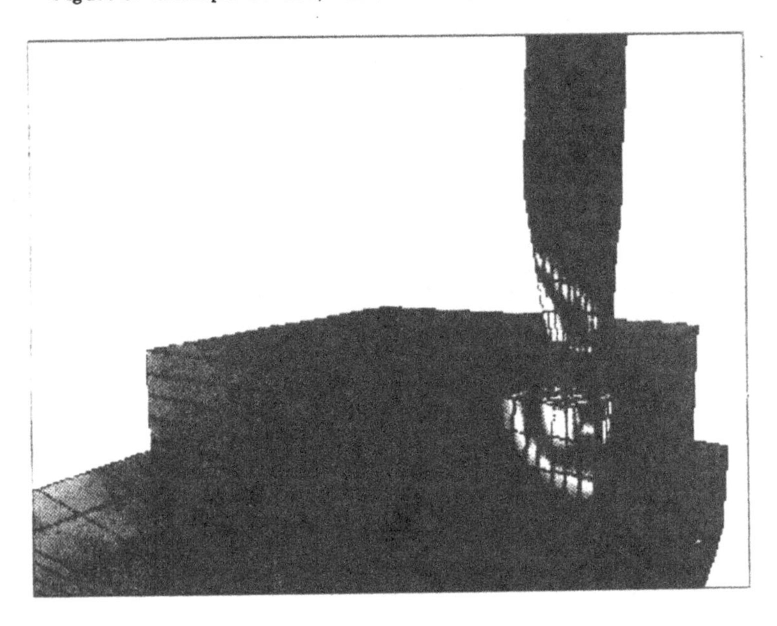

Figure 6. Example 4 Plastic Strain Contours with Mesh Outline.

Crosshole Tomography Animation in X Windows with Convex-2/IBM RS 6000 Client-Server Display Connection

O.G. Johnson

Computer Science Department, University of Houston, Houston, TX 77204-3475, U.S.A.

ABSTRACT

Recent emphasis in scientific visualization has centered on CRT displays which occur in real time as the simulation proceeds. In such visualizations, scientists can "see" intermediate answers as they are computed. This helps researchers conceptualize the phenomenon more completely. They can also adjust parameters, boundary conditions or other pieces of input data more quickly and can obtain immediate feedback on the response of the simulation to these changes.

Since a fundamental objective of animated modeling is to see something happen, the speed of both the client and server systems is important. The marriage of advanced computer models simulated on number crunching hardware with sophisticated visual displays on bitmapped graphic devices is the order of the day. The developer of scientific animation software must have an awareness of what can be expected from computers of different capabilities. Such awareness will necessarily translate itself into proper choices for model complexity and numerical algorithms as well as graphics and program environment.

Here, we use a particular application area, the movement of waves originating from a downhole acoustic source and propagating through short distances between wells in an oil or gas reservoir. As applied to such tomographic modeling, visualization consists of "seeing" the recorded values of receivers in a second well as they occur in the simulation while simultaneously seeing the acoustic waves travel back and forth between the wells interacting with the various geological strata.

Computer Visualization and Animation

Computer visualization has revolutionized such divergent enterprises as television commercials, architectural modeling and art. It has been called the second computer revolution and has been thought of as a process which "fuses the computer with the 'preconscious' process of visual perception and ... fundamentally changes the way we realize creative ideas and solve complex problems." [Friedhoff, 1989] In science and engineering, computer visualization is the art of displaying the results of computer simulations in graphic form. Recent emphasis in this field has centered on CRT displays which occur in real time as the simulation proceeds. In such visualizations, the scientist can "see" intermediate answers as they are computed. This helps the researcher conceptualize the phenomenon more completely. It also allows him to abort simulations which are proving to be useless or meaningless early in the process. He can also adjust parameters, boundary conditions or other pieces of input data more quickly and he can obtain immediate feedback on the response of the simulation to these changes.

Techniques for displaying results visually in the best way can present a challenge in themselves. Historically, computers generated tables of numbers which had to be studied and analyzed after the computer had finished its run. These tables were replaced with graphs and contours by using plotters. These in turn have been replaced by pictures on color CRT devices. These pictures may be wire-frame drawings or solid figures. The user may require the ability to rotate, translate or scale the objects in the picture as the simulation proceeds. The user may wish to alter the lighting sources, the colors, the texture or the translucence of an object. The user may wish to open and close "windows" with various attributes of the simulation showing simultaneously.

Animation involves the rapid change of visualization images so that a cinematographic effect is achieved. Typically 30 images per second might be used in the spiraling logos of television commercials but often only one or two images per second will suffice for scientific studies.

The partial differential equations which govern transitory physical phenomena equate the change in value of a quantity over time with its change in value from place to place. At a given time, the spatial values of the quantity at discrete, equally spaced intervals can be represented by a vector (in one space dimension) or a matrix (in two space dimensions) or a higher dimensional array (in three space dimensions usually) of real numbers. These numbers can be color coded by range to form a picture of the variation in values with respect to space at a given time. Each image is thus a snapshot of the phenomenon in time. The color coded matrix can be thought of as a raster of data, and the colors as pixels which can be displayed on a CRT screen. Snapshots of the phenomenon at succeeding time intervals can be computed from those of the past by using the partial differential equation as the regulator of change. If the rasters for succeeding time intervals are displayed on the CRT successively, the result is the same as an animated movie.

Examples of the successful utilization of this technique in science are now quite plentiful. They include: ion interactions, stress wave propagation, fluid dynamics, Lorenz attractors, hydrogen diffusion on platinum surfaces, enzyme reactions and fluorpolymer simulations to name only some of the areas of application. The videotape has become an adjunct to the scientific journal. Indeed some journals now have videotaped supplements. [Volume Visualization/ State of the Art, Issues 44-50], [International Journal of Supercomputer Applications, special issue on computer visualization], [ACM SIGGRAPH Video Review, vols 1-35].

Visualization in Tomographic Seismology

Tomographic mappings of geological strata are produced by creating an acoustic source in a well bore and subsequently recording the reflected energy at several points in a neighboring well. Typically, the recordings take place over a 3 to 6 second period following the initiation of the source. The recordings are taken simultaneously at small time intervals of 1 to 4 milliseconds. While there may be as many as 100 such receivers, most instruments at present have only a few, less than 10, such receivers.

Computer visualization of this process would consist of "seeing" the recorded values of the receivers as they occur in the simulation. Simultaneously, one would like to see the acoustic waves move between the wells interacting with the various strata. As they reach the receiver well, one would like to see the amplitudes of these waves transferred from the wavefield picture to the picture of the recordings.

In order to see a wavefront bounce back from encountering a change in geological formation, one must of course be able to see the formation as well as the waves which travel in the formation.

Typically, the models use finite differences in time, t. For second order differences, three arrays are kept: p (the value recorded by the surface instruments, for instance this may be pressure for off-shore exploration), pf (p's father, i.e. the value of the array at the previous time step and pgf (p's grandfather). Visualization and animation can occur simultaneously with the computations in the simulation provided there is a reasonable balance between compute time per time step and display time per time step. If so, one time step consists of:
a) display pf b) compute p and c) replace pgf by pf and replace pf by p.

At present, a PC can produce a 1D animation in real time. A 2D model of size 100x100 needs a workstation of the Sun-3 or Sun-4 class in order to achieve a reasonable display rate. Animation of 3D models requires a supercomputer such as the Convex-2 or a high-end workstation such as the IBM RS6000. Here we use both. Animation for 2D models is performed by rasterizing pf. In monochrome mode, the rasters are displayed as wiggle traces. In color mode, there are typically 16 to 256 colors available at any one time for color coding the amplitudes by range.

A Convex-2/IBM RS6000 Animation System

In two space dimensions, 2D, the wavefield $p(x, z, t)$ is a function of position along the earth's surface (x), depth into the earth (z) and time (t). At a given time t_{j+1} we have the matrices $p = p(x, z, t_{j+1})$, $pf = p(x, z, t_j)$ and $pgf = p(x, z, t_{j-1})$. At the surface of the earth we will compute another matrix $ps = p(x, 0, t)$. For animation purposes, we wish to view the matrices, pf and ps simultaneously as the animation proceeds.

In viewing pf for successive values of t, we will see snapshots of the wavefield as it moves in the earth. The pf raster values show the amplitudes of p only. In order to see the subsurface geology as well, polygonal lines must be drawn at the interfaces of strata of different acoustic velocity. These lines are drawn so that they overlay the wavefield to scale. Also, since the color coding of matrix data does not give the same level of detail as a graph of amplitude versus time at a given receiver position, (x, z) a separate graphical display may be needed.

Axes for both x and z are needed for the pf raster and axes for x and t are needed for the ps raster. Initially, in the simulation, the ps array will be zero. Lines of the array will appear as the simulation progresses and the values of ps for each t value displayed are computed in turn. The values for all time steps will not be displayed due to the size of the screen. Hence, only enough samples can be viewed as will fit the terminal. In viewing ps we will see the simulated record, or tomographic section, as it develops.

The programs discussed here were developed for workstations with color monitors. In order to show figures in this paper, a special black and white version was developed from which screendumps could be taken and printed with a laser printer. The resolution is, of course, not as good, but the figures show the type of images produced.

Figures 1 through 6 show snapshots of the screen for a simulation of a two layer model. Here the acoustic velocity of the upper layer is 6000 ft/sec and that of the lower is 6200 ft/sec. The two wells are 600 ft. apart and both the source and receiver are at a depth of about 700 ft. The instrument being simulated in this case is the **Bender** produced by Southwest Research Institute. This instrument can be controlled to broadcast at a given frequency for a specific period of time. In the program the user can select a menu item to set the frequency and time interval. This causes a sub-window to appear as in Figure 2. The user moves the mouse to any frequency-time combination and then clicks the mouse. This model consists of an (x, z) array of size 116x208. Interpolations are made to smooth out the raster display and to increase the display array size to approximately 464x832. The initial shot location is selected by the mouse anywhere in the source well.

This is accomplished by warping and grabbing the pointer At time $\delta t = 0$ milliseconds, point source occurs at the source point. At time $\delta t = .004$ milliseconds, a 2D surface with centers at the circle determined by outward motion from the shot point with the local velocities is computed. These two arrays form the initial pf and pgf matrices and drive the simulation.

Figure 1 shows the screen as the simulation begins. The menu is used to select the desired action, in this case, the firing of the *Bender* instrument for a specific time interval at a specific frequency. A customized cursor, in the form of the instrument, is then created for manipulation by the mouse to select the location of the shot (Figure 2). The shot is activated by pressing the rightmost button of the mouse. This causes an asterisk to appear, denoting the shot. The simulation then starts. Figure 3, shows the wavefield early in the simulation. The direct wave from the shot spreads out toward the receiver well increasing its spherical radius as time proceeds. The time section is void at this point. In Figures 4 and forward we see snapshots as the simulation proceeds. The wavefield continues until it arrives at each point in turn. The acoustic velocity of the upper layer in this case is smaller than that of the lower. Hence we see the wavefield slow down upon entering the lower velocity material. In Figures 4 through 6, we see that the time section is much more filled out. The graph of the wave amplitude at the receiver is superimposed on the output tomographic section.

AVS II

In addition to being more powerful than than most workstations, the Convex-2 supercomputer is accompanied by a special visualization package which interfaces with the user's compute modules to display results in a more sophisticated way. This system, The Automatic Visualization System II (AVS), provides an interactive viewing application for the user which runs concurrently with his modeling program. This system handles all graphic input and output for the model including handling mouse interactions, defining hierarchical 3D objects, calculating transformation matrices for complex motions and specificing colors and other surface properties [Stardent, 1989].

For instance, the user can export the matrix pf to AVS for display. AVS will create a 3D version of pf. The length and width of this structure are determined by the size of the matrix. The height is determined by the values of each element of the array. The height can further be displayed as a surface via triangularization and color shading. Further, the structure can be scaled or rotated in real time as the simulation proceeds.

A separate 3D structure can be created for the velocity values at each point. This structure, in effect, shows the various geologic strata in the model. The two structures can be overlaid so that the user sees one on top of the other. The top array can be made translucent to various degrees, interactively, so that the geological model can be seen through the moving wavefield. Figures 7 and 8 show 3D versions of the wavefield.

The power of AVS as an animation device is equaled by its ease of use. The time required to learn AVS and to convert the base X system to the Convex was only one weekend. AVS is an open system based on PHIGS+ and X Windows.

References

[1] Etgen, J. and Dellinger, J.,"Accurate Wave Equation Modeling", *S.E.G. Expanded Abstracts*, v. 1, pp. 494-497 (1989).

[2] Friedhoff, R., *Visualization*, Harry N. Abrams, Inc., New York (1989).

[3] Loewenthal, D., Lu, L., Robertson, R. and Sherwood, J., "The Wave Equation Applied to Migration", *Geophysical Prospecting*, v. 24, pp. 380-399.

[4] Nelson, R., *New Technologies in Exploration Geophysics*, Gulf Publishing Co., Houston (1983).

[5] Stardent Computer, Inc., *Application Visualization System, User's Guide*, Newton, Mass. (1989).

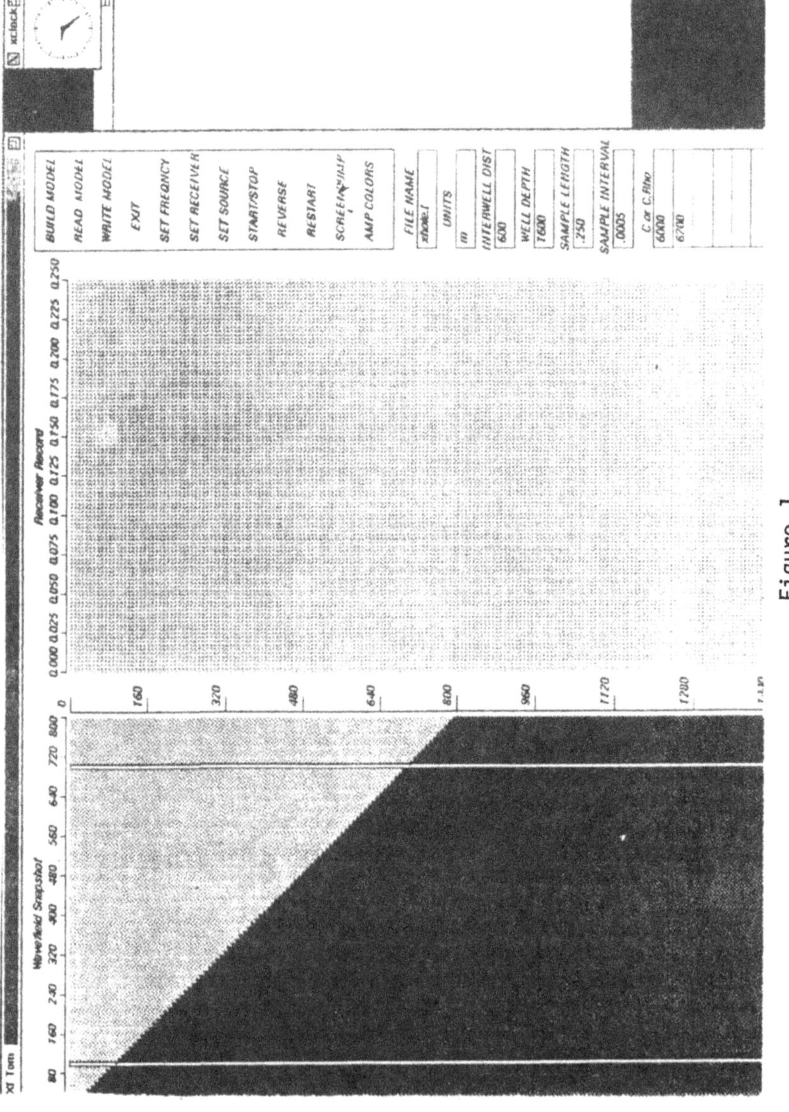
xclock
Wavefield Snapshot
Receiver Record
0.000 0.025 0.050 0.075 0.100 0.125 0.150 0.175 0.200 0.225 0.250
80 160 240 320 400 480 560 640 720 800
0 160 320 480 640 800 960 1120 1280
BUILD MODEL
READ MODEL
WRITE MODEL
EXIT
SET FREQNCY
SET RECEIVER
SET SOURCE
START/STOP
REVERSE
RESTART
AMP COLORS
FILE NAME
UNITS
m
INTERWELL DIST
600
WELL DEPTH
1600
SAMPLE LENGTH
.250
SAMPLE INTERVAL
.0005
6000
6700

Figure 1

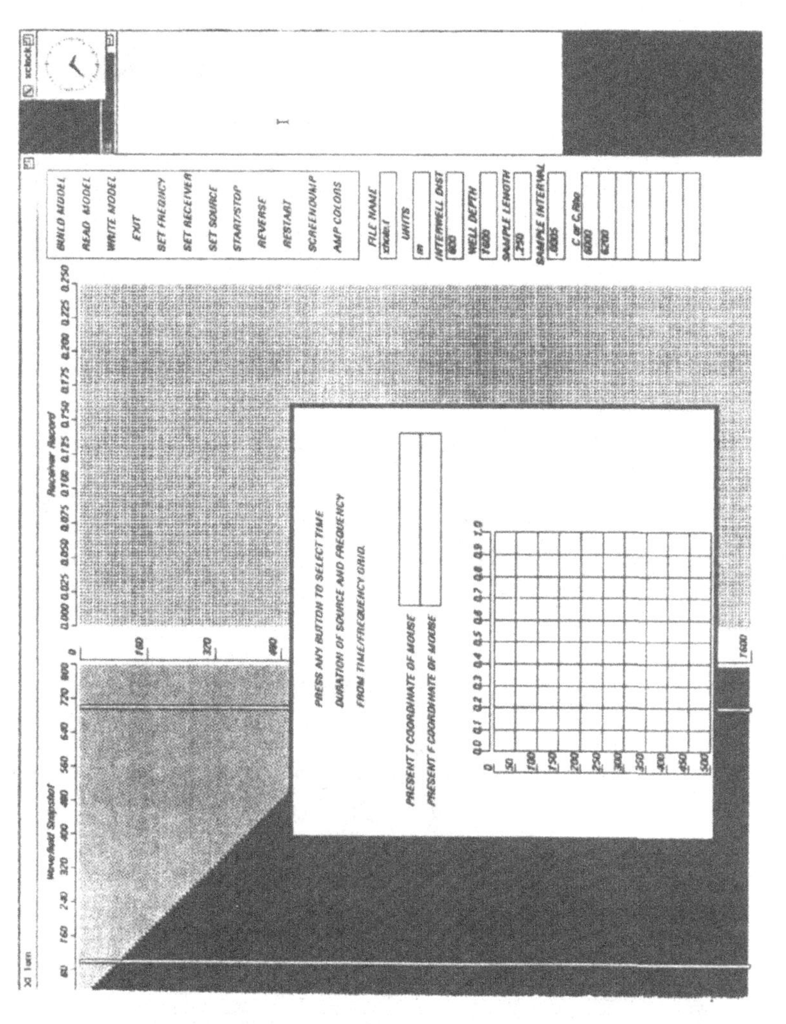
Wavefield Snapshot
Receiver Record
BUILD MODEL
READ MODEL
WRITE MODEL
EXIT
SET RECEIVER
SET SOURCE
START/STOP
REVERSE
RESTART
SCREENDUMP
AMP COLORS
FILE NAME
UNITS
INTERWELL DIST
WELL DEPTH
SAMPLE LENGTH
SAMPLE INTERVAL
PRESS ANY BUTTON TO SELECT TIME
DURATION OF SOURCE AND FREQUENCY
FROM TIME/FREQUENCY GRID.
PRESENT T COORDINATE OF MOUSE
PRESENT F COORDINATE OF MOUSE

Figure 2

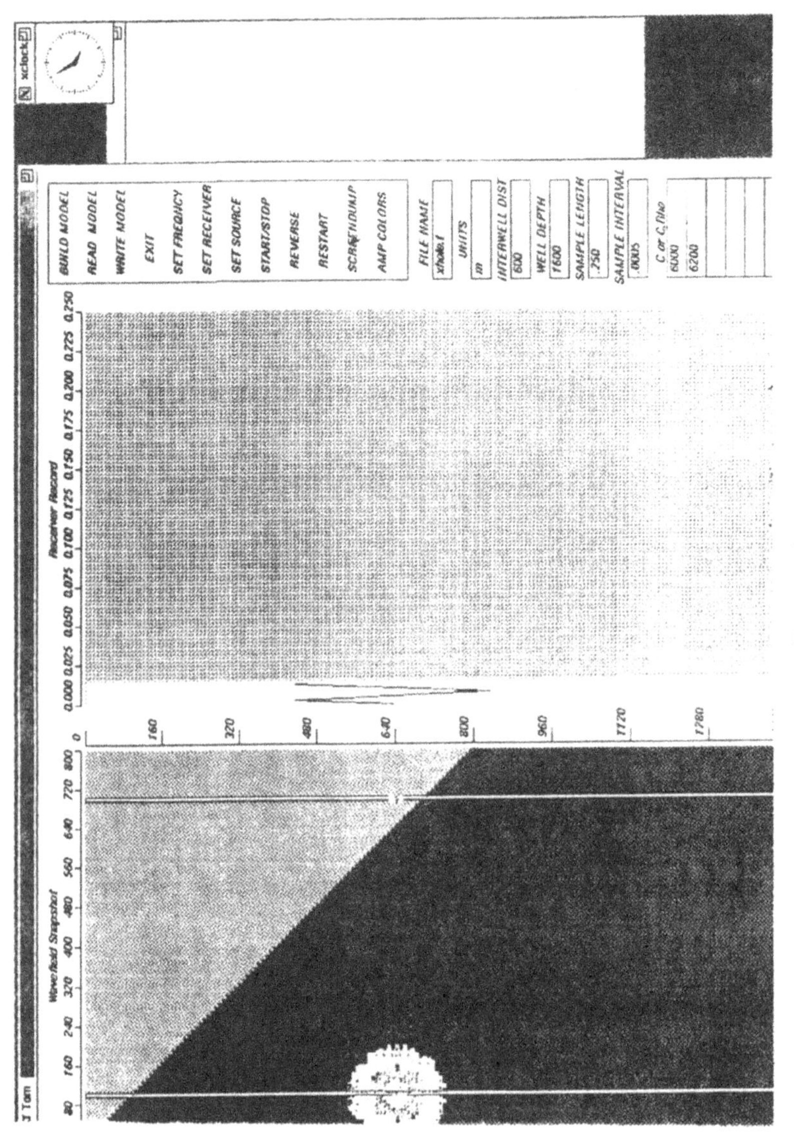
xclock
Wavefield Snapshot
Receiver Record
BUILD MODEL
READ MODEL
WRITE MODEL
EXIT
SET FREQNCY
SET RECEIVER
SET SOURCE
START/STOP
REVERSE
RESTART
AMP COLORS
FILE NAME
UNITS
INTERWELL DIST
600
WELL DEPTH
1600
SAMPLE LENGTH
.250
SAMPLE INTERVAL
.0005
6000
6200

Figure 3

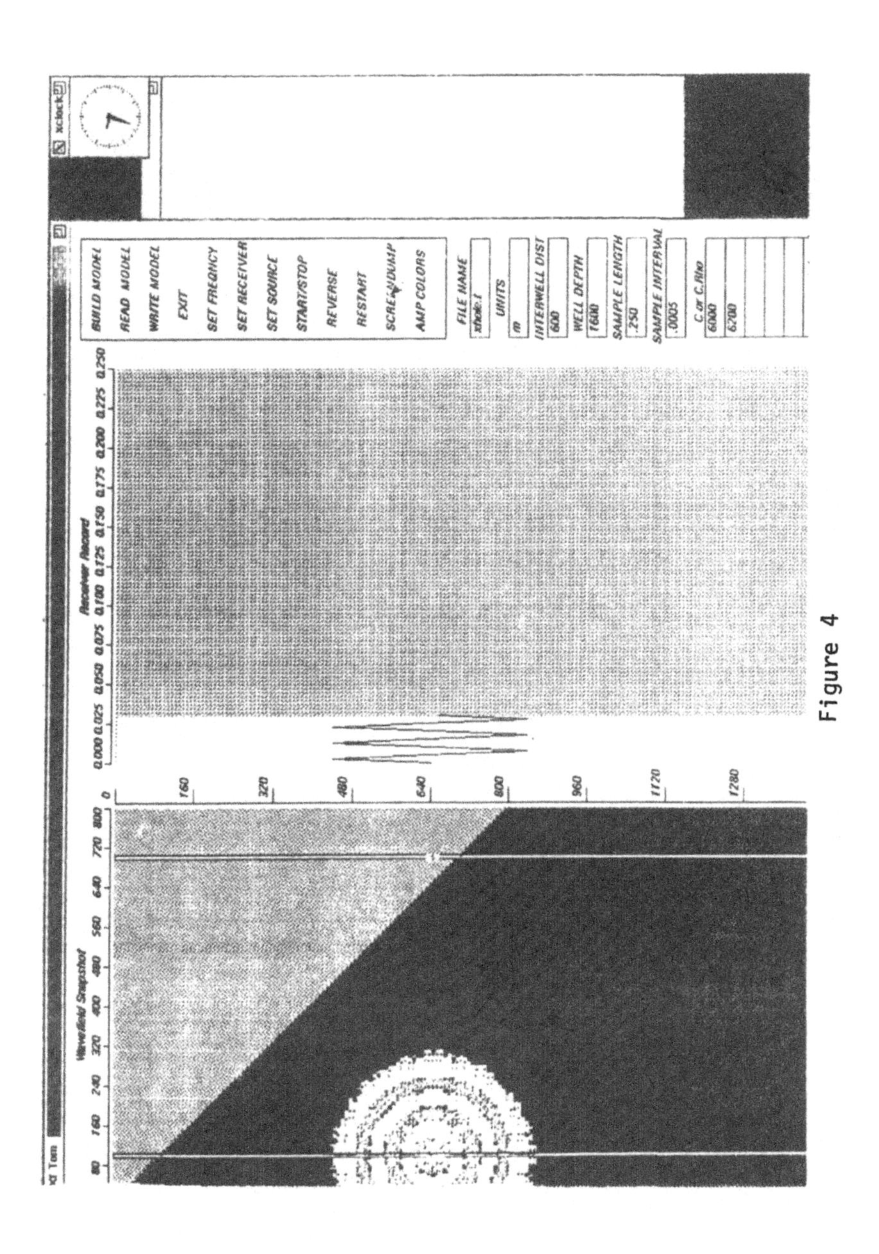
xclock
Wavefield Snapshot
80 160 240 320 400 480 560 640 720 800
Receiver Record
0.000 0.025 0.050 0.075 0.100 0.125 0.150 0.175 0.200 0.225 0.250
0 160 320 480 640 800 960 1120 1280
BUILD MODEL
READ MODEL
WRITE MODEL
EXIT
SET FREQNCY
SET RECEIVER
SET SOURCE
START/STOP
REVERSE
RESTART
AMP COLORS
FILE NAME
UNITS
m
INTERWELL DIST
600
WELL DEPTH
1600
SAMPLE LENGTH
.250
SAMPLE INTERVAL
.0005
C or C,Rho
6000
6700

Figure 4

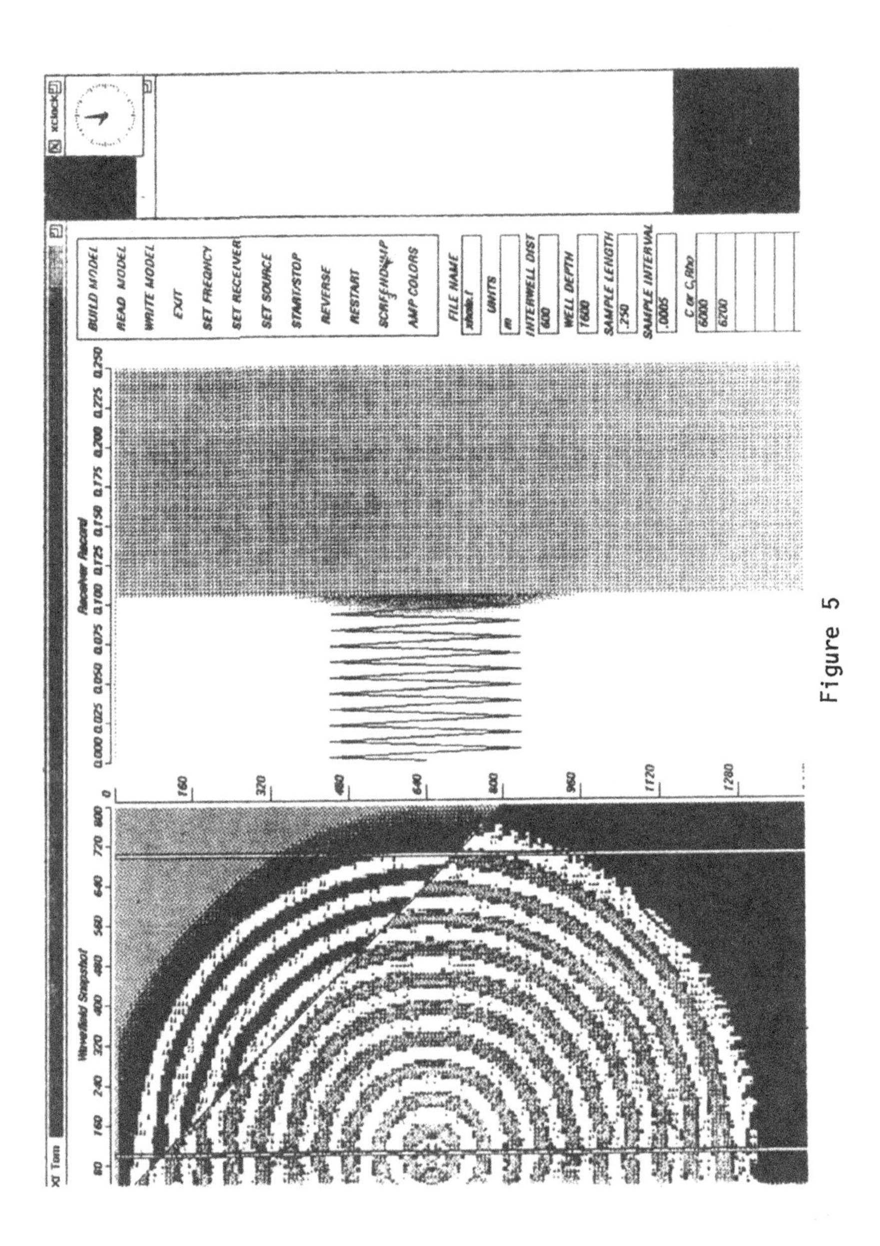
xclock
Xterm
Wavefield Snapshot
Receiver Record
BUILD MODEL
READ MODEL
WRITE MODEL
EXIT
SET FREQNCY
SET RECEIVER
SET SOURCE
START/STOP
REVERSE
RESTART
AMP COLORS
FILE NAME
UNITS
m
INTERWELL DIST
600
WELL DEPTH
1600
SAMPLE LENGTH
.250
SAMPLE INTERVAL
.0005
6000
6200

Figure 5

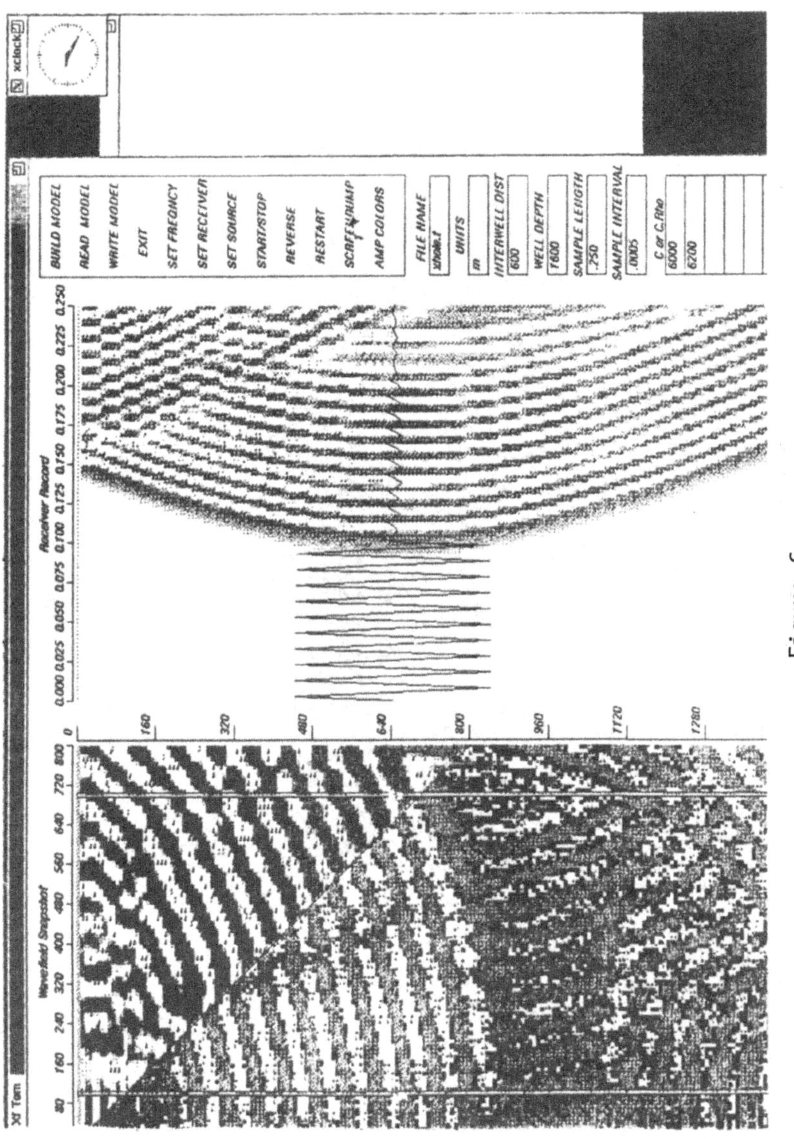
xclock
Wavefield Snapshot
Receiver Record
BUILD MODEL
READ MODEL
WRITE MODEL
EXIT
SET FREQNCY
SET RECEIVER
SET SOURCE
START/STOP
REVERSE
RESTART
AMP COLORS
FILE NAME
UNITS
WELL DEPTH
SAMPLE LENGTH
SAMPLE INTERVAL

Figure 6

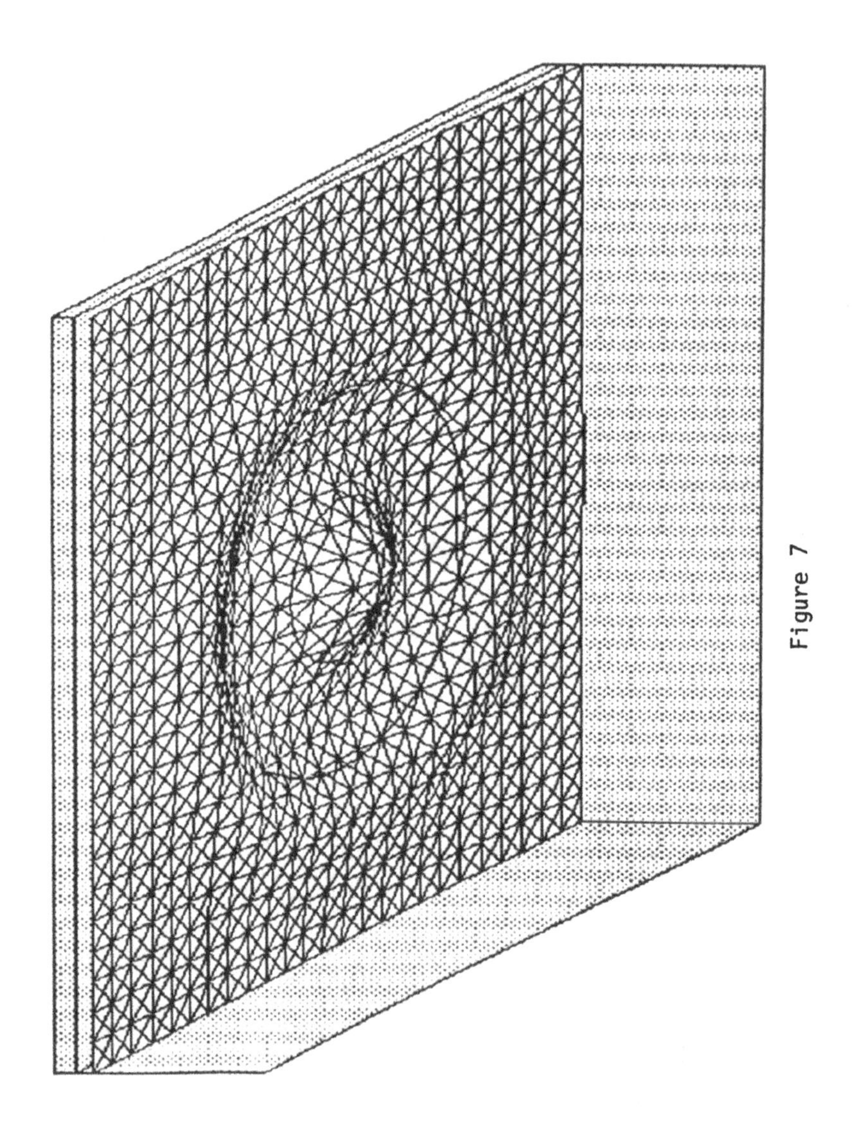

Figure 7

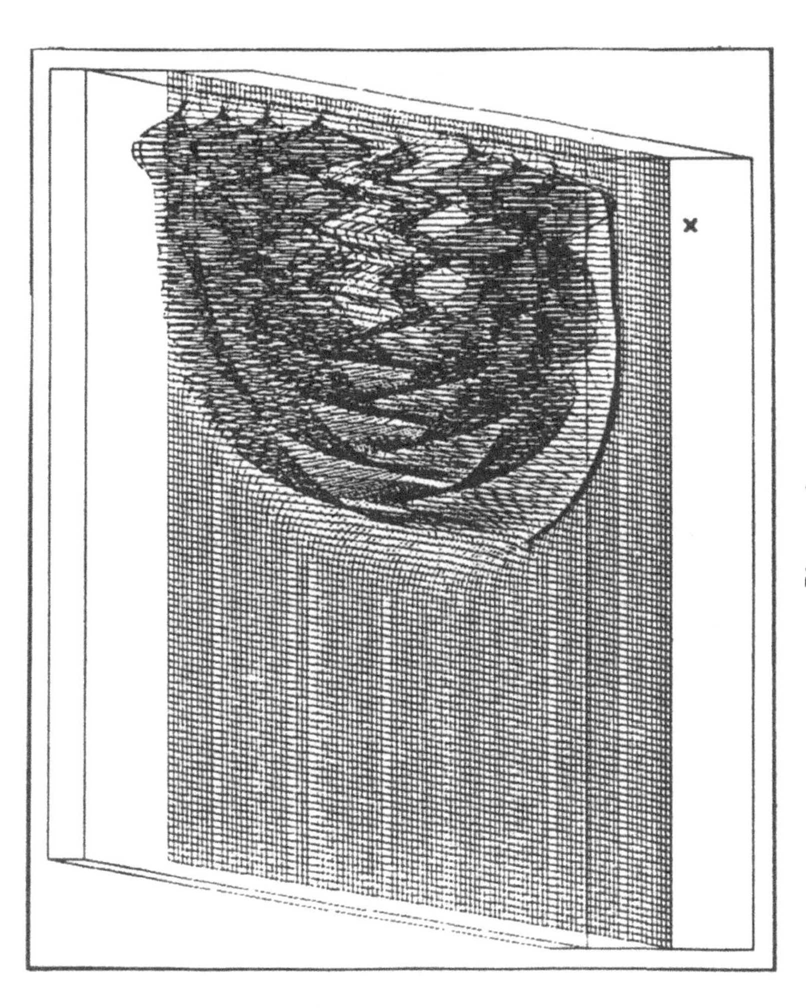

Figure 8

SECTION 3: PERFORMANCE

A Critical Discussion of Supercomputer Architectures for Engineering Applications

W. Schönauer, H. Häfner

Rechenzentrum der Universität Karlsruhe, D-7500 Karlsruhe 1, Germany

ABSTRACT

The main purpose of this paper is to explain the large gap between theoretical peak performance and real performance for present supercomputers. From this knowledge result valuable hints for the design of supercomputer architectures and of supercomputer algorithms.

1. THE PROTOTYPE VECTOR COMPUTER AND ARITHMETIC OPERATIONS

In Fig. 1.1 the prototype vector computer is depicted. The technology is characterized by the cycle time τ in nsec. The floatingpoint units are pipelines, thus have a startup time and need long vectors for efficient use. Chainig means coupling and overlapped operation of different pipelines. For basic references see [1,2]. In this paper we denote by "word" 64 bits = 8 bytes and by "pipe group" an addition and multiplication pipe. Critical points of vector computers are: memory bandwidth in words per cycle and pipe group (there are vector computers with multi-track pipelines); memory size in Mwords (million words); size and bandwidth of extended memory (the two sizes limit the problem size); i/o bottleneck: disks are extremely slow compared to the pipeline speed.

For the discussion of the memory bandwidth we present four arithmetic operations:

$$a_i = b_i + c_i, \tag{1.1}$$
$$a_i = b_i + c_i * d_i, \tag{1.2}$$
$$a_i = b_i + s * c_i, \tag{1.3}$$
$$a_i = a_i + s * b_i. \tag{1.4}$$

Equation (1.1) is a dyadic operation and needs two loads and one store per cycle and pipe group. Equation (1.2) is the vector triad or full triad that needs three loads and one store per cycle and pipe group. This is the most important operation. For parallel operation of the addition and multiplication pipeline two results per cycle and pipe group are obtained (supervector speed). A vector computer with the one-word memory bottleneck per cycle and pipe group delivers for (1.2) only one quarter of the possible pipe performance. The linked triad (1.3) and the repeated contracting linked triad (1.4) are special cases of (1.2) with less memory references. For vector computers not only additions and multiplications, but "operations" like (1.1) to (1.4) should be counted.

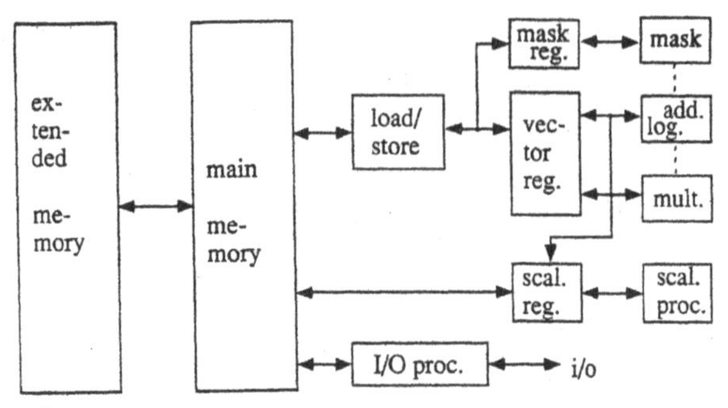

Fig. 1.1 Prototype vector computer.

Table 1.1 Performance of three vector computers in MFLOPS.

performance in MFLOPS	theoret. peak	vector triad n = 10 000	MMUL(Fortran) n = 500	MMUL (library) n = 500
CRAY Y-MP (one proc.) τ = 6 nsec	333.3	148.9 - 150.2	199.5 - 204.4	308.5
IBM 3090S VF (one VF) τ = 15 nsec	133.3	17.7 - 19.0	32.5	102.3
Fujitsu VP 2600 (Siemens S600) τ = 3.2 nsec, 8-track pipes	5000	1111.6	2813.6	4422.7

In Table 1.1 the performance for three vector computers is presented. Our rating of these computers is the performance of the vector triad that corresponds astonishingly well to benchmark ratings. The matrix multiplication (MMUL) with a columnwise algorithm in Fortran (the compiler selects which loop is to be vectorized) and with a library routine demonstrates the wide gap between vectorized Fortran and assembler code, the basic operation is equation (1.4), i.e. an "exceptional" operation.

Up to now we have discussed memory access by contiguous vector elements. If we access the memory with a stride, memory access and memory bank conflicts may occur, above all if the stride is a power of 2. With a stride the cache of the IBM VF quickly becomes useless. The least efficient access is by indirect addressing, either on the right hand side (gather) or on the left hand side (scatter) of an assignment; index vectors must be loaded and addresses must be computed. The consequence of these properties is the

$$\text{\underline{Essential rule:} Separation of the selection and of the processing of the data.} \qquad (1.5)$$

This means that we must at first select an optimal data structure and then process the data in order to keep the pipes continuously busy, see [1]. Not a code optimization but an optimal data structure is the key for an efficient use of a vector computer.

2. SOME BASIC ALGORITHMS

We want to discuss some exemplary algorithms for full and sparse linear algebra problems with respect to data structure and related types of operations. At first we discuss the matrix multiplication (MMUL) C = A * B for full (square) matrices. In Fortran columns (first index) are in contiguous storage locations, therefore a columnwise algorithm as indicated

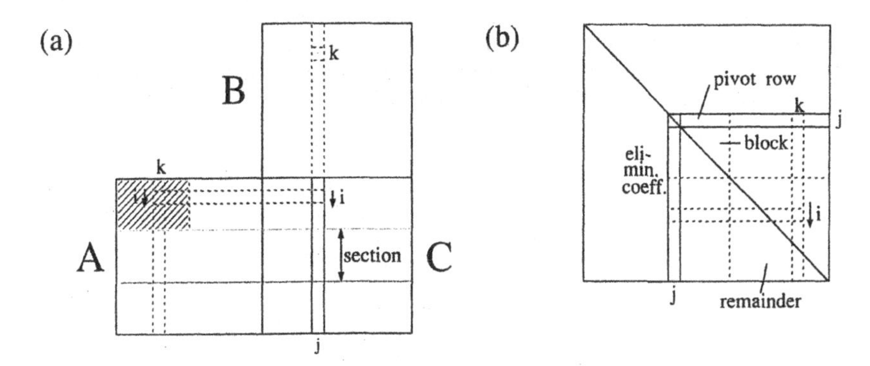

Fig. 2.1(a) Columnwise algorithm for MMUL, (b) illustration for Gauss elimination.

in Fig. 2.1a, that computes a whole column of C, is optimal. The two innermost loops of the algorithm (see [1]) are:

```
      do 20  k = 1,n
             do 20  i = 1,n                              (2.1)
20           c(i,j) = c(i,j) + a(i,k) * b(k,j)
```

Here b(k,j) is a scalar for the i-loop and line 20 is a contracting linked triad of type (1.4). A compiler that does not optimize over at least two do-loops will process whole columns of C with many unnecessary load/stores of c(i,j). Then on the Fortran level only unrolling of the k-loop could improve the situation. A better compiler or an assembler program would process a whole "section stripe" of A (see the dotted lines in Fig. 2.1a), during which the section of C is fixed in the vector register and only one load per cycle and pipe group is needed. Then the next section stripe is processed etc. But for an IBM VF with a cache we have even here for this data structure to apply the essential rule (1.5): To avoid cache rolling we fix in the cache a part of the section stripe of A that just fits into the cache (shaded part in Fig. 2.1a), and process now all parts of the section stripe of C. In order to use the cache completely we must not use the shaded part of A itself but a copy of it in a dense array because only data that is dense in the memory is also dense in the cache. Now the performance values of the last two columns of Table 1.1 are cleared up.

For the LU-decomposition of a matrix A = LU by Gauss elimination, Fig. 2.1b, the two innermost loops of a columnwise algorithm (see [1]) are:

```
      do 20  k = j + 1, n
             do 20  i = j + 1, n                         (2.2)
20                  a(i,k) = a(i,k) - a(i,j) * a(j,k)
```

Here a(i,j) denotes the column with the elimination coefficients, a(j,k) is a scalar for the i-loop. We have again a repeated linked triad of type (1.4), but it is not a contracting one and thus unrolling of the k-loop is not possible. But an assembler program could fix in a section-oriented algorithm the section of the elimination coefficients a(i,j) in a vector register so that for the i-loop one load and one store per cycle and pipe group is needed. A step towards the contracting linked triad could be made by a block-oriented algorithm, see Fig. 2.1b. In the small square part denoted by "block" and the rectangular parts below and right of it, a "usual" algorithm is executed. Then in the large square block denoted by "remainder" all elimination columns below the small square block can be applied "simultaneously", leading (for section stripes) to contracting linked triads over the number of elimination columns. The transition to block-algorithms is also the means for an efficient use of the cache for the IBM VF. Such algorithms must be optimized for each architecture. This is done presently for all supercomputers in the form of BLAS (Basic Linear Algebra Subprograms) [3].

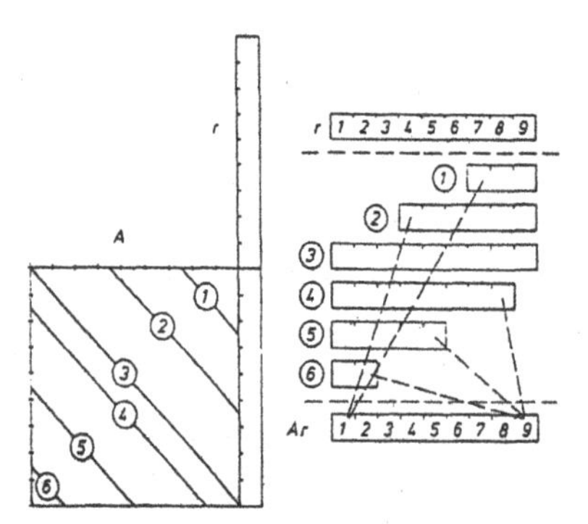

Fig. 2.2 Illustration of the matrix-vector multiplication in diagonal form.

The extremely large and spars linear systems of equations that result from the discretization of PDEs by the FDM or FEM usually are solved iteratively, because the fill-in produced by a direct solution method is too expensive. The basic "operation" of all iterative solvers is the matrix-vector multiplication (MVM) A * r. There are many possibilities to store sparse matrices. Because usually those matrices have a "band structure" we decided to store them by diagonals, i.e. as a "file fo diagonals". But the diagonals themselves may be sparse, thus we store them as a file of packed diagonals with nonzero elements and corresponding index vectors for the indices in the diagonals. This is a typical example for the choice of an optimal data structure for vector computers. Now the MVM must be formulated in diagonal form, see Fig. 2.2. For the multiplication of the elements the upper/lower diagonals are arranged to end/start with the elements of r and the products are added to the result vector that they start/end with A * r. There are different possibilities to optimize this algorithm and to select between masked operation, unpacking of diagonals or sparse arithmetic for the selection of the active elements. The latter method proved to be the best one [4].

The development of our black box PDE solver FIDISOL (see [1]) started with this data structure of packed diagonals for the matrix of the linear system and then the data structure of the whole program package has been selected to support this structure, i.e. to generate the matrix directly by diagonals. It is not surprising that this procedure resulted in an optimal data structure also for other parts of the program package because this is a type of "natural" ordering. Presently we are developing the FEM kernel program VECFEM in which the element computations, the mounting of the global matrix and the iterative solution of the linear system is oriented towards packed diagonal storing and has resulted in a complete vectorization [5].

3. MICRO-MEASUREMENTS REVEAL SOURCES FOR LOST CYCLES

In order to clear up the large gap between theoretical peak performance and the real performance for the vector triad in Table 2.1, we may ask the question: Where are the lost cycles? This question has been investigated for the IBM VF, the Fujitsu VP400-EX (predecessor of VP2600) and briefly for the CRAY Y-MP in [6] and for the latter in more detail in [7], further measurements will be published in [8]. Here we report about the VF and Y-MP. The VP600 that has been installed only recently, will be treated in a forthcoming paper. Without losses we needed n cycles (one chime) with supervector speed to execute a

all data in cache

load d, sectioning

$$t_{mic}(n)=\tau\Big(\underbrace{2\big(n+\lceil \tfrac{n}{vrl}\rceil\cdot 28\big)}_{\text{load b,c}}+\underbrace{n+\lceil \tfrac{n}{vrl}\rceil\cdot(28+22)+5+}_{\text{multiply and add , d from cache}}$$

$$\underbrace{+n+\lceil \tfrac{n}{vrl}\rceil\cdot 31+\lceil \tfrac{n}{cll}\rceil}_{\text{store a}}\Big)$$

all data in main storage

d, sect.

$$t_{mic}(n)=\tau\Big(\underbrace{2\big(n+\lceil \tfrac{n}{vrl}\rceil\cdot 38+\lceil \tfrac{n}{cll}\rceil\cdot 6}_{\text{load b,c from m.s.}}+\underbrace{n+\lceil \tfrac{n}{vrl}\rceil(38+22)+}_{\text{multiply and add , d from m.s.}}$$

$$\lceil \tfrac{n}{cll}\rceil\cdot 6+5+\underbrace{n+\lceil \tfrac{n}{vrl}\rceil\cdot 32+\lceil \tfrac{n}{cll}\rceil\cdot 18}_{\text{store a into m.s.}}+\underbrace{(\lceil \tfrac{n}{ps}\rceil-1)\cdot 95}_{\text{3 ld. 1 st.over p.s.}}\Big)$$

Fig. 3.1 Result of micro-measurements for the IBM VF for $a_i = b_i + c_i * d_i$, all data in cache and all data in main storage.

vector triad (1.2) for a vector of n elements. But if we look at the corresponding assembler code we recognize that in reality several scalar instructions to prepare the operation and to organize the section loop for the strip-mining of long vectors, and vector load/store operations to load/store the operands to/from the vector registers are needed. In Fig. 3.1 for the IBM 3090S VF the time $t_{mic}(n)$ to process n elements of the vector triad is depicted, that gives all the details of the needed cycles. In front of the main parentheses is the cycle time τ, thus inside the parentheses we count cycles. For the IBM 3090S VF we must distinguish if the data is in the cache or if the data is in the main storage, i.e. if the cache is rolling. By $\lceil x \rceil$ we denote the next integer $\geq x$. We have the parameters cycle time τ = 15 nsec, vector register length vrl = 256 words (à 64 bits), cache line length cll = 16 words, page size ps = 512 words, n = vector length to be processed.

If we add up all terms of the same type we get the full model $t_{step}(n)$, and if we replace the step function by a "mean" straight line (see [6]) we get the mean model $t_{mean}(n)$:

data in cache:

$$t_{step}(n)=\tau\left(4n+137\left\lceil\frac{n}{vrl}\right\rceil+1\left\lceil\frac{n}{cll}\right\rceil+5\right), \qquad (3.1)$$

$$t_{mean}(n)=\underbrace{\underbrace{\left(4+\frac{137}{vrl}+\frac{1}{cll}\right)}_{W}\tau}_{\tau_{eff,mean}}\Bigg(n+\underbrace{\frac{73.5}{W}}_{n_{\frac{1}{2}eff,mean}}\Bigg), \qquad (3.2)$$

data in main storage:

$$t_{step}(n) = \tau\left(4n + 146\left\lceil\frac{n}{vrl}\right\rceil + 36\left\lceil\frac{n}{cll}\right\rceil + 95\left(\left\lceil\frac{n}{ps}\right\rceil - 1\right) + 5\right), \tag{3.3}$$

$$t_{mean}(n) = \underbrace{\underbrace{\left(4 + \frac{146}{vrl} + \frac{36}{cll} + \frac{95}{ps}\right)}_{W}\tau}_{\tau_{eff,mean}}\left(n + \underbrace{\frac{48.5}{W}}_{n_{\frac{1}{2},eff,mean}}\right). \tag{3.4}$$

If we look at $t_{mean}(n)$, equs. (3.2) and (3.4), we see in the second parantheses the number n of useful cycles. But the first parentheses constitute a wasting factor W that increases the hardware cycle time τ to $\tau_{eff,mean}$. With $\eta_a = 1/W$ we define an architectural efficiency. For infinitely long vectors the effective real performance in MFLOPS is theoretical peak performance times η_a. Half of this performance is obtained at a vector length $n_{1/2,\ eff,\ mean}$ (see (3.4)) that is Hockney's half performance length, see [1,2]. The computed values of the performance parameters for data in cache/data in main storage are: W = 4.60/7.01 (ideal value is 1), $\tau_{eff,\ mean}$ = 69.0/105.1 nsec (ideal value is 15 nsec), η_a = 0.212/0.143 (ideal (value is 1), $r_{\infty,\ eff}$ = 29.0/19.0 MFLOPS (ideal value is 133.3 MFLOPS), $n_{1/2,\ eff,\ mean}$ = 16/7. If we interpret the values for data in main storage we can say: For each useful cycle 6.01 additional wasted cycles must be paid, instead of the performance of a 15 nsec technology we get the performance of a 105.1 nsec technology, the architectural efficiency is 14.3 %, instead of 133.3 MFLOPS we get 19 MFLOPS for infinitely long vectors, and half this performance is attained at a vector length n = 7. This explains the 17.7 -19.0 MFLOPS in the third column of Table 1.1 for the VF, the difference to $\bar{r}_{\infty,\ eff}$ is due to the "mean"

$$t_{mic}(n) = \tau\Big(n + \underbrace{\left\lceil\frac{n}{2\ vrl}\right\rceil * \min(n,vrl) + 6\left\lceil\frac{n}{2\ vrl}\right\rceil}_{\text{additional cycles because of third chime per 128 elements}} + \overset{\text{section loop organization}}{6\left\lceil\frac{n}{vrl}\right\rceil} + \underset{\text{programming of a port for load and termination of add and store pipe}}{23} + \overset{\text{startup}}{29}\Big)$$

Fig. 3.2 Result of micro-measurement for the CRAY-Y-MP (one processor) for $a_i = b_i + c_i * d_i$.

value. If we make the same measurements of the vector triad for one processor of a CRAY Y-MP (that of the Y-MP832 at Jülich) we must observe that we need three loads and one store, but the Y-MP has only two loads and one store available. Therefore double sections of 2 * 64 elements are processed in an interleaved manner to deliver in 3 chimes the necessary 6 loads, full chaining leads to efficient overlapping of load/store and arithmetic operations. The result of the micro-measurement is presented in Fig. 3.2, the sources of the lost cycles are explained by comments. These measurements have been arranged that no memory contention occurs. But in realistic operation the simultaneous two loads and the store lead to self-induced memory access contention, lost cycles per useful cycle are represented by a coefficient c_s, and the simultaneous access of several processors in a Y-MP832 leads to external memory contention by bank conflicts, lost cycles per useful cycle are represented by a coefficient c_e. The value of c_s depends on the relative location of the operand and result vectors, the value of c_e depends on the actual job profile, see [7]. With "mean" values for the step functions and c_s, c_e we get the mean model

$$t_{mean}(n) = \underbrace{\underbrace{\left(\frac{3}{2} + c_s + c_e + \frac{9}{vrl}\right)}_{W} \tau}_{\tau_{eff,mean}} \left(n + \underbrace{\frac{58 + vrl/2}{W}}_{n_{\frac{1}{2},eff,mean}}\right). \qquad (3.5)$$

We compute the performance parameters for $\tau = 6$ nsec, vrl = 64, $c_s = 0.4229$, $c_e = 0.2123$ to W = 2.28, $\tau_{eff,mean} = 13.7$ nsec (ideal 6 nsec), $\eta_a = 0.439$, $\bar{r}_{eff,\infty} = 146.4$ MFLOPS (ideal 333.3 MFLOPS), $n_{1/2,eff,mean} \approx 40$. Here we have much more favorable values of W and η_a compared to the IBM VF. These better values result above all from the higher memory bandwidth of the Y-MP. Now the 148.9 - 150.2 MFLOPS for the Y-MP in the third column of Table 1.1 is explained, the difference to $\bar{r}_{eff,\infty}$ results from the character of the "mean" value.

4. MORE MFLOPS, PERFORMANCE FORMULA

Today (March 1991) the Fujitsu VP2600 has a cycle time of $\tau = 3.2$ nsec, the NEC SX-3 has $\tau = 2.9$ nsec, for the CRAY-3 $\tau = 2$ nsec is expected, perhaps in 1995++ we shall have $\tau = 1$ nsec. For $\tau = 1$ nsec we get 2 GFLOPS supervector speed for a single pipe group. For higher performance parallelism is the only means to increase performance. This can be done by external parallelism like the 6 processors of an IBM 3090/600 with 6 VFs or a CRAY Y-MP8 with 8 processors, or it can be done by internal parallelism like in the Fujitsu VP 600 with two 4-track pipe groups, i.e. internally 8 pipe groups, or it can be done by external and internal parallelism like for the NEC SX-3, Model 44 with 4 processors, each with internally 8 pipe groups. But if we want to use the whole performance of a multiprocessor supercomputer, we must distribute the workload onto several processors in a multitasking style and we are immediately in the area of parallel computing. We shall discuss the related problems in a more general way.

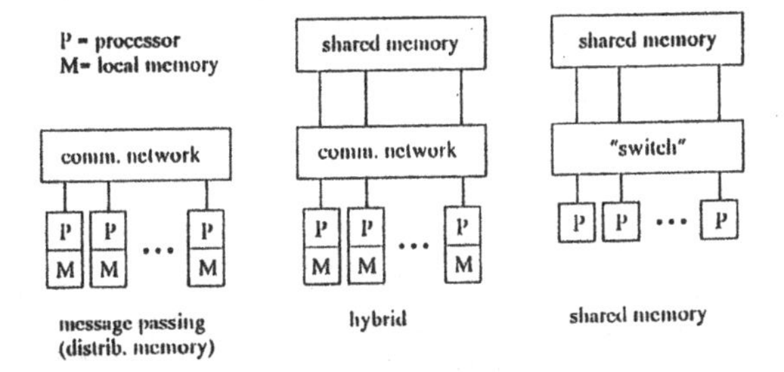

Fig. 4.1 Three basic type parallel architectures.

In Fig. 4.1 three basic type parallel architectures are depicted, all are of MIMD type. The shared memory system has the hardware problem of memory bandwidth and memory contention with increasing number of processors, but the user has a central storage and must distribute only the workload onto the processors. The message passing system or distributed memory system offers in the ideal case an "infinite" number of processors, but the processors can operate only with the data in their (local) memory. Thus the user must distribute workload and data onto the processors and select data structures and algorithms with a high ratio of computation to communication. The real problem for message passing systems is that a dynamic central resource management to use idling processors is basically not possible and thus these are GFLOPS-PCs. Hybrid systems with local and shared memory combine the disadvantages of the two inherent systems and add the severe problem of data coherence. The argument of the manufacturers of massively parallel systems is, that they deliver cheap MFLOPS or GFLOPS. But the price/performance relation depends strongly on the usage of the processors, counted over a whole year. A comparison of the price/performance for supercomputers and parallel computers is presented in [9]. It turns out that the massively parallel computers are not "cheap".

A brief remark to the development of programs for parallel computers: Try to apply the principle of "divide and conquer" in order to break up the whole problem into optimal "chunks" for the available processors. Large problems always result from a nest of loops. If we have e.g. a nest of three loops

```
do  10  i = ........  (parallelize)
 do  20  j = ........
  do  30  k =........  (vectorize)
```

we should try to parallelize the outermost loop and distribute the data correspondingly to the local memories, and try to vectorize the innermost loop with contiguous elements. Usually the processors will have vector pipes for the arithmetic units, e.g. for the INTEL iPSC/860.

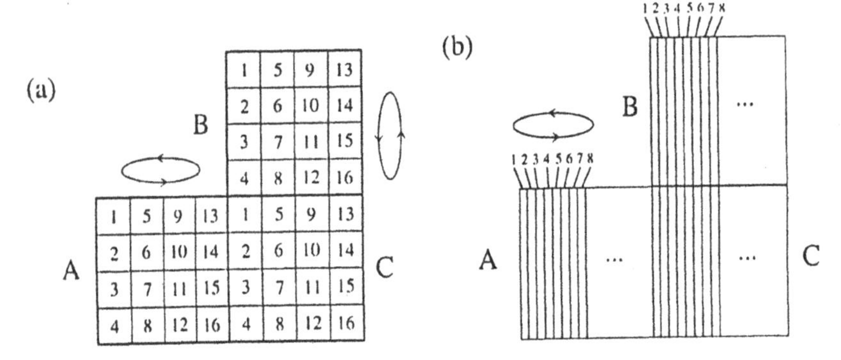

Fig. 4.2 Matrix multiplication on 16 processors, (a) with square blocks, (b) with column blocks.

We want to discuss the matrix multiplication C = A * B on 16 processors of a message passing system. If we distribute the matrices in square blocks onto the processors, see Fig. 4.2a, we must shift the matrices A and B through all processors, if the processors indicated in C are considered to be fixed. We have 2 n^2 transfers if n is the dimension of the matrices, n^3 linked triads, a vector length of n/4 and a storage overhead of 5/3 = 1.67 because for parallel execution and transfer in a processor 5 blocks are needed (two for the transfer buffers). If we distribute the matrices in column blocks, see Fig. 4.2 b, we must shift only matrix A through all processors. We have n^2 transfers, n^3 linked triads, vector length n and a storage overhead of 4/3 = 1.33. Thus the column blocks are much more efficient. It should be mentioned that in the papers about parallel algorithms usually the storage overhead is not mentioned. For the individual blocks the columnwise algorithm for vector computers will be applied, if the processors have vector pipelines.

Seemingly we can increase the performance of our supercomputer arbitrarily by increased parallelism. But what we get really out of the computer may be very poor. In [9] a performance formula has been presented in which "the losses" (i.e. only the most significant ones) have been factorized. Here we give only a brief summary of these investigations, a more detailed discussion will be given in [10]. A supercomputer with a total of P pipe groups will give a real performance

$$\bar{r}_{real} = \underbrace{\frac{1000}{\tau[nsec]} * 2 * P}_{\textit{theoret. peak}} * \underbrace{f_1 * f_2 * \ldots * f_N}_{\textit{reduction factors}} \text{ [MFLOPS]} . \tag{4.1}$$

The $f_i \leq 1$ are reduction factors that account for special losses. A selection of such factors could be: $f_1 = 1/(1+d)$ gives losses for section loop organization (z * d lost cycles for z useful ones), $f_2 = m/4$ gives losses due to the memory bottleneck for the full triad which is our key operation (m = memory bandwidth in words per cycle and pipe group), $f_3 = n/(n + n_{1/2})$ gives losses by finite vector length n ($n_{1/2}$ is Hockney's half performance length [2]), $f_4 = 1/((1-v) * w + v)$ is Amdahl's law (v = vectorizable part of operations, w = ratio of vector to scalar performance). For multiprocessor computers further losses occur: $f_5 = 1/((1-q) * p + q)$ is Amdahl's law for parallelization (q = parallelizable part of operations, p = number of processors), for a shared memory computer we have $f_{6,sh} = 1/(1+2 * (1-c)/a)$ as a "model" for memory contention (c is part of operations with contiguous, (1-c) with random elements (indirect addressing), a is an empirical factor, e.g. a = 2), for a message passing system we have $f_{6,mp} = 1/(1+b)$ for additional cycles by non-overlapping communication (z*b lost cycles for z useful ones), $f_7 = (\sum_{i=1}^{p} t_i)/(p * \max_i t_i)$ accounts for load balancing (a user has reserved p processors, e.g. his "subcube", t_i is the time that processor i is active). Finally f_8 = (hours of usage per year)/8760 defines a long-range utilization factor.

For reasonable assumptions you may get $f_1 * \ldots * f_8 = 0.01$, i.e. then you get just 1% of the theoretical peak performance of your "supercomputer". Note that then 99% of the performance is lost, a 5 GFLOPS computer then delivers poor 50 MFLOPS! Try to evaluate the f_i for your computer and class of problems and you will recognize: The architecture of the present supercomputers is not user-friendly. Thus we may ask the question:

5. COULD USER-FRIENDLY SUPERCOMPUTERS BE DESIGNED?

This is the title of the paper [11] which we summarize here briefly. There is an inherent coupling of sustained performance and memory size. In order to use a certain performance of sustained MFLOPS or GFLOPS we must be able to store the necessary operands to keep the pipes busy. An investigation for large FDM and FEM problems gave a relation of 64 Mwords (à 64 bits) for 100 MFLOPS sustained performance. This means 64 Gwords for 100 GFLOPS, thus memory size is the real problem for a balanced system. To observe the essential rule (1.5) we need efficient data transfer operations (pack, unpack, merge, gather, scatter). To escape Amdahl's law as far as possible we should have the fastest cycle time because this is the only possibility to speed up scalar operations. Quite naturally our supercomputer must be programmable in portable Fortran to preserve software investments.

The goal of our supercomputer must be to come as close as possible to the theoretical peak performance. This is only possible by a memory bandwidth of three loads and one store per cycle and pipe group. But experience tells us that only 10% of the memory are needed with this high bandwidth as a main memory. Therefore 90% of the memory should be an extended memory, but with a bandwidth of one word per cycle and pipe group to the main memory, transferred in blocks. Thus one operand can be obtained via buffer from the extended memory. Let us assume that we have N = 64 pipe groups. Then for a 128/12.8/1.28 nsec technology our supercomputer will deliver 1/10/100 GFLOPS. About 1995 we can expect a 1.28 nsec technology and also the 64 Mbit chip that allows a 64 Gword total memory. For marketing considerations the whole spectrum of these three technologies with N = 1, 2, 4, ..., 64 processors should be offered.

The crucial problem is, how to organize such an extremely large memory bandwidth. This is only possible for a synchronous access, i.e. by SIMD (never by MIMD). In Fig. 5.1 the

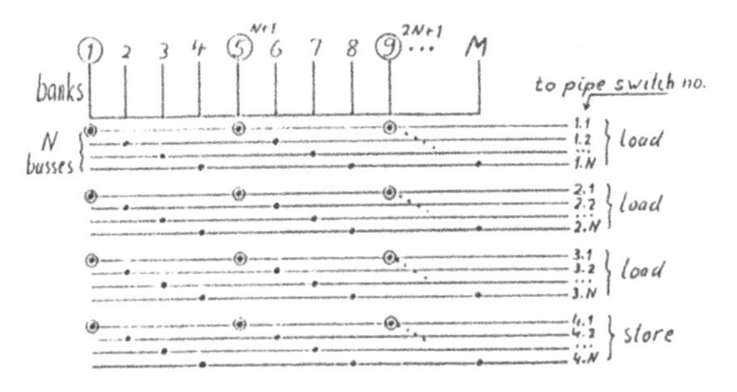

Fig. 5.1 Organization of memory banks and busses.

organization of the memory banks and of the four bus groups, each with N busses, is depicted for N = 4. In order to get the operands for a synchronous execution in the correct cycle to the correct pipe entry, a delay register and a "pipe switch" in the form of a pipelined register transfer must be inserted between each bus group and the pipe entries. In Fig. 5.2a

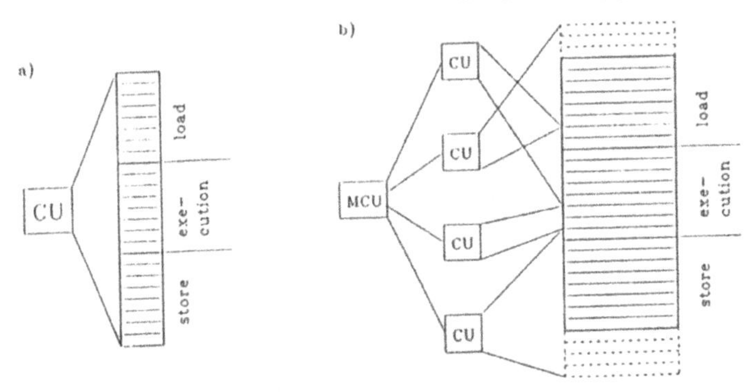

Fig. 5.2 (a) "Classical" vector computer with a single control unit (CU), (b) Continuous Pipe Vector Computer with master control unit (MCU) and many CUs.

the "classical"vector computer with a single control unit (CU) is depicted. In Fig. 5.2b the "Continuous Pipe Vector Computer" (CPVC) with many CUs and a master control unit (MCU) is presented. Each CU guides "its" vector operation through the "bundle" of N pipelines. Each CU has time to prepare its vector operation, there is no gap and thus no lost cycles between vectors except the N/2 lost operations in the mean for the last "section" of each vector and the loss for the very first startup. A problem arises for dependent vector operations whose vector length is larger than $n_{critical}$ = "volume" of the bundle of pipes. For such operations a register file is introduced only for intermediate results which allows a flexible transition between memory-to-memory and register computer. There are further new ideas, e.g. a combination pipeline with add, multiply, add parts, several parallel scalar units, synchronization and condition flugs that allow an early preparation of operations, and above all there is a proposition for indirect addressing by a two-stage priority-controlled memory access with nearly vector speed, for details see [11].

The architecture of the CPVC has been designed to minimize the number of lost cycles, i.e. to come as close as possible to the theoretical peak performance. All real problems are shifted to the hardware because cycles that are lost on the hardware level can never be regained on the software level. Only increased internal parallelism leads to a user-friendly design.

6. CONCLUDING REMARKS

We wanted to explain the wide gap between theoretical peak and real performance for supercomputers. The reason for the gap are lost cycles. We have revealed the losses for an individual operation, the vector triad which we consider as a key operation, by micro-measurements. We have explained the total losses by the discussion of the reduction factors in the performance formula (4.1). All these losses are ultimately caused by a lack of internal parallelism. This is the weak point of all present supercomputer architectures. If all "auxiliary" cycles are fully overlapped by the "useful" cycles, no cycles are lost. This lack of internal parallelism makes life so difficult for the software, for the compiler and for the user. The most critical point is the memory bandwidth. For a bandwidth of four memory references per cycle and pipe group we had not to distinguish between vector triad and contracting linked triad; special section stripes or data locality in a cache would not be needed.

In order to give a positive criticism we have presented the ideas for a Continuous Pipe Vector Computer that minimizes the lost cycles and makes life as easy as possible for the compiler and for the user, including parallel pipelines. The problem of "parallel computing" can never be solved satisfactorily by MIMD parallel architectures. It must be solved internally on the hardware-level, i.e. user-transparent. But we may not end this paper without recalling the responsibility of the user for optimal data structures. Any architecture that is based on pipelines (also a form of parallelism) needs data structures with long and contiguous vectors that allow a type of "data flow algorithm", for details see [1].

Acknowledgement: This research has been supported in part by the Stiftung Volkswagenwerk in the frame of the project "Vektorrechner-Praktikum für Ingenieure".

REFERENCES

1. W. Schönauer, Scientific Computing on Vector Computers, North-Holland, Amsterdam 1987.
2. R.W. Hockney, C.R. Jesshope, Parallel Computers 2, Adam Hilger, Bristol 1988.
3. J. Dongarra, J. Du Croz, I. Duff, S. Hammarling, A sect of level 3 basic linear algebra subprograms, ACM Trans. Math. Software 16 (1990), pp. 1 - 17.
4. R. Weiss, H. Häfner, W. Schönauer, Tuning the matrix-vector multiplication in diagonal form, in D.J. Evans et al. (Eds.), Parallel Computing '89, North-Holland, Amsterdam 1990, pp. 93 - 98.

5. L. Groß, P. Sternecker, W. Schönauer, Optimal data structures for an efficient vectorized finite element code, in H. Burkhart (Ed.), CONPAR 90/VAPP IV, Lecture Notes in Computer Science 457, Springer-Verlag, Berlin 1990, pp. 435 - 446.
6. W. Schönauer, H. Häfner, Supercomputers: Where are the lost cycles, to appear in the proceedings of the 1991 ACM Internat. Conf. on Supercomputing, Cologne, Germany, June 17 - 21, 1991.
7. H. Häfner, W. Schönauer, Micro-measurements of the CRAY Y-MP and models for memory contention, submitted to Sympos. for High Performance Computation, Montpellier, France, Oct. 1991.
8. W. Schönauer, H. Häfner, Micro-measurements for the IBM 3090/VF, the Siemens/Fujitsu VP 400-EX and the CRAY Y-MP, Interner Bericht des Rechenzentrums der Universität Karlsruhe, in preparation.
9. W. Schönauer, H. Häfner, Supercomputers: The hardware, the architecture, to appear in the proceedings of the 13th IMACS World Congress on Computation and Applied Mathematics, Dublin, July 22 - 26, 1991, edited by J.J.H. Miller.
10. W. Schönauer, H. Häfner, Performance estimates for supercomputers: The responsibilities of the manufacturer and of the user, to appear in Parallel Computing.
11. W. Schönauer, R. Strebler, Could user-friendly supercomputers be designed?, in J.T. Devreese, P.E. van Camp (Eds.), Scientific Computing on Supercomputers II, Plenun Press, New York, pp. 99 - 122.

A High Performance Vector-Parallel Skyline Solver Using Level 3 BLAS

D.F. Lozupone

IBM ECSEC (European Center for Scientific and Engineering Computing), via Giorgione 159, 00147 Roma, Italy

Abstract

This paper deals with a parallel solver for matrices stored following the the skyline storage pattern. The algorithm performs the Gaussian elimination by using the Level 3 BLAS blocked approach. Large grain parallelism is achieved performing operation on separate blocks in parallel.
The performance of the solver interfaced with the finite element program ADINA(*) is presented and discussed.

1 - Introduction

The factorization of a linear system of equations represents typically the compute intensive section of finite element codes performing implicit nonlinear analyses. As the system of equation has a variable banded structure these codes are used to store the matrix using the skyline storage pattern in order to save storage and floating point operations during its factorization. This storage technique was proposed more than 20 years ago and is nowadays widely used in many industrial packages.
We have considered the problem of developing versions of the skyline solver that would run efficiently over a wide range of architecture. To do this we have implemented a code based on standard kernels, extending ideas that have been originally developed for dense linear algebra codes.
On computers provided with hierarchical levels of storage, in order to avoid having to access the slower main memory more than is necessary,

it is essential to make as much use as possible of data stored in the high speed memory, thus avoiding costly data movements. A set of Level 3 BLAS (Basic Linear Algebra Subprograms) routines [4] [5] has been proposed so that operations on matrices partitioned into blocks can be performed by calls to a few basic kernels with standard interfaces.
The LAPACK library has been recently developed for solving linear algebra problems. LAPACK [2] functions are written in terms of calls to the BLAS. The code is thus modular, and very easy to maintain. Performance is achieved by using the version of the BLAS optimized for the machine the code is currently executing on. Many experiments over the past few years indicate that with this approach LAPACK routines attain a performance with is very close to that which is reached by a code entirely tuned for a particular architecture.
The Engineering and Scientific Subroutine Library (**ESSL**), supplied by IBM to provide high performance on IBM ES/9000 Vector Multiprocessors, already contains some Level 3 BLAS routines, thus giving to the user the capability of developing many linear algebra algorithms using highly tuned versions of these kernels.

The Cholesky factorization of a symmetric positive definite skyline matrix was faced using Level 3 BLAS kernels. The unsymmetric version of the method was afterwards developed.
Results obtained on IBM ES 9000/VF show speeds very close to those obtained for dense matrices, even for very irregular skyline profiles. We will describe in the following the parallelization of the method carried on through a large grain approach obtained by splitting computations between blocks to different processors.
Performance data and validation tests are discussed, and comparisons with the skyline solver used in the ADINA(*) finite element system are also presented.

2 - The Skyline Storage Pattern

The weighted residual formulation at the base of the Finite Element Method, discretizes the partial differential equation so that the single nodal unknown is not coupled with all the others. Hence the resulting stiffness matrix is substantially sparse, with non-zero elements grouped around the diagonal with a variable band profile.
The skyline storage pattern (or **envelope pattern**, proposed by A. Jennings in 1966) stores and processes (e.g. in the case of a symmetric matrix) for each column of the matrix, only those elements that lie between the first non-zero element and the diagonal element. Zero elements outside the skyline are not stored because they are not

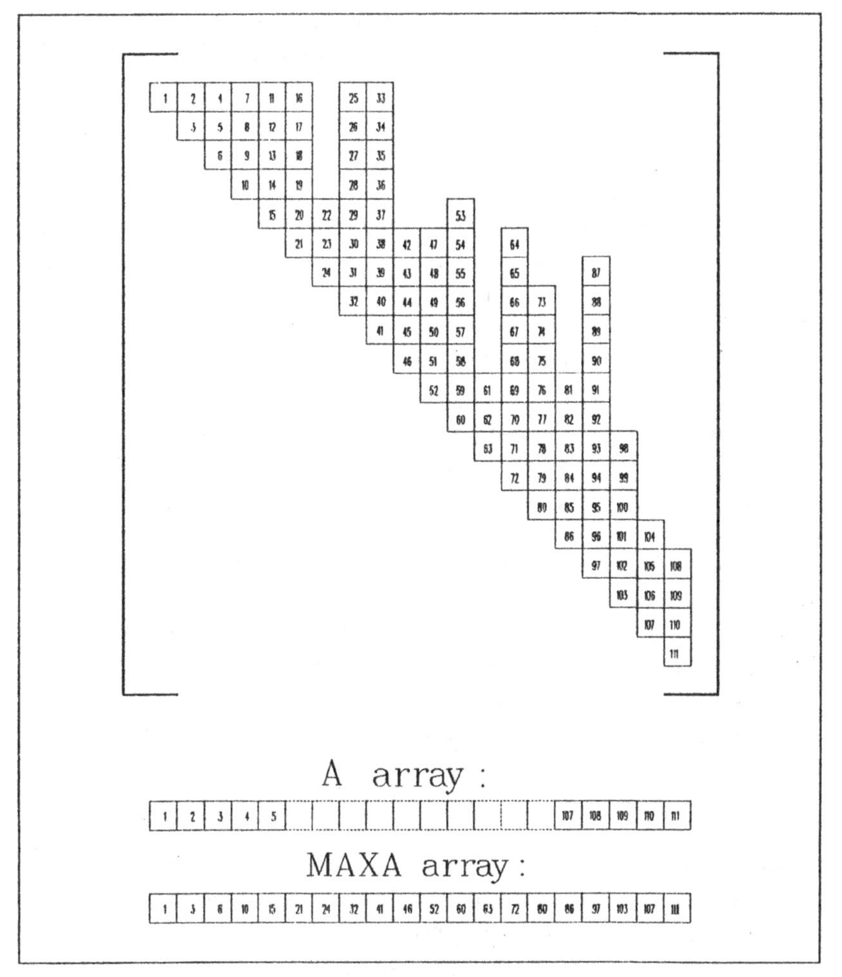

Figure 1: A Typical Skyline Storage Scheme for Symmetric Matrices

reduced during the factorization process. On the other hand, zero elements inside the skyline change, generally, to non-zero during the reduction.

In mathematical terms, following George and Liu [6], if a_{ij} is the generic entry of the matrix A, let

$$f_i(A) = \max\,(j \mid a_{ij} \neq 0)$$

and

$$\beta_i(A) = f_i(A) - i$$

be respectively the column subscript of the last non-zero element and the bandwidth of the i-th row. Then the **bandwidth** of the matrix will be defined as:

$$\beta(A) = \max\{\beta_i(A) \mid 1 \leq i \leq N\}$$

Along with other authors, we will use the term bandwidth to mean the quantity $2\beta(A)+1$, and refer to the quantity $\beta(A)$ defined above as the **half-bandwidth**. The skyline (or envelope) of A is defined by:

$$Env(A) = \{\{i,j\} \mid 0 < j - i \leq \beta_i(A)\}.$$

This storage scheme is normally implemented using two arrays (see fig. 1): the first one (A) contains columns of the matrix from the first to the last. For each column only the elements from the profile to the diagonal are stored; the second one (MAXA) contains one element for each column of the matrix, giving the position of the diagonal element of that column in the array A. Using these arrays it is possible to compute the bandwidth of each column and the row containing the first non-zero element in each column. The formula connecting the position of the generic element $A(I,J)$ in a dense matrix with the corresponding element in a skyline storage pattern, is the following:

$$A(I,J) = A(I + MAXA(J) - J)$$

Referring to fig. 1, showing a skyline matrix with N=20 nodal unknowns, the skyline storage obviously reduces the number of storage location required. In fact, using a dense scheme we need $N \times (N+1)/2 = 210$, whereas with the skyline scheme we use only $111 + 20 = 121$ storage locations (having counted both A and AMAX). In a constant banded matrix with N equations and half bandwidth M, the number of operations to perform is:

$$O(N \times M^2 - \frac{2}{3}M^3)$$

This estimate may be quite inaccurate for small N or very small M [3]. The above formula can be employed also for the skyline pattern to estimate the approximate number of operations required if a **mean** band is used [1]. Since the operation count for a symmetric dense matrix is $O(\frac{1}{3}N^3)$, the saving in arithmetic using the skyline storage scheme is even more dramatic than the saving in storage, as it varies **with the square of the mean bandwidth.**

Different schemes to that shown in fig. 1 are also used. For example the elements may be stored backwards, from the diagonal up to the first non-zero element in each column. Note that the formula given above refers to the scheme shown in fig. 1. The code described in the following implements both the possibilities and can be easily modified to treat allocation methods implemented in third parties finite element packages.

When dealing with unsymmetric matrices, similar schemes are used to store the matrix. For example the above described method is implemented for the upper triangle and an other array is used to store the lower one with the same skyline structure. The profile is normally considered symmetrical thus only one MAXA array is used in the scheme. Using this approach the position of the element I,J in the upper triangle and of the element J,I in the lower one are determined by the same operation.

Since factorization time increases quadratically with respect to the bandwidth while storage requirements increases linearly, care has therefore to be taken in order to chose a good equation numbering scheme; various automatic schemes known as **bandwidth minimizers** have been developed to renumber the mesh in order to decrease the bandwidth, or the number of elements stored under the skyline. The most widely used is probably the Reverse Cuthill-McKee Method (1969) implemented, for example, in the ADINA(*) finite element system.

3 - Dense matrix factorization using Level 3 BLAS

If we consider the LU decomposition of a dense matrix A, we may derive a block variant by partitioning the matrices so that:

$$\begin{bmatrix} A_{11} & A_{12} & A_{13} & \cdots \\ A_{21} & A_{22} & A_{23} & \cdots \\ A_{31} & A_{32} & A_{33} & \cdots \\ \cdots & \cdots & \cdots & \cdots \end{bmatrix} = \begin{bmatrix} L_{11} & 0 & 0 & \cdots \\ L_{21} & L_{22} & 0 & \cdots \\ L_{31} & L_{32} & L_{33} & \cdots \\ \cdots & \cdots & \cdots & \cdots \end{bmatrix} \begin{bmatrix} U_{11} & U_{12} & U_{13} & \cdots \\ 0 & U_{22} & U_{23} & \cdots \\ 0 & 0 & U_{33} & \cdots \\ \cdots & \cdots & \cdots & \cdots \end{bmatrix}$$

$$= \begin{bmatrix} L_{11}U_{11} & L_{11}U_{12} & L_{11}U_{13} & \ldots\ldots \\ L_{21}U_{11} & L_{21}U_{12} + L_{22}U_{22} & L_{21}U_{13} + L_{22}U_{23} & \ldots\ldots \\ L_{31}U_{11} & L_{31}U_{12} + L_{32}U_{22} & L_{13}U_{13} + L_{23}U_{23} + L_{33}U_{33} & \ldots\ldots \\ \ldots\ldots & \ldots\ldots & \ldots\ldots & \ldots\ldots \end{bmatrix}$$

The approach, as outlined below, consists of factorizing the current diagonal block, then using this result to reduce elements of the current block row and column and to immediately update the remaining submatrices below.
Thus, in the first stage:

1. Factorizing the current diagonal block
   $A_{11} \rightarrow L_{11}U_{11}$

2. Computing block row and column $\rightarrow$ CALL DTRSM(...)
   $U_{12} = (L_{11})^{-1}(A_{12})$
   $L_{21} = (A_{21})(U_{11})^{-1}$
   ....

3. Rank-k Update $\rightarrow$ CALL DGEMM(...)
   $A'_{22} \leftarrow A_{22} - L_{21}U_{12} = L_{22}U_{22}$
   $A'_{23} \leftarrow A_{23} - L_{21}U_{13} = L_{22}U_{23}$
   $A'_{32} \leftarrow A_{32} - L_{31}U_{12} = L_{32}U_{22}$
   ....

DTRSM performs the solution of a triangular system of equations, DGEMM performs general matrix-matrix multiplications.

At the end of this stage, the whole matrix has been reduced. The second stage begins with the factorization of the new diagonal block $A'_{22} \rightarrow L_{22}U_{22}$ and so on.
When dealing with symmetric matrices, simple modifications have to be applied to the above outlined description, for example by modifying L_{ij} blocks into U^T_{ij} blocks if the symmetric matrices are stored in the upper triangle.

4 - Block Skyline Factorization

The basic idea to face the skyline storage pattern with a blocked version, is to perform computations working on rectangular arrays so that, at each stage, parts of the skyline array must be copied into these blocks before computations and copied back to the skyline array after the work on the blocks has been completed. Subroutines have been written to copy chunks of a skyline array to rectangular matrices and vice-versa. Time spent in these copies is actually negligible and these subroutines are fully vectorized.

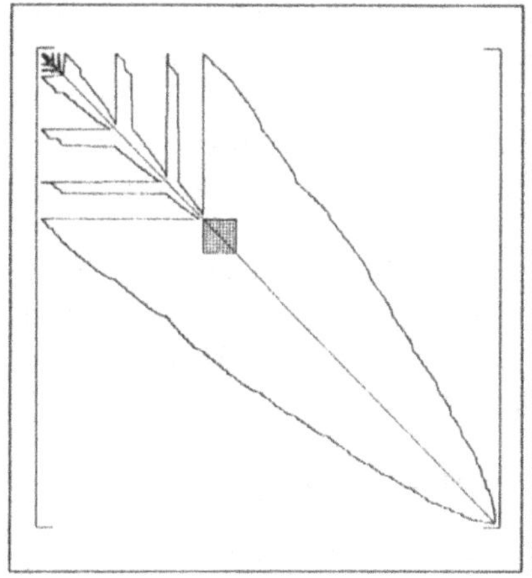

Figure 2

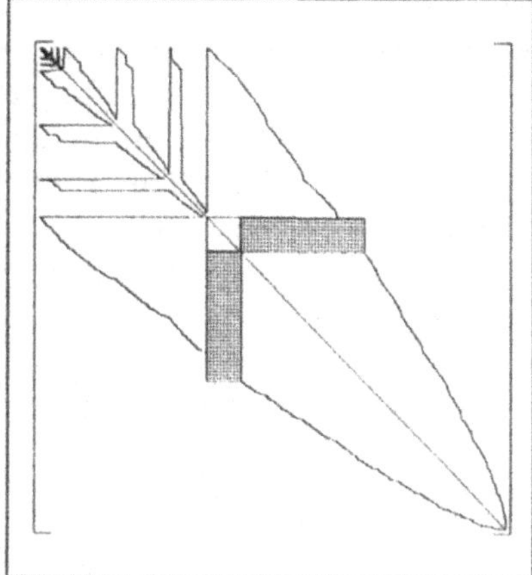

Figure 3

The basic approach involves the copying into a rectangular working array of a diagonal section of the skyline array, its factorization (fig. 2) and its usage to reduce the block row and column (fig. 3) in the DTRSM phase.

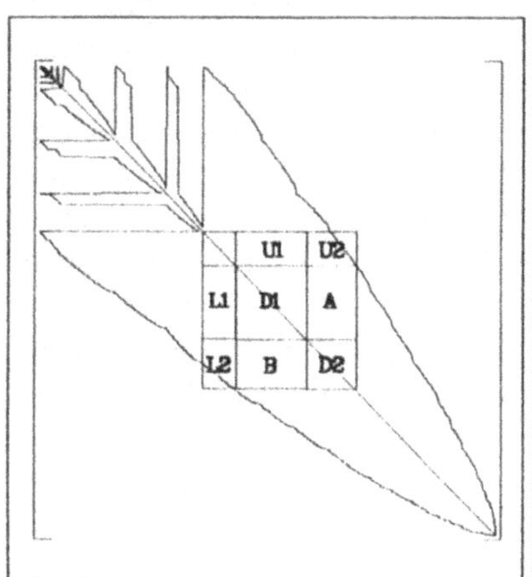

Figure 4

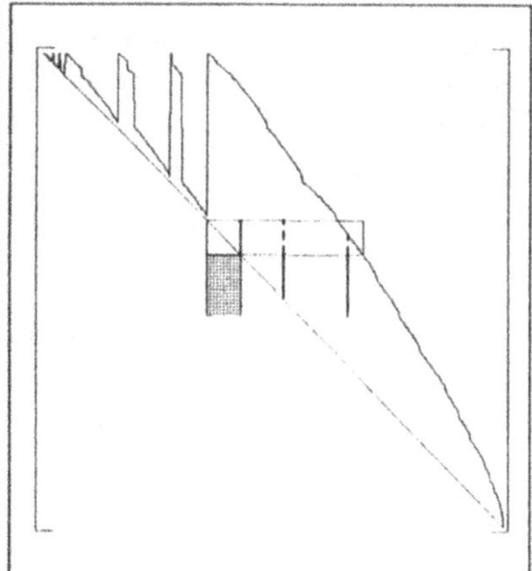

Figure 5

The compute intensive section is represented by the rank-k update phase. In fig. 4 we describe how to perform this linear algebra operation onto a skyline array, using rectangular work arrays.

Some modifications have been added to the basic bersion outlined improving the effectiveness of the method. In order to avoid many of the copies between skyline storage and rectangular arrays during the rank-k phase, the calls to the Level 3 BLAS routine DGEMM have

been replaced by calls to DGEMV (general matrix-vector product from the Level 2 BLAS), and the vector arguments to DGEMV are addressed directly in the skyline array. (fig. 5). It is important to remark that although a Level 2 BLAS kernel has been used, we are still essentially using a Level 3 BLAS operation, as the blocks are used as much as possible before the next block is accessed. Using this approach, DGEMV performs substantially as DGEMM, with the great advantage of avoiding a number of copies into the work array and back to the skyline array.
Further improvements have consisted in tuning the method for very irregular skyline profiles. A detailed description of these strategies can be found in [8] where the source of the Level 3 BLAS skyline factorization for symmetric matrices is also included.

5 - Parallel Version of the Blocked Skyline Factorization

In this section we describe a parallel version of the skyline solver for shared memory multiprocessors. Shared memory parallelism consists in a parallel execution where each processor accesses data from a common area. IBM ES/9000 multiprocessors are a typical example of shared memory multiprocessor and parallelism can be exploited on this family of computers through the IBM Fortran compiler.
This compiler support Fortran language extensions to exploit, for example, the execution of PARALLEL DO LOOPS, the origination and termination of tasks, the scheduling of Fortran subroutines.
Critical sections of a parallel program that have to be executed one at a time by all processors can be efficiently treated by PARALLEL LOCK managements. The EVENT handling can be used to have a synchronization of scheduled subroutines with very small overhead.
We have chosen the approach of producing a parallel version of the outlined blocked algorithm, by using a large grain parallelism approach basically splitting computations between blocks to different processors.
Referring to a constant banded matrix, addressed by a skyline storage pattern, with 5000 equations and a half bandwidth of 1050, we compute the following CPU time percentages spent in three main sections:

1. Factorizing the diagonal block 0.37 sec. (0.23%)
2. DTRSM Phase 11.90 sec. (7.10%)
3. Rank-k update phase 155.23 sec. (92.67%)

where the entire factorization takes 167.50 sec. on an IBM 3090-60E/VF. All these sections include time spent to copy chunks of the skyline in and out of working arrays.

We have parallelized phases 2 and 3 since the time spent in factorizing the diagonal block is actually negligible.

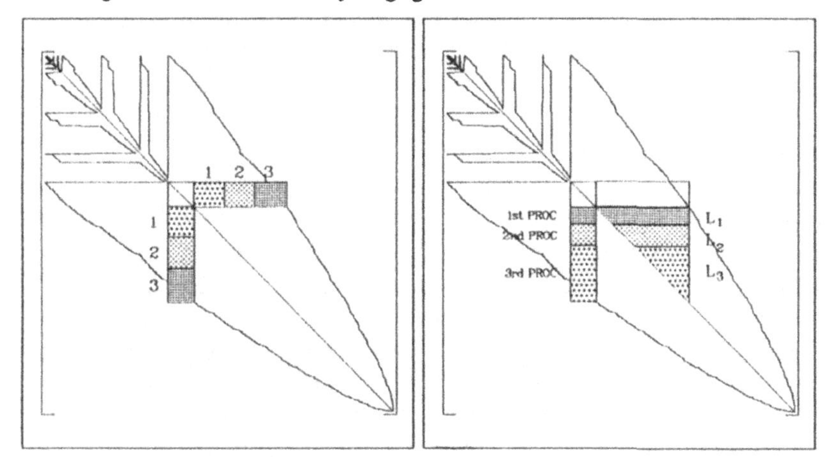

Figure 6 Figure 7

DTRSM is a section easily parallelizable by dividing equally the block row (and or column) between available processors. For constant banded matrices the load is well balanced between processors.

The parallelization of the rank-k phase is more complex. because the issue of balancing the load becomes important. The approach used is illustrated in fig.7. The section of the upper triangle to be reduced is subdivided into a number of strips equal to the number of processors. Each processor loads a section of the block column on a local working array and performs the rank-k update using this private array and the shared skyline array.

We set the areas of the strips to be equal to guarantee that the number of floating point operations is equal for each processor. That however does not imply that the load is well balanced between each processor. In fact the CPU time used by a processor to perform a certain operation is given by: $T_i = \frac{FLOPS}{v_i}$ where T_i is the CPU time in the $i-th$ processor and FLOPS is the number of floating point operations. The term v_i indicates the mean execution speed that the $i-th$ processor achieves during the execution.

From fig. 7 it appears that we separate equal areas by using different lengths L_i. Assuming e.g. a system with a constant half-band of 200 we want to factorize using 3 processors, we subdivide the triangular area of 200 by 200 elements into a triangle and two trapezes where $L_1 = 37$, $L_2 = 48$, $L_3 = 115$. Thus the Rank-k Update performs the same number of operations on each processor but the vectorization on L_i

leads to different speeds being different lengths L_i of array involved in the vectorization.
Thus the approach of balancing the number of floating point operations between different processors leads to a correct balance of CPU times spent in each processor only when the L_i are large enough to be processed more or less in the same time in each processor.
A more accurate position of the problem, by involving the variation of speed with dependance on the length of arrays can be faced, but it appears dependent on the section size of the computer used and on the band of the system. So we decided to use definitely the method outlined above to preserve the portability of the Level 3 BLAS approach.
From an implementation point of view it may be interesting to note that, we dispatch N tasks (being N the number of processors) only at the beginning of the factorization. At each stage one task factorizes the diagonal block while the others wait for the completion. Then all tasks begin to reduce the diagonal block and column working onto a section of it. Another synchronization is set immediately before the rank-k update phase and, after that, at the termination of the stage. Thus only 3 barriers at each stage are set into the code and the speed of the synchronization through events managements reduced greatly overheads leading to an asymptotic speed-up of 5.2 on 6 processors on a constant band reference test case (see fig. 10).

6 - The ADINA Finite Element System

ADINA(*) is a state-of-the-art computer system for finite element analysis of structural, fluid, heat transfer and other field problems. The program has been developed to provide an effective tool for three dimensional linear and non-linear analysis; emphasis has been placed on the use of reliable and efficient finite element techniques and user conveniences for modeling and performing the analyses of complex industrial problems.
The ADINA system consists of the following programs: ADINA for displacements and stress analysis; ADINA-T for analysis of heat transfer and field problems; ADINA-F for fluid dynamics problems; ADINA-IN for free format command-language input, generation, and display of input data; ADINA-PLOT, for graphical a numeric display of program results, developments of cracks and contact zones during dynamic analyses.
The ADINA system is applicable to general analyses and can be employed in many different technologies such as civil, mechanical and aerospace, nuclear, offshore, automobile, geomechanics and manufacturing technologies.

7 - Performance and Validation Tests

Tests on ADINA. A number of tests have been carried out with cases produced by the ADINA(*) finite element system either for structural and fluid-flow problems. Measurements have been carried out using the IBM 3090-60E/VF model of ECSEC, where a version of ADINA highly tuned for the IBM 3090 has been produced.
In this section we present a fluid dynamic test case that was analysed with ADINA-F in its original version and with the optimized version produced at ECSEC using the parallel Level 3 BLAS solver, vectorization and parallelization of the assembly and optimization of the I/O [9].
The test case treats an incompressible viscous fluid flow in a schematic hydraulic press (fig. 9). made by a vertical pipe with varying cross section connected to an horizontal tube terminating with a non-symmetric T-shaped pipe. The branches of the "T" turn backward and the cross section of one of this increases on its final path which turns clock wise. We studied this case increasing the geometry discretization in order to augment the mean half bandwidth of the problem and the number of equations. The test case was prepared with the graphic pre-processor ADINA-IN. The material type was selected to be a constant properties model and the selected element type was a 3-D fluid flow elements with 27 nodes each. A pressure load on the top of the pipe and no-slip conditions were defined as boundary conditions. The outlet was characterized by defining the velocity parallel to the axis of the branches.
Results are summarized in Table 1 for the three cases obtained from the basic model described above, with different mesh discretization and for different size of the cross section of the horizontal pipe. Elapsed times on the table are expressed in seconds; for each test case total timings are outlined and partial timings related to the factorization. For each case the number of equations (NEQ), the mean half-bandwidth (MHB) and the number of finite elements (NEL) in each model are indicated. Each analysis consists of 1 time step with 2 iterations to convergence. No bandwidth minimizer is used and that leads to an irregular skyline profile. With reference to the T4 test case we note that the lowest factorization time is achieved with 4 procs (62*). Using 6 procs the factorization requires a longer time because the MHB is short. In the total time we report a mixed execution where the factorization is executed on 4 processors and the assembly on 6.

Test case	Original version	1 proc	3 proc	4 proc	6 proc
T4: NEQ = 11528; MHB = 310; NEL = 576					
Total	*2456*	*534* (4.6)	*230* (10.7)	*194* (12.7)	*164* (15.0)
Fact.	*1093*	*157* (7.0)	*72* (15.1)	*62* (17.6)	*62** (17.6)
T5: NEQ = 15754; MHB = 410; NEL = 768					
Total	*4642*	*862* (5.4)	*380* (12.2)	*317* (14.6)	*272* (17.1)
Fact.	*2517*	*343* (7.3)	*153* (16.5)	*124* (20.3)	*119* (21.2)
T6: NEQ = 32040; MHB = 659; NEL = 1500					
Total	*19188*	*2962* (6.5)	*1371* (14.0)	*1171* (16.4)	*1001* (19.1)
Fact.	*13400*	*1784* (7.5)	*760* (17.6)	*632* (21.2)	*541* (24.8)

Table 1

A reference test case. In order to estimate the performance of the proposed method we tested the factorization of a constant banded matrix of 5000 equations addressed using the skyline storage pattern. This reference test case gives an estimate of the performance on skyline matrices arising from application problems using the mean halfbandwidth of the skyline matrix. This estimate is in good agreement with measured data for matrices with a smooth skyline profile (obtained e.g. with a bandwidth minimizer). In fig. 10 we show the speed-up of the parallel version using up to 6 processors. For the T6 problem with a mean halfbandwidth of 659, for example, we obtain a speed-up of 3.8 using 6 processors. This figure fits well with our measured data.

Fig. 8 shows the performance in terms of Megaflops of the algorithm on 1 CPU. This performance results very close to the performance obtained on dense matrices.

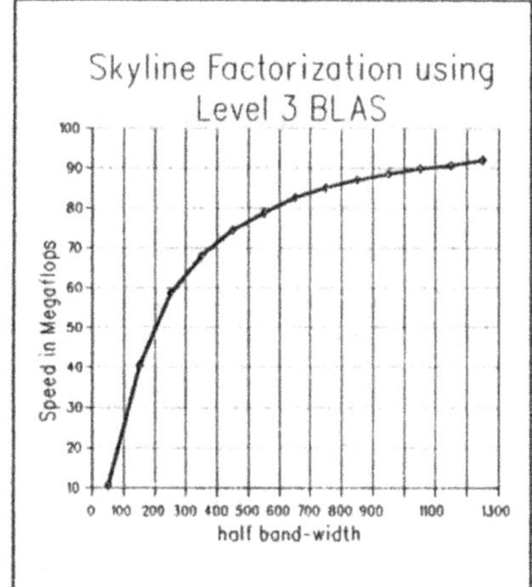

Figure 8

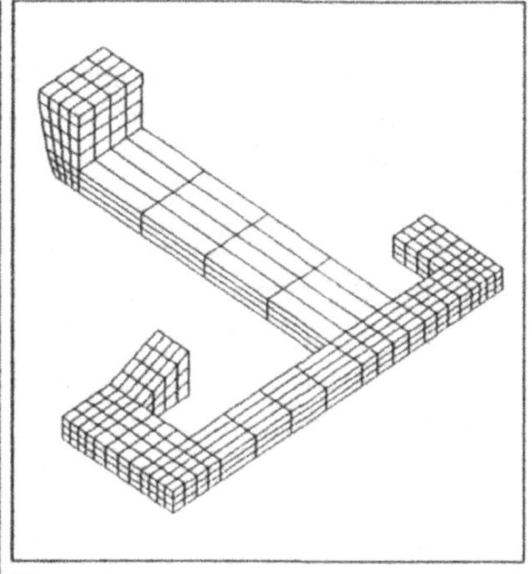

Figure 9

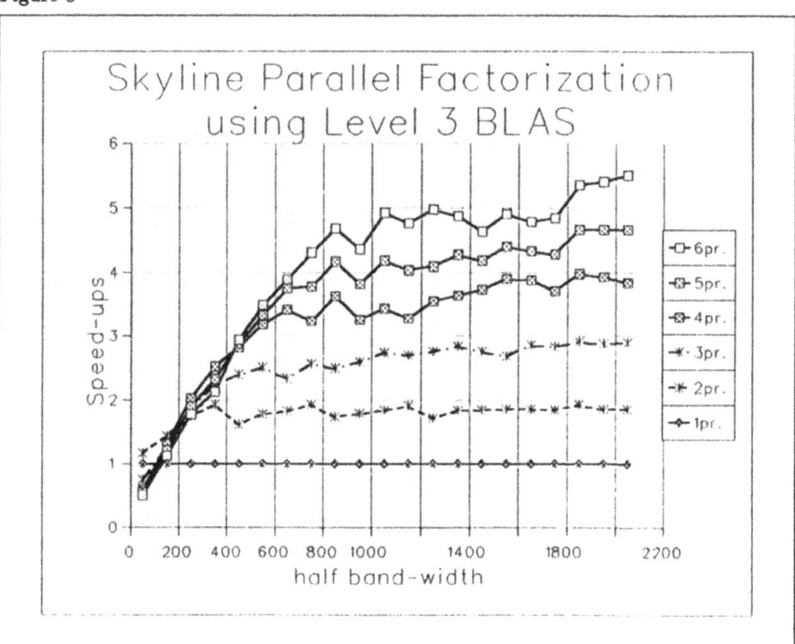

Figure 10

8 - Conclusions

In many industrial finite element codes, the largest percentage of the time is spent in the skyline factorization. That motivates the development of highly tuned versions of this kernels for vector multiprocessors.
An efficient parallel solver based on a blocked algorithm (Level 3 BLAS) for the factorization of skyline matrices has been presented. Using this solver with the ADINA(*) finite element system we have obtained a speed-up of 25 in the solver that leads to a speed-up of 19 on the entire run.

(*) **ADINA** is a trademark of ADINA R&D Inc. 71 Elton Avenue Watertown, MA 02172 U.S.A.

References

[1] Bathe, K.J. **Finite Element Procedures in Engineering Analysis** *Prentice-Hall,* 1982

[2] Demmel, J., Dongarra, J.J., Du Croz, J.J., Greenbaum, A., Hammarling, S.J. and Sorensen, D. **Prospectus for the Development of a Linear Algebra Library for High-Performance Computers.** *Technical Memorandum No. 97 MCS Division, Argonne National Laboratory,* 1987

[3] Dongarra, J.J., Bunch, J.R., Moler, C.B., Stewart, G.W. **LINPACK User's Guide** *SIAM Philadelphia,* 1979

[4] Dongarra, J.J., Du Croz, J.J., Duff, I.S., Hammarling, S.J. **A Set of Level 3 Basic Linear Algebra Subprograms** *Technical Memorandum No. 88 (1) MCS Division, Argonne National Laboratory,* 1988

[5] Dongarra, J.J., Du Croz, J.J., Duff, I.S., Hammarling, S.J. **A Set of Level 3 Basic Linear Algebra Subprograms: Model Implementation and Test Programs.** *Technical Memorandum No. 122 MCS Division, Argonne National Laboratory,* 1988

[6] George, A., Liu, J.W-H **Computer Solution of Large Sparse Positive Definite Systems** *Prentice-Hall,* 1981

[7] Lozupone, D.F. **A Skyline Solver for Symmetric Matrices on the IBM 3090 Vector Multiprocessor** *ASE89 Proceedings Computational Mechanics Publications and Elsevier,* 1989

[8] Lozupone, D.F., Mayes, P.,Radicati di Brozolo, G. **Skyline Cholesky Factorization using Level 3 BLAS.** *ICE-0037 "IBM ECSEC Technical Report", Roma(Italy)*, 1990

[9] Lozupone, D.F., Piccolo, F. **The Use of ADINA-F for Large Model Fluid-Flow analyses.** *8th ADINA Conference Proceedings International Journal of Computers and Structures* 1991 (in print)

[10] **ADINA User's Manuals** *ADINA R&D Inc. Watertown, MA(USA)*, 1987

Exploitation of Data in Memory in the FIDAP Code on the IBM ES/3090

V. Zecca

IBM ECSEC European Center for Scientific and Engineering Computing, Via Giorgione 159, 00147 Rome, Italy

ABSTRACT

The IBM ESA/370 architecture incorporates unique features that are of benefit in numerical intensive applications. One of these features, as implemented in the MVS/ESA operating system, is the hiperspace, an extension to the addressing scheme available on the base System/370-XA, that allows indirect access to Expanded Storage. Access to hiperspaces has been implemented through usage of the Data Window Services library of MVS/ESA.

The present paper discusses a hiperspace exploitation for the IBM 3090 of the FIDAP (*) Finite Element code, that analyses 2-D and 3-D incompressible viscous fluid flow.

In FIDAP, if the storage available is not sufficient for solving the problem in-core, the solver splits the system into blocks that are put into a FORTRAN logical unit. For very complex geometries, the limit of two gigabytes virtual storage in IBM System/370-XA can well be surpassed. In that case, heavy I/O activity can reduce performance of the solver. MVS/ESA permits fast access to hiperspaces through windows in virtual storage; FIDAP takes advantage of that feature by transforming access to the spill unit into access to a hiperspace, thus gaining a reduction of up to three in elapsed time.

The purpose of the work was to demonstrate the capability of the standard IBM VS FORTRAN compiler and MVS/ESA operating system to optimally compile and execute an industrial code like FIDAP on a supercomputer exploiting its storage hierarchy, and particularly virtual and Expanded Storage, with small intervention on the programmer's part.

The paper also presents performances and results concerning an industrial test case involving thermocapillary convection.
(*) FIDAP is a registered trademark of Fluid Dynamics International.

Keywords
Large Systems, Hiperspaces, Finite Element Method

INTRODUCTION

The constant increase in performance and reliability of modern mainframes allows the solution of larger problems in several areas such as engineering, physics, and more generally in scientific and technical computing. Problems that were insolvable a few years ago are now handled efficiently by computers such as the IBM ES/3090. In the past years we have seen a consistent upgrade of computer architectures to enable execution of industrial and academic applications requiring a larger range of addressing.

The IBM System/370 Extended Architecture (XA) increased the address range on IBM mainframes by a factor of 128 over the earlier IBM System/370 architecture. Today, we are beginning to feel a need to go beyond the two gigabytes addressing scheme of MVS/XA for very large applications in fluid dynamics, aerodynamics, and similar areas, where solutions for problems like the flow around an airplane, or a space shuttle, are sought. This is a predictable evolution from earlier studies concerning local flow around an airfoil or inside a nozzle, that were fully contained within the MVS/XA addressability limitations.

The IBM Enterprise Systems Architecture (ESA/370) [1] provides a virtually unlimited number of two-gigabyte address spaces to user applications in an MVS Version 3 (MVS/ESA) environment. In our study, we used MVS/ESA's extensive addressing capability on the fluid dynamics Finite Element industrial code FIDAP [2]. In particular, we used the Data Window Services (DWS) [3] of MVS/ESA to exploit addressing in a large FORTRAN application.

The value of DWS is obvious for applications where the storage demand is greater than the two gigabytes address space limit. Measurements show that even applications smaller than two gigabytes may benefit from using DWS, when their storage demand is greater than the physical storage available on the machine. Applications that can fit into available storage are not good candidates for implementing DWS.

The intent of this paper is to show how ESA was applied to an industrial code like FIDAP, using DWS. The findings and the experience gained are detailed in the following sections. The following section describes in more detail the hiperspace concept in MVS/ESA, for application programmers.

HIPERSPACES AND DATA WINDOW SERVICES

A hiperspace (high-performance space) provides an explicit interface to the Expanded Storage feature of the IBM ES/3090, which, prior to MVS/ESA, was only available to the operating system for swapping, paging, and temporary Virtual Input/Output (VIO) data sets. The use of hiperspaces allows an application to take direct advantage of Expanded Storage in an industrial application, both to improve its performance, and to have access to a large addressable data range. If we consider that a page fault handling from disk requires approximately 10 ms of elapsed time for processing, while 100 μs are needed from expanded storage, it is easy to understand the advantages of using electronic storage rather than disk storage to store bulk data. The recently announced IBM ES/9000 can have up to eight gigabytes of Expanded Storage thus allowing processor storage of unprecedented amounts of data.

A hiperspace can be seen from a user's point of view as a data buffer. This buffer is backed either by Expanded Storage, or by both Expanded Storage and auxiliary storage. The hiperspace is block-addressable storage that cannot be directly addressed by a program. Data must be moved in four-kilobyte blocks from the hiperspace to virtual storage, and vice-versa. The application can use the data only while it is in its address space. The address space area provides a "view" into the hiperspace. Handling of hiperspaces and views can be done in FORTRAN through access to Data Window Services. The maximum size of a hiperspace is two gigabytes: Data Window Services allow concatenation of hiperspaces into an "object" of virtually arbitrary length, only limited by the auxiliary space available on disk for paging.

By using Data Window Services a FORTRAN program can:

- Access an object.
- Open a view in virtual storage to an area of an object.
- Save in the object changes made in the view.
- Terminate access to an object.

Objects can be permanent or temporary (in FIDAP temporary objects were exploited). The maximum size of a temporary object is 16 terabytes. This means that DWS will automatically allocate as many hiperspaces as needed by the application.

Access to the data in the object, also called viewing the data, can be done only through an intermediate area. This area, called a window, is defined by the user in the program and must begin on a page (four-kilobyte) boundary. The concept of data boundary, or alignment, is foreign to most high-level language programmers, and is perhaps the most difficult obstacle to implementing DWS in existing programs. An alleviating factor is that dynamic COMMONs defined in IBM VS FORTRAN are page-aligned, so that they can be immediately used as windows for an object.

The window not only must start on a page boundary, but its size must be a multiple of a page. A feature of DWS is the ability to transfer multiple pages from an object to the window. Every time a page in the window is needed, that page and the successive 15 are read from the object. This is called *Read-ahead*. A speed up in elapsed time of up to two has been reached with Read-ahead.

FIDAP

Introduction to FIDAP

The FIDAP fluid dynamics package [4], along with similar packages such as ADINA-F, is a general purpose computer program that uses the Finite Element Method (FEM) to simulate many classes of viscous incompressible fluid flows. Two dimensional axi-symmetric, and 3-dimensional steady state or transient simulations in complex geometries, including the effects of temperature, are possible. The analysis is limited only by practical considerations of computer time and the capacity of auxiliary storage devices for problems that cannot be solved in core. Versions of FIDAP exist for a wide range of computers under many operating systems.

The Finite Element Method, while enjoying widespread use in structural problems, has a relatively short history in computational fluid dynamics. In recent years, however, both academic research and industrial practice have shown convincingly that FEM is becoming as powerful a tool in fluid mechanics as it is in structural analysis.

In FEM, the flow region is subdivided into a number of small regions, called Finite Elements. The partial differential equations of fluid mechanics covering the flow region are replaced by ordinary differential or algebraic equations in each element. The system of these equations is then solved simultaneously by sophisticated numerical techniques to determine the velocities, pressures, and temperatures throughout the region. The great advantage of FEM over other methods is its inherent flexibility in treating complex flow domains and boundary conditions.

FIDAP Structure

The FIDAP package is comprised of three programs: FIPREP, FIDAP, and FIPOST, with the following functions:

1. FIPREP: mesh generation, input of physical properties, input of boundary and initial conditions, specifications of class of problem, specification of solution procedures, and file conversion.
2. FIDAP: transforms governing partial differential equations into algebraic equations, and solves them with iterations and time-stepping.
3. FIPOST: handles graphical post-processing of output field variables, and computation and graphical display of derived output quantities.

All model definition data is prepared using FIPREP, while analysis of the results is performed using FIPOST. The FIDAP module was optimized for ESA by exploiting the DWS feature of MVS/ESA.

Solution Techniques

Several nonlinear iterative solution techniques are available to solve the nonlinear system of equations that arises when a steady-state analysis or a transient analysis with an implicit time integrator is being performed. These techniques include:

- Successive substitution
- Newton-Raphson
- Quasi-Newton, Broyden's update
- Matrix-free iterative solver
- Combinations of the above
- Incremental approach to Newton-Raphson or Broyden's update
- Segregated solver

If the storage available is not sufficient for solving the problem in-core, the solver splits the system into blocks that are put in an external data set. For very complex geometries, the limit of two gigabyte virtual addressing in MVS/XA can well be surpassed. In that case, heavy I/O activity can reduce performance of the FIDAP solver.

As explained in the introduction, MVS/ESA permits fast access to large objects through windows in virtual storage; FIDAP can take advantage of that feature by transforming access to the spill data set into access to a DWS object.

FIDAP Test Case

The first step in enabling FIDAP for MVS/ESA was to select a large test case to check the correctness of the modifications applied to FIDAP for MVS/ESA support. The test case selected was EX26, from the test case suite delivered along with the FIDAP package by Fluid Dynamics International. EX26 simulates thermocapillary convection; that is, convective motion occurring in a liquid having a free surface and subject to differential heating as a result of the temperature dependence of the surface tension. This effect, often termed Marangoni flow, is important in a variety of processes, such as welding, crystal growth, and so on.

In this example, a liquid is located in a 2-dimensional container and has one free surface. The liquid is heated by applying a constant temperature along one side of the container. The force due to gravity is set equal to zero in this simulation; the motion is then entirely due to the temperature gradient surface tension. The solver selected for this case is the Quasi-Newton, Broyden's update method. The mesh of the problem, shown in Figure 1, is a square with 25×25 points in the original formulation.

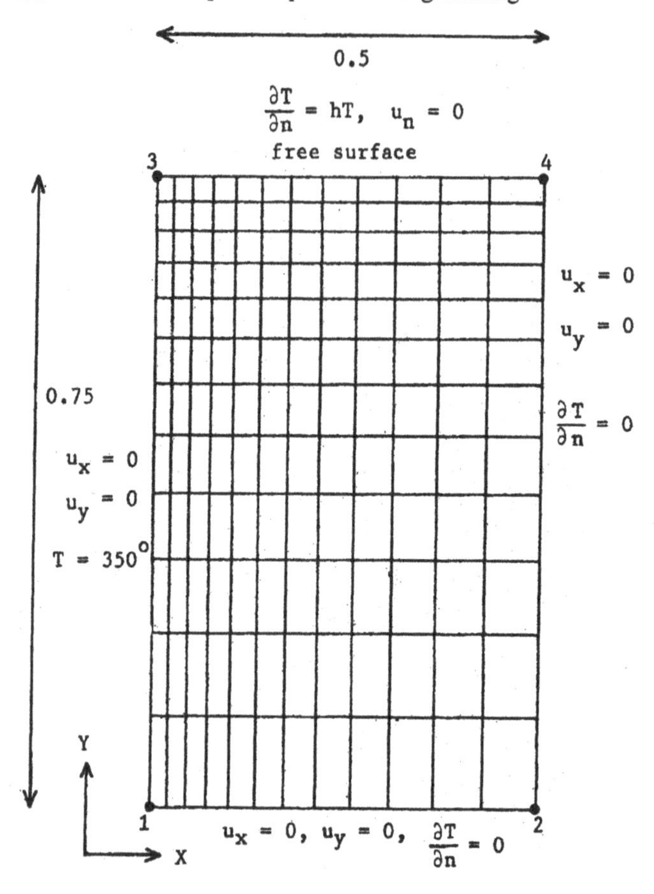

Figure 1. Boundary Conditions and Mesh for Thermocapillary Convection

Such a mesh generates a linear system with 2163 equations, 641,487 elements, and a mean half-bandwidth of 148. The mean half-bandwidth (MHB) of the problem is the average width of the band containing non-zero elements of the matrix to be solved; it is a key indicator of the computational complexity of the solution. A good approximation of the MHB is six times the number of points on the edge of the mesh, in our case 25. Experimenting with finer grained meshes has shown that if the linear system remains in virtual storage the CPU time in seconds required for LU decomposition, which dominates the total CPU time for large linear systems, is approximated by $1.2 \times 10^{-5} \times N^4$, where N is the number of points. These measurements were done on an IBM ES/3090 E supercomputer. If the linear system dimension exceeds the virtual storage available, FIDAP is forced to split the matrix into blocks which are transferred back and forth between virtual storage and the spill data set. This situation is an ideal show case for MVS/ESA; the spill data set was transformed into an MVS/ESA object and access was handled through DWS.

Figures 2 and 3 illustrate the velocity vector field and streamline contours of the flow, respectively. The motion driven by the surface tension gradient is clearly evident. In the absence of gravity, the flow is basically confined to the region near the free surface, with quiescent flow near the bottom. The deformation of the free surface, although not very large, can also be seen in these Figures.

Figure 4 shows the computed isotherms of the field.

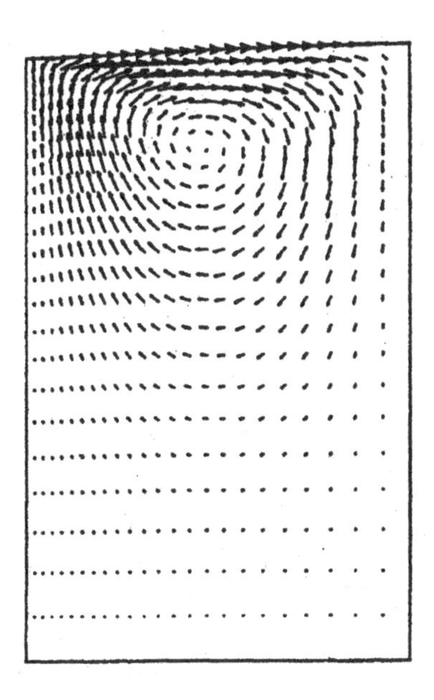

Figure 2. Velocity Vector Field for Thermocapillary Convection

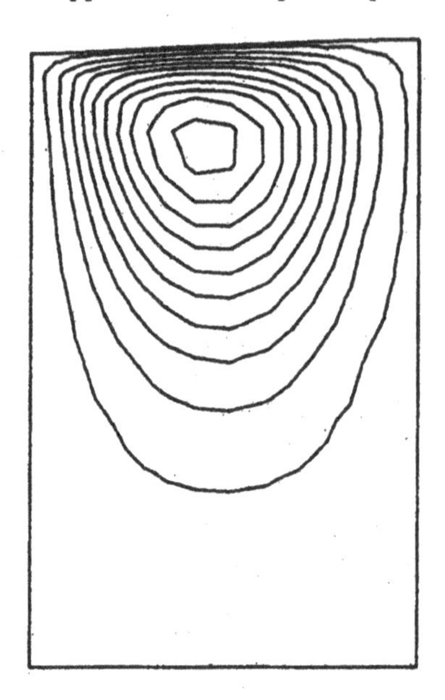

Figure 3. Streamlines for Thermocapillary Convection

Enabling FIDAP for MVS/ESA

Once the test case was selected, the source code was searched for I/O statements on unit IMAT, the spill unit for storing the problem matrix blocks. The I/O statements were located in subroutines INRAN for input and OUTRAN for output. A sample call for I/O routines is:

```
CALL INRAN(IMAT, IFWA, NWORDS, NREC)
```

where NWORDS words of the array IFWA are written on unit IMAT starting at record number NREC. For an easy and optimal usage of DWS, the FIDAP blocks were adjusted in such a way that their length was a multiple of four kilobytes, the unit of access to MVS/ESA objects.

The FIDAP blocks could not be used as windows to the object, however, because they are not, in general, page aligned. Also on write operations to the object the contents of the window are discarded, which is unacceptable for FIDAP, which reuses the block. Therefore, an eight megabytes area was defined in a dynamic COMMON as the window. All data transfers pass through this buffer which is page aligned, and can be safely discarded after transfer to the object. In addition, data transfers were accomplished in vector mode, thus exploiting the vector processing capabilities of the 3090.

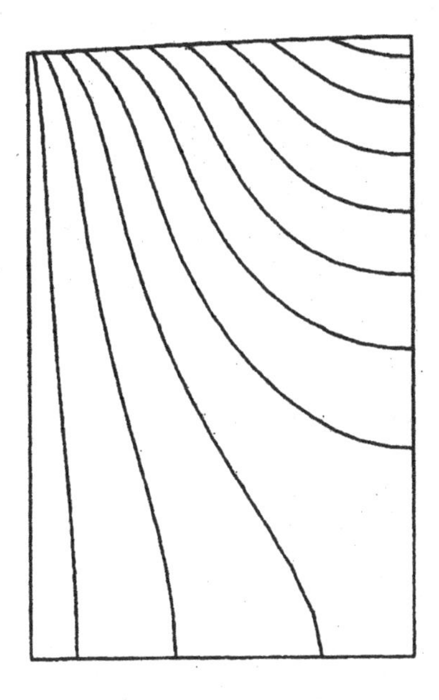

Figure 4. Isotherms for Thermocapillary Convection

Performance Results

After correct execution of smaller test cases, a large case was tried. This case had a mesh with 111 × 111 points, with 45,378 equations, 62,270,592 elements, and a mean half-bandwidth of 686. An area of 400 megabytes was allocated for FIDAP buffers, and three blocks with 200 megabytes each were required to spill the matrix to be solved onto auxiliary storage.

The table summarizes the performance results of the original (I/O) version and the enabled (DWS) version. Times are given in minutes. The EXCP (Execute Channel Program) count is a measure of the amount of disk I/O done by FIDAP. With hiperspace exploitation all the I/O done to the spill unit is avoided, thus greatly decreasing the total I/O count.

Table 1. Performance of FIDAP in MVS/ESA

Version	CPU Time	Elapsed Time	EXCP Count
I/O	26.2	150.2	334000
DWS	25.4	51.8	28000

The hardware configuration was comprised of an IBM ES/3090 model 300J with 128 megabytes of Central Storage, and one gigabyte of Expanded Storage.

While the CPU time is nearly the same, which can be expected since the algorithms of FIDAP were not changed, a great decrease in elapsed time was achieved using hiperspaces.

The high overhead of the original version is due in part to the disk formatting at OPEN time required by VS FORTRAN access methods for direct access units, and in part to lower performance of disk I/O as compared to data transfer between Central Storage and hiperspaces. For this case, 20 minutes were required to format the disk unit. To avoid the overhead incurred in formatting a direct access unit, a preformatted dataset could be used. In this case the OPEN routine would recognize that the unit is already formatted, thus skipping the onerous formatting process. However, this procedure puts a limit to the size of the problem to be solved, because larger problems would need a larger dataset. On the other hand, if the formatted dataset is very large, it wastes disk space when it is not used.

CONCLUSIONS

The IBM Enterprise Systems Architecture (ESA) incorporates unique features that are of benefit in Numerically Intensive Computing (NIC) applications. Features like Data Window Services (DWS), which are available to high level languages, like FORTRAN, in the MVS/ESA Operating System, enable NIC applications to take advantage of the full power of the Enterprise Systems Architecture. Our experience shows that DWS is best utilized when the data model of the program is coherent with the services provided by DWS. Such a case arises when the application has to access a storage object, such as a large array or a large data set, in blocks that are not too large compared with the size of the full object.

An ideal case for DWS is the Finite Element industrial code FIDAP. Its solver tries to maintain the matrix, representing the linear system to be solved, in storage. If the problem is very large, this is not possible and blocks of the matrix are spilled on a data set. By transforming access to that data set into access to an MVS/ESA object, using the facilities of DWS, the traditional XA barrier of two gigabytes for virtual storage addressing can be overcome with better performance than having the spill object mapped onto a data set.

The faster paging native I/O of MVS/ESA, as compared to the relatively slower I/O of VS FORTRAN access methods, makes the major difference. Another factor to consider is that, as far as possible, MVS/ESA backs the object on Expanded Storage, thus achieving a high transfer data rate between the object and the application virtual storage, while DASD I/O is limited by the 4.5 MBps rate of current IBM ES/3090 parallel channels. On the other hand, limitations in the data model handled by DWS limit its usability, so that careful planning is needed to decide if DWS is suited for the application at hand.

A major limitation is that access to an MVS/ESA object is made on a page basis, while VS FORTRAN accesses storage on a word, or double-word, basis. This makes accessing MVS/ESA objects very cumbersome, unless the application buffers are page aligned and adjusted to a length that is a multiple of a page, 4096 bytes.

Optimum usage of Data Window Services involves cases where there is a definite need for relieving virtual storage constraints and when migration to DWS can be accomplished with a few man-weeks of effort. A larger effort would require a careful evaluation of the performance benefits that could result from DWS exploitation.

BIBLIOGRAPHY

[1] **Special issue on Large Systems,** IBM Systems Journal 28 (1989).

[2] **ESA Exploitation by NIC Applications,** IBM Order No GG24-3539.

[3] **MVS/ESA Callable Services for High Level Languages,** IBM Order No GC28-1843.

[4] **FIDAP General Introduction Manual,** Fluid Dynamics International, September 1987.

A Recursive Database Query Language on an Object-Oriented Processor

L. Natanson, W. Samson, A. Wakelin

Department of Mathematical and Computer Sciences, Dundee Institute of Technology, Dundee, DD1 1HG, U.K.

1 Abstract

Applications of database technology, such as Computer Assisted Design and Computer Assisted Software Engineering, demand expressive and powerful database query languages capable of supporting, for instance, recursive queries [6, 9].

The performance of such query languages on conventional von Neumann architectures is so poor that well understood enhancements such as integrity constraint checking, are often not incorporated so as not to further weaken performance.

The REKURSIV processor [3] has a microprogrammable object-oriented architecture. The microcode level is powerful enough to support recursion. The object store is persistent and much of the overhead of garbage collection is carried out in hardware. An object-oriented language, Lingo, which is similar to Smalltalk is available for the REKURSIV in source form, together with the source for its microcoded instruction set, which is thus extensible.

The paper describes the implementation (in Lingo, on the REKURSIV system) of a recursive query language, DEAL [2], from a formal specification [5].

2 Introduction

Traditionally, database management systems were designed to meet needs from business data processing applications. Areas such as Computer Assisted or Automated Design are better supported by languages of Turing equivalent power and with expressivity at least as high as that of modern programming languages.

For several years,a group at Dundee Institute of Technology had been involved with the development and utilisation of a relational query language, DEAL [2]. Some of this work was directed at using DEAL to show that its enhanced expressivity made problems in certain application areas more tractible. These areas included -

- History - relational databases with an inbuilt model of time [6] allow selection predicates to involve temporal relations. This is of use in a wide variety of areas including, within engineering, design version and configuration control.

- Graphics - CAD systems necessitate the integration of the Database Management System with the ability to view and operate on database objects graphically within the same language [9].

The usefulness of the language DEAL in real applications at the institute was limited by the efficiency of the implementation and the language was very much used as a research model against which to test ideas for further language development [8] and to carry out experiments in algebraic specification of the Relational Algebra [7]. The group won an award, under the Department of Trade and Industry's Awareness Initiative in Object-oriented programming, with a proposal entitled 'The REKURSIV as a fast database engine' to implement the language DEAL on the REKURSIV.

The implementation of DEAL on the REKURSIV proceeded by developing a formal specification of the relational algebra in the functional language Standard ML and then transforming the specification into the language Lingo. This layer was then used as the target virtual machine for an interpreter, written in Lingo, which translated DEAL queries into an intermediate code consisting of calls to the relational algebra layer. The interpreter was constructed by first implementing a parser/compiler generator which takes as input a B.N.F. like description of a language (along with a specification of semantic actions) and produces a Lingo class that is the required compiler.

The REKURSIV architecture is described briefly in section 3 and the language Lingo in section 4. The process of specifying the relational algebra and transforming the specification into Lingo is covered in section 5. The implementation of DEAL is discussed in section 6 and some conclusions are drawn in section 7.

3 The REKURSIV

The use of a conventional processor, which views memory as a linear array of identical cells, for any high level task leads to several mismatches.

- data types – at the programmer level complex data types are used to maximise expressivity. At the machine level, however, these are implemented by complex mechanisms involving several accesses to memory and much processing.
- virtual memory – programmers deal with aggregates of objects of various sizes whereas operating systems attempt to maintain a regime of memory residency optimal in relation to *page* access.

The REKURSIV addresses memory by unique *object* identifiers which are the only programmer-available memory access points. Each object has associated with it a **type** word, a **size** word and as many **representation** words as are required. The type and size of an object are fetched at the same times as its representation. Object identifiers are allocated by hardware and are 38 bits long.

The virtual memory manager employs strategies based on *objects* rather than *pages*. These strategies are implemented through microcode and supported in hardware via a chip called OBJEKT. High level data structures are supported by OBJEKT which has several ALUs to handle address modification and indexing through aggregate objects.

The REKURSIV's control is provided by LOGIK, the microcode sequencer. This has available to it a large stack allowing recursion and highly powerful microcode. The REKURSIV's ALU, NUMERIK, provides 32 bit arithmetic and also has its own stack and stack addressing logic.

Since the REKURSIV has so many memories, over forty 40 bit registers and many internal 40 bit data paths, the von Neumann bottleneck is minimised.

4 The language Lingo

The language Lingo supports the class concept in the sense of Simula and the message passing terminology of SMalltalk [4]. This means that Lingo allows the creation of objects that contain executable statements and a persistent state. Every object belongs to a class and the behaviour of objects is determined by the class they belong to. The equivalent to the notion of *program* in Lingo is the **Class definition** : class definitions are like templates from which instantiations (objects belonging to the class) are created.

The message passing terminology describes what is essentially a remote procedure calling mechanism. A statement such as

```
stack new;
```

can be interpreted in various, equivalent, ways –

- send the message *new* to the object *Stack*.
- invoke the **method** *new* encapsulated in the object *Stack*.
- call the procedure *new* encapsulated in the object *Stack*.

In general, message sending elicits a response. In the above example, it can be inferred that since the (receiver) object's name starts with an uppercase character it is actually a class (this is a syntactically enforced rule of Lingo) and it is probable that the message *new* has semantics **create a new instance of the receiving class (Stack) and respond with the new object's identifier**. The above is more likely to be used in the context of an assignment statement as in

```
myStack := Stack new;
```

which would assign to the variable myStack the object identifier of the newly created instance of the class Stack.

The names of an object's procedures are known as **selectors**. The executable statements invoked by a send message, the procedure bodies, are known as **methods**. Methods can also take arguments. For example –

```
myStack push: 24 ;
```

can be thought of, in Pascal terms, as a procedure call like push(24). Selectors that require an argument, such as push:, must end with a colon and are termed **keyword selectors**.

Methods can take more than one argument. For instance

```
today setDay:31 setMonth:"January" setYear:1991 ;
```

Sends the three arguments – the integer 31, the string January and the integer 1991 – to the method with selector setDay: setMonth: setYear:.

Class definitions, as well as defining explicit behaviour for instances also specify a superclass from which behaviour can be inherited.

5 Formal methods

5.1 Specification of the Relational algebra in SML

A **Relation** is a set whose elements are drawn from a Cartesian product of **domains**. From a database point of view, it is important to associate a relation with a description of the domains involved (termed the **scheme** of the relation). We choose to regard a Relation as a pair (S,T) where S is a scheme and T is a set of tuples drawn from a set of domains described within the scheme S.

SML was chosen to represent formally our description of the relational algebra. Relations (and the other classifications of entities within the relational algebra) were represented by **abstypes** so that these entities could

- only be constructed by using the declared constructor functions
- only be manipulated by calling functions declared within the abstype's interface.

For example, the abstype **relation**'s declaration starts

```
abstype relation = rel of ( scheme * tupset )
```

This expresses exactly that a relation is a pair drawn from the product of the set of schemes and the set of tupsets (both of which are defined elsewhere). An individual relation can be constructed using the constructor function, rel, as in, for example -

```
val supprel = rel( supsch, suppliers)
```

Schemes are defined to contain information about the domains involved in a relation. In addition, the domains are named (as well as typed) and a list of names of a scheme's domains are defined (these are to hold any key fields) —

```
abstype scheme = sch of ((name list) * ((domain * name) list))
```

Tupsets hold the body of data within a relation as a list of tuples —

```
abstype tupset = set of ( tuple list)
```

Tuples (the rows of a relation viewed as a table) contain individual atomic values known as attributes —

```
abstype tuple = tup of ( attribute list)
```

Attributes can either be string or integer values —

```
abstype attribute = ival of int | cval of string
```

Domains represent the set from which attribute values are drawn. These are represented by strings ("string" for strings, "int" for integers).

```
abstype domain = dom of string
```

The validity of a domain is checked with a function in domain's interface, **validdom** with signature : domain $\rightarrow$ bool.

The abstype declarations are completed by defining functions. For the abstype relation, the key functions are the operators of the relational algebra, whose signatures are given —

- select : relation $\times$ (tuple $\rightarrow$ bool) $\rightarrow$ relation
- project : relation $\times$ (name list) $\rightarrow$ relation
- cartprod : relation $\times$ relation $\rightarrow$ relation
- union : relation $\times$ relation $\rightarrow$ relation
- difference : relation $\times$ relation $\rightarrow$ relation
- intersection : relation $\times$ relation $\rightarrow$ relation

A detail to notice here is select's signature — this allows the application of any tuple $\rightarrow$ bool function to the elements within the tupset (body) of a relation, giving a great degree of freedom in selection predicates.

5.2 Implementation from a functional specification

The authors' view is that a program is essentially a mapping (of input values to output values) and so the **effect** of a program can be naturally specified as a function. An implementation, in an imperative, non-functional programming language, can be derived by **transforming** such a functional specification into a program. The mapping approximates to —

abstract data type $\mapsto$ **class**
constructor functions $\mapsto$ **class methods**
functions $\mapsto$ **instance methods**
pattern matching $\mapsto$ **specialisation and inheritance**
higher order functions $\mapsto$ **Lingo modules**

We start with an approach based, using the language SML, which is applicable to any imperative implementation language, and not especially Lingo, on the following stages —

- Consider all the data types that the program will encounter and specify all these as abstract data types using constructor functions.
- define each operation as an operator on these abstract data types, using pattern matching in the usual way to provide case analysis
- define **destructor** functions — these are used to extract the arguments of any nonconstant constructor functions
- eliminate pattern matching in all function definitions (except destructors)
- choose an implementation language representation for each abstract data type

- code the constructors and destructors as result returning functions in the implementation language
- the definition of all the operators is now expressed entirely in terms of function application. Coding these is simply a matter of transliteration.

5.2.1 An example — sequences

As an example, consider the representation of sequences of integers and the implementation of methods to reverse the sequence and to append an integer to the right hand end of a sequence.

- an abstract datatype —

```
abstype seq = empty | cons of int * seq
```

  This says that a seq (a sequence) is either the constant sequence, **empty**, or can be constructed from an application of the function **cons** to an integer, sequence pair.

- the operators —
  - right append —

    ```
    fun rap(empty,i) = cons(i,empty)
      | rap(cons(h,t),i) = cons(h,rap(t,i))
    ```

    Notice the use of pattern matching here. The first line states that right appending an integer to the empty sequence results in a sequence containing just that integer, which is obtained by applying **cons** to the integer and **empty** (which is a valid sequence).

    The second line matches the case where an integer is being right appended to a nonempty sequence; this sequence, being nonempty, must be constructible by an application of **cons** to some integer, h say, and some sequence, t say. Clearly this recursion will come to an end.
  - the reverse function, similarly —

    ```
    fun rev(empty) = empty
      | rev(cons(h,t)) = rap(rev(t),h)
    ```

- Define destructor functions.In both the function definitions above pattern matching has been used to break a constructed item into its component parts. Here we would define constructor functions —

```
    – fun head(empty) = raise seqFault
        | head(cons(h,t)) = h
    – fun tail(empty) = raise seqFault
        | tail(cons(h,t)) = t
```

- Eliminate pattern matching by using destructor functions —

```
fun rap(s,i) = if s = empty then
                   cons(i,empty)
               else
                   let val h = head(s) in
                       let val t = tail(s) in
                            cons(h,rap(t,i));
                        end
                    end;
```

  and similarly for rev.

- Choose a representation in the implementation language. In general, we declare a class of objects, each with a single instance variable as in —

```
Seq is Object
  [ sequence ]
```

  and we code instance methods to access this data —

```
sequence [] sequence.
sequence:s [] { sequence := s ; }.
```

- Code the constructor and destructor functions. The constructors, which create new items of the class, are naturally coded as class methods —

```
empty [] { ^ ( super new) ; }.
cons: anInteger with: aSequence []
{ ^ (super new sequence:
  (Pair of:anInteger and:aSequence));}.
```

  Notice the use of a class Pair. This is appropriate, pairs (indeed ntuples) are a built-in type in SML which has no direct correspondent in Lingo so we code it. Pair has two instance methods: **first** and **second**; these are destructor functions for extracting the components of a pair. With these, the destructor functions for Seq are coded as instance methods —

```
- head [] { if (sequence = nil ) then
                  raise seqException
            else
                  ^ (sequence first);
          }.
- tail [] { if (sequence = nil ) then
                  raise seqException
            else
                  ^ (sequence second);
          }.
```

- Code the operators. Again, we use instance methods since one of their arguments is always an instance of this class.

```
self rap: anInteger [ h t ]
    {
    if sequence = nil then
        ^ Seq consOf: anInteger with: self
    else
    {
       h := self head; t := self tail ;
       ^ Seq consOf:h with:( t rap: anInteger);
    }
  }.
```

Which is clearly in one to one correspondence with the original SML.

5.2.2 Making more use of Lingo

One source of inefficiency in an implementation derived as above comes from pattern matching. When translating into a language such as Pascal or C, pattern matching would make use of tag fields within a variant record or a union type. In Lingo, we can do better by using inheritance. Consider again the SML for the seq abstype —

```
abstype seq = empty | cons of int * seq
```

We can represent sequences by a class Seq, as before, and have two specialisations Empty and Cons. The constructor functions **empty** and **cons** become instance creating class methods of Seq's subclasses Empty and Cons respectively —

```
Seq is Object
 [ ]
```

Which needs no class methods since only its subclasses should ever be instantiated.

```
Empty is Seq
 [ ]
```

Which needs no class methods — simply creating an object of this type with it's inherited new method is enough.

Cons, whose instances are nonempty is coded along with destructors head and tail as instance methods —

```
Cons is Seq
 [ sequence ]
 {
 of:anInteger and:aSeq []
  {
    ^ ((super new) sequence: (Pair of:anInteger with:aSeq));
  }.
 head [] { ^ sequence first; }.
 tail [] { ^ sequence second; }.
 sequence [] { ^ sequence; }.
 sequence:s [] { sequence := s; }.
```

Now if again we consider the SML function rev

```
fun rev(empty) = empty
 | rev(cons(h,t)) = rap(rev(t),h)
```

we can view rev as having two specialisations, one a function has domain only those instances of the abstype that are empty, the other has the complementary domain of all nonempty seqs. We add an instance method within the Empty class as

```
  rev [] { ^ Empty new }.
```

and an instance method to the Cons class as

```
  self rev [] { ^ ((( self tail) rev) rap:(self head)); }.
```

Which is concise. Notice this facility cannot be mirrored in SML since SML has no facilities to specialise a type.

5.3 Higher order functions

As has already been pointed out, the SML specification of the relational algebra that we use makes use of higher order functions (consider the signature of the select operation on the abstype relation). These can be catered for in our transformation strategy. By way of example, consider a function called map that takes as arguments a function with integer domain and range and a sequence of integers. Its result is the sequence of integers obtained by applying the int $\rightarrow$ int function to each integer of the given sequence and constructing the sequence of results. That is, the signature of map is

map : (int $\rightarrow$ int $\times$ seq) $\rightarrow$ seq

and its implementation is

```
fun map(f,empty) = empty
  | map(f,cons(h,t)) = cons(f(h),map(f,t))
```

This is transformed into Lingo into separate methods, one for the Empty class and one for the Cons class as

- For the empty class —

  ```
  map: f [] { ^ Empty new }.
  ```

- For the Cons class —

  ```
  self map:f []
   { ^ Cons of:(f performWith:(self head))
                  and:((self tail) map:f);}.
  ```

6 Implementing the language DEAL

6.1 The target: an overview of DEAL

In this section we give an overview of the salient details as they affect the analysis and synthesis strategies we selected.

DEAL (DEductive ALgebra) is a relational language. It attempts to provide a unified framework for both conventional and deductive database processing. Rather than supporting knowledge based systems by providing Prolog, say, with an interface to an underlying relational database, in DEAL the relational language is extended. Deductions are regarded as the generation of new facts from existing facts (extensional database) using deduction rules (intensional database).

An interactive session with the interpreter started by creating an instance of a Deal class object. For example, assuming x is a Lingo process variable:

```
x := Deal new;
```

will initiate an interactive session that will end when a <ctrl> D is entered for end of file. The object x still exists, with whatever environment the Deal object had, and can be queried as in:

```
x ask:"parts where weight > 14" ;
```

(This query returns an object of Relation class, the relation being the subset of the parts relation whose weight attribute is greater than 14.

Within an interactive session, the DEAL interpreter recognises two categories of statement.

- **Queries** – the purpose of the language is to answer relational algebra queries such as

  ```
  parts where weight > 14
  ```

  but the interpreter will accept any expression – relational or arithmetic – and print its result.

- **Environmental statements** – these change the context in which queries are evaluated. Functions that take parameters and return results, for example can be defined; Relations can be loaded from text files and so on.

We will concentrate on the translation and evaluation of queries and the compilation of function definitions.

6.1.1 The statements

We have chosen to take the view that (relational) queries are merely a special case within the general class of expressions, specialised by their type. We follow a conventional syntax for expressions, using syntactic classes such as factor and term to express operator precedence and associativity. The operators of the relational algebra, set operators and numeric operators (such as addition and multiplication) are treated identically in terms of parsing – the appropriateness of an operator in a context is determined by typechecking later in the analysis phases.

This strategy was not followed by Sadeghi [6] who used syntax to effect some typechecking within the parser and differentiated syntactically between relational queries and arithmetic expressions. Sadeghi's syntax was not amenable to recursive descent parsing (our chosen tactic) without manipulating the grammar into an unnatural state. Delaying this differentiation to the typechecking phase eases our treatment of arbitrarily complex selection predicates within the **where** section of a query. For example, the query

```
parts where weight * quantity <= vancapacity / 2
```

asks for all the parts whose weight multiplied by quantity in stock does not exceed half the carrying capacity of a van (a global, integer typed variable). The selection predicate can involve logical **and** and **or**, a relation's attribute names, global variables, function results and constants.

Function definitions include declarations of the function's formal parameters and their types – but not the returned type. The body of a function definition can consist of assignment, selection (if) and repetition (while) statements combined in sequence and nested. A function cannot contain a bare expression to be evaluated, a function definition or a housekeeping statement.

An example function definition –

```
func factorial( n : int)
{
  if (n=0) then
     factorial := 1
  else
    factorial := n * factorial(n-1);
};
```

6.1.2 Variables

Functions have no local variables (apart from their parameters, which can be assigned to having only a local effect.) Global variables are introduced by being assigned to — there is no explicit declaration.

6.2 The interpreter's analysis phases

Lexical analysis was achieved by defining a class Scanner that could be instantiated to form a lexical analyser for most conventional languages. To ease the parsing burden, Scanner was then specialised by a subclass DEALScanner.

The recursive descent method of syntax analysis was employed – this entails writing a method for each nonterminal of the language much as one would in a conventional language [1] At first this was coded manually and then the parser generation process was automated by creating a tool which the authors call YAKK. YAKK allows a BNF-like grammar description to contain semantic, synthesizing actions. More information about the Scanner and the parser generation is available from the authors.

6.2.1 Synthesis

The parser creates an abstract syntax tree whose leaves are either constants of some type (**Relations, Integers, Strings**) or variables from some scope.

It can be determined during the parse whether an encountered variable is global or the formal parameter of a function definition. If an identifier is encountered that does not fit into this category, it is assumed to be an attribute name — which cannot be checked at parse time. There are three classes representing variable type leaves — **Global, Local** and **AttrVar**.

The nodes of the abstract syntax tree are instances of classes representing operations; for instance an addition node is represented by an instance of the class Plus.

All operators either have or inherit an instance variable which, at instantiation is assigned to a module that carries out the required operation. This allows, for example, the select operator to be instantiated with as complex a predicate as the arithmetic expressivity of DEAL will allow — each select node carries with it a module to evaluate its own predicate.

At run time, these abstract syntax trees are evaluated within two contexts - one holds global bindings, the other holds bindings local to the current invocation of a function. Since DEAL uses tuple predicates the currently active tuple and currently active relation are held in the global context.

7 Conclusions

The work described in this paper continues. The next phase will see a qualitative evaluation of the performance of the DEAL system and the coding of some demonstration applications in it.

What are the hypotheses that the work attempts to test ?

- expressivity – some comparison can be made between the work in this project and an earlier implementation of DEAL in the language C [6]. The expressivity of the language Lingo certainly increases programmer productivity (after an initial learning curve) for projects of this kind. Moreover, certain aspects of Lingo, such as polymorphism, allow features in this implementation of DEAL that were considered and rejected by [6] – the atomic attribute values in a tuple within a relation can easily belong to any Lingo class *without altering significantly the DEAL interpreter itself.* Indeed, attributes may themselves be relations. Selection predicates can be as complex as one likes – as long as atomic values have a behaviour encompassing (boolean) relational operators such as = and <.

- object-oriented memory - unfortunately the only REKURSIVs to be manufactured have been hosted on Sun workstations, using communication over the VME bus to the Sun processor to simulate the action of a dedicated disk processor. This complicates testing the hypothesis that an object-oriented store squeezes the performance difference between low level fast languages and highly expressive slow languages. Early results suggest that this is true up to the point where the REKURSIVs real memory is full. Thereafter, the VME penalty is severe.

References

[1] Davie, A.J.T. and Morrison, R. Recursive Descent Compiling, Ellis-Horwood, 1981.

[2] Deen, S.M. DEAL - a Relational language with Deductions, Functions and Recursion, Data and Knowledge Engineering, Volume 1, 1985.

[3] Harland, D.M. REKURSIV - an Object-oriented architecture, Ellis-Horwood, 1988.

[4] Goldberg, A. Smalltalk80 : the language and its implementation, Addison-Wesley, 1983.

[5] Natanson, L.D., Samson, W.B. and Wakelin, A.W. Object-oriented implementations from a functional specification, in proceedings of Software Quality Workshop, Dundee, 1990.

[6] Sadeghi, R.S. HQL - a Historical Query Language, in proceedings of British National Conference on Databases 6, Cambridge University Press, 1988.

[7] Samson, W.B, Deen, S.M. Wakelin, A.W. and Sadeghi R. Formalising the Relational Algebra — Some specifications, observations, problems and suggestions, presented at Formal Methods Workshop, Teesside Polytechnic (UK), 1987.

[8] Samson,W.B. and Wakelin, A.W. PEARL — a database query language for the integration of data and knowledge bases, Proc. Int. Conf. on AI in industry and government, Hyderabad, India, ed P. Balagurusamy, Macmillan 1989.

[9] Wakelin, A. A database query language for operations on graphical objects. Ph.D thesis, Dundee Institute of Technology, 1988.

A Reverse Communication Interface for "Matrix-free" Preconditioned Iterative Solvers

M.A. Heroux

Mathematical Software Research Group, Cray Research, Inc., 655F Lone Oak Drive, Eagan, MN 55121, U.S.A.

Abstract

This paper describes an interface for a set of "matrix-free" preconditioned iterative routines for solving a sparse linear system $Ax = b$. These routines are matrix-free in the sense that only dense linear algebra operations are performed internally. By using reverse communication, all sparse linear algebra operations, i. e. , those that depend on the matrix A or preconditioners for A, are computed outside these routines. Thus, these routines can be used regardless of the representation of A since only the action of A on a given vector is needed.

Along with its description, we also discuss the advantages and disadvantages of this interface and demonstrate its use by providing an example. A brief comparison with other common interfaces is also given.

Introduction

Standard interfaces for numerical linear algebra software have been shown to be very useful if the interface is simple yet flexible enough for a wide variety of users. Examples of successful interfaces are the Level 1, Level 2 and Level 3 BLAS [10, 9, 5, 4], LINPACK [3] and EISPACK [14], and more recently, LAPACK [2]. That these standards have been accepted by a large number of users is a testimony to their simplicity and flexibility. However, with the exception of perhaps the LINPACK banded storage scheme, widely adopted software standards for numerical linear algebra have been almost exclusively for dense linear systems.

Several attempts have been made at standards for sparse linear systems but the task is much more difficult. At the heart of this difficulty is the fact that there is no single widely accepted storage format for sparse matrices. In fact, there are many well-known storage formats for sparse matrices and for each basic format there is typically several minor variations. Any standard which would allow only a single storage format would severely limit the number of users that could effectively utilize the standard. Therefore, attempts at standards for sparse linear algebra software have tried to accommodate a variety of storage formats. In this paper we describe one such attempt.

Interface Description

The interface presented in this paper is intended for preconditioned conjugate gradient (PCG) routines. These routines are used to solve linear systems $Ax = b$ where b is a given right-hand-side vector, A is a given sparse matrix and x is an unknown solution vector. The operations of PCG routines can be placed in two categories:

1. Dense linear algebra (A-independent) operations:
   - Vector updates
   - Vector dot products
   - Dense matrix-vector products
2. Sparse linear algebra (A-dependent) operations:
   - Sparse matrix-vector products
   - Preconditioners:
     - Sparse factor/solve
     - Diagonal scalings
     - Polynomial expansions

The dense linear algebra operations, or A-independent operations, can be computed independent of the storage format used for A. However, the sparse linear algebra operations, or A-dependent operations, are entirely dependent on the storage format of A. Because of this, it is desirable to separate the A-independent and A-dependent operations.

One technique for separating A-independent and A-dependent operations is *reverse communication*. Reverse communication is a fairly common technique in numerical software. In particular, it has been a popular technique in software for optimization and ordinary differential equations[7]. The basic idea of our use of reverse communication is this: When the PCG routine reaches a step in the algorithm that depends explicitly on the matrix A, it sets a flag indicating the type of operation needed, (e. g. , sparse matrix-vector product), sets pointers to the input and output vectors, and return to the calling routine. The calling routine must then perform the requested operation on the input vector, place the result in the output vector, and then re-call the PCG routine. The PCG routine then jumps to next step in the algorithm and continues.

Thus, these PCG routines are matrix-free in the sense that only the A-independent operations are performed internally. By using reverse-communication, all A-dependent operations are computed by the calling routine. This provides flexibility in several ways. It means that the PCG routines can be used regardless of the representation of A (and its preconditioners) since only the action of A on a given vector is needed. This gives the user great freedom in cases where the matrix A cannot be explicitly built and only the action of A is available. Also, the user only needs to provide the matrix-vector product and preconditioning routines in order to have several preconditioned iterative solvers. However, there is a price to pay for this flexibility. The user must write a driver routine similar to the one shown in the following section. Also, it can be cumbersome to write a PCG routine using reverse communication, especially if there are nested subroutine calls and A-dependent operations are required several levels deep.

Another important feature of these PCG routines is compatible calling sequences. By clustering integer parameters into **iparam**, scalar parameters into **sparam**, vectors into **swork** and pointers into **ipntr**, (where **iparam**, **sparam**, **swork** and **ipntr** are defined in the man page below) we have a calling sequence which is the same for all PCG routines even though the required lengths of these arrays may vary from method to method. If two iterative methods are very dissimilar, then this clustering is of no benefit. However, many PCG methods are similar in their parameter and workspace requirements. Thus, by generalizing the parameter list for each method to include the parameters for every other method, we can interchange calls to different PCG routines with very few code modifications.

Below is a man page description of the reverse communication interface. This present interface is general enough to express implementations of a variety of preconditioned iterative methods including the following: the standard preconditioned conjugate gradient method[6], PCG for the normal equations, the bi-conjugate gradient and bi-conjugate gradient squared methods[15] and GMRES[12].

REVCOM **(ido, m, n, x, b, iparam, sparam, ipntr, swork, lswork, ierr)**

ido **Integer [input/output]** The value of **ido** determines the type of matrix-dependent operation needed. The input vector x starts at **swork(ipntr**(1)) and the output vector y starts at **swork(ipntr**(2)). The meaning of **ido** is as follows:

On entry: Whenever REVCOM is called with **ido** = 0, initialization occurs. **ido** must be zero on the first call to REVCOM, and should not be zero on subsequent calls.

On exit:

1 - Compute $y = Ax$. (Matrix times vector)

2 - Compute $y = A^T x$. (Matrix transpose times vector)

3 - Compute $y = P_L^{-1} x$. (Left preconditioning)

4 - Compute $y = P_R^{-1} x$. (Right preconditioning)

5 - Compute $y = P_L^{-T} x$. (Left transpose preconditioning)

6 - Compute $y = P_R^{-T} x$. (Right transpose preconditioning)

9 - Compute error estimate and store in variable **sparam**(2) = *err*.

ido also indicates when execution is completed:

99 - all done

m **Integer [input]** Row dimension of the linear system.

n **Integer [input]** Column dimension of the linear system.

x **Scalar(n) [input/output]** On entry, **x** contains the initial guess to the solution of the linear system. On exit, **x** contains the solution.

b **Scalar(m) [input]** Right-hand-side vector.

iparam **Integer(*) [input/output]** On entry, **iparam** contains the initial values of the integer parameters used by REVCOM. On exit, some of these values are modified to report execution statistics. The integer parameters are as follows:

- **iparam**(1) *ktest* **[input]** Flag indicating the type of convergence criterion to use. If *ktest* = 0 the 'natural' (cheapest) convergence criterion is used. Otherwise, the user must compute the error via reverse communication with **ido** = 9.

- **iparam**(2) *maxit* **[input]** Maximum number of iterations to be performed.

- **iparam**(3) *kpre* **[input]** Indicates how to apply preconditioner.

1 - Apply left preconditioning

2 - Apply right preconditioning

3 - Apply 2-sided preconditioning

- **iparam(4)** *kunit* **[input]** Unit number for reporting results.
- **iparam(5)** *iter* **[output]** Number of iterations performed.
- **iparam(6)** *ktrunc* **[input]** Krylov subspace dimension for truncated methods.
- **iparam(7)** *krstrt* **[input]** Number of iterations between restarts.

sparam Scalar(*) **[input/output]** On entry, **sparam** contains the initial values of the scalar parameters used by REVCOM. On exit, some of these values are modified to report execution statistics. The scalar parameters are as follows:

- **sparam(1)** *tol* **[input]** Error tolerance used to test for convergence.
- **sparam(2)** *err* **[output]** Current estimate of solution error.

ipntr Integer(*) **[output]** Used to mark the starting locations in the array **swork** of vectors used by the iterative method. Also, **ipntr(1)** and **ipntr(2)** point to the starting address in **swork** of the operand and result vectors, respectively, for matrix-dependent operations specified by **ido**. The exact meaning of **ipntr(3)**,... may vary from method to method. For the standard preconditioned conjugate gradient method the meanings are:

- **ipntr(1)** - pointer to current operand vector.
- **ipntr(2)** - pointer to current result vector.
- **ipntr(3)** - pointer to residual vector.
- **ipntr(4)** - pointer to direction vector.
- **ipntr(5)** - pointer to pseudo-residual vector.

Thus, at any point in the computation, the residual vector can be found in **swork(ipntr(3))** through **swork(ipntr(3)+m-1)**.

swork Real/Complex(**lswork**) **[output]** Work array.

lswork Integer **[input]** Length of work array.

ierr Integer **[output]** Error flag.

Sample Program

In this section we illustrate how to build a driver for the interface defined in the previous section. The following program `pcgsol` solves $Ax = b$ via the standard PCG method with left preconditioning. A is stored in `a` and the preconditioner is denoted by `p`. All other notation is the same as above. We assume that `revcom` is the reverse communication implementation of the standard PCG method. Note that the only possible values for `ido` are 1, 3 and 99. Also, `iparam(6)` and `iparam(7)` are not used.

```
      subroutine pcgsol (n, x, b, a, p, swork, lswork, ierr )
      integer n, lswork, ierr, iparam(5), ipntr(5), ido
      real x(n), b(n), a(*), p(*), sparam(2), swork(lswork)
c
c.....Call setup to compute preconditioner.  Store in p.
      call setup(n, a, p )
c
c.....Define iparam and sparam (iparam(6) and iparam(7) not used).
      iparam(1) = 0       ! ktest = 0:     Use default error test.
      iparam(2) = 50      ! maxit = 50:    Do maximum of 50 iterations.
      iparam(3) = 1       ! kpre = 1:      Use left preconditioning.
```

```
      iparam(4) = 6      ! kunit = 1:      Print output to Fortran unit 6.
      sparam(1) = 1.0E-6 ! tol = 1.0E-6:  Error tolerance.
c
c.....Initialize ido equal to zero.
      ido = 0
 1000 continue
c
c.....Call reverse communication routine.
      call revcom(ido, n, n, x, b, iparam, sparam, ipntr, swork,
     &            lswork, ierr)
      if (ierr.ne.0)  return  ! Check if any error.
c
      iinp = ipntr(1) !  Pointer to input vector.
      iout = ipntr(2) !  Pointer to output vector.
c
c.....Determine what to do based on value of ido.
      if (ido.eq.1) then
        call matvec(n, swork(iinp), a, swork(iout))
      else if (ido.eq.3) then
        call preconl(n, swork(iinp), p, swork(iout))
      else
        return ! The only other possible value is ido = 99.
      endif
      goto 1000 ! Re-call revcom.
      end
```

Other Interfaces

There are two other common interfaces for sparse iterative solvers. One is the *black box* interface which provides routines for A-dependent operations for one or more common storage formats. Thus, users need not supply any executable code if their matrices are stored in one of supported formats and can treat the iterative solver like a black box. However, it is practically impossible to provide support for all storage formats.

The other common interface is *direct communication*. Like reverse communication, this interface requires that the user provide routines for the A-dependent operations. However, it is also required that these routines conform to pre-defined calling sequences. Thus, because the calling sequences are known to the iterative routine, direct calls can be made to perform the A-dependent operations and there is no need to use reverse communication. A well-developed interface of this type is described in [1]. Some examples of packages which use direct communication are found in [8], [11] and [13].

One can certainly argue that the black box and direct communication interfaces are simpler to use than reverse communication. But, this simplicity comes at the cost of restricting the user to a fixed set of storage formats (black box) or to a fixed calling sequence (direct communication) which may be cumbersome or even impossible for the user to conform to. However, it is not necessary to use only one of the three interfaces since it is quite easy to integrate all three interfaces into a single package in a very natural way.

In fact, a fully developed software package for solving sparse linear systems can provide a three-level hierarchy of user interfaces. The first level is the black box interface which would provide routines for A-dependent operations for one or more common storage formats. This level could meet the needs of a large number of users. The second level is the direct communication interface where users provide their own routines for A-dependent operations and these routines conform exactly to given calling sequences.

Finally, the third level is the reverse communication interface. This level is used if neither the black box nor direct communication interfaces is appropriate.

One should note that the reverse communication subroutines can also serve as kernels for performing the A-independent operations for all three interfaces. This is not so important for algorithms like the standard conjugate gradient method where the A-independent operations are few and simple. However, for algorithms like GMRES, the A-independent operations are much more complex and having them in a single subroutine that can be used by all interfaces is very beneficial.

Conclusions

This paper describes a reverse communication interface for sparse linear systems solvers. This interface is meant to be flexible and easy to use in several ways:

1. By placing the A-dependent steps of the solution process outside the iterative routine, the user only needs to provide the matrix-vector product and preconditioning routines in order to have several preconditioned iterative solvers. Also, the user has great freedom in cases where only the action of A is available.

2. By clustering parameters, work vectors and pointers, we have a calling sequence which is the same for all PCG routines making it possible to interchange calls to different PCG routines with very few code modifications.

3. Finally, even if a black box or direct communication interface is provided, the reverse communication routines can serve as kernels which perform the A-independent operations for all interfaces.

References

[1] Steven F. Ashby and Mark K. Seager. A proposed standard for iterative linear solvers. Technical report, Lawrence Livermore National Laboratory, January 1990.

[2] Chris Bischof, James Demmel, Jack Dongarra, Jeremy DuCroz, Anne GreenBaum, Sven Hammarling, and Danny Sorenson. LAPACK working note #5: Provisional contents. Technical Report ANL-MCS-TM-122, Argonne National Laboratory, July 1988.

[3] J. J. Dongarra, J. Bunch, C. Moler, and G. Stewart. *LINPACK Users' Guide*. SIAM Pub., 1979.

[4] Jack Dongarra, Jeremy DuCroz, Iain Duff, and Sven Hammarling. A proposal for a set of level 3 basic linear algebra subprograms. Technical Report ANL-MCS-TM-88, Argonne National Laboratory, April 1987.

[5] J.J. Dongarra, J. DuCroz, S. Hammarling, and R. Hanson. An extended set of fortran basic linear algebra subprograms. *ACM Transactions on Mathematical Software*, 14, 1988.

[6] M. R. Hestenes and E. Stiefel. Methods of conjugate gradients for solving linear systems. *J. Res. National Bureau of Standards*, 49:409–436, 1952.

[7] Alan Hindmarsh. Reverse communication in general–purpose fortran math software. *Tentacle*, 4(11):2–7, November 1984.

[8] David R. Kincaid, Thomas C. Oppe, John R. Respess, and David M. Young. ITPACKV 2C user's guide. Technical Report CNA-191, Center for Numerical Analysis, The University of Texas at Austin, November 1984.

[9] C. Lawson, R. Hanson, D. Kincaid, and F. Krogh. Algorithm 539: Basic linear algebra subprograms for fortran usage. *ACM Transactions on Mathematical Software*, 5, 1979.

[10] C. Lawson, R. Hanson, D. Kincaid, and F. Krogh. Basic linear algebra subprograms for fortran usage. *ACM Transactions on Mathematical Software*, 5, 1979.

[11] Thomas C. Oppe, Wayne D. Joubert, and David R. Kincaid. *NSPCG User's Guide.* Center for Numerical Analysis, The University of Texas at Austin, December 1988.

[12] Youcef Saad and Martin H. Schultz. GMRES: A generalized minimal residual algorithm for solving nonsymmetric linear systems. *SIAM J. Sci. Stat. Comput.*, 7(3):856–869, July 1986.

[13] Mark K. Seager. A SLAP for the masses. Technical report, Lawrence Livermore National Laboratory, December 1988.

[14] B. T. Smith, J. M. Boyle, J. J. Dongarra, B. S. Garbow, Y. Ikebe, V. C. Klema, and C. B. Moler. *Matrix Eigensystem Routines - EISPACK Guide*, volume 6 of *Lecture Notes in Computer Science*. Springer–Verlag, New York, second edition, 1976.

[15] Peter Sonneveld. CGS, a fast lanczos-type solver for nonsymmetric linear systems. *SIAM J. Sci. Stat. Comput.*, 10(1):36–52, January 1989.

Knowledge Modeling and Model-Based Problem Solving - Towards a Multi-Use Engineering Knowledge Base

H. Ueno (*), Y. Yamamoto (*), H. Fukuda (**)

(*) Department of Systems Engineering, College of Science and Engineering, Tokyo Denki University, Hatoyama, Saitama 350-03, Japan

(**) RAS Design No.2, Yamato Laboratory, IBM Japan, Yamato, Kanagawa, Japan

ABSTRACT

This paper presents a way of systematic knowledge modeling representing objective engineering knowledge on a digital system and reasoning for shooting troubles as an example of model-based problem solving. Since the knowledge of an object itself is separated from the knowledge for specific problem solving the knowledge base can be used for multiple purposes. The object model consists of the knowledge on structure and that on behavior. The knowledge on structures is represented by means of combining ISA and PARTOF relations in a two dimensional three-layer hierarchy. The knowledge on behaviors is attached to associated unit in the hierarchy. By means of the extended inheritance, control knowledge description is minimized. This knowledge scheme results in several benefits: 1) The generalized model at the top layer allows users to define the knowledge of similar objects easily. 2) Since the model includes objective knowledge in a general form on the target object the knowledge base can be used for a variety of problem solvings in a shared use. 3) Since the modeling procedure is systematic an automated knowledge acquisition could be achieved from CAD systems. 4)A problem solving of the model-based trouble shooting shows the usefulness of the approach.

1 Introduction

We have learned many things from the experiences of developing expert systems in the last decade. One of the most important issues should be the use of deep knowledge in order to achieve high performance and high flexibility in problem solving, especially in the domain of engineering. This is understandable because human experts have the knowledge about basic theories and principles extracted from education and research, in addition to expertise obtained through a long time experience in a specific domain. In another case we know that plans and design specifications are very useful for engineers who are trying to find failures within an engineering system, or are trying to understand documents describing complex devices.

It should be noted that these types of knowledge which are referred to as "*deep knowledge*" are common and domain independent, and can be shared by a variety of problem solvings. The knowledge about principles and theories and the knowledge about structures and behaviors are known as typical deep knowledge, e.g. Davis[2], Hart[11] and Ueno[5]. Deep knowledge is objective, while surface one is subjective. We know that experienced engineers have enough objective knowledge as well as subjective, and that they are applying both in solving complex engineering problems. In this paper we deal with the second type of deep knowledge, i.e., the knowledge about structures and behaviors, which are useful to manipulate complex structured systems. In this paper we discuss the knowledge representation schema to represent structures and behaviors by means of "*object model*" in the domain of digital systems, and model-based problem solvings for trouble shootings.

An important feature of deep knowledge should be universality in interpretation which results in high flexibility and performance in problem solving, while surface knowledge which has been applied to most currently available expert systems is dedicated to specific interpretation in specific problem solving within a narrow domain, and therefore resulting in limited flexibility and performance.

We know also that an engineer who has the knowledge about one system can easily acquire the knowledge of a similar system using that. This means that appropriate modeling methodologies are required to achieve knowledge acquisition as well as knowledge independence. We are trying to provide an useful paradigm for systematic knowledge acquisition in engineering domain by means of the object model-based approach.

The object model-based knowledge representation schema are considered to be shared by a variety of applications from simple analysis-oriented problems to complex synthesis-oriented problems by many users. "*Knowledge sharing*" must become one of key issues in the next decade. We believe that the knowledge modeling and reasoning by means of the object model will provide a useful methodology in this direction.

In this paper we show a systematic modeling method of a digital system and trouble shooting by means of model-based reasoning in order to examine the flexibility of this approach as an example of applications. A feasibility study was done about a keyboard system of a microcomputer system. The experimental system has demonstrated some interesting features under a limited situation. The system was developed using frame-based knowledge engineering environment ZERO with an extension on inheritance control on a SUN workstation. Although the examination was tried on a virtual keyboard system an essential structure must be applicable to most microcomputer system as well as digital systems. The extension was accomplished so that the concept and methods of the object model can be realized smoothly. The ZERO system is written in the COMMON LISP language.

2 Motivations

The idea of knowledge modeling on a complex structured system is not new. Many advanced expert systems by means of object modeling have been developed and reported already. However ways of modeling which have been proposed and applied seem for us not to be useful enough from the point of view of such as knowledge acquisition, knowledge base maintenance and multi-use of knowledge in computer-based problem solvings. We have been trying to develop a systematic method for modeling of engineering knowledge to meet them. Already we have proposed the idea and method of an object model-based approach and developed several experimental expert systems, by Oomori[3] and Kato[8]. We have leaned several hints through the efforts. The method of modeling and reasoning which is discussed in this paper is an advanced version.

Most currently available expert systems based on surface knowledge can only work for the specific problem solving for a specific problem in a narrow domain. Meanwhile, engineers who have the knowledge about principle and the structure of a target system can solve a variety of problems related to the system using the same knowledge. One of the issues of the deep knowledge should be the multiple interpretability of the knowledge. Most of this type of knowledge is objective and can be shared in use. We need a knowledge modeling scheme which allow us to use a knowledge base for multiple purposes in shared use.

From observation of expert engineers' tasks we can obtain useful hints. Some of them are listed in the following which resulted in our way of knowledge modeling and model based problem solvings.

1) A person who has the knowledge of one system can easily obtain the knowledge about similar systems.
2) Human beings can solve a variety of problems using the same knowledge.
3) In the field of engineering objective knowledge seems to be much more important than subjective knowledge.
4) In problem solvings on complex structured systems human beings seem to control a focus-of-attention along the structural and functional connections of the system.
5) Structures and functions seem to have strong relationships.

We have been trying to develop our modeling method by keeping these hints in mind.

In this paper we discuss a systematic knowledge modeling and model-based trouble shooting for a computer keyboard system. Major motivations of this work are as follows. In the field of production support engineering we need a unified system combining a data base and a knowledge base. The data base is used for maintaining such as design specifications, documents, plans and data for supporting both productions and field services. The knowledge base should support a variety of problem solvings related to the products, although current knowledge base technologies are used in limited tasks as separated systems. There are strong relationships between the data base and the knowledge base.

It is hard to keep the consistency of the contents of the data base and the knowledge base since the design of products are modified occasionally. We think that the knowledge base based on the concept of object model could support such tasks as redesign of products, maintenance of information, maintenance of knowledge and problem solvings. Advanced problem solvings could be achieved by means of knowledge base and data base as a total system. The object model-based knowledge base could work as a knowledge base and also as a bridge to a data base in order to support a variety of tasks in productions and services. This is the major motivation and the long range goal of the project.

3 Object Model

By "*object*" we mean the objective of problem solving by means of knowledge-based approaches. It includes a physical object, a conceptual object and an engineering system. We call a representation model of such objects an "*object model*", by Ueno[5] and [10]. For example, the computer model of a computer is the object model when the computer is an objective of knowledge-based problem solving.

The structure of a knowledge base to meet these circumstances should have a two-layer hierarchy as shown in Fig.1a, where the lower layer consists of the object model ,i.e., deep knowledge, and the upper layer surface knowledge, i.e., such as a set of production rules. A usual knowledge base has a flat structure as in Fig.1b. The two-layer hierarchical knowledge model could allow us to organize a multi-use engineering knowledge base as shown in Fig.2. In this scheme common knowledge is maintained by means of the object model and each application could be achieved by attaching additional surface knowledge on it respectively.

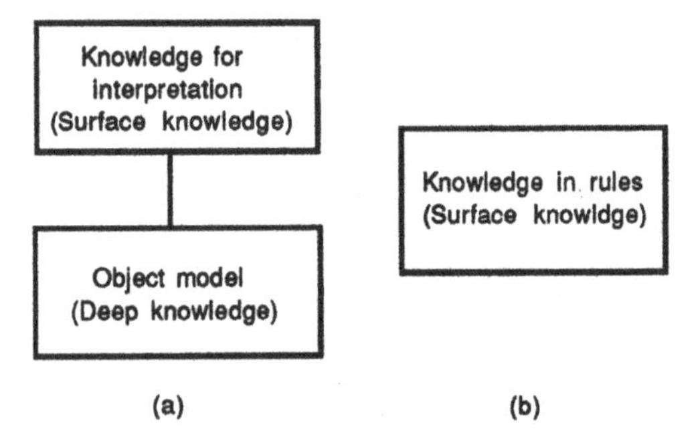

Fig.1 Two-layer knowledge base(a) and single layer knowledge base(b).

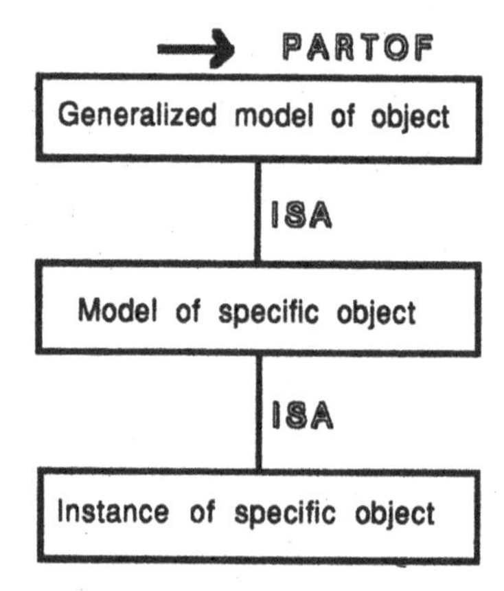

Fig. 2 Three-layer model to represent objects.

Next, we need a suitable data structure to represent the object model. It should be noted that a complex structured system can only be represented by a combination of abstract-specific relations and whole-part relations. In general a complex system consists of a number of subsystems, which in turn consist of a number of sub-subsystems, and can be decomposed until every subsystem becomes a primitive component. The whole-part hierarchy is suitable to represent this kind of structure. In addition there exist semantic structural relations among the subsystems or elements. The semantic relations should be embedded within each unit in the hierarchy in order to encapsulate the knowledge. While, components themselves which include system, subsystems and primitive components can be organized by abstract-specific hierarchies. It should be stressed that the inheritance control of the attributes are inevitably important in order to maintain the knowledge in this situation. However, usual inheritance mechanisms can only manage the abstract-specific relations by the nature. We developed an extended inheritance control mechanism to solve this problem.

As the conclusion in our modeling a two-dimension hierarchy with the extended inheritance control mechanism is used to represent the object model within the frame-based knowledge representation schema as shown in Fig.2, by Ueno[5]. The horizontal hierarchy is used to represent the structure of the object by means of whole-parts(PARTOF) relations and the vertical hierarchy is to represent abstract-specific(ISA) relations of each component of the object. Functional(behavioral) attributes are attached to each associated structural node of the hierarchy. We have developed the systematic knowledge modeling procedure in the frame-based schema as shown in the following. By means of this knowledge modeling schema encapsulation of knowledge can be accomplished, by Ueno [9].

In the following the systematic modeling procedure for complex structured objects is outlined, from Ueno[5]:

1) The principal objects, i.e., top-level objects, of a target system must be represented in a skeleton hierarchy, i.e., the left-most hierarchy, by means of ISA slots.

2) The components are organized in an abstract-specific hierarchy using ISA slots, while the structures are organized in whole-part hierarchies using a combination of HASPARTS, PARTOF and RELATIONS slots.

3) HASPARTS slot has a list of subsystem element names as the slot value. The semantic relations between the elements for representing local structure are described by a set of predicates which is stored as the value of the RELATIONS slot of the frame.

4) The behavioral attributes of each component are described using the BEHAVIORS slot of an associated frame. Additional attributes are described using additional slots.

As shown in the procedure the knowledge modeling is considered according to the functional structure of the object. The knowledge about the behaviors is embedded within each associated unit of the structure. This is the key point in our modeling. This modeling method is applicable under the extended frame-based knowledge representation environment. It should be noted that this knowledge modeling procedure does not include any heuristics, and that the contents and structure of the knowledge are application independent. It should also be stressed that this modeling procedure gives the basic idea. Actual modeling procedure should be modified according to the characteristics of the problem to be handled. In the following chapter an example of actual knowledge modeling is discussed.

4 Modeling of Keyboard Unit

In this chapter we discuss how the knowledge is modeled and represented in detail using an example of digital systems.

4.1 Description of Structures

The basic idea of representing a structural object according to our object model methodology is that each behavioral attribute is strictly connected to each associated substructure of an object. In this scheme the behavioral attributes should be represented within the frame which represents a substructure. This strict relationship seems not to be directly applicable to a microcomputer system. In case of manipulating an object which is consisted of highly generalized functional elements it is difficult to define the relationships among specific functions, i.e., behaviors, and specific substructures because the relationship between chunks of functions assigned to specific units and structures for realizing them are unclear.

For example, on a keyboard unit which is our target object a single processing chip is used for key scanning, parallel-to-serial data conversion, command handling between a system unit and the keyboard unit, and so on. We need a suitable extension on the modeling policy so that such multi-functional systems achieved by combining a microprocessor and microcodes can be represented transparently. We have solved this problem by introducing a combined model representation scheme. That is, the conceptual representation level for an object is divided into two sub-models which are

- logical configuration model, and
- physical configuration model.

The logical configuration model is the model based on functional blocks, while the physical configuration model is the model which is directly associated to the physical configurations of the object system. In the former the model is represented by means of functional units. Physical devices are not considered in this modeling. While, in the latter components are physical devices such as a processor, switches and connectors. Associated components between two models are linked together as shown is Fig.3.

In the logical configuration model the function of the system is decomposed into subfunctions which are decomposed into sub-subfunctions. These functional relations are structured in a PARTOF hierarchy. The model elements in this hierarchy are functional units such as a scanner and a serial-to-parallel converter, which are independent from physical structure of the system.

On the other hand, since the physical configuration model is a representation model strictly connected to the physical structure of the system, model elements are hardware units and devices such as switching devices and connectors. The physical structure is also represented in a PARTOF hierarchy. In the case of objects which include processors as parts, microcodes can be treated as abstract elements. A circuit diagram, a timing chart, a wiring diagram, a shape diagram, a flowchart, and a program list are examples of elements of the model. These are seen in system documents.

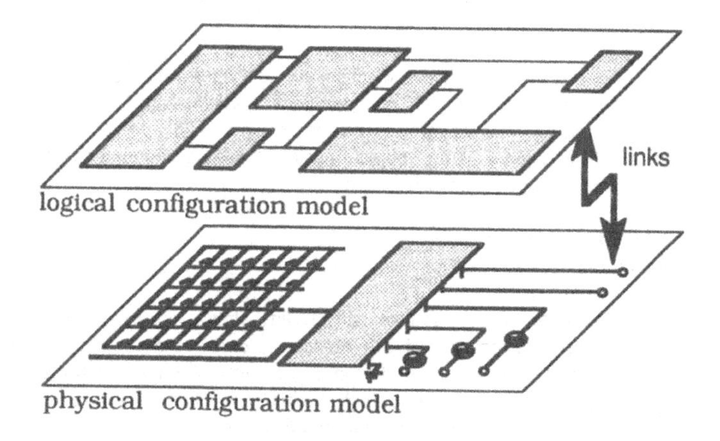

Fig.3 An image of logical configuration model, physical configuration model and links between them.

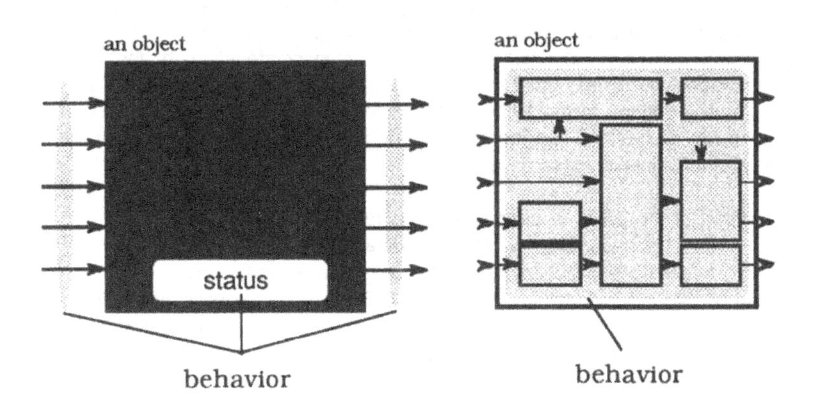

Fig.4 An image of black box model(BBM) and gray box model (GBM) to represent an object.

4.2 Description of Behaviors

In this section we concentrate the discussion on the representation of functions, i.e., behaviors, in the logical configuration model which consists of a set of functional elements. The issue should be to obtain a flexible and powerful method of describing behaviors of an object system. We introduced the two-way view modeling as shown in Fig.4, which are

- Black box model(BBM), and
- Gray box model(GBM).

In BBM the behavior of an object is described simply by the relations between input and output informations of the object. In case that objects have storage functions behaviors are required to be represented by the relations of input, output and internal states of the object system. The BBM modeling is a kind of an external specification-oriented representation since the internal structure of the object is not considered.

On the other hand, in the GBM model the behaviors of a forcusing object is considered to be achieved by the combination of the sub-behaviors which are achieved by the sub-objects of it. Therefore the outline of the internal structure of the object can be seen from outside. By the GBM modeling the relations between behaviors and sub-behaviors and the causal relations among sub-behaviors can be treated. By combining two model descriptions it becomes possible to control a focus of attention during problem solving reasoning.

The structural and functional attributes are represented by a set of slots. Table 1 shows a list of the slots and their roles. For example, as in the table the behavioral attributes of a unit is defined by a combination of a 'behaviors' slot and '!behavior' slot(s). In the 'behavior' slot a list of behavior identifiers of the unit is defined. Each behavior gives the constraint among input, output and internal state of the object and is defined in the following '!behavior' slot(s). In case of the 'kbs-unit-ps2' unit three behavioral attributes are defined which are '!generate-code', '!indicate' and '!communicate'(see Fig.7).

The format of the behavior description is as follows:

((input-port-identifier -> output-port-identifier)
((condition1 -> action1) ... (condition-n -> action-n)))

An example of detailed descriptions are shown in Fig.7. It should be stressed that by combination of the slots in Table 1 the detailed knowledge about logical structures and behaviors of the object can be represented. Similar descriptions are used to represent the physical configuration model.

4.3 Outline of keyboard system

The target system is a keyboard unit of a microcomputer system. The experimental system was specially designed for this project according to the document of the IBM PS/2 system, by IBM [1]. The system was simplified in order to examine the feasibility of the approach.

The three major functions of the keyboard unit are outlined as follows:
(1) the unit has 101 keys, and generates a specific serial code according to a specific key press,

Table 1 Additional system slots and their roles.

attribute name	data type	value
BLACK BOX DESCRIPTIONS		
a-kind-of	frame	pointer to the abstruct object in the ISA hierarchy.
d-descendents	flist	a set of pointers to the child objects in the ISA hierarchy.
partof	frame	parent component in the PARTOF hierarchy.
ports	list	a set of input/output ports of this object.
.*port*	list	property of the port named *port*.
states	list	internal states of this object.
behaviors	list	a set of identifiers of the behaviors of this object.
!*behavior*	list	definition of each behavior of this object.
GRAY BOX DESCRIPTIONS		
variables	list	a set of global variables for the following description.
hasparts	flist	a set of directly connected components of this object.
links	list	a set of links including this object.
_*link*	list	property of the link named *link*.
functionalpaths	list	a set of identifiers of the functional paths for the behaviors.
!!*functionalpath*	list	definitiion of each functional path in the Gray Box Model.

(2) the associated indication lamp is turned on/off according to key-pressing for Caps Lock, Scroll Lock and Num Lock, and
(3) the unit sends the result to the cpu unit according to acceptance and interpretation of the command which is sent from the cpu unit.

Fig.5 shows the functional block structures and the functional paths for typical functions of the keyboard system we are dealing with. As shown in the figure the keyboard system has four major functional blocks which are (1) "keys" to detect user's input operations, (2) "kbd main unit" to generate scan codes according to a specific key press and so on, (3) "indication unit" to handle indicator units, and (4) "cbd cable" to accept and send data between the keyboard unit and the cpu unit. In addition, each functional block is composed of several sub-functional blocks repeatedly.

4.4 Knowledge Representation in Frames

Fig.6 shows the skeleton of the knowledge structure for the keyboard system. As shown in the figure the model is represented in a three-layer ISA hierarchy. At the top layer the general structure of a keyboard system is defined with identifier "kbd-unit". At the middle layer the model for target system "kbd-unit-ps2" is described as a subclass of "kbd-unit". The bottom level knowledge is

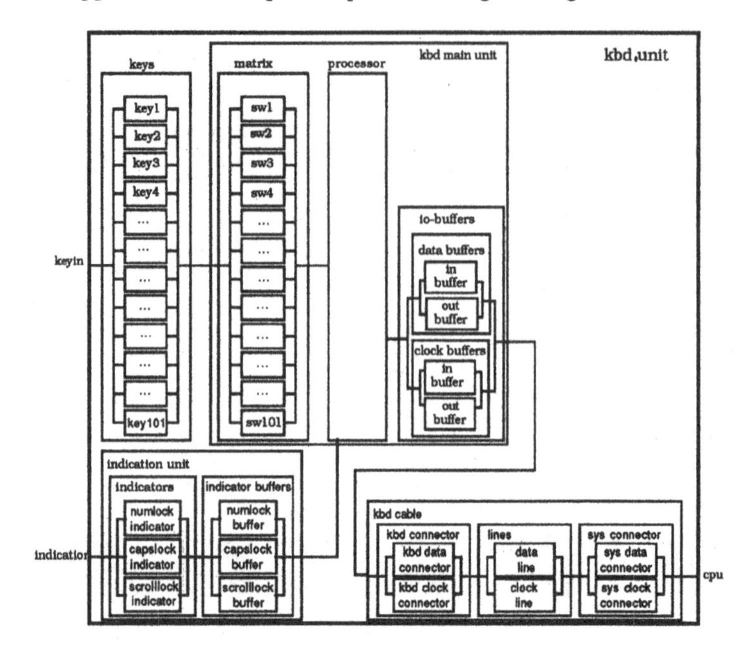

Fig.5 A simplified functional structure of a keyboard unit.

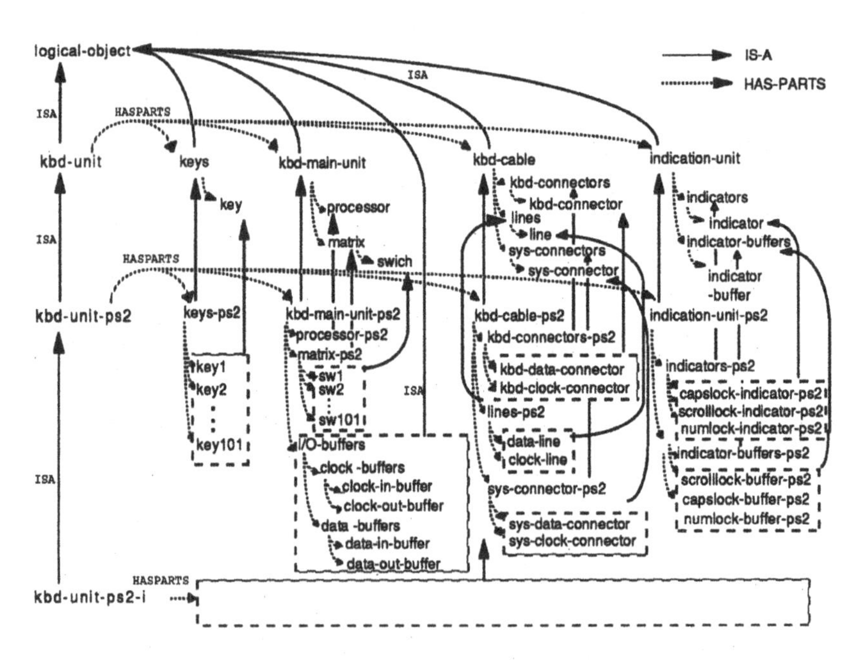

Fig.6 The skeleton of the two dimentional hierarchical frame structure to represent the knowledge of the keyboard unit.

```
frame name:  kbd-unit-ps2      frame type: class
a-kind-of            frame       kbd-unit
d-descendents        flist       (kbd-unit-ps2-i)
...
::: BLACK  BOX  DISCRIPTIONS
part-of              frame       ps2-system
ports                list        (.keyin .indication .system)
.keyin               list        ((/action :value ("push" "release"))
                                  (/key-number :value (&range 1 101)))
.indication          list        ((/num-lock :value ("bright" "dark"))
                                  (/caps-lock :value ("bright" "dark"))
                                   (/scroll-lock :value ("bright" "dark")))
...
states               list        (@indication)
%scan-code-set       table       scan-code-set
behavior             list        (!generate !indication !communicate)
!generate-code       list        ((.keyin --> .system)
                                  ((/key-number  %scan-code-set/key-number)
                                   (/action "push")
                                   --> (/data/output %scan-code-set/make-code))
                                  ((/key-number  %scan-code-set/key-number)
                                   (/action "release")
                                   --> (/data/output %scan-code-set/break-code)))
!indicate            list        ((.keyin -[ @indicate -> @indicate ]-> .indication)
                                  ((/key-number 98) (/action "push")
                                   -[ (/num-lock "bright") -> (/num-lock "dark") ]->
                                  (/num-lock "dark"))
                                  ((/key-number 98) (/action "push")
                                   -[ (/num-lock "dark") -> (/num-lock "bright") ]->
                                  (/num-lock "bright")) ... )
...
::: GRAY  BOX  DESCRIPTIONS
has-parts            flist       (keys-ps2  kbd-main-unit-ps2
                                             kbd-cable-ps2  Indication-unit-ps2)
links                list        (_type-in _stroke _io _system
                                         _connect _Indicator _control _light)
_type-in             list        ((.keyin keys-ps2.operation)
                                         :data-categories ((/action :value ("push" "release"))
                                                          (/key-number :value (&range 1 101))))
_stroke              list        ((keys-ps2.response  kbd-main-unit-ps2.operation)
                                  :data-categories ((/action :value ("down" "up")
                                                          (/key-number :value (&range 1 101))))
_Indicator-control
                     list        ((kbd-main-unit-ps2.Indication  Indication-unit-ps2.Input)
                                  :data-categories ((/num-lock :value ("on" "off"))
                                                   (/caps-lock :value ("on" "off"))
                                                   (/scroll-lock :value ("on" "off"))))
functional-path      list        (!!generate-code !!indicate !!communicate)
!!generate-code      list        (.keyin _type-in
                                         keys-ps2!be-hit _stroke
                                         kbd-main-unit-ps2!indicate _indicator-control
                                         indication-unit-ps2!indicate _light
                                         .system)
...
```

Fig.7 An image of the detailed content of the frame for representing the "kbd-unit-ps2" unit which is the principal class object in the knowledge model as shown in Fig.6.

instantiated during the model-based reasoning of a specific keyboard system, i.e., the reasoning for shooting trouble in this case. The whole-part relations are used to represent the whole-part relations of the functional blocks of the system and represented within each layer from left to right by menas of PARTOF links.

An example of a detailed frame structure is shown as in Fig.7. We introduced several additional system slots to manipulate the two dimensional model representation in addition to basic system slots as shown in Table 1. These slots are installed with the original ZERO system. By system slots we mean the slots which are maintained by the ZERO system itself in editing and/or inheriting the knowledge. These additional system slots are attached to the original ZERO system by means of the user extensible function the system has.

As shown in the figure a frame consists of two parts which are the black box descriptions and the gray box descriptions. In the black box descriptions such as the pointer to the parent unit(by the 'partof' slot), a list of ports and and their attributes(by the 'port' and following slots), a set of behaviors(by the 'behavior' and following slots) are described. In the gray box descriptions a list of the pointers to the sub-elements(by 'hasparts' slot), a set of links and their attributes(by 'links' and following slots) and a set of functional paths(by 'functional-path' and following slots) are described. The functional path shows the sequence of the sub-units for achieving a specific behavior which is described within the black box description part of the frame. In this example the functional path for the '!generate-code' behavior is described. This information is used for such as focus control in the model-based trouble shooting.

5 Model Based Reasoning for Trouble Shooting

In this chapter we discuss the way of model-based problem solving and the method of model-based trouble shooting as an example.

5.1 Model-Based Reasoning

By model-based reasoning we mean a reasoning which is achieved by means of the object model. We suppose that a part of problem solvings done by engineers seems to be achieved by imaging the conceptual model in mind. By this method they are able to solve a variety of problems about a specific object, such as understanding a picture, a drawing and specifications, designing a system and shooting troubles. By combining the model-based reasoning and other reasoning methods such as a rule-based reasoning the performance of a knowledge system must be increased greatly.

Problem solving for trouble shooting by means of the model based reasoning should be defined at first as in the following: "The cause of a trouble occurred within a target system may result in abnormal behavior of one of subsystem of it. The shooting trouble is the problem solving to identify the smallest component of the system which shows the abnormal behavior."

The limitation is that we suppose a single and permanent fault at this time. Multiple faults and temporal faults should be considered in the future. In addition, the target of faults is limited to those which could be identified at conceptual level of reasoning. In other words, time components such as shown in a timing chart is ignored here. As mentioned above we introduced a "functional path" which is the path linking a set of subunits where an influence of a behavior passes through. The functional path is the representation of behaviors in the GBM model.

The procedure of identifying a faulty component by object model-based trouble shooting is as follows.
Step 1: Examines the appropriateness of the behaviors of a focused unit by checking the adequacy of the input and output information of it as a black box model. This is done to make sure whether a target system performs the functions completely or not by means of the knowledge of behaviors. If abnormality is not detected then it is understood that the focused unit does not have a failure and the process terminates. If abnormality is detected and no subunits are defined to this unit then it is concluded that the failure is included within this unit and the process terminates, otherwise the process proceeds to step 2.
Step 2: Detects and divides the functional paths of the focused unit into the normal functional paths and the abnormal functional paths using the gray-box model, if any inadequacy has been detected by the test of step 1.
Step 3: Ruleouts the subunits which have no faulty components by overlapping the normal functional paths and abnormal ones. This is applicable by the single fault assumption.
Step 4: Repeats the above procedures for each subunit along the abnormal functional path.

5.2 Trouble Shooting System

An image of the reasoning process of the model-based trouble shooting is shown in Fig.8. The reasoning begins with unit "kbd-unit-ps2" which is the first focusing

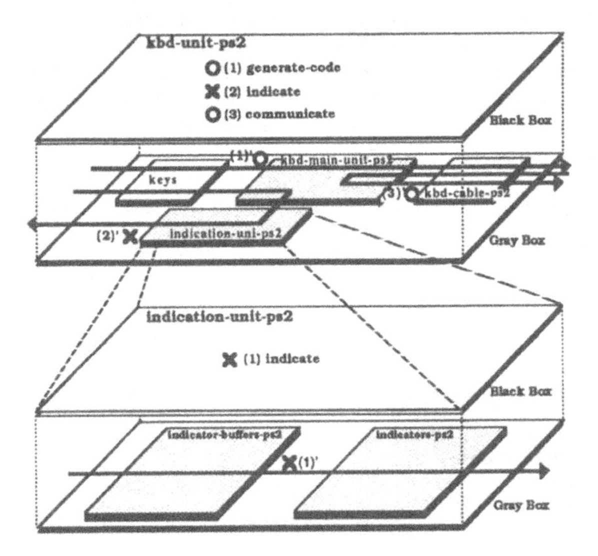

Fig.8 An image of focus control for trouble shooting reasoning by means of the object model of the keyboard unit.

unit assigned by the system. This unit has three behaviors: (1), (2) and (3). If behavior (2) is detected as abnormal one due to the data then the system assumes that unit "kbd-unit-ps2" has a failure. Next, using the gray box information three functional paths can be divided into one abnormal functional path (2') which may include the failure subunit and two normal functional paths (1', 3'). Next, by overlapping all functional paths normal subunits which are belonging to the normal functional paths are deleted from the subunits belonging to the abnormal functional path. The rest of these subunits which are belonging to the abnormal functional path are the candidates of faulty subunits. The figure shows an example that "kbd-main-unit-ps2" and "indication-unit-ps2" are decided as the candidates. Same procedure is applied repeatedly up to detecting the faulty unit which does not have subunits within it. The trouble shooting reasoning terminates when the smallest faulty unit is identified.

The trouble shooting is executed in an interactive way, that is, the system make questions to a user who replies by measuring a target unit. The interactive screen consists of multiple windows as shown in Fig.9. The interaction about functional validity of specific unit is done through Function Validity Check window. The user is requested to reply to questions displayed within this window. Currently focused unit is displayed within the FOCUSING UNIT window in the left hand side of the screen as a black box unit. In the PARENT UNITS OF THE FOCUSING UNIT window in the right hand side the gray box structure which includes the focusing unit as an element is shown. Currently focused unit is indicated by shading within the gray box description. The history of the focused units is displayed here also. In the lower part of the screen interactions for trouble shooting by means of the physical configuration model are displayed.

Fig. 9 shows an example of trouble shooting session using the logical configuration model of the key board unit. The result is sent to the trouble shooter by the physical configuration model, where the reasoning starts with this information by focusing a physical unit. The similar reasoning method is applied using the knowledge on the physical structure of the keyboard unit. Fig.10 shows the final decision of the trouble shooting.

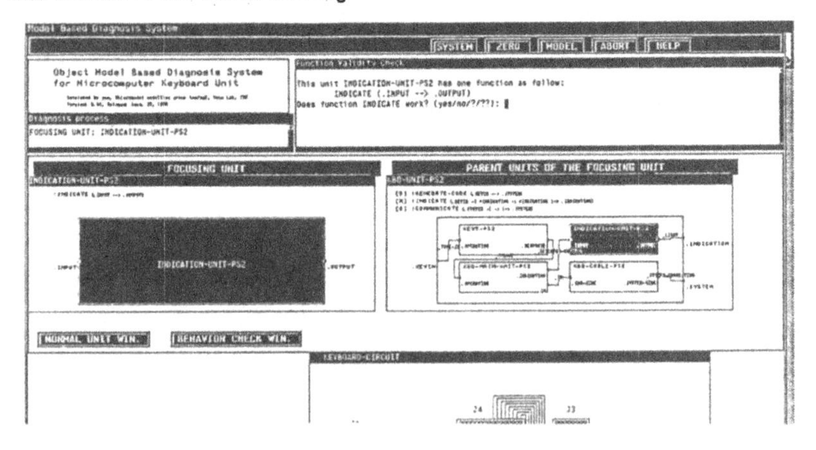

Fig.9 An example of a display of model-based trouble shooting session. Finding the candidate subunits by GBM and testing the functions of kbd-main-unit-ps2 by BBM.

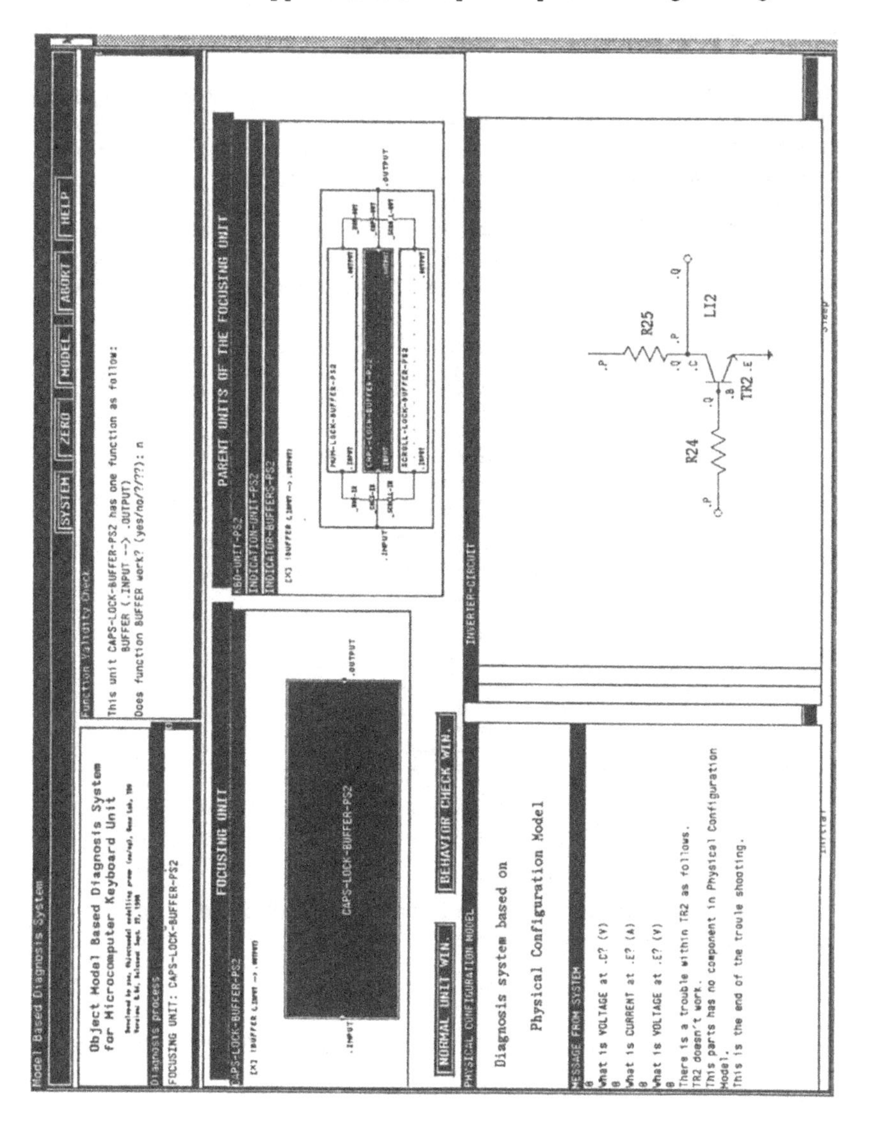

Fig.10 An example of a display of model-based trouble shooting session. The final conclusion has been produced by means of the physical configuration model.

6 Concluding Remarks

In this paper we have presented the concept of the object model, the modeling method for a digital system based on the concept, and the method of model-based reasoning for trouble shooting. The modeling is suitably achieved by means of the frame-based knowledge representation formalism with some extensions. Since the knowledge of an object, i.e., the keyboard system in this case, is separated from a specific problem solving, i.e., a trouble shooting, and the form of representation is not linked to a specific style of reasoning, the knowledge base could be used for a variety of problem solvings as shown in Fig.11. This is our major issue in this work.

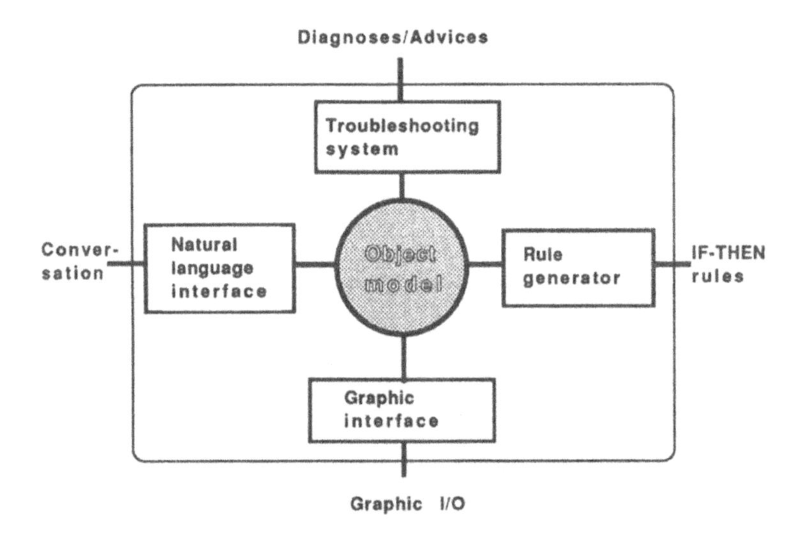

Fig. 11 Ideal view of a multi-use knowledge system based on the concepts of object model.

The way of describing the knowledge by a combination of BBM and GBM models seems to be useful to encapsulation of engineering knowledge and should be familiar for domain experts who are solving problems using such documents as system specifications and manuals. This means that it will be possible to translate the information written in the documents to the knowledge base smoothly.

The model-based trouble shooting reasoning seems to be quite natural and simple. No expertise, no if-then rules were used here. Use of physical configuration model, modeling of micro-programs, and examination of the model from the point of view of a multi-use knowledge base are issues for next study.

Since the modeling is systematic and needs no expertise, the concept and the methodology discussed here seem to be applicable to automatic/semi-automatic knowledge acquisition from the documents produced by design systems. We are trying to develop an (semi)automatic knowledge acquisition system from a CAD system. We are also trying to develop a system which generates IF-THEN rules for describing a specific situation to identify a specific problem. This would be achieved by means of fault simulation methods and be useful to assist field engineers who are working for customers.

In addition, modeling of micro-programs, revision of the system for supporting problem solvings in actual domains, extensions of the system such as for linking to database systems are topics for future study.

ACKNOWLEDGMENTS

The works shown in this paper are supported in part by Science Research Foundation of Ministry of Education, IBM and Tokyo Denki University Research Center. The authors would like to thank members of Ueno group of Tokyo Denki University for system implementations and discussions.

REFERENCES

Book

1. IBM, IBM: Personal System/2 Model1130 Technical Reference, IBM, 1987

Paper in a journal

2. Davis,R., Expert Systems: Where Are We and Where Do We Go From Here, The AI Magazine,3(2),1982
3. Oomori,Y. and Ueno,H., Hybrid Trouble Shooting System Based on Object Model - Object Modelling and Model-Based Reasoning -, J. of JSAI, vol.5, No.5, pp.604-616, 1990(in Japanese)

Chapter in a book

4. Minsky,M., A Framework for Representing Knowledge, in The Psychology of Computer Vision (P.H.Winston, ed), McGraw- Hill, 1975
5. Ueno,H, Object Model for a Deep Knowledge System, Lecture Notes in Echonomics and Mathematical Systems 286(Ed. by Y.Sawaragi, K.Inoue and H.Nakayama), pp.11-19, Springer-Verlag, 1986
6. Ito,H. and Ueno,H., ZERO: Frame + Prolog, Lecture Notes in Computer Science 221(Ed. by E.Wada), pp.78-89, Springer-Verlag,1986

Paper in Conference Proceedings

7. Feigenbaum,E., The Art of Artificial Intelligence- Themes and Case Studies of Knowledge Engineering, Proc. NCC78, 1978
8. Katou,H., Oomori,Y. and Ueno,H., Trouble Shooting Expert System of a Petroleum Refinery Based on Object Model - Model-Based Reasoning by Using the Knowledge about Behavior -, Proc. JSAI'89, pp.239-242, 1989(in Japanese)
9. Ueno,H., Expert Systems Based on Object Model, Proc.ISME-AI'90, pp.29-36
10. Ueno.H and Oomori,Y, Expert Systems Based on Object Model, Proc.ISAI90 Nagoya, pp.25-32,1990

News letter

11. Hart,P.E.,, Direction for AI in the Eighties, SIGART, 79, 1982

Empirical Results of a Hybrid Monte Carlo Method for the Solution of Poisson's Equation

R.E. Hiromoto, R.G. Brickner

Computing and Communications Division, Los Alamos National Laboratory, Los Alamos, New Mexico, U.S.A.

INTRODUCTION

The application of Monte Carlo techniques have considerable importance in solving computational problems that exhibit complex physical interactions and irregular spatial geometries [1,2]. Yet equally important is the large computational parallelism that is inherent in these methods. With the current interest and emphasis on massively parallel computer systems, Monte Carlo methods would seem to be a likely candidate for parallel implementation. However parallelism in itself is only one metric for parallel computational performance. Measures such as load-balancing, data locality, communication latency, and numerical accuracy are equally important metrics for the evaluation of a parallel numerical algorithm.

In this paper, a highly parallel Monte Carlo technique is modified and used in estimating the solution to Poisson's equation. This simple modification incorporates a more efficient use of boundary data and appears to yield a better estimate to the exact solution. As a by product, the new approach is a more efficient sequential algorithm but at the expense of sacrificing parallelism.

DESCRIPTION

We consider the solution of a two-dimensional Poisson equation with Dirichlet boundary conditions given by:

$$\begin{aligned} \nabla^2 \phi(x,y) &= -\rho(x,y), \\ \phi(x_0,y_0) &= f(x_0,y_0). \end{aligned} \tag{1}$$

Only problems with the Dirichlet boundary conditions are discussed here, but the method may be extended to other boundary conditions [3]. Using the five-point finite difference stencil, the discretized form of Equation (1) is given by

$$\phi_{ij} = \frac{1}{4} (\phi_{i-1j} + \phi_{ij-1} + \phi_{i+1j} + \phi_{ij+1} + \frac{\rho_{ij}}{h^2}). \quad (2)$$

Here, the spacing in x and y are both equal to some small grid spacing h.

For Eqn. (2) and correspondingly higher order differencing schemes, the potential at location (i, j) is determined from the average value of the potentials at the appropriate neighboring points. Since the coefficients of the right hand ϕs sum to 1, a probabilistic formulation would prescribe an estimate of ϕ_{ij} to be made by randomly sampling points within the solution domain with a given probability (in this case a uniform probability of 1/4 in each of four directions) to move from one point to any other neighboring point. There are several Monte Carlo techniques discussed in the literature [4,5,6,7,8] that may be used to solve Poisson's equation. We choose to examine a standard Monte Carlo technique [1] that requires sampling the solution domain by means of discrete random walks that score the value of the boundary where the walks terminate. For non-zero ρ, one also scores the value of $\frac{\rho_{ij}}{4h^2}$ at every interior point (i, j) reached during the walk. This entails what is known as a "primary estimate" of the value of ϕ at point (i, j). For N primary estimates, the arithmetic mean of the N primary estimates gives the final estimate.

In this standard method, estimates for the solution of Poisson's equation at the point (i, j) requires a "particle" or "random walks" to be initialized with its position being equal to the cell indices (i, j) and two tally bins. One bin scores $\frac{\rho_{kl}}{4h^2}$ at each interior point (k, l) of the walk, the second scores the value of $\phi(x_0, y_0)$ when the particle reaches a boundary. The interior tally is initialized with ρ at (i, j); the boundary tally is set to zero. As in other Monte Carlo simulations, the use of a pseudo-random number generator determines which of four possible directions the particle is to take for its "step." If the step takes the particle to an interior point, the value for the charge density ρ is tallied at this point. The particle then continues its random walk from this point. If the new cell is a boundary point, the value of ϕ on the boundary is scored in the boundary tally and the particle history terminates. The tallies are multiplied by appropriate factors, giving an estimate for ϕ at this point. To achieve sufficient accuracy to the solution, a large number N of particles are issued from point (i, j).

The expression for ϕ_{ij} is given as follows:

$$\phi_{i,j} = \sum_{n=1}^{N} \phi_{I(n),J(n)} + \frac{1}{4h^2} \sum_{n=1}^{N} \sum_{m=1}^{M(n)} \rho_{I'(n,m),J'(n,m)} \quad (3)$$

where N is the number of random walks started from (i, j), $M(n)$ is the number of cells the N'th walk comprises, $(I(n), J(n))$ are the cell coordinates of the boundary on which the N'th walk terminates, and $(I'(n, m), J'(n, m))$ are the cell coordinates of the M'th cell through which the N'th walk passes.

The standard Monte Carlo (MC) method may be regarded as a Jacobi-like scheme where the estimate for each interior grid point is determined without reference to estimates obtained at other points. The standard MC method, therefore, exhibits a high degree of parallelism, and appears as a rather good candidate for implementation on a massively parallel system. Yet as a Jacobi-like scheme, some inefficiencies exist in converging the solution to some acceptable accuracy. Standard iterative schemes such as Gauss-Seidel mix both old and new iterate values while other schemes in addition use gradient information to significantly reduce the number of iterations required for convergence over the pure Jacobi scheme. Without major modifications to the standard MC scheme described above, a Gauss-Seidel-like scheme is developed with the hope of minimizing both the number of random walks required to reach the boundary and the total number of particles required to obtain a comparable or better converged solution.

The new technique "propagates" the boundary information into the interior of the solution domain. The procedure consists of initiating random walks from only those grid points adjacent to the boundary (see Fig. 1). When all random walks have terminated, the original boundary is logically replaced by the boundary formed by those points where the solution has just been estimated (i.e., the adjacent points). The procedure continues in this manner where the boundary, incrementally replaced by the solution estimates at the corresponding adjacent points, is in effect "propagated" into the center of the solution domain. We refer to this as a Monte Carlo boundary propagation (MCBP) method. Figure 1 shows the "propagation" of the boundary for a few "iterations." Because each step of the boundary propagation depends on the result of the previous boundary calculation, it is also clear that MCBP exhibits less computational parallelism as compared to the standard MC technique described above.

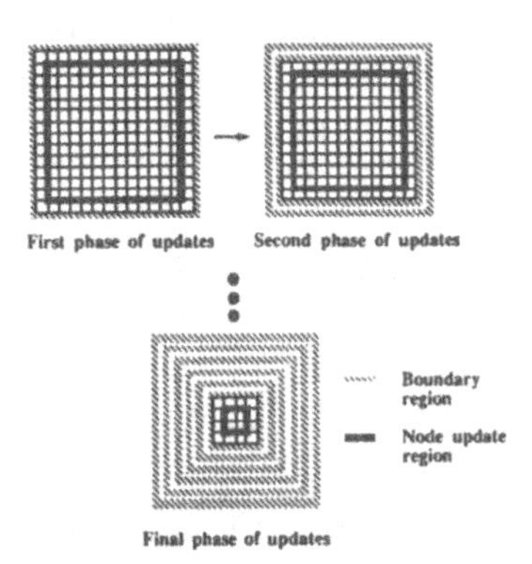

Fig. 1. Phases of the boundary propagation method.

RATE OF CONVERGENCE

The average number of steps per random walk per mesh point (ω) required to exit a 2-dimensional (2-D) solution domain of size $L \times L$ is determined empirically for both MCBP and the standard MC method. The values of mesh length L range from four (4) to 256. From this data, a least square fit is applied to ω as a function of L.

Following the argument made by R.P. Feynman [9], it can be shown that the average number of steps required by the standard MC method to exit the domain is of $O(L^2)$. A least squares fit of ω applied to αL^2 gives:

$$\omega_{MC} = 0.14\, L^2. \tag{4}$$

For a sample size of $N = 50k$ random walks per mesh point, Table I. lists both the measured (ω) and computed (ω_{MC}) values for the average number of steps required to reach the mesh boundary.

TABLE I: Standard MC						
L	4	8	16	32	64	128
ω	2.00	8.19	33.98	139.46	566.30	2282.76
ω_{MC}	2.23	8.91	35.65	142.60	570.38	2281.52

For the boundary method, the functional form of ω is observed to be directly proportional to the length of the mesh size L. Using this fact, a least squares fit of ω applied to αL gives:

$$\omega_{MCBP} = 0.37\, L. \tag{5}$$

From Eqn. 5, Table II. lists both the measured (ω)and computed (ω_{MCBP}) values for the average number of steps required to reach the mesh boundary.

TABLE II: MCBP							
L	4	8	16	32	64	128	256
ω	1.44	3.05	6.12	12.14	24.08	47.97	95.63
ω_{MCBP}	1.50	2.99	5.98	11.97	23.94	47.87	95.74

RESULTS

For the test problem, the charge density on the right hand side of Poisson's equation is chosen to be

$$\rho(x,y) = \sin(x)\sin(y),$$
$$\phi(x_0,y_0) = 0. \tag{6}$$

The convergence for the boundary propagation method is studied by running the test problem with different number of samples (or particles) per grid point. The estimated value of the potential is compared against the exact solution. The test problems are solved with mesh sizes of 32×32 and 64×64 (where the boundaries are included in the mesh size). Table III. shows the execution times for the standard MC and the MCBP method on a Sun 3 workstation. The times are measured for an example using 100 random walks initiated from each mesh point.

TABLE III: Execution times for MC vs. MCBP.

L	MC time (s)	MCBP time (s)	Ratio Mc to MCBP
32	3,981.0	349.6	11.39
64	67,675.6	2,967.8	22.80

A least squares fit of the function $\alpha/\sqrt{N}$, is applied to the maximum error measured for the various grid sizes. Figure 2 shows the correspondence (for $\alpha = 0.82$) between the measured data and the fitted curve. Figures 3 and 4 show the difference in maximum error between the two methods for the same number of random walks, and the norm of the errors for the respective mesh sizes.

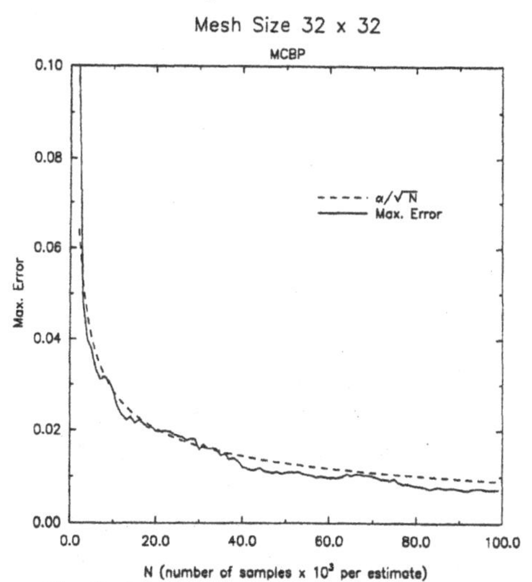

Fig. 2. Maximum error as a function of N.

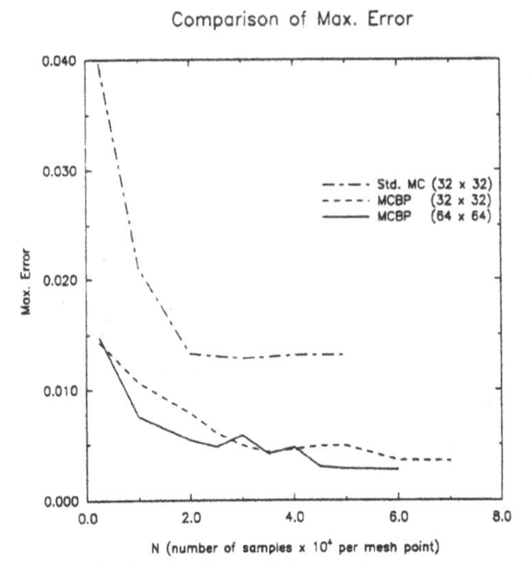

Fig. 3. Comparison of the maximum errors.

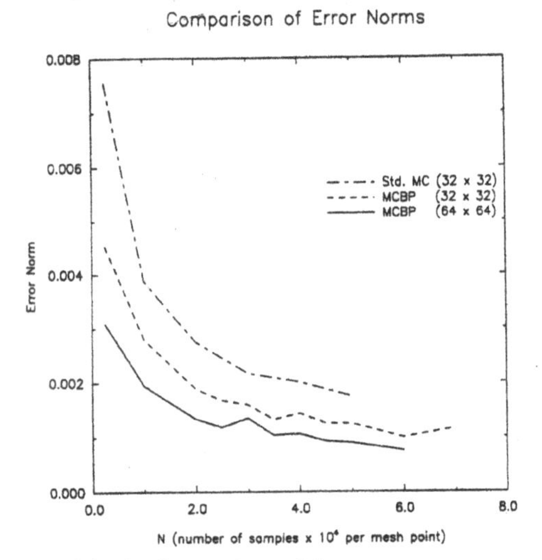

Fig. 4. Comparison of the error norms.

Noting that the longest duration of random walks for the MCBP method corresponds to the shortest duration of random walks for the standard MC method (that is, for those walks that start at the grid points adjacent to the original boundary), the boundary method is significantly more efficient than the more parallel (Jacobi-like) method. In fact, in this study the MCBP method is found to be of O(L) times faster. Furthermore for the same number of samples per grid point, the boundary propagation method achieved better estimates of the solution for the example problem studied.

REFERENCES

1. Hammersley, J.M. and D.C. Handscomb, Monte Carlo Methods, Methuen, London (1964).

2. Buslenko, N.P., et al., The Monte Carlo Method: The Method of Statistical Trails, edited by Yu.A. Shreider, Pergamon Press, New York, (1966).

3. Bevensee, R.M. Applications of Probabilistic Potential Theory (PPT) to the Solutions of Electromagnetic Problems, Preprint UCRL-82760, Lawrence Livermore Laboratory, June 1, 1979.

4. Haji-Sheikh, A. Application of Monte Carlo Methods to Thermal Conduction Problems, Ph.D. thesis, University Microfilms, Inc., Ann Arbor, Michigan (1965).

5. DeLaurentis, J.M. and L.A. Romero, A Monte Carlo Method for Poisson's Equation, Journal of Computational Physics, Vol. 90, No. 1, pp 123-140 (Sept. 1990).

6. Booth, T.E. Exact Monte Carlo Solution of Elliptic Partial Differential Equations, Journal of Computational Physics, Vol. 39, No. 2, pp 396-404 (Feb. 1981).

7. Booth, T.E. Regional Monte Carlo Solution of Elliptic Partial Differential Equations, Journal of Computational Physics, Vol. 47, No. 2, pp 281-290 (Aug. 1982).

8. Mascagni, M. A Monte Carlo Method Based on Wiener Integration for Solving Elliptic Boundary Value Problems, submitted to the SIAM Journal on Scientific and Statistical Computing.

9. Feynman, R.P. Lecture on Physics, Vol.I, Section 41-4, Addison-Westley, Reading, MA., (1963).

SECTION 4: APPLICATIONS IN FLUID FLOW

Implementation and Application of the Multiblock NSFLEX Navier-Stokes Solver

K.M. Wanie, M.A. Schmatz

Messerschmitt-Bölkow-Blohm GmbH, FE211, Postfach 80 11 60, D-8000 München 80, Germany

ABSTRACT

The need of design aerodynamics to calculate viscous flow past complex configurations like complete aircraft with wings, fins, flaps and external stores represents one of the major challenges for computational fluid dynamics. If a structured grid topology is to be applied, which has proven to be very advantageous for viscous calculations at least near solid surfaces, a block decomposition of the space around the vehicle can simplify grid generation considerably. Furthermore storage requirements can be reduced by such a decomposition. For this purpose the present paper focuses on the implementation of a multiblock technique into the Navier–Stokes code NSFLEX. The governing equations and the solution method are shortly reviewed. The fundamental details of the multiblock incorporation like indirect addressing and use of boundary indicators are described thoroughly. The results achieved for hypersonic flow past several configurations demonstrate the capabilities of the approach.

INTRODUCTION

The design of new aerodynamic shapes requires a detailed investigation of the flow properties past entire configurations. With the advent of modern high speed computers computational fluid dynamics (CFD) is more and more applicable for these investigations and has become an important tool of design aerodynamics. It is commonly used today to support the experimental work.

Advantages of CFD as compared to experiments are lower cost and lower turn–around times. This holds especially for the design of hypersonic cruise aircraft or re-entry vehicles. Due to the high enthalpies of the free–stream flows windtunnel experiments are very expensive in the hypersonic flow region. A tremendous technical effort is necessary to simulate effects like radiation, catalytic walls or chemical reactions correctly. CFD is in principle suited to overcome these problems, provided a suitable mathematical modelling for the particular effects is available. The development of these models can, of course, be a very difficult task.

In order to get valuable contributions to the design process it will always be necessary to compute the flow past entire complex configurations including wings, fins, flaps and external stores. For viscous calculations it is furthermore mandatory to use a very good spatial resolution in the boundary–layer region and it has proven to be advantageous to use a structured grid at least in this part of the flow field. Bearing these constraints in mind it is obvious that the generation of a single block grid can be very difficult or even impracticable.

The above consideration lead to a multiblock decomposition technique for the discretisation of the flowfield in a natural way. Advantages of such a multiblock technique are considerable simplifications of the grid generation procedure and a reduction of the storage requirements, which is very important especially with regard to large three–dimensional problems.

The present paper focuses on the implementation of a multiblock technique exploiting these advantages into the Navier–Stokes code NSFLEX. The single-block version of NSFLEX was widely used and verified in the last years for a large number of different applications ranging from low Mach number subsonic flow past cars to very high Mach number hypersonic flow past re-entry vehicles [1,2]. The code is able to solve both the full or thin–layer Navier-Stokes and the inviscid Euler equations.

The overall goal of the present work is to end up with a numerical method suitable especially for engineering applications. For this purpose the code must be flexible enough to allow changes in flow conditions, geometry, grid topology or boundary conditions without a large amount of additional programming. Furthermore storage requirements have to be minimized in order to allow either for a fine spatial resolution, if desired, or to keep cost for the calculations small. As a last point it is desired to incorporate cases like two–dimensional or axisymmetric flow into the three–dimensional code. In this way the effort for the program maintenance can be kept as small as possible. The multiblock technique incorporated fulfills all these requirements due to a very flexible treatment of the block boundaries, a one–dimensional storage technique and generalized approach for 3D–, 2D– and quasi–2D–flows.

In the following the governing equations and some details of the fundamental computational method are given. After this details of the multiblock implementation are discussed thoroughly. Features like indirect addressing and treatment of block boundary conditions by boundary indicators are described. In the remainder of the paper some representative applications are given.

GOVERNING EQUATIONS AND SOLUTION PROCEDURE

Navier–Stokes equations

The general governing equations for the present investigations are the time dependent Reynolds-averaged compressible Navier-Stokes equations in conservation law form. In terms of body-fitted arbitrary coordinates ξ, η, ζ using Cartesian velocity components u, v, w they read

$$\frac{\partial}{\partial t}\vec{U} + \frac{\partial}{\partial \xi}\vec{E} + \frac{\partial}{\partial \eta}\vec{F} + \frac{\partial}{\partial \zeta}\vec{G} = 0 \tag{1}$$

where

$$\vec{U} = (\rho, \rho u, \rho v, \rho w, e)^T \tag{2}$$

is the solution vector of the conservative variables and

$$\vec{E} = J\left(\tilde{\vec{E}}\xi_x + \tilde{\vec{F}}\xi_y + \tilde{\vec{G}}\xi_z\right), \tag{3}$$

$$\vec{F} = J\left(\tilde{\vec{E}}\eta_x + \tilde{\vec{F}}\eta_y + \tilde{\vec{G}}\eta_z\right), \tag{4}$$

$$\vec{G} = J\left(\tilde{\vec{E}}\zeta_x + \tilde{\vec{F}}\zeta_y + \tilde{\vec{G}}\zeta_z\right) \tag{5}$$

are the flux vectors normal to the $\xi = const., \eta = const., \zeta = const.$ faces. The Cartesian fluxes therein are

$$\tilde{\vec{E}} = \begin{pmatrix} \rho u \\ \rho u^2 - \sigma_{xx} \\ \rho uv - \sigma_{xy} \\ \rho uw - \sigma_{xz} \\ (e - \sigma_{xx})u - \sigma_{xy}v - \sigma_{xz}w + q_x \end{pmatrix}, \tag{6}$$

$$\tilde{\vec{F}} = \begin{pmatrix} \rho v \\ \rho uv - \sigma_{xy} \\ \rho v^2 - \sigma_{yy} \\ \rho vw - \sigma_{yz} \\ -\sigma_{xy}u + (e - \sigma_{yy})v - \sigma_{yz}w + q_y \end{pmatrix}, \tag{7}$$

$$\tilde{\vec{G}} = \begin{pmatrix} \rho w \\ \rho uw - \sigma_{xz} \\ \rho vw - \sigma_{zy} \\ \rho w^2 - \sigma_{zz} \\ -\sigma_{xz}u - \sigma_{yz}v + (e - \sigma_{zz})w + q_z \end{pmatrix}. \tag{8}$$

Accepting Stokes' hypothesis for Newtonian fluids, $3\lambda + 2\mu = 0$, the stress tensor reads

$$\sigma_{xx} = -p - \frac{2}{3}\mu\left(-2\frac{\partial u}{\partial x} + \frac{\partial v}{\partial y} + \frac{\partial w}{\partial z}\right), \tag{9}$$

$$\sigma_{xy} = \sigma_{yx} = \mu\left(\frac{\partial u}{\partial y} + \frac{\partial v}{\partial x}\right), \tag{10}$$

$$\sigma_{xz} = \sigma_{zx} = \mu\left(\frac{\partial u}{\partial z} + \frac{\partial w}{\partial x}\right), \tag{11}$$

$$\sigma_{yy} = -p - \frac{2}{3}\mu\left(\frac{\partial u}{\partial x} - 2\frac{\partial v}{\partial y} + \frac{\partial w}{\partial z}\right), \tag{12}$$

$$\sigma_{yz} = \sigma_{zy} = \mu\left(\frac{\partial v}{\partial z} + \frac{\partial w}{\partial y}\right), \tag{13}$$

$$\sigma_{zz} = -p - \frac{2}{3}\mu\left(\frac{\partial u}{\partial x} + \frac{\partial v}{\partial y} - 2\frac{\partial w}{\partial z}\right). \tag{14}$$

For the heat flux vector Fourier's law

$$q_x = -k\frac{\partial T}{\partial x}, \quad q_y = -k\frac{\partial T}{\partial y}, \quad q_z = -k\frac{\partial T}{\partial z} \tag{15}$$

is valid.

The letters ρ, p, T, μ, k denote density, pressure, temperature, viscosity and heat conductivity. e is the total inner energy (inner energy + kinetic energy) per unit volume. The indices $()_\xi, ()_\eta, ()_\zeta$ denote partial derivatives with respect to ξ, η, ζ except for the stress tensor σ and the heat flux vector $\vec{q}$. The definition of the metric can be found for example in [1].

For turbulent flows effective transport coefficients are introduced with the BOUSSINESQ approximation. The equations are closed with a turbulence model for the turbulent viscosity. For the present investigations the algebraic model [3] was used.

Numerical method

The Navier–Stokes method used for the solution of the above equations is called NSFLEX (Navier–Stokes solver using characteristic flux extrapolation) [4,2]. NSFLEX is written in finite volume formulation. To reach the steady state solution asymtotically the equations are solved in time–dependent form. For the time integration an implicit relaxation procedure for the unfactored equations is employed which allows large time steps [4]. In the following only a coarse survey of the method will be given. A detailed description can be found for example in [1,2].

Time integration Starting point for the time integration is the first order in time discretized implicit form of Eq. 1,

$$\frac{\vec{U}^{n+1} - \vec{U}^{n}}{\Delta t} + \vec{E}_{\xi}^{n+1} + \vec{F}_{\eta}^{n+1} + \vec{G}_{\zeta}^{n+1} = 0. \tag{16}$$

A Newton method can be constructed for $\vec{U}^{n+1}$ by linearizing the fluxes of Eq. 16 about the known time level n,

$$\vec{E}^{n+1} = \vec{E}^{n} + A^{n}\Delta\vec{U}, \quad \vec{F}^{n+1} = \vec{F}^{n} + B^{n}\Delta\vec{U}, \quad \vec{G}^{n+1} = \vec{G}^{n} + C^{n}\Delta\vec{U}, \tag{17}$$

leading to

$$\frac{\Delta\vec{U}}{\Delta t} + A_{\xi}^{n} + B_{\eta}^{n} + C_{\zeta}^{n} = -(\vec{E}_{\xi} + \vec{F}_{\eta} + \vec{G}_{\zeta})^{n} = RHS. \tag{18}$$

Therein matrices A, B, C are the Jacobians of the flux vectors $\vec{E}, \vec{F}, \vec{G}$,

$$A = \frac{\partial\vec{E}}{\partial\vec{U}}, \quad B = \frac{\partial\vec{F}}{\partial\vec{U}}, \quad C = \frac{\partial\vec{G}}{\partial\vec{U}}. \tag{19}$$

$\Delta\vec{U}$ is the time variation of the solution vector and the update is

$$\vec{U}^{n+1} = \vec{U}^{n} + \Delta\vec{U}. \tag{20}$$

The system of equations resulting from the discretisation of Eq. 18 is solved approximately at every time step by a point Gauss–Seidel relaxation, see [1]. The time step is calculated with the maximum of the eigenvalues of the inviscid Jacobians. The CFL number is typically about 150 to 200 for moderate Mach numbers and may reduce to lower ones for hypersonic applications. Typically, three Gauss–Seidel steps are performed at every time step. The point–Gauss–Seidel technique is fully vectorized. The speed–up factor on a Siemens/Fujitsu vector computer is about 25 compared to a scalar run on the same processor.

Flux calculation To evaluate the inviscid fluxes $\vec{E}, \vec{F}, \vec{G}$ a linear locally one–dimensional Riemann problem is solved at each finite–volume face up to third order accurate in the computational space. A hybrid local characteristic (LC) and Steger–Warming (SW) type scheme is employed, which allows the code to work for a wide range of Mach numbers [5]. Van Albada type sensors are used to detect discontinuities in the solution.

The flux in ξ–direction at cell face $i + 1/2$ is found by

$$\vec{E}_{i+1/2} = (\vec{E}_{LC}(1 - a) + a \cdot \vec{E}_{SW})_{i+1/2} \tag{21}$$

with $a = SW \cdot s_{\rho} \cdot (M_{r} - M_{l})^{2}$. s_{ρ} is the sensing function for the density, M_{r} and M_{l} are the Mach numbers on the left– and on the right–hand side of the volume face. SW is an input constant to be specified greater or equal to zero. The fluxes in the η– and ζ–directions are calculated in a similar way. To calculate the local characteristic fluxes, the conservative variables on either side of the volume face are extrapolated up to third order in space (MUSCL type extrapolation). This scheme guarantees the homogeneous property of the Euler fluxes, a property which simplifies the evaluation of the true Jacobians of the fluxes for use on the left-hand side [2,6].

Because this local characteristic flux is not diffusive enough to guarantee stability for hypersonic flow cases especially in the transient phase, where shocks move, at regions of high gradients a hyper-diffusive flux is used locally. This is a modified Steger-Warming type (flux vector splitting) flux [7] which gives good conservation of the total enthalpy.

Diffusive fluxes at the cell faces are calculated with central differences [4].

Boundary conditions At the farfield boundaries non-reflecting boundary conditions are inherent in the code since the code extracts only such information from the boundary which is allowed by the characteristic theory. At outflow boundaries, as long as the flow is supersonic, the code does not need information from downstream. In the part of the viscous regime where the flow is subsonic the solution vector is extrapolated constantly. No upstream effect of this extrapolation could be observed up to now.

At solid bodies the no-slip condition $u = v = w = 0$ holds. Several temperature and heat flux boundary conditions are incorporated. It is possible to prescribe adiabatic wall ($q_w = 0$), given wall heat flux q_w or given wall temperature T_w. Radiation of solid bodies can be taken into account as well. In this case the heat-flux vector at the wall is calculated with

$$q = -k\frac{\partial T}{\partial n} + \epsilon\sigma T_w^4, \tag{22}$$

where n is the wall normal direction. The second term in Eq. 22 is the radiation term. σ is the Stephan-Boltzmann constant with $\sigma = 5.67 \cdot 10^{-8} W/(m^2K^4)$. ϵ is the emissivity factor of the surface material ($0 \leq \epsilon \leq 1$). T_w is the temperature at the wall.

Equilibrium real gas incorporation A simple approach to account for equilibrium real gas effects is incorporated which allows the Riemann solver and the left–hand side of the flow solver to stay unchanged.

In the inviscid fluxes the ratio of the specific heats $\gamma_r = \gamma_r(p, \rho)$ appears only in the energy equation. Following EBERLE [5] a new total energy e is defined with a reference ratio of specific heats γ which is the freestream γ. Introducing this definition into the energy equation a new source term is obtained. If only steady state solutions are of interest the time derivative of the source term can be set identically to zero. This means that the left-hand side of the energy equation is the perfect gas equation and the real gas influence can be formulated as a source term on the right-hand side. In the code only a few lines are necessary to calculate the source terms. Note, however, that this approach is restricted to steady flow simulations.

To account for equilibrium real gas effects in the viscous fluxes additional thermodynamic subroutines for the temperature $T = T(p, \rho)$ and the transport coefficients $\mu = \mu(p, \rho)$, $k = k(p, \rho)$ are necessary.

Fully vectorized curve fits [8] are used for the calculation of the thermodynamic and transport properties. For this reason the computer time for real gas simulations is only about 20 % higher than for perfect gas runs.

IMPLEMENTATION OF THE MULTIBLOCK TECHNIQUE

Block structure and boundary conditions

The new mulitblock code is designed to allow the use of arbitrary multiblock meshes. A principle sketch of a possible block decomposition is given in Fig. 1 for a cross section of the HERMES reentry vehicle. The flow field in the cross section is divided into a total number of five blocks for the present example.

All around each block two dummy cell rows are used to define the boundary condition for each cell of the block boundaries separately (Fig. 2). In these cells boundary indicators generated in the grid generation system are stored. It is either possible to specify an adjacent cell in the same or an other block or to prescribe various other boundary conditions by an identification code. Implemented are solid body boundary conditions as described in the preceeding chapter, symmetry conditions, singular line conditions and various inflow and outflow conditions. Note that the use of two dummy cell rows is mandatory in order to maintain third-order accuracy for adjacent blocks also across the block boundaries.

Due to the general treatment of the block boundaries a high amount of flexibility is achieved. Adjacent blocks need not to have the same number of grid points (block 3 and 5, Fig. 1). At each block face an arbitrary number of neighbouring blocks is possible (block 2,3,4). It is possible to specify an arbitrary number of different boundary conditions on each block face (solid wall and overlapping, block 2). The i, j, k–indices in an adjacent block do not need to have the same direction in computational space (block 4 and 5).

The use of three blocks (1,2,3) instead of one larger block can reduce storage requirements considerably. This is due to the fact that most of the three–dimensional arrays used in the flux routines and in the implicit part of the flow solver need not to be stored for the whole flowfield. They are temporary arrays which may be overwritten when calculating the next block.

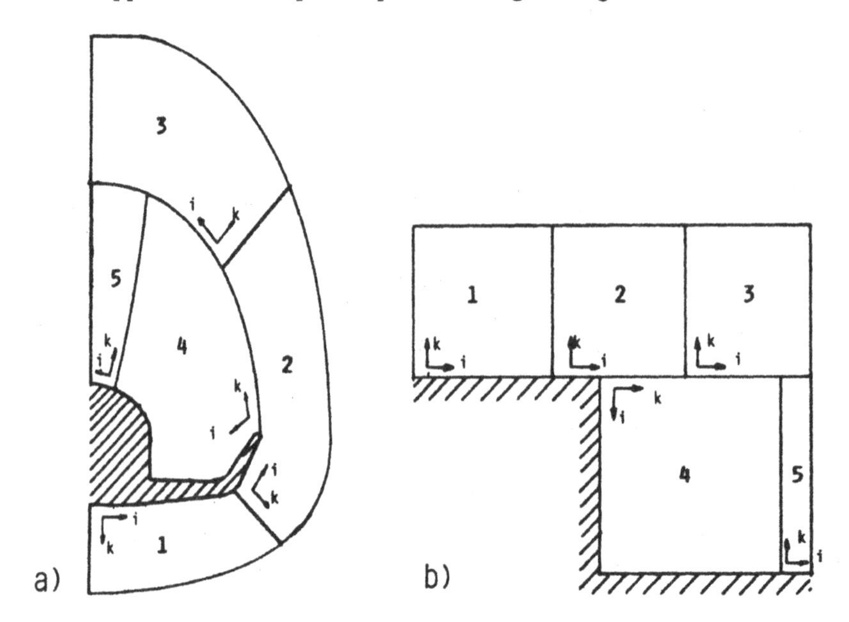

Figure 1: Principle of block decomposition a) physical space, b) computational space

The subroutines for the boundary conditions are completely separated from the flux routines and the implicit part. This simplifies the implementation of new boundary conditions considerably. Furthermore it enhances the use of arbitrarily overlapping meshes or grid refinement by doubling the points in one coordinate direction across a block boundary. In the subroutine for the boundary conditions only an interpolation has to be introduced. The coefficients for this interpolation can be calculated once for all in the grid generation system. Results for an example of grid refinement by doubling the number of points will be given in the next chapter.

As was mentioned before an algebraic turbulence model [3] is incorporated up to now. It is possible to specify for each cell at solid walls if the flow in the cell row above it is laminar or turbulent. Of course, the usual restrictions and limitations of the model are present in the new code as well.

Storage technique

As gets obvious from Fig. 1 block 5 of the present example is very small compared to the other blocks. This situation will almost always be present in one or the

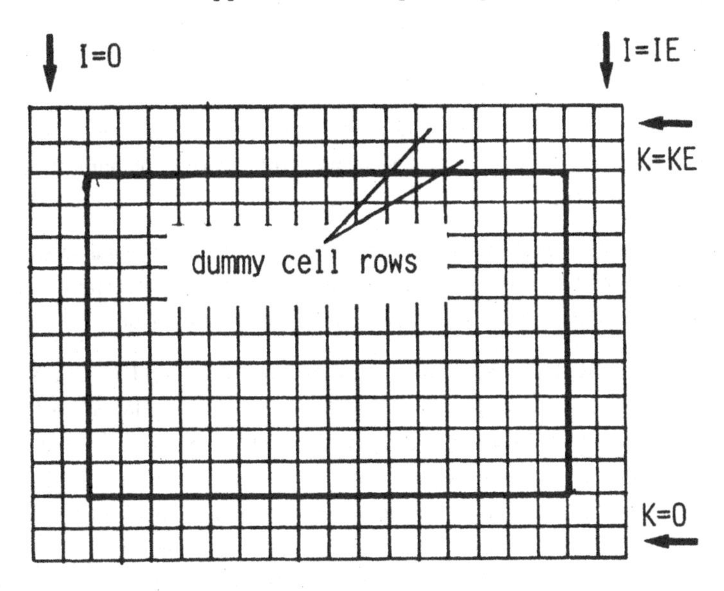

Figure 2: Dummy cell rows at block boundaries

other way if for example complex configurations with features like flaps or similar things are to be calculated. Since for such calculations usually a large number of blocks is required a specification of each block dimension is impracticable. A specification of the dimensions of the largest block together with the number of blocks on the other hand would result in an unjustifiable waste of storage and therefore is impracticable as well.

For this purpose a special solution technique is applied to minimize the storage requirements no matter how many blocks are used. The solution vector of all blocks is sequentially stored in an one–dimensional array. Within the subroutines indirect addressing is used. The actual one-dimensional index NN is calculated by the function statement

$$NN(i,j,k) = IB(MM) + (I+1) + K*(IE+1) + J*(IE+1)*(KE+1) \quad (23)$$

from the i,j,k–indices of the current block. MM is the number of the current block, IE, JE, KE are its dimensions (Fig. 2) and $IB(MM)$ is its starting address.

Advantage of this indirect adressing technique beside an optimum use of storage is that only a few very simple modifications are necessary in the flux routines and in the implicit part of the code when migrating from a singleblock to a multiblock method. The DO–loops of the singleblock routines may stay almost unchanged execpt that the three–dimensional indices have to be replaced by the call for the function statement Eq. 23.

It should be emphasized that the indirect addressing technique used does not hamper vectorization at all on the Siemens/Fujitsu vector computer used for the present calculations.

Generalized version A two-dimensional and a quasi–two–dimensional version of the code are inherent in the three-dimensional one. A special technique is used to achieve this without a large amount of additional programming.

For two–dimensional calculations a three–dimensional grid is established from the two–dimensional one. For this purpose $j = const$ planes are found from the two–dimensional grid by shifting it in y–direction. In this way a total number of three cell planes are generated (Fig. 3). The middle cell plane is used for the calculation, while in the two outer planes boundary conditions are specified. For 2D–flows the variables are simply extrapolated constantly into the boundary cells. For axisymmetric flow the velocity vectors have to be rotated around the axis. Also other quasi–2D–flows like conical flows or infinite swept wing flows can be calculated very easily. For these cases only modifications of the boundary conditions are to be implemented.

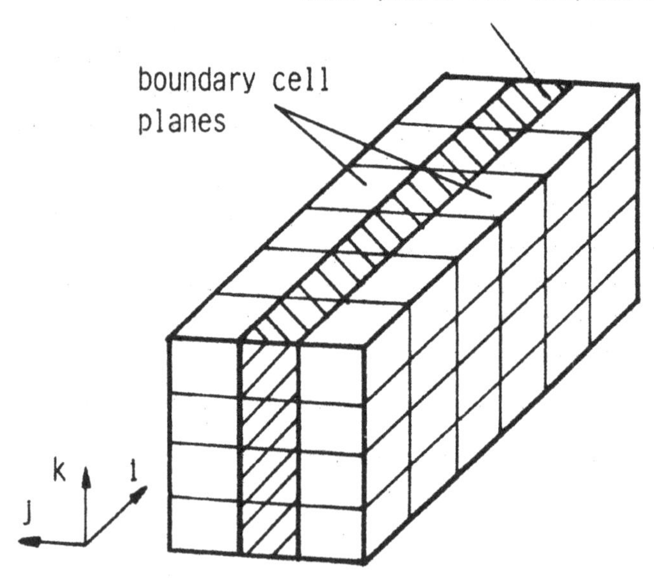

Figure 3: Principle of two–dimensional calculations

On the grid described actually a three-dimensional calculation is performed in only one cell plane. For this reason no changes in the flux routines or in the implicit part are necessary. The fluxes and the implicit terms, of course, are calculated only in the necessary directions. The overhead introduced compared to a two-dimensional code is very small and is more than justified by the increased flexibility and the reduced effort for program maintenance.

APPLICATIONS

The monoblock NSFLEX method was validated in the last years for a great variety of configurations and a large Mach number range. Results can be found for example in [1,4,9,10]. For this purpose the applications presented in the following have been chosen especially to validate the multiblock implementation.

In Fig. 4 an application of the multiblock code in its two-dimensional mode is given. Shown are results for the hypersonic laminar flow past a two-dimensional ramp at $M_\infty = 5.0$, $Re/m = 6 \times 10^6$, $T_\infty = 80K$, $T_{wall} = 288K$. In Fig. 4a the computational grid and the computed Mach contours for a calculation using only one block are given. For the results presented in Fig. 4b the grid was adapted to the solution by the grid embedding technique described in the preceeding chapter. For this purpose the flowfield was divided into a total number of three blocks. In the region of strong interaction the points were doubled in each coordinate direction. The boundary condition between the blocks were set by interpolation as mentioned before. No influence of the block boundaries is discernable from the Mach contours. The solution using three blocks is as smooth as that one on a single block.

Another test case is the laminar flow past a delta wing at $M_\infty = 8.7, \alpha = 30°$, $Re/m = 2.25 \times 10^6$, $T_\infty = 55K$, $T_{wall} = 300K$. For the calculation perfect gas was assumed. The skin friction pattern on the lower side is given in Fig. 5a and Mach contours are shown in Fig. 5b. For the calculation the flow field was divided into a total number of six blocks in order to keep the storage requirements small enough to allow calculations during day time working hours. The decomposition gets obvious from Fig. 5b. In streamwise direction three blocks were introduced, each of them again being split in two blocks normal to the surface. Note that each block contains approximately the same number of grid points. The boundary between the inner and the outer blocks is, however, quite close to the surface, which is due to the fine resolution near the solid wall.

The next example is the laminar perfect gas flow past a double ellipsoid at $M_\infty = 8.15$, $\alpha = 30°$, $Re/m = 1.67 \times 10^5$, $T_\infty = 56K$, $T_{wall} = 288K$. Fig. 6 shows the Mach contours and the skin friction pattern. The solution is perfectly smooth across the block boundaries as can be seen from the Mach contours. Also from the skin friction lines no influence of the block boundaries is discernable.

For the HERMES reentry vehicle a laminar equilibrium real gas calculation was performed for $M_\infty = 20$, $\alpha = 30°$, altitude $h = 75km$, $T_{wall} = 1500K$ and surface radiation ($\epsilon = 0.85$). In Fig. 7 the grid and the isobars are given. The same block topology as for the delta wing was used. Clearly vissible from Fig. 7a is the use of a special nose grid which helps to avoid the singular line of O–O–type grids. Again the solution is smooth across the block boundaries.

Another application calculated recently is the turbulent flow in the supersonic NASA P8 intake. The results can be found in [11], where two– and three–dimensional calculations are reported, the quality being compareable to the solutions presented above.

CONCLUSIONS

The implementation of a multiblock technique into the Navier–Stokes code NS-FLEX was described. Advantages of the multiblock technique are both considerable simplifications of the grid generation procedure for viscous flow past complex configuations and reductions in storage requirements. The latter item can be essential if a very good spatial resolution is required.

Particular emphasis was layed upon the versatility of the multiblock program with regard to modifications in grid topology and boundary conditions. Therefore all around the blocks dummy cell rows are stored containing the i,j,k indices of the adjacent cell or boundary indicators. The boundary indicators allow to specify a variety of boundary conditions separtely in each cell of a block face. Due to this treatment of the block boundaries every grid topology and boundary condition can be used without changes in the Navier–Stokes code. A further important point is use of indirect addressing which allows to store the solution vector in a one–dimensional array. In this way storage requirements are as small as possible even for very different size of the single blocks.

By means of the results presented at the end of the paper the approach is verified. The solutions show an almost perfect smoothness across the block boundaries. The reductions in storage requirements due to the block decomposition enabled the calculation past an entire reentry vehicle. To sum up the present multiblock approach has proven to be a very valuable tool for viscous flow calculations past complex configuration. One of its major benefits with regard to engineering applications is the flexibility in boundary conditions and grid topology.

References

[1] Schmatz, M.A.: *Hypersonic three-dimensional Navier-Stokes calculations for equilibrium air.* AIAA-Paper 89-2183, 1989.

[2] Schmatz, M.A.: *NSFLEX - An implicit relaxation method for the Navier-Stokes equations for a wide range of Mach numbers.* Hackbusch, W., Rannacher, R. (eds): Proc. 5th GAMM-Seminar 'Numerical treatment of the Navier-Stokes equations', NNFM Vol. 30, Vieweg, 1989.

[3] Baldwin, B.S., Lomax, H.: *Thin layer approximation and algebraic model for separated flow.* AIAA-Paper 78-257, 1987.

[4] Schmatz, M.A.: *Three-dimensional viscous flow simulations using an implicit relaxation scheme.* Kordulla, W. (ed.): Numerical simulation of compressible viscous-flow aerodynamics. NNFM Vol. 22, Vieweg, 1988, pp. 226-242.

[5] Eberle, A.: *Characteristic flux averaging approach to the solution of Euler's equations.* VKI lecture series, Computational fluid dynamics, 1987-04, 1987.

[6] Eberle, A., Schmatz, M.A., Schaefer, O.: *High-order solutions of the Euler equations by characteristic flux averaging.* ICAS-Paper 86-1.3.1, 1986.

[7] Steger, J.L., Warming, R.F.: *Flux vector splitting of the inviscid gasdynamic equations with application to finite difference methods.* J. Comp. Phys., Vol. 40, 1981, pp. 263-293.

[8] Mundt, Ch.; Keraus, R.; Fischer, J.: *New, accurate, vectorized approximations of state surfaces for the thermodynamic and transport properties of equilibrium air*, paper accepted for publication in the ZfW, 1989.

[9] Wanie, K.M., Schmatz, M.A.: *Numerical analysis of viscous hypersonic flow past a generic forebody.* ICAS-Paper 90-6.7.2, 1990.

[10] Wanie, K.M., Schmatz, M.A.: *Verification and application of the NSFLEX method for hypersonic flow conditions.* 1st European Symposium Aerothermodynamics, ESTEC, 1991.

[11] Brenneis, A., Wanie, K.M.: *Navier-Stokes results for hypersonic inlet flows.* Paper submitted to AIAA, 1991.

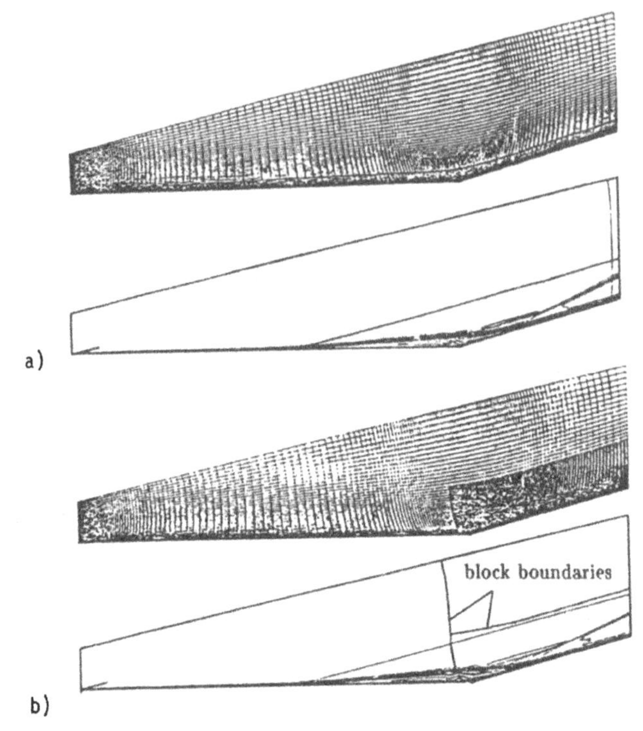

Figure 4: Grid and Mach contours for 2D ramp flow ($M_\infty = 5$, $Re/m = 6 \times 10^6$, $T_\infty = 80K, T_{wall} = 288K$, laminar); a) without adaption; b) with grid embedding

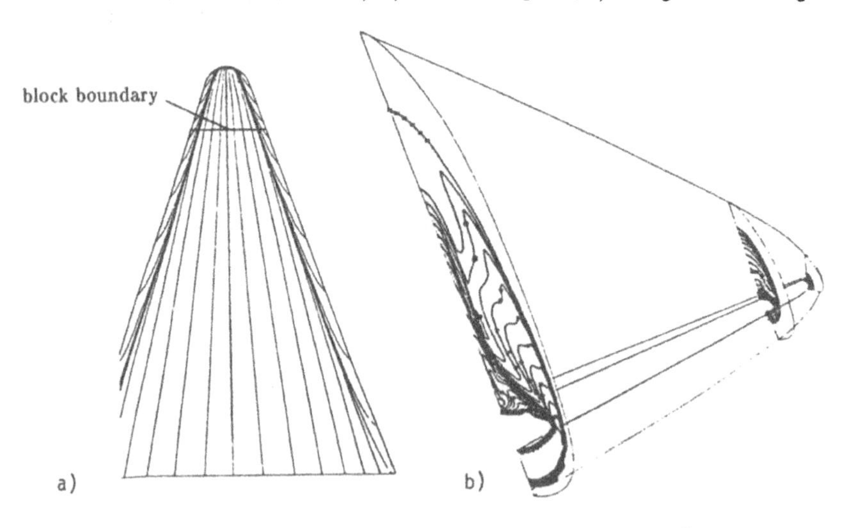

Figure 5: Delta wing flow ($M_\infty = 8.7$, $\alpha = 30°$, $Re/m = 2.25 \times 10^6$, $T_\infty = 55K$, $T_{wall} = 300K$, perfect gas, laminar); a) Skin friction pattern on lower side; b) Mach contours

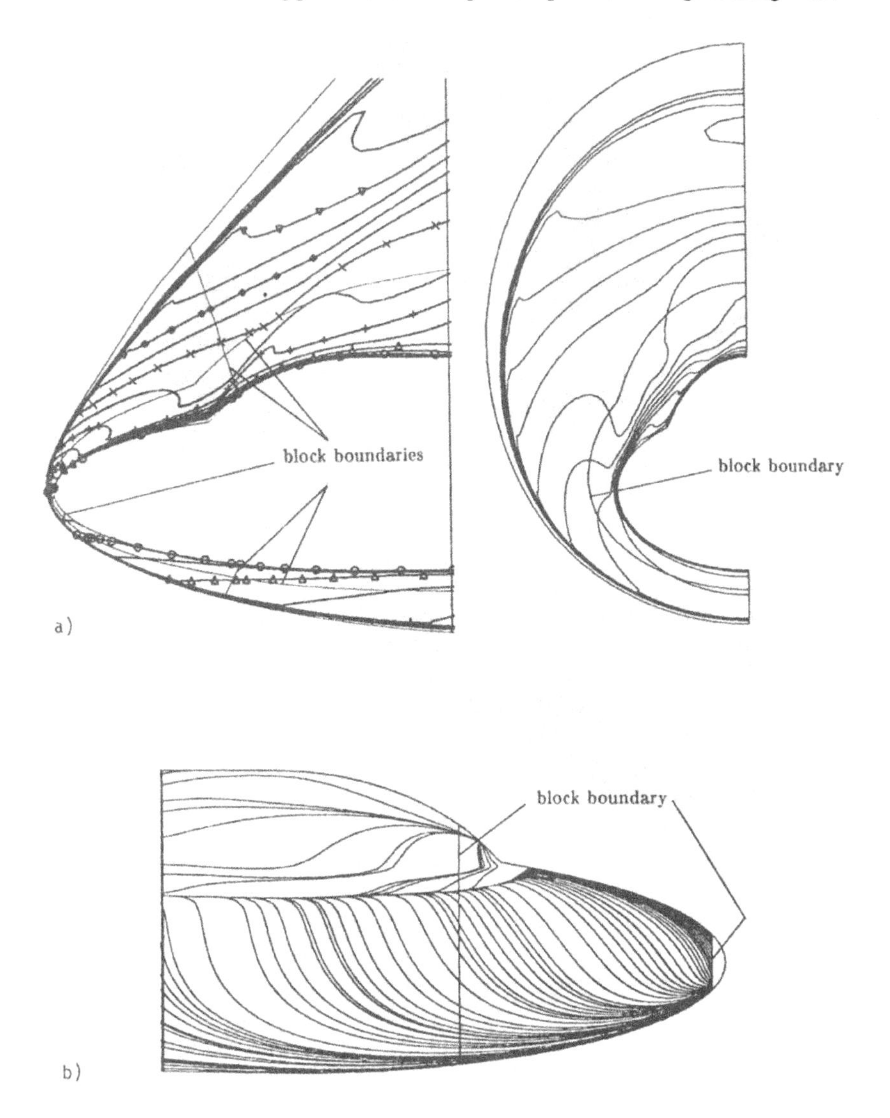

Figure 6: Double ellipsoid flow (M_∞ = 8.15, α = 30° , Re/m = 1.67 × 10^5, T_∞ = 56K, T_{wall} = 288K, perfect gas, laminar); a) Mach contours; b) Skin friction pattern

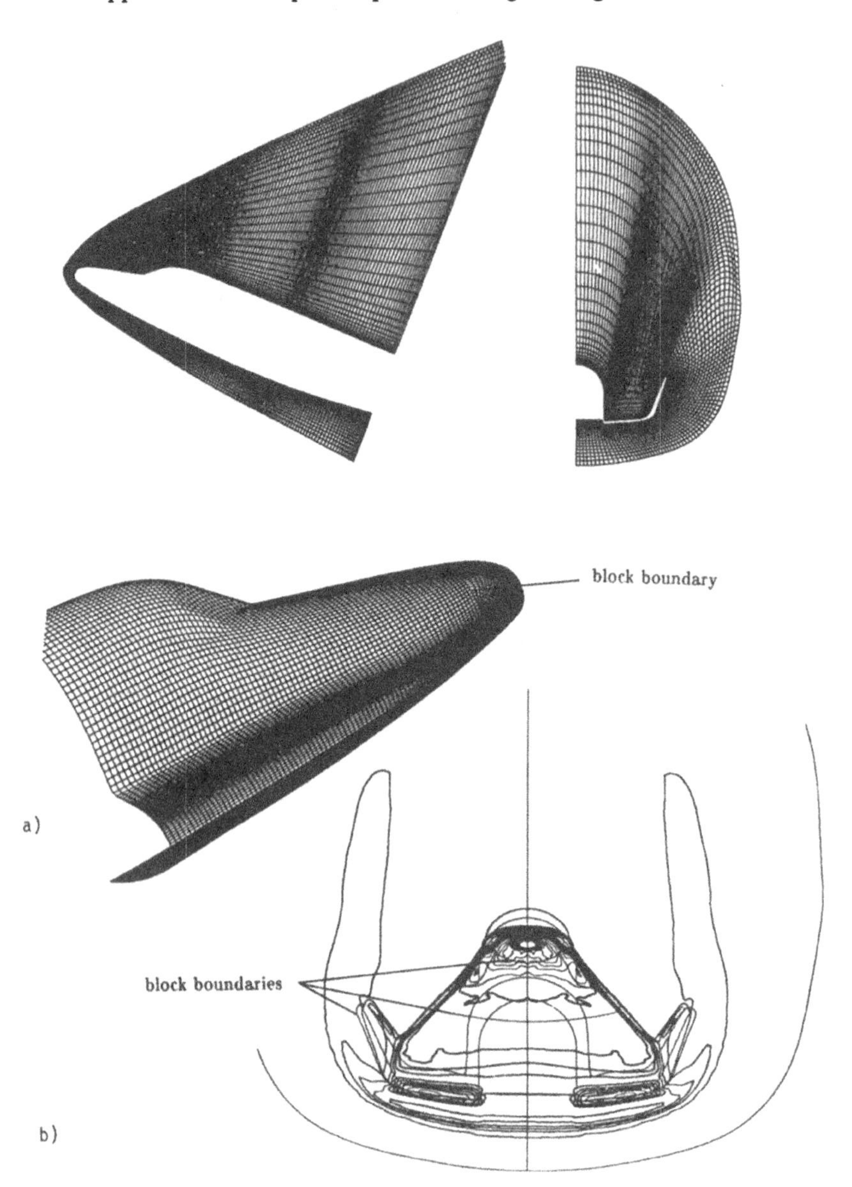

Figure 7: Hermes calculation ($M_\infty = 20$, $\alpha = 30°$, altitude = $75km$, $T_{wall} = 1500K$, equilibrium real gas, surface radiation, laminar); a) Grid; b) Isobars

A Comparison of Differential Systems and Numerical Methods for the Computation of Smooth Oceanographic Flows

G.L. Browning (*), W.R. Holland (*), H.-O. Kreiss (**)
(*) National Center for Atmospheric Research, P.O. Box 3000, Boulder, CO 80307, U.S.A.
(sponsored by the National Science Foundation)
(**) U.C.L.A., Department of Mathematics, Los Angeles, CA 90024, U.S.A.
This work was supported by the Office of Naval Research under contract ONR-N00014-90-F-0057

ABSTRACT

Recently, a new system of equations which can be used to accurately describe approximately hydrostatic and incompressible oceanographic flows has been developed. The "approximate system" is derived by slowing down the speed of the fast waves instead of increasing their speed to infinity as in the primitive equations. While the initial-boundary value problem for the primitive equations is always ill posed, boundary conditions can be chosen for the approximate system so that the initial boundary value problem is well posed. And the new system is not restricted to the midlatitudes as is the quasi-geostrophic system.

In this paper a model based on the proper mathematical limit of the approximate system, the "reduced system," is compared with models based on the primitive and quasi-geostrophic systems for a fixed grid and numerical method. This illuminates some of the advantages of the reduced system compared to the traditional systems. Then a series of reduced models based on different numerical methods are compared in terms of their accuracy and efficiency.

INTRODUCTION

Many of the dynamical systems of geophysical fluid dynamics admit motions which can vary on different time scales. In many situations, the low frequency class contains the majority of energy and is therefore the class of main interest. Computation of this class of oceanographic motions using a model based on the unmodified Eulerian equations is prohibitive due to severe accuracy and time step restrictions (Browning et al [4]). To alleviate these restrictions, a number of modifications of these equations have been proposed. Originally "filtered systems," such as the quasi-geostrophic equations, removed all fast waves by solving the three dimensional elliptic equation satisfied by the Rossby motions at each time step (Charney [6]). Because of a number of shortcomings of the filtered systems, e.g. their restriction to the midlatitudes, the "primitive equations," which only employ the globally accurate hydrostatic and incompressible assumptions, were developed (Bryan and Cox [5]). Recently, the "approximate system," which slows down the speed of the fast waves instead of increasing it as in the primitive equations, has been introduced (Browning et al [4]). Analytically the proper mathematical limit of the approximate system, the "reduced system," has a number of possible advantages over previously proposed systems. In Section 3 these advantages will be demonstrated by comparing the relative abilities of models based on the different systems to recreate an analytic solution which is developed in Section 2. In Section 4 reduced models employing a number of different numerical methods are compared in terms of accuracy and efficiency. The point of this test is to select the best numerical approximation for the development of a global ocean model based on the reduced system.

ANALYTIC SOLUTION

It is generally accepted that the Eulerian equations correctly describe the hyperbolic evolution of the oceanographic flows of interest. However, because of the reasons stated in the previous section, these equations are not used as the basis for current numerical models. Instead, simplifications of the Eulerian equations, e.g. the hydrostatic approximation, are made in order to develop systems with more reasonable numerical approximation requirements. Clearly there are then two possible sources of error in the numerical solution of the simplified system. In order to measure both the continuum and discrete errors made by models based on the different systems discussed in the previous section, an analytic solution of the unmodified Eulerian equations would be the most desirable basis of comparison. However, the only known analytic solutions are time independent and therefore not general enough to ensure all errors will be taken into consideration (Browning et al [3]). Here the alternative method of choosing specified representative functions for all of the dependent variables will be used. The given functions are solutions of the forced Eulerian system

$$\frac{ds}{dt} - \bar{s}' w = F_s , \tag{2.1a}$$

$$\frac{du}{dt} + R^{-1} p_x - f_0 v = F_u , \tag{2.1b}$$

$$\frac{dv}{dt} + R^{-1} p_y + f_0 u = F_v , \tag{2.1c}$$

$$\frac{dw}{dt} + R^{-1} (p_z + gs) = F_w , \tag{2.1d}$$

$$\frac{dp}{dt} + c^2 R (u_x + v_y + w_z) = F_p , \tag{2.1e}$$

where $s = \rho - c^{-2} p$ is essentially the potential density, $\mathbf{V} = (u, v, w)^*$ is the velocity, p is the pressure, $d/dt = u\partial/\partial x + v\partial/\partial y + w\partial/\partial z$ is the total derivative, $\mathcal{S}$ is the stratification parameter, R is the mean density, f_0 is the Coriolis parameter evaluated at $\pi/4$ radians, g is the gravitation, and c is the speed of sound. The forcing function on the right-hand side of a particular equation of (2.1) is determined by substituting the representative solution into the corresponding expression on the left-hand side of that equation. For example, if a prime is used to indicate the representative solution, then $F_s = ds'/dt - \mathcal{S}\,w'$. It is important to realize that the forcing functions play no role in the local error equations and that it is only the derivatives of the representative solution that determine the local error. Also note that several simplifications of the original Eulerian system have been made, e.g. deletion of some Coriolis terms in the u and w equations and the assumption that the remaining Coriolis terms are constant. The simplifications are only for presentation purposes and will not be present in the final model.

The best choice for representative dependent variables appears to be the selection of a stream function and the determination of the remaining dependent variables from bounded derivative constraints. For this study, the selection of the streamfunction will be based on the fact that one of the important smooth flows in the ocean is the mesoscale eddy. Although the stream function can be chosen arbitrarily, there is an equation which gives the Laplacian of the velocity potential in terms of the stream function that is possible to solve analytically if the stream function is representable in the form

$$\psi'(x,y,z,t) = \tag{2.2}$$

$$-u_0 y + \psi_0 \nabla_2^2 [F(x - u_0 t)\, G(y)]\, H(z) = -u_0 y + \psi_0 (F_{xx} G + F G_{yy}) H \ ,$$

where u_0 is the magnitude of the background flow and translation speed of the eddy (assumed constant), ψ_0 is a normalization constant (discussed later), $\nabla_2^2 = \partial^2/\partial x^2 + \partial^2/\partial y^2$, and F, G, and H are known functions of their respective arguments.

To reproduce the essentially circular nature of an eddy, a Gaussian function with an argument which is proportional to the square of the distance from the center of the eddy will be employed to represent the product FG, i.e.

$$F(x)G(y) = e^{-(x/r_e)^2} e^{-(y/r_e)^2} = e^{-|(x,y)|^2/r_e^2} , \tag{2.3}$$

where r_e is the e folding factor. Note that because of (2.2) this means that the stream function itself is not a pure Gaussian, but nevertheless a circular function. For simplicity only, the ocean will be assumed to have a rigid lid with no bottom topography which implies that the vertical velocity w must vanish at the top and bottom of the basin. Since w vanishes at the vertical boundaries, it can be expanded in a Fourier sine series in the vertical coordinate z and the simplest choice is to use only the first term in the series, i.e. to assume the vertical dependence of w is half a sine wave. The solution will be assumed to be incompressible, so the corresponding functional dependence of the stream function must be half a cosine wave, i.e.

$$H(z) = \cos(-\pi\, z/z_b) , \tag{2.4}$$

where $z_b = -3\ km$ is the right-handed coordinate value of the ocean bottom.

The normalization constant ψ_0 is determined so that the maximum value of the perturbation speed on the background flow has the magnitude u_1. The maximum occurs at a distance given by $r_m^2 = (7 - 33^{1/2})/(4\ c_d)$ where $c_d = r_e^{-2}$. Then the normalization factor is

$$\psi_0 = \{u_1 e^{c_d r^2}/[8c_d^2 r(2 - c_d r^2)]\}_{r = r_m} \ . \tag{2.5}$$

The eddies typically have a diameter on the order of $10 - 100\ km$ and can be advected along at approximately $.10\ m\ s^{-1}$ so $r_e = 30\ km$, $u_0 = .10\ m\ s^{-1}$, and a perturbation velocity given by $u_1 = .05\ m\ s^{-1}$ are reasonable values for the stream function parameters.

The remaining dependent variables will be determined from the equations of the linear quasi-geostrophic system. Given the stream function defined by (2.2-2.5), the geostrophic pressure p' can be determined from the geostrophic equation

$$\nabla_2^2 p' = Rf_0\zeta = Rf_0\nabla_2^2\psi' , \tag{2.6}$$

where $R = 10^3\ kg\ m^{-3}$ is the mean density, $f_0 = 2\ \Omega \sin(\pi/4)$ is the constant Coriolis parameter, and $\Omega = 7.292 \times 10^{-5}\ s^{-1}$ is the rotation rate of the Earth. With a constant Coriolis parameter the vorticity equation in the quasi-geostrophic system is

$$\frac{\partial}{\partial t}[\nabla_2^2\psi + f_0^2 (g\bar{s})^{-1}\psi_{zz}] = 0 , \tag{2.7}$$

where $\bar{s}$ is the mean stratification (assumed constant). Under the given assumptions, the only solution of the linear quasi-geostrophic f-plane equations is time independent. In order to have a time dependent solution, all equations of the system will be satisfied with the exception of the vorticity equation which will have a forcing term generated by the time dependency of the stream function. The linear quasi-geostrophic potential density and hydrostatic equations are

$$s'_t - \bar{s}\ w' = 0 , \tag{2.8}$$

$$p'_z + g\ s' = 0 , \tag{2.9}$$

where $g = 9.8\ m\ s^{-1}$ is the gravitational acceleration. Differentiating (2.8) with respect to z, using (2.9) to replace s' by p' and (2.6) to replace p' by ψ', an equation for the horizontal divergence in terms of the stream function,

$$w'_z = -\nabla_2^2\chi' = -Rf_0(g\ \bar{s})^{-1}\psi'_{zzt} = Rf_0(g\ \bar{s})^{-1}u_0\psi'_{xzz} , \tag{2.10}$$

is obtained and since ψ' was selected appropriately this can be solved in the form

$$\chi' = -Rf_0 u_0 \psi_0 (g\,\bar{s})^{-1} F_x G H_{zz} \,. \tag{2.11}$$

Since R, f_0, u_0, ψ_0 and g are known values, the size of the divergence is clearly determined by the size of $\bar{s}$. If $\bar{s}$ is too large (e.g. infinite), then the divergence is unrealistically small and the vertical coupling term in the vorticity equation (2.7) can be neglected. The behavior of the system in this case is exactly that of the barotropic vorticity equations. To ensure full three dimensional flow, the value of $\bar{s}$ was chosen so that the vertical term had the same magnitude as the horizontal term. The value $\bar{s} = 1.4 \times 10^{-4}$ indicates a cancellation between the two terms of $\bar{s}$ of approximately two digits which corresponds to a length scale of 10 *km* (Browning et al [4]). Even though the test eddy has a lateral extent on the order of 100 *km*, it clearly consists of waves on a shorter scale so the above length scale is reasonable.

Given the stream function ψ' and the velocity potential χ', the horizontal velocity (u', v') can be computed from the relations

$$u' = -\psi'_y + \chi'_x \,, \tag{2.12}$$

$$v' = \psi'_x + \chi'_y \,, \tag{2.13}$$

the pressure from the geostrophic equation (2.6), $p' = Rf_0\psi'$, the potential density from the hydrostatic equation

$$s' = -g^{-1}p'_z = -Rf_0 g^{-1}\psi'_z \,, \tag{2.14}$$

and the vertical velocity from (2.8)

$$w' = Rf_0 u_0 (g\bar{s})^{-1} \psi'_{xz} \,. \tag{2.15}$$

Thus known functions for all the dependent variables have been determined. These functions are a solution of the forced nonlinear system (2.1) and can be used to determine both the continuum and discrete errors in numerical approximations of systems based on (2.1).

COMPARISON OF DIFFERENTIAL SYSTEMS

The traditional systems of oceanography were developed to overcome several deficiencies of the Eulerian equations. To clarify a number of points, the non-dimensional version of (2.1) will be used in the remainder of the paper. For mid-latitude oceanographic flows which are geostrophic, hydrostatic, and fully three dimensional the scaled version of (2.1) is (Browning et al [4])

$$\frac{ds}{dt} - \bar{s}w = F_s \,, \tag{3.1a}$$

$$\frac{du}{dt} + \varepsilon^{-1}(p_x - fv) = F_u \,, \tag{3.1b}$$

$$\frac{dv}{dt} + \varepsilon^{-1}(p_y + fu) = F_v \,, \tag{3.1c}$$

$$\frac{dw}{dt} + 10^{4-n}\varepsilon^{-2}(p_z + g\ s) = F_w \,, \tag{3.1d}$$

$$\frac{dp}{dt} + 10^{4+n}\varepsilon^{-1}(u_x + v_y + \varepsilon w_z) = F_p \tag{3.1e}$$

where ε is the Rossby number and the variable n, which is used to ensure that $\bar{s}$ is of order unity, determines the local length scale with $n=0$ ($n=2$) corresponding to $10^5\ m$ ($10^4\ m$). For the representative solution in the previous section, the Rossby number can be approximated by $\varepsilon = u_0/(f_0 L) \approx 10^{-1}/(10^{-4} \cdot 10 \approx 10^{-1}$. From the discussion of the length scale of the representative solution, reasonable values for n for the representative solution would be either $n=1$ ($L = 30\ km$) or $n=2$ ($L = 10km$).

A number of alternative systems to overcome the difficulties mentioned above have been developed. In this section the accuracy of models based on three of these systems, namely the reduced, primitive, and quasi-geostrophic equations, will be compared for a fixed mesh size and second order accurate numerical methods. For all of the numerical tests to follow the flow will be assumed to be periodic in x with rigid walls at the remaining boundaries.

Reduced Equation Model

The reduced system of equations corresponding to (3.1) is

$$\frac{ds}{dt} - \bar{s}w = F_s \,, \tag{3.2a}$$

$$\frac{du}{dt} + \varepsilon^{-1}(p_x - f\ v) = F_u \,, \tag{3.2b}$$

$$\frac{dv}{dt} + \varepsilon^{-1}(p_y + f\ u) = F_v \,, \tag{3.2c}$$

$$\frac{dw}{dt} + \alpha 10^{4-n}\varepsilon^{-2}(p_z + gs) = F_w \,, \tag{3.2d}$$

$$u_x + v_y + w_z = 0 \,, \tag{3.2e}$$

where $\alpha = 10^{n-4}$.

System (3.2) is similar to the standard incompressible Navier-Stokes system and pressure can be determined in both systems by differentiating (3.2e) with respect to time to obtain an elliptic equation for the pressure. A traditional method (Williams [12]) to solve the incompressible system is to use a staggered mesh in a manner that guarantees that the discrete divergence will be zero to the order of the roundoff error. In this section this technique will be applied although the ultimate goal will be to develop a nonstaggered mesh model (see next section). There are a number of staggered meshes in use. Here s, u, v, and w were defined on the mesh $M_V = \{(x_i, y_j, z_k) \mid x_i\ (i-1)\Delta x\ ,(j-1)\Delta y\ ,(k-1)\Delta z\}(1 \le i,j \le I+1, 1 \le k \le K+1)$ and p was defined on the mesh $M_p = \{(x_i, y_j, z_k) \mid x_i = (i-.5)\Delta x\ ,(j-.5)\Delta y\ ,(k-.5)\Delta z\ \}(1 \le i,j \le I, 1 \le k \le K)$, where Δx, Δy, and Δz are the grid spacings in the x, y, and z directions. The corresponding centered second order finite difference approximation may be written as

$$D_{0t}s + uD_{0x}s + vD_{0y}s - \bar{s}w = F_s\ , \tag{3.3a}$$

$$D_{0t}u + uD_{0x}u + vD_{0y}u + \tag{3.3b}$$

$$\varepsilon^{-1}[D_{-x}(p + p_{j-1} + p_{k-1} + p_{j-1,k-1})/4 - f\ v] \ = F_u\ ,$$

$$D_{0t}v + uD_{0x}v + vD_{0y}v + \tag{3.3c}$$

$$\varepsilon^{-1}[D_{-y}(p + p_{i-1} + p_{k-1} + p_{i-1,k-1})/4 + f\ u] \ = F_v\ ,$$

$$D_{0t}w + uD_{0x}w + vD_{0y}w + \tag{3.3d}$$

$$\alpha\, 10^{4-n}\varepsilon^{-2}[D_{-z}(p + p_{i-1} + p_{j-1} + p_{i-1,j-1})/4 + gs] = F_w\ ,$$

$$D_{+x}(u + u_{j+1} + u_{k+1} + u_{j+1,k+1})/4 + D_{+y}(v + v_{i+1} + v_{k+1} + v_{i+1,k+1})/4 + \tag{3.3e}$$

$$D_{+z}(w + w_{i+1} + w_{j+1} + w_{i+1,j+1})/4 = 0\ .$$

The requirement that (3.3e) apply at time level $n+1$ leads to an equation for the pressure in the form

$$D_{+x}D_{-x}A_yA_zp/16 + D_{+y}D_{-y}A_xA_zp/16 + D_{+z}D_{-z}A_xA_yp/16 = B\ , \tag{3.4}$$

where $A_s = T_{+s} + 2I + T_{-s}$ and B is only a function of the grid variables at time level n. The left-hand side of (3.4) is a finite difference approximation of the three dimensional Laplacian, but more complicated than ones arising from other staggered grids. Nevertheless, it can be solved using fast Fourier quarter wave transforms in the y and z directions and a tridiagonal solver in the x direction. One of the main difficulties with this approach is the rigid wall boundaries. A singular system of equations for each plane $x = constant$ must be solved. For odd numbers I and K, this system can be solved recursively.

To determine the dependence of the error on the value of α, a fixed resolution was selected and numerical solutions computed for different values of α. The relative l_2 errors $e(\mathbf{V}_h) \equiv ||\mathbf{V}'_h - \mathbf{V}_h|| / ||\mathbf{V}'_h||$ ($\mathbf{V}_h = (u,v)^*$ and $||\cdot||$ denotes the l_2 norm) and $e(w) \equiv ||w' - w|| / ||w'||$ in the horizontal and vertical velocities for the case $I = 25$, $K = 5$, $\Delta t = 7200\,s$, and various values of α at five days are shown in Table 3.1.

Table 3.1
RM error as a function of α

I	K	Δt (s)	α	$e(boldV_h)$	$e(w)$
25	5	7200	10^{-4}	.38	.10
25	5	7200	10^{-3}	.36	.40
25	5	7200	10^{-2}	.38	1.47

To have all terms of the elliptic equation for pressure be of the same size, it is necessary that $\alpha = (\frac{\Delta z}{\Delta x})^2 \approx 3.6 \times 10^{-3}$. The mathematical theory says that for the correct value of α, the error in $\mathbf{V}_h$ and w should be of the same order and the numbers in Table 3.1 support this conclusion. The constant value of $\alpha = 10^{-3}$ will be chosen instead of the value above for the following reason. If the value is allowed to the error is being affected by both changes in the resolution and (possible) changes in α. By freezing the value of α the latter effect is removed.

A number of tests were run to determine the relative sizes of the various factors contributing to the discretization error (at this level of total error the continuum error is negligible as will be seen later). Tests similar to those in Table 3.1 but with a time step $\Delta t = 3600\,s$ were also run and the results were essentially the same, i.e. the time truncation error is not dominant at this error level. Five day runs were also run with fixed horizontal resolution ($I = 25$, $\Delta t = 7200\,s$) and varying vertical resolution. The results in Table 3.2 indicate that for $K = 5$, the vertical truncation error is roughly balanced with the horizontal truncation error.

Table 3.2
RM error as a function of K

I	K	Δt (s)	$e(\mathbf{V}_h)$	$e(w)$
25	3	7200	.42	.44
25	5	7200	.36	.40
25	7	7200	.34	.38

To check that the model is second order accurate, the number of horizontal and vertical points were doubled and the time step halved. From Table 3.3 it can be seen that he relative error $e(w)$ at five days for $I = 25$ and $K = 5$ was 40%, while for $I = 49$ and $K = 9$ it was 10%. Thus the error correctly decreased by a factor of four when the mesh size was halved.

Table 3.3
RM error as a function of resolution

I	K	Δt (s)	$e(\mathbf{V}_h)$	$e(w)$
25	5	7200	.36	.40
49	9	3600	.11	.10

The stability criterion for the staggered numerical approximation is that

$$|\tilde{\lambda}_{\max}|\Delta t \leq 1,$$

where $\tilde{\lambda}_{\max}$ is the numerical frequency with the largest magnitude $|\tilde{\lambda}_R| \approx u/\Delta x \approx 10^{-5} s^{-1}$. Since $\alpha g\tilde{s}R^{-1} \approx f_0^2$, $|\tilde{\lambda}_G| \approx f_0 \approx 10^{-4} s^{-1}$. Thus the Coriolis terms lead to the largest numerical frequency and the time step must be less than $10^4 s$. The model was stable for $\Delta t = 7200 s$, but not for $\Delta t = 10800 s$. Now the Coriolis terms could be treated implicitly to increase the time step by a factor 10 with only a small increase in computation. However, in the real ocean the advection velocities can be on the order of $1 m s^{-1}$ and then all frequencies will be comparable in magnitude. In that case the explicit reduced model can take the same time step as the quasi-geostrophic model (see discussion below).

Primitive Equation "Model"

The primitive equations are derived from the Eulerian system by neglecting the substantial derivatives in the vertical velocity and pressure equations. To be consistent, the corresponding forcing terms are also neglected so the forced system in this case is

$$\frac{ds}{dt} - \tilde{s}w = F_s, \tag{3.5a}$$

$$\frac{du}{dt} + \varepsilon^{-1}(p_x - f v) = F_u, \tag{3.5b}$$

$$\frac{dv}{dt} + \varepsilon^{-1}(p_y + f u) = F_v, \tag{3.5c}$$

$$p_z + gs = 0, \tag{3.5d}$$

$$u_x + v_y + w_z, \tag{3.5e}$$

The stability criterion for a primitive equation model will depend on the particular numerical approximation. In contrast with the reduced equations, the gravitational portion will tend to be the dominant term (in fact unbounded as $\Delta x, \Delta z \to 0$) and one would expect that a time step smaller than that required by the Coriolis frequency alone (or advection speeds on the order of $1 m s^{-1}$) would be necessary. At current resolutions a time step 10 times smaller than that required by advection alone is necessary.

Numerical problems at the boundary of a limited area model based on the meteorological primitive equations were the impetus for the original mathematical analysis of the initial-boundary value problem for that system. The analysis showed that the initial-boundary value problem for the primitive equations of meteorology is always ill posed (Oliger and Sundström [9]; Browning and Kreiss [2]) and that was the source of the observed boundary oscillations. Later a similar problem with the primitive equations of oceanography was demonstrated (Browning et al [4]). In contrast, the reduced system is the proper mathematical limit of a hyperbolic system so that with reasonable boundary conditions (e.g. the limiting form of well posed boundary conditions for the approximate system) the initial boundary value problem for the reduced system will be well posed. There is another concern with the primitive equations. The vertical velocity is essentially computed by integrating the incompressible equation in the form

$$w = -\varepsilon^{-1}\int_{z_b}^{z} (u_x + v_y)\, d\tilde{z} \ , \tag{3.6}$$

and this results in a considerable loss of accuracy (Browning et al [4]). This loss can be seen in the first iteration.

On the staggered mesh the horizontal divergence can be approximated by the quantity

$$\delta = D_x u + D_y v \ ,$$

and then the integral in (3.6) can be approximated by the trapezoidal rule (a second order finite difference scheme can also be used). Using the analytic solution from the previous section, the relative error $e(w)$ can be computed at the first time step for different resolutions.

Table 3.4
PEM Error as a function of resolution

I	K	$e(w)$
25	5	.56
49	9	.16

From Table 3.4 when $I = 25$ and $K = 5$ $e(w) = 56\%$, while when $I = 49$ and $K = 9$ $e(w) = 16\%$. Thus, as theoretically predicted, the error reduces roughly by a factor of four as the mesh size is halved. But the error at the first time step in the lower resolution case is on the order of 56% while that level of error was not exceeded in the corresponding reduced model until the 14th day of integration.

It should be mentioned that the staggered mesh used in this study aids in the computation of the pressure gradient terms in the reduced system since the truncation error mesh size for those terms is effectively $\Delta x/2$ while the truncation error grid size for the computation of the divergence is Δx. Other staggered grids compute the divergence with an effective truncation error mesh size of $\Delta x/2$. If the error in the computation of w exactly reduced by a factor of four (when using those meshes there is considerable averaging that can result in a larger truncation error coefficient and added dissipation), then the error in the two methods would be comparable at the first iteration. It is not the intent here to review all possible staggered schemes, only to show that the reduced system is at least as good as the traditional systems, but without any disadvantages. That has effectively been done for the primitive equations even in the best scenario for a model based on those equations.

Quasi-geostrophic model

The best comparison of the models would be to assume that the solution is periodic in all three directions so that boundary conditions do not cloud the issues. However, because of the desire to have a mean background flow (handled by rigid walls for the southern and northern boundaries) and the assumption of rigid vertical boundaries, the solution has only been assumed to be periodic in x. But it is still possible to remove the question of boundary stability questions for the quasi-geostrophic system by neglecting all advection terms. In this case the quasi-geostrophic equation for the stream function is

$$\frac{\partial}{\partial t}[\nabla_2^2\psi + f^2(g\bar{s})^{-1}\psi_{zz}] = F_q \, , \tag{3.7}$$

where the forcing term F_q is obtained by substituting the representative stream function into the left-hand side of (3.8), i.e. in this case the advection terms have been dropped on both sides of (2.1). The equation was approximated by

$$D_t E = F_q \, , \tag{3.8}$$

where

$$E = [D_{+x}D_{-x} + D_{+y}D_{-y} + f^2(g\bar{s})^{-1}D_{+z}D_{-z}]\psi \, .$$

Note that in the general case f and $\bar{s}$ would be functions of y and z, respectively, and E would be an approximation of a nonseparable variable coefficient elliptic equation. The quantities E and ψ are known at time $t = -\Delta t, 0$ from the analytic solution. Then equation (3.8) can be used to update E to time level $t = \Delta t$ at all interior points. Finally ψ can be updated to the new time level using a standard direct solver since ψ is periodic in x and has known constant values on the rigid walls. This process can then be repeated to integrate further in time.

The errors after five days from two runs of the linear quasi-geostrophic model are shown in Table 3.5.

Table 3.5 LQGM error as a function of resolution			
I	K	Δt (s)	$e(\mathbf{V}_h)$
25	5	7200	.08
49	9	7200	.02

At five days with $I = 25$ and $K = 5$, the relative error in the horizontal velocity $e(\mathbf{V}_h)$ was 8% in(advection terms dropped on both sides of (3.2)), i.e. the models produce essentially equivalent results. At five days with $I = 49$ and $K = 9$, the error was 2% in the quasi-geostrophic model so the error reduced by the correct factor.

It might at first seem surprising that the two models produce essentially equivalent results since the truncation error terms in the reduced model have large additional multiplicative factors, e.g. $\varepsilon^{-1}(\Delta x/2)^2 p_{xxx}$ while there are no such large factors in the quasi-geostrophic equation. The fact that the effective mesh size in the truncation error expression for the pressure term of the reduced model is half that of those in the truncation error terms of the quasgeostrophic model is a partial explanation. It has been shown (Kreiss [8]) that the error is really of order Δx^2, i.e. a factor of ε is gained.

Looking at the advection terms in the quasi-geostrophic equation more closely, it is also possible to include the advection term in the x direction on both sides of (3.7) without running into any boundary questions. Table 3.6 shows the errors after five days for this model.

Table 3.6
NQGM error as a function of resolution

I	K	Δt (s)	$e(V_h)$
25	5	7200	.28
49	9	7200	.11

In this case with $I = 25$ and $K = 5$ the velocity error was 28% (which is comparable to the full nonlinear reduced model error discussed earlier) while for $I = 49$ and $K = 9$ the error was 11%. Thus the error appears not to reduce by the correct factor. At first this might be thought to be due to a coding or numerical error (one reason for the previous example). That this is not the case becomes more apparent when considering $t = 0$ where the respective errors are 14.1% and 7.2%. The initial velocities are computed from centered difference formulas applied to the analytic stream function. Thus the error must reduce by the proper factor if the stream function were the only component of the analytic velocity. But this is not the case. The continuum error in the quasi-geostrophic system is on the order of the ratio of the divergence over the vorticity since the divergence has been ignored in the advection terms. For the current solution this ratio is on the order of 7% and it can not be expected that the error of a quasi-geostrophic model will be below this level for any resolution. Clearly to reduce the continuum error requires the extension of the quasi-geostrophic system to an intermediate system that includes the divergence in the advection terms. This means that minimally an additional K two dimensional elliptic systems need to be solved for χ (see 2.10) at each time step which considerably increases the computational expense of the associated model. Also the system has not been generalized to the equatorial region and attempts to do so further increase the complexity of the system. On the other hand, the globally applicable reduced system only requires the computation of a single constant coefficient three dimensional elliptic equation even for two digits of accuracy. For example, at five days with $I = 99$ and $K = 21$, the reduced model velocity error was only 2.6% and later it will be seen that this can even be reduced even further.

COMPARISON OF NUMERICAL METHODS

In the comparisons of the previous section it was essential to keep the numerical methods employed to approximate the various differential systems as close to one another as possible. Now the goal is to study a number of numerical methods for a single differential system, namely the reduced system. While the staggered mesh method works and has one distinct advantage discussed below, it is difficult to use with the composite mesh method which uses overlapping meshes to cover an irregular region (Starius [10]). Thus a number of methods on a nonstaggered mesh will be developed and compared. In particular, the approximate solution of the elliptic equation at each time step will be computed using a highly vectorized direct method and the multigrid iteration. A fourth order method will also be investigated.

In this section all variables will be defined on the mesh M_V. The centered second order approximation of the reduced equations (3.2a-d) on this mesh is

$$D_{0t}s + uD_{0x}s + vD_{0y}s - \tilde{s}w = F_s \, , \tag{4.1a}$$

$$D_{0t}u + uD_{0x}u + vD_{0y}u + \varepsilon^{-1}(D_{0x}p - f\ v) = F_u \, , \tag{4.1b}$$

$$D_{0t}v + uD_{0x}v + vD_{0y}v + \varepsilon^{-1}(D_{0y}p + f\ u) = F_v \, , \tag{4.1c}$$

$$D_{0t}w + uD_{0x}w + vD_{0y}w + \alpha 10^{n-4}(D_{0z}p + gs) = F_w \, . \tag{4.1d}$$

At the solid wall boundaries, e.g. at $j = 1$, there are no derivatives in the normal direction except in the equation for the normal velocity which is not needed. Thus the difference formulas can be used at every gridpoint. Instead of approximating (3.2e) and then forming a discrete equation for the pressure, the continuum equation for the pressure is determined by applying the total differential operator to (3.2e) to obtain the elliptic equation

$$\left(\frac{\partial^2}{\partial x^2} + \frac{\partial^2}{\partial y^2} + \alpha 10^{n-4}\frac{\partial^2}{\partial z^2}\right)p = M \, , \tag{4.2a}$$

where

$$M = f(-u_y + v_x) \tag{4.2b}$$

$$-\alpha 10^{n-4} g s_z - (u_x u_x + v_x u_y + u_y v_x + v_y v_y + u_z w_x + v_z w_y)$$

contains at most first order derivative terms.

For a channel the solution is only periodic in the x direction. The natural boundary condition at the remaining boundaries involves the corresponding normal derivative of the pressure, e.g. at $y = 0$ the boundary condition on the pressure is

$$p_y = -fu \, . \tag{4.3}$$

All of the current software packages that approximately solve the elliptic equation (4.2) use the centered difference approximation of (4.3)

$$D_{0y}p = -fu \tag{4.4}$$

on the boundary at $j = 1$. This is unfortunate because energy techniques require the more natural approximation

$$D_{+y}p = -fu \ , \tag{4.5}$$

which is second order accurate if the boundary is located between grid lines. Since the approximation (4.4) requires the right-hand side of (4.2) at the boundary and second order accuracy must be maintained, the centered difference operator D_{0y} in (4.2) was replaced by the difference operator

$$D_y = \begin{cases} [-3I + 4T_{+y} - T_{+y}^2]/(2\Delta y)\ , & j = 1 \\ D_{0y}\ , \quad 2 \le j \le J \\ [\ 3I - 4T_{-y} + T_{-y}^2]/(2\Delta y)\ , & j = J+1 \end{cases}$$

and a similar substitution for D_{0z}. Since the reduced system analysis of Gustafsson and Kreiss [7]) is not applicable to this case, the normal mode analysis of the stability of this method will be analyzed in a separate note.

A method which reduces the amount of coding was also tested. The equations are first updated in the normal manner except that the pressure terms are neglected. For example,

$$\tilde{u}^{n+1} = u^{n-1} / 2\Delta t + [uD_{0x}u + vD_{0y}u - f\ v - F_u\ ]. \tag{4.6}$$

Then the right-hand side of (4.2) is determined by the approximation

$$M = -[D_{0x}(\tilde{u}^{n+1}) + D_{0y}(\tilde{v}^{n+1}) + D_{0z}(\tilde{w}^{n+1})] \tag{4.7}$$

The expression is clearly derived from the requirement that the divergence at time level $n+1$ vanish. This requirement would lead to the nonstandard approximation of the Laplacian

$$(D_{0x}^2 + D_{0y}^2 + \alpha 10^{n-4} D_{0z}^2)p = M \ ,$$

so the standard 7 point approximation is used instead. In this case there are approximations of higher order derivatives in M and it is interesting to see if this causes any difficulties in practice. Assuming approximations of s, u, v, and w are known at time level n and $n+1$, either of the approximations of (4.2) can be used to compute the pressure at time level n and then the remaining variables can be updated to time level $n+1$ using (4.1). Clearly this process can be repeated. The actual errors for the model based on (4.7) were slightly less than for the one based on (4.2b) so those errors will be used. However it must again be emphasized that at this point there is only preliminary computational evidence for the stability of either method.

The first model used the CRAYFISHPACK routine H3GCSS (Sweet [11]) to solve the approximation of the elliptic equation given in (4.2). Here also a number of tests were run to determine the relative sizes of the various factors contributing to the discretization error. In Table 4.1 the errors at day 5 for $I = 24$, $K = 4$ and various time steps are shown.

Table 4.1 DM error as a function of Δt			
I	K	Δt (s)	$e(\mathbf{V}_h)$
24	4	7200	.94
24	4	3600	.88
24	4	1800	.77
24	4	900	.66
24	4	450	.60
24	4	225	.58

Direct solvers work the best with a number of grid points that have low prime numbers as factors ($I = 2^3\,3$ and $K = 2^2$) and this is the reason for the slightly different number of points than in the previous section. The results in Table 4.1 are rather disturbing. They indicate improved results with decreasing time step (although the amount of improvement decreases). From earlier results it is clear that such small time steps are not necessary for accuracy in time. Instead the improvement is obtained by an increase in the amount of the damping of the divergence. Whereas the staggered mesh was used to ensure that the discrete divergence is zero (to the order of the roundoff error) this is not the case for a nonstaggered approximation. The terms in M at time level $n-1$ can be shown to provide a damping mechanism which controls the growth of the divergence. A smaller time step leads to larger damping of the divergence and this can be used to explain the results in Table 4.1.

Table 4.2 shows the errors at five days for $I = 48$, $\Delta t = 1800\,s$ and two values of K.

Table 4.2 DM error as a function of K			
I	K	Δt (s)	$e(\mathbf{V}_h)$
48	4	1800	.38
48	8	1800	.39

This shows that the vertical truncation error is not dominant.

Table 4.2 shows that although the number of points compared to the run in 4.1 was doubled, the error only reduced by a factor of two. This is due to the fact that the resolution is insufficient, i.e. the asymptotic range has not been reached, as now will be shown. In Table 4.3, the errors at five days for two resolutions are shown.

Table 4.3 DM error as a function of resolution			
I	K	Δt (s)	$e(\mathbf{V}_h)$
48	4	900	.28
96	8	900	.07

When sufficient resolution is reached, the error reduces by the correct factor. Clearly approximately twice the resolution used in the staggered mesh reduced model is required in the nonstaggered reduced model to obtain the same level of error. This is because the largest term in the truncation error for any of the equations arises from the pressure gradient term because of the ε factors. In the staggered model this is approximated over a step size half of that in the corresponding nonstaggered model and so a factor of four increase in the error is to be expected. Although more resolution is required in the non-staggered case to obtain the same error as in the staggered case, the nonstaggered model only took 53% more time than the staggered model. This is due in part to the extra boundary complications in the staggered model.

The second model used the MUDPACK multigrid routine MUD3SP (Adams [1]) to solve the matrix equation (4.2). At day five with $I = 48$, $K = 4$ and $\Delta t = 1800\ s$ the error $e(V)$ in the multigrid model was 44%. The number of points is slightly different than the staggered model because the multi-grid routine works best when the number of points on the coarsest grid is a small prime number. For this choice, the coarsest grid is a $3 \times 3 \times 2$ mesh. Note that the error in this case is approximately the same as for the direct method (as it should be) but the multigrid model took approximately 6 times longer. Thus on rectangular basins and lower resolutions the direct solver is more efficient [multigrid is more efficient at sufficiently high resolutions because it has an $O(I^2K)$ operation count versus $O(I^2K \ln I)$ for the direct solver]. But the real ocean is not rectangular. If the capacitance matrix method is employed, then minimally two direct solves per iteration instead of one must be performed. Then the two methods should be roughly within a factor of three (or less) of each other and the choice is not as clear. Because of the curved boundaries and boundary layers, a composite mesh is a natural candidate. There the multigrid method will be the elliptic solver of choice, but there will clearly be an increase in the cost of the solution relative to that in a rectangular basin no matter which solver is used.

MUDPACK (Adams [1]) also contains a fourth order solver. The initial call to MUD3SP computes the second order approximation and then a call to MUD34SP raises the approximation to fourth order. Table 4.4 and 4.5 show the errors at five days as functions of Δt and K.

Table 4.4
FOM error as a function of Δt

I	K	$\Delta t\ (s)$	$e(V_h)$
48	6	3600	.10
48	6	1800	.05
48	6	900	.03

Table 4.5
FOM error as a function of K

I	K	$\Delta t\ (s)$	$e(V_h)$
48	4	3600	.10
48	6	3600	.08

Table 4.4 shows that the sensitivity to the time step is not as great as before (since the solution is more accurate there is less generation of spurious divergence so less damping is necessary) and Table 4.5 show that the vertical truncation error is not dominant.

Table 4.6 shows the errors at five days as a function of resolution.

Table 4.6 FOM Error as a function of resolution			
I	K	$\Delta t\ (s)$	$e(V_h)$
48	6	1800	.05
72	12	1800	.01
96	12	900	.005

The error reduced by the correct factor $[(2/3)^4 \approx 1/5]$ when going from $I = 48$ and $K = 6$ to $I = 72$ and $K = 12$ so clearly the continuum error at this point is still not dominant. But the error when going from $I = 48$ $I = 96$ and $K = 12$ does not decrease by a factor of 16 and the continuum error is now evident. Note that the error is on the order of $\max(\varepsilon^{-2}, 10^{-2}) = 10^{-2}$ as predicted by theory (Browning et al [4]) up to five days.

The fourth order model took approximately 3 times less to achieve a 1% error than the staggered model took to achieve a 3% error. If additional accuracy is required, the fourth order model will clearly be superior to all of the other models since even at the 1% level its higher accuracy has more than offset the modest increase in computational cost. The accuracy gain of going from a second order to a fourth method helps to offset the loss of accuracy of going from a staggered to a nonstaggered grid, i.e. the nonstaggered second order model has approximately four times the error of the staggered second order model at roughly the same resolution. At the same resolution the fourth order model is clearly more expensive per time step than the second order model. But for a given accuracy requirement, the fourth order method requires less resolution and for higher accuracy requirements, e.g. to maintain two digits of accuracy for longer integrations, the fourth order model is clearly more efficient.

REFERENCES

1. Adams, J. MUDPACK: Multigrid Fortran Software for the Efficient Solution of Linear Elliptic Partial Differential Equations, Applied Mathematics and Computations, Vol.34, pp. 113-146, 1989.
2. Browning, G.L. and Kreiss, H.-O. Scaling and Computation of Smooth Atmospheric Motions, Tellus, Vol.38A, pp. 295-313, 1986.
3. Browning, G.L., Hack, J.J. and Swarztrauber, P.N. A Comparison of Three Numerical Methods for Solving Differential Equations on a Sphere, Monthly Weather Review, Vol.117, pp. 1058-1075, 1989.
4 Browning, G.L., Holland, W.R., Kreiss, H.-O. and Worley, S.J. An Accurate Hyperbolic System for Approximately Hydrostatic and Incompressible Oceanographic Flows, Dynamic Atmospheric Oceans, Vol.14, pp. 303-332, 1990.
5. Bryan, K. and Cox, M.D. A Numerical Investigation of the Oceanic General Circulation, Tellus, Vol.19, pp. 54-80, 1967.
6. Charney, J. G. The Use of the Primitive Equations of Motion in Numerical Prediction, Tellus, Vol.7, pp. 22-26, 1955.
7. Gustafsson, B. and Kreiss, H.-O. Difference Approximations of Hyperbolic Problems with Different Time Scales I; The Reduced Problem,

SIAM Journal of Numerical Analysis, Vol.20, pp. 46-58, 1983.

8. Kreiss, H.-O. Problems with Different Time Scales for Ordinary Differential Equations, SIAM Journal of Numerical Analysis, Vol.16, pp. 980-998, 1979.
9. Oliger, J. and Sundström, A. Theoretical and Practical Aspects of Some Initial-boundary Value Problems in Fluid Dynamics, SIAM Journal of Applied Mathematics, Vol.35, pp. 419-446, 1978.
10. Starius, G. On Composite Mesh Difference Methods for Hyperbolic Difference Equations, Numerical Mathematics, Vol.35, pp. 241-255, 1980.
11. Sweet, R. CRAYFISHPAK: A vectorized Fortran package to solve Helmholtz equations. Recent Developments in Numerical Methods and Software for ODEs/DAEs/PDEs, (Ed. G. Byrne and W. Schiesser), World Scientific Publishing Company, 1991.
12. Williams, G.P. Numerical Integration of the Three-dimensional Navier-Stokes Equations for Incompressible Flow, Journal of Fluid Mechanics, Vol.37, pp. 727-750, 1969.

Study of Incompressible Flow with 3D Finite Elements

D. Howard

Numerical Analysis Group, Oxford University Computing Laboratory, Oxford, OX1 3QD, U.K.

Abstract

This study investigates a number of u-P, primitive variables, finite element algorithms based on 3D Lagrangian brick elements for simulation of steady incompressible confined fluid flow. These were applied to flow through a tight 90° bend of circular cross section, and compared to available experimental measurements on this geometry for laminar flows at Re=500 and Re=1093, and for turbulent flow at Re=43000. Three distinct pressure approximations were investigated: (1) discontinuous linear, (2) discontinuous tri-linear, and (3) continuous tri-linear, all of these combined with a 27 noded tri-quadratic approximation for velocity. Also studied were the penalty method, Galerkin weighting, and an inconsistent version of the SUPG method. Turbulence was simulated with a $k-\varepsilon$ model. Computations were carried out on a Cray XMP supercomputer with an SSD for efficient I/O. Discontinuous pressure elements were more stable in the presence of convection than continuous pressure elements for both laminar Reynolds numbers studied. The discontinuous approximation is less restrictive than the continuous approximation for the modelling of strong pressure gradients present inside the 90° bend — a fact which may account for such an observation. Flow simulations at Re=1093 gave an estimate of the distortion of the test space required for good non-linear convergence and accuracy, when implementing the inconsistent SUPG method. The turbulence simulations revealed the importance of a near wall model able to capture strong secondary flows near pipe walls.

1 Introduction

1.1 Partial differential equations studied

Consider the equations describing the behaviour of a Newtonian incompressible fluid. These can be derived from an equilibrium of forces acting on an element of fluid as:

$$\rho \frac{Du_i}{Dt} = \frac{\partial \sigma_{ij}}{\partial x_i} + b_i \tag{1}$$

where ρ is the density of the fluid, u_i are the velocity vector components, σ_{ij} are the stress tensor components, and b_i is a body force. When a uniform external field such as a driving pressure is superimposed on a fluid, equilibrium is accommodated by requiring its equations to equal the resulting total change of velocity with time. This change is known as the Convected, Material or Substantial derivative and is expressed mathematically as

$$\frac{D}{Dt} = \frac{\partial}{\partial t} + u_i \frac{\partial}{\partial x_i} \tag{2}$$

where x_i are components of the orthogonal space coordinate system, u_i the components of velocity in this system, and t the time. Assuming the existence of a unique steady state allows for the solution of the steady form of the equilibrium equations:

$$\rho u_i \frac{\partial}{\partial x_i} = \frac{\partial \sigma_{ij}}{\partial x_i} + b_i \tag{3}$$

The stress tensor for a Newtonian fluid can then be expressed in terms of the ***primitive variables***, velocity and pressure, as

$$\sigma_{ij} = \mu \left(\frac{\partial u_i}{\partial x_j} + \frac{\partial u_j}{\partial x_i} \right) - \left(\frac{2}{3} \mu \, \nabla \cdot u + P \right) \delta_{ij} \tag{4}$$

where δ_{ij} is the Kronecker delta. The stress tensor is related both to friction, through the fluid viscosity μ and to pressure P through its normal components. The divergence, or incompressibility equation is given by

$$\frac{\partial \rho}{\partial t}+\frac{\partial(\rho u_i)}{\partial x_i}=0 \tag{5}$$

For a constant density fluid under steady flow assumptions it reduces to

$$\frac{\partial u_i}{\partial x_i}=0 \tag{6}$$

Perhaps the most popular statistical method for describing turbulence is the two equation or $k-\varepsilon$ model [11]. The statistical nature of the model results in the appearance of an additional term in the momentum equations which is known as the Reynolds Stress tensor: $-\rho\langle u_i u_j\rangle$. The Boussinesq approximation [1] which relates the Reynolds stresses to the mean rate of strain through a turbulence viscosity, also known as the Reynolds Stress Hypothesis, is mathematically stated as

$$-\rho\langle u_i u_j\rangle=\mu_t\left(\frac{\partial \overline{u_i}}{\partial x_j}+\frac{\partial \overline{u_j}}{\partial x_i}\right)-\frac{1}{3}\langle u_i^2\rangle\delta_{ij} \tag{7}$$

where the term $\frac{1}{3}\langle u_i^2\rangle$ is also known as the time averaged turbulence pressure and is absorbed into the $\bar{P}$ variable, and where μ_t is the turbulence or eddy viscosity, a parameter which varies in time and space, and one which must be related to other flow variables in a turbulence closure. It was the combined ideas of Prandtl and Kolmogorov to let the turbulence viscosity depend on the *turbulence kinetic energy*, or the total energy per unit mass of turbulence, denoted by k and defined as $k=\langle u_i^2\rangle/2$. The Prandtl–Kolmogorov relationship is given by

$$\mu_t=C_\mu \rho l_\mu \sqrt{k} \tag{8}$$

It introduces C_μ, a nondimensional constant, and a length scale l_μ which can be specified algebraically, one equation model, or through a differential equation for a variable of the form $z=k^m l^n$ where m and n are constants, eg. $k-\varepsilon$ model where ε corresponds to the choice $m=\frac{3}{2}$ and $n=-1$. The kinetic energy of turbulence is obtained from a differential equation of transport which assumes isotropy of turbulence and applies to the contracted form of the Reynolds Stress closure [13] . A steady incompressible $k-\varepsilon$ model as derived by Launder and Spalding [11] and others [17, 16] can be summarised as:

$$u_j\frac{\partial u_i}{\partial x_j}-\frac{\partial}{\partial x_j}[(\frac{\mu+\mu_t}{\rho})(\frac{\partial u_i}{\partial x_j}+\frac{\partial u_j}{\partial x_i})]+\frac{1}{\rho}\frac{\partial P}{\partial x_i}=0 \tag{9}$$

$$\frac{\partial u_i}{\partial x_i}=0 \tag{10}$$

$$u_j\frac{\partial k}{\partial x_j}-\frac{\partial}{\partial x_j}[(\frac{\mu+\frac{\mu_t}{\sigma_k}}{\rho})\frac{\partial k}{\partial x_j}]+C_D\varepsilon=\frac{\mu_t}{\rho}\frac{\partial u_i}{\partial x_j}(\frac{\partial u_i}{\partial x_j}+\frac{\partial u_j}{\partial x_i}) \tag{11}$$

$$\textit{and} \tag{12}$$

$$u_j\frac{\partial \varepsilon}{\partial x_j}-\frac{\partial}{\partial x_j}[(\frac{\mu+\frac{\mu_t}{\sigma_{\varepsilon t}}}{\rho})\frac{\partial \varepsilon}{x_j}]+(\frac{C_{2t}\varepsilon^2}{kC_{\mu t}})=(\frac{\mu_t C_{1t}\varepsilon}{k\rho})\frac{\partial u_i}{\partial x_j}(\frac{\partial u_i}{\partial x_j}+\frac{\partial u_j}{\partial x_i}) \quad \textit{or} \tag{13}$$

$$u_j\frac{\partial \varepsilon}{\partial x_j}-\frac{\partial}{\partial x_j}[(\frac{\mu+\frac{\mu_t}{\sigma_{\varepsilon l}}}{\rho})\frac{\partial \varepsilon}{x_j}]+(\frac{C_{2l}\varepsilon^2}{k})=(\frac{\mu_t C_{1l}\varepsilon}{k\rho})\frac{\partial u_i}{\partial x_j}(\frac{\partial u_i}{\partial x_j}+\frac{\partial u_j}{\partial x_i}) \tag{14}$$

where subscripts l and t, with the symbol μ_t as an exception, stand for Launder and Thomas respectively, and help distinguish between the parameters of both models. The

Table 1: Constants in the k-ε model

Model	C_D	C_μ	C_1	C_2	σ_k	σ_ε
Thomas	0.417	0.22	1.43	0.18	1.53	1.0
Launder	1.0	0.09	1.44	1.92	1.0	1.3

constants in these models have been calibrated by fits to experimental data and to idealised isotropic turbulence conditions. Recently constraints between the values of these constants have been derived in order to make these consistent with asymptotic conditions in isotropic turbulence. However, the numerous assumptions in the derivation of these equations means that the calibration of these constants should be thought of as an engineering tool designed to give a reasonable fit to experimental data. Standard acceptable values for these constants are summarised in table 1.

1.2 Velocity-pressure discretisation

Application of the Weighted Residuals method to (9) and (10) can be interpreted as a Virtual Power Principle [2]. Immediately we recognise the incompressibility constraint as an additional term in the Virtual Power Principle and given by:

$$\int_V P^\diamond \epsilon_{kk} dV \tag{15}$$

where ϵ_{ij} is the strain tensor. Incompressibility is a kinematic constraint which gives rise to a reaction stress. This reaction is an arbitrary hydrostatic pressure, a virtual pressure $P^\diamond$, which can be superimposed on the stress field without causing deformation. This virtual pressure does no work in any actual deformation which satisfies the incompressibility constraint. Hence, the term can be safely added to the principle as it does no power for an arbitrary virtual pressure. A dual role is associated with the pressure variable, it a) is a solution variable in the equations of motion, and b) acts as a Lagrange multiplier imposing the continuity constraint. This can also be seen more clearly when interpreting the Weighted Residuals method as a functional minimisation [9]. The stationarity of this functional leads to the following matrix equation system:

$$\begin{bmatrix} K_n + K_l & -C \\ -C^T & 0 \end{bmatrix} \begin{bmatrix} u \\ P \end{bmatrix} = \begin{bmatrix} f_n \\ 0 \end{bmatrix} \tag{16}$$

where the submatrices and vectors with subscript indices l and n are linear and non-linear parts respectively. K_n will contain the convective terms present in the Navier-Stokes case, and C^T is a finite element discretisation of (10) and similar to (15).

Alternatively, the incompressibility constraint can be incorporated into the functional through a penalty method. This avoids P as a solution unknown. The stationarity of this penalised functional results in the following matrix equation system:

$$[K_n + K_l + \Upsilon CC^T]\,[u] = [f_n] \tag{17}$$

with Υ the penalty number or large constant whose size in Stokes flow depends on the fluid viscosity — a turbulent implementation would necessitate a distinct Υ for each integration point or individual finite element. Note that the reduced integration penalty method is equivalent to (17) in the discontinuous tri-linear pressure case [12] but this is more difficult to show for the discontinuous linear pressure case. In this study all terms in (17) were integrated with a 3x3x3 Gauss rule. Apart from its ability to handle the 4 noded discontinuous pressure space case quite naturally, (17) under a 3x3x3 rule is likely

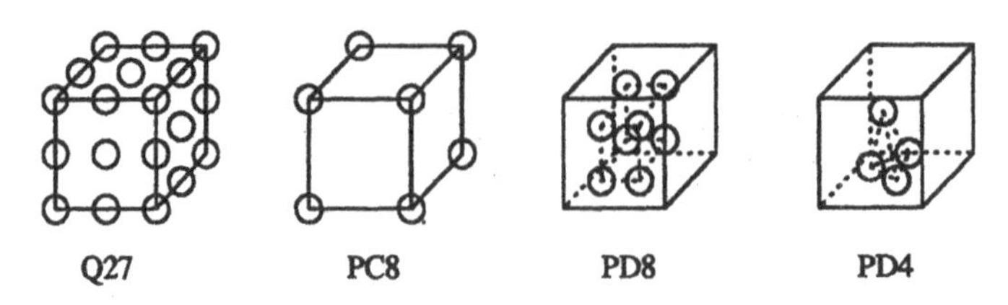

Figure 1: Velocity and pressure discretisations used.

to have accuracy advantages over other integration rules when the mapping between the finite element and its canonical or parent finite element becomes quite distorted.

The duality of the pressure variable in the Lagrange Multiplier approach leads to the possibility of spurious solutions or non-uniqueness of the pressure solution. A careful balance between the pressure space and the velocity space can avoid the presence of such spurious solutions but generally at the expense of some 'compressibility' effect. The combinations of pressure and velocity space investigated are illustrated in Fig. 1. These 3D shape functions are tri-quadratic 27 noded Lagrangian functions for the velocity variables, with either: (a) 8 noded continuous tri-linear (PC8), or (b) 8 noded discontinuous tri-linear (PD8), or (c) 4 noded discontinuous linear (PD4) functions for the pressure variable, and a superparametric mapping for the element using the 27 noded tri-quadratic functions (Q27).

With discontinuous pressure, cases (b) and (c) above, the discontinuous pressure variables were solved for directly, or alternatively, recovered from the penalty formulation using $P = -\Upsilon C^T u$. Once obtained they were smoothed onto the continuous tri-linear pressure node locations for plotting purposes. This smoothing was accomplished with a global least squares method. It requires solution of a global matrix problem, solved by the following stationary iteration:

$$p_c^{n+1} = p_c^n + \hbar\mathcal{K}(f_d - \mathcal{M}p_c^n) \tag{18}$$

$\hbar$ is a relaxation factor, n is an iteration count, $\mathcal{K}$ is an approximation to the inverse of the mass matrix $\mathcal{M}$, also denoted by $\mathcal{M}_l^{-1}$ or *lumped mass matrix*. The iteration in equation (18) if convergent is exactly equivalent to the solution of the original least squares matrix system:

$$\mathcal{M}p_c = f_d \tag{19}$$

where p_c is the solution vector of continuous pressure, and f_d the right hand side vector containing the discontinuous pressure solution. Five iterations of (18) sufficed for convergence.

1.3 Incompressibility measures, treatment of non-linearities

Picard linearisation, or successive substitution, was chosen over Newton's method to linearise the momentum and turbulence transport equations. The Picard iteration converges more slowly than the Newton iteration, but converges with a wider choice of starting points than the Newton iteration.

Picard's linearisation was adopted particularly because turbulence results were obtained with a Galerkin weighting. The rate of non-linear convergence was sensitive to the ratio of velocity to turbulence viscosity. Decoupling of the incompressible Navier-Stokes equations from the turbulence transport equations, required because of limited memory and CPU, degraded even further the rates of non-linear convergence of all solution variables. The incompressible momentum equations were solved first, followed by k transport, followed by ε transport. Convergence rates also deteriorated with the implementation of complex near wall velocity boundary conditions intended to capture the strong recirculating flow just adjacent to the pipe walls.

At the beginning of a computation the turbulent viscosity was set to zero, but the inlet velocity boundary conditions produced a very fast flow upon solution of the first Picard iteration — Stokes flow. Hence, although at convergence the turbulence viscosity was hundreds of times larger than the laminar viscosity in most locations, the ratio of convection to diffusion remained very large during the initial Picard iterations. Although at convergence Galerkin weighting was appropriate, the intermediate Picard linearisations solved a system of equations where the convection or inertia terms overpowered the diffusive terms because the turbulence viscosity was still quite small compared to the size of the velocity vectors. The Galerkin formulation therefore failed to converge unless measures were taken to balance the non-linear rate of growth of the velocity and turbulence viscosity. Originally, this convergence difficulty was overcome with a Petrov-Galerkin method which faded away into a Galerkin formulation at convergence. However, it was found that a small background level of Petrov-Galerkin distortion was necessary at all times. It was just not possible to restore the full Galerkin formulation and at the same time to avoid divergence.

A more successful approach was to simulate a continuation scheme in Reynolds number, or stage by stage solutions to increasingly faster flows, with a non-physical continuation parameter multiplying the convection terms in all equations and which assumed values between 0.0 and 1.0. Small values for this parameter helped to delay the growth of the inertia terms relative to the growth of the turbulence viscosity. Another similar non-physical continuation parameter premultiplied the sink terms in the k transport equation, and helped to keep the k variable positive at all times. These continuation parameters were varied in an ad-hoc way to give converged Galerkin solutions to the turbulent Navier-Stokes equations when their values were restored to 1.0.

In order to depict convergence of the non-linear iterative process the following two error estimators were chosen

$$\Delta_1 = \max_j \frac{\max_i |\,{}^j\varphi_i^n - {}^j\varphi_i^{n-1}|}{\max_i |\,{}^j\varphi_i^{n-1}|} \qquad \text{and} \qquad \Delta_2 = \max_{i,j} \left| \frac{{}^j\varphi_i^n - {}^j\varphi_i^{n-1}}{{}^j\varphi_i^{n-1}} \right| \tag{20}$$

where φ stands for a degree of freedom of j type, (ie. velocity, pressure, etc.), and n is the current iteration step. It is worth noting that Δ_2 is unstable in the presence of a discrete zero value of the solution, whereas the Δ_1 is only so for the trivial case of zero global solution. Δ_2 found application when the precise orientation of relatively small velocity vectors in some corner of the flow domain was of interest.

Another useful indicator was the incompressibility norm described in Zienkiewicz, Villotte and Toyoshima [19]. It is a measure of the incompressibility condition by the finite element velocity solution. At element level, an incompressibility semi-norm $\|\cdot\|_{2,V_e}$ can be defined such that

$$\|\nabla \cdot u\|^2_{2,V_e} = \int_{V_e} \left(\frac{\partial u_i}{\partial x_i}\right)^2 dV_e \tag{21}$$

or in an average sense over the entire mesh, a norm $\|\cdot\|_V$ can be defined such that

$$\|\nabla \cdot u\|^2_V = \frac{1}{\text{ne}} \sum_{e=1}^{\text{ne}} \|\nabla \cdot u\|^2_{2,V_e} \tag{22}$$

where ne is the total number of elements in the mesh. These norms cannot be used to compare the true level of incompressibility of a continuous pressure formulation to that of a discontinuous pressure formulation. Different norms favour one type of pressure approximation over another, eg. discontinuous over continuous. Both (21) and (22) were useful in ascertaining the stability of a given pressure approximation at various Reynolds numbers. They were also useful for investigating the size of the penalty number by computing the 'incompressibility' of the penalty method and comparing it to that of a discontinuous pressure formulation which solved for the pressure variables explicitly using (16).

1.4 Advection dominated flow

Methods for dealing with a mixture of convection and diffusion are based on analysis for the 1D homogeneous convection-diffusion equation with two Dirichlet boundary conditions. In this context a Petrov-Galerkin function can be found which gives an optimal solution for any cell Peclet number. It is called the Hemker function.

The literature is full of schemes for the multidimensional situation but none of these uses the optimal function (like Hemker in 1D) because finding it is as difficult as solving the equations. Finite volume/difference 'upwind' schemes are analysed in terms of the order of truncation of the scheme, and finite element Petrov-Galerkin methods in terms of the 'optimality' of the method in some norm. In the former case the incorrect assumption is made that the order of truncation of the error is the same as that of the scheme, and in the latter case there is little attempt to give numerical values to the constants in the error estimates and to qualify what the norms represent as far as pointwise accuracy, accuracy in the infinity norm, is concerned. In multidimensions, schemes are constructed from one dimensional arguments or tensor products of 1D functions such as the Hemker 1D function. These must be tested for accuracy against standard known solutions.

In this study, a scheme similar to the SUPG method [3] was used to restore stability to the solution of the Navier-Stokes equations in the convection dominated laminar flow regime. The scheme was an 'inconsistent' version of SUPG because neither the diffusion nor the pressure gradient terms were weighted using the SUPG function. The accuracy of the method under strong pressure gradients or under appreciable diffusion was therefore not expected to be so good. This inconsistent SUPG was implemented at every Gauss point, at every element, and at every Picard iteration.

The technique will now be described in more detail. Denoting the velocity solution from the previous Picard iteration at a Gauss point G by u_G, v_G, and w_G, the distortion of the weighting function in the direction of flow is accomplished by the addition of a convective term which uses the derivatives of the global weighting function at node I and velocities at the Gauss point G; it is given by:

$$W^I = N^I + \frac{1}{2}\gamma\frac{\alpha_o \ell}{2u_C}\left(u_G\frac{\partial N^I}{\partial x} + v_G\frac{\partial N^I}{\partial y} + w_G\frac{\partial N^I}{\partial z}\right) \tag{23}$$

where ℓ and u_C are quantities to be defined shortly, γ is a constant which indicates how much of the 'optimal' SUPG or α_o to give to the distortion. There is an extra $\frac{1}{2}$ term since the distance between nodes in a 1D quadratic element is half of the element size. SUPG schemes applied to quadratic functions were first developed by Nakazawa [14]. The term α_o simulates the Allen and Southwell operator which is given by a hyperbolic cotangent fitting factor:

$$\alpha_o = coth\left(\frac{Pe_\ell}{2}\right) - \frac{2}{Pe_\ell} \tag{24}$$

The magnitude, or L_2 norm, of the velocity, the characteristic velocity u_C, is given by:

$$u_C = \|u\| = \sqrt{u_G^2 + v_G^2 + w_G^2} \tag{25}$$

The characteristic element length ℓ was derived for the brick element SUPG implementation by extending the definition of Yu and Heinrich [18] for a two dimensional quadrilateral into three dimensions. This gave the following relations:

$$l_{(1)}^{(1)} = \frac{\partial x}{\partial \xi} \quad l_{(2)}^{(1)} = \frac{\partial y}{\partial \xi} \quad l_{(3)}^{(1)} = \frac{\partial z}{\partial \xi}$$

$$l_{(1)}^{(2)} = \frac{\partial x}{\partial \eta} \quad l_{(2)}^{(2)} = \frac{\partial y}{\partial \eta} \quad l_{(3)}^{(2)} = \frac{\partial z}{\partial \eta} \tag{26}$$

$$l_{(1)}^{(3)} = \frac{\partial x}{\partial \zeta} \quad l_{(2)}^{(3)} = \frac{\partial y}{\partial \zeta} \quad l_{(3)}^{(3)} = \frac{\partial z}{\partial \zeta}$$

$$hh^{(i)} = \frac{1}{u_C}\left(u_G l_{(1)}^{(i)} + v_G l_{(2)}^{(i)} + w_G l_{(3)}^{(i)}\right) \tag{27}$$

(28)

$$\ell = |hh^{(1)}| + |hh^{(2)}| + |hh^{(3)}| \tag{29}$$

Thus, the characteristic element length, ℓ, is the element length through the Gauss point in the direction of flow.

1.5 Boundary Conditions

The form of the Navier-Stokes equations considered was the one in which the stress tensor is integrated by parts with the weighting function and the resulting boundary traction integrals are updated concurrently with non-linear iteration. It is known as the *update traction boundary condition* [15]. The stress terms appear on the right hand side vector in the Navier-Stokes equations, f_n in (16) or (17), and upon substitution of (4) are constructed with values of velocity, pressure and turbulence viscosity from the previous non-linear iteration (the hat values), and with direction cosines t_j evaluated from the finite element geometry. For turbulence flow, the surface terms in the momentum equations are given by:

$$\int_{\Gamma_2} N^I \, [\, (\mu + \hat{\mu}_t)(\frac{\partial \hat{u}_i}{\partial x_j} + \frac{\partial \hat{u}_j}{\partial x_i})t_j - \hat{P}\delta_{ij} \,] \, d\Gamma \tag{30}$$

where N is the Galerkin test function.

The advantages of this type of non-linear updating boundary condition over the more traditional Neumann, or even a prescribed zero traction boundary condition, are significant for cost and accuracy of approximation: (a) the erroneous assumptions of developed flow conditions at outflow can be abandoned in favour of more realistic developing flow conditions which ensue in an automatic fashion, (b) they do not propagate as much numerical noise upstream as other more restrictive boundary conditions, (c) updating tractions permits the use of a smaller computational domain, reducing CPU and memory requirements.

All these advantages are relevant to the flow though a 90° bend considered next because of (a) the presence of strong secondary flows which do not decay easily, (b) the instability of the Galerkin method in the presence of a mixture of advective and diffusive operators which manifests both as wiggles in the solution and in slow non-linear convergence, (c) the CPU demands of three dimensional simulations.

Inlet Dirichlet velocity boundary conditions in laminar or turbulent flow were gathered by interpolating data from the experiments. The Law of the Wall was used in conjunction with the van Driest damping factor to prescribe Dirichlet velocity data on the near wall surface for the turbulence flow simulations. The computational domain was placed so that $6 \leq y^+ \leq 15$. Shear stresses at the near wall were then evaluated by an analytical least squares fit of the three velocity values perpendicular to the wall, eg. along a brick finite element edge [7].

It was not necessary to prescribe any pressure reference boundary condition. Pressure boundary conditions are not required when the gradient of pressure is integrated by parts, and when in addition there exists a Neumann pressure boundary condition (eg. an update traction boundary condition).

Turbulence Dirichlet k and ε boundary conditions at the flow inlet were obtained iteratively so as to be compatible with the prescribed velocity data there. Downstream boundary conditions for the k and ε variables where of Neumann type. Near wall boundary conditions for k and ε were of Dirichlet type and assumed local equilibrium in the log law region.

1.6 Solution of matrix equations

As in other applications of computers to Engineering it was both economies in computing time and I/O requirements that presented an equal challenge for the solution of large 3D problems with the finite element models studied.

Both direct and iterative methods can achieve the solution of the matrix equations (16) and (17). The most effective technique for this study was a direct method based on unsymmetric Gauss elimination with diagonal pivoting — the unsymmetric frontal method [6]. Optimised on a Cray XMP/48 and running in a multiuser environment it achieved in excess of a 100 megaflop rate, while the average 'well vectorised' code in that environment achieved just over a 70 Mfl rate. The most important area of code requiring optimisation performed the Gauss elimination around the pivotal row and column. Other important areas for code optimisation were the pre-front loops and the integration of the finite element matrices with forced vectorisation. The SSD device proved very helpful to store and retrieve the upper triangular matrix resulting from the frontal elimination algorithm.

The frontal method is as competitive as the preconditioned Unsymmetric Conjugate Gradient (UCG) methods [8] in the context of pipe flows because, except for impracticably large problems on a Cray-XMP supercomputer, such geometries give a minimum maximum front width small enough to make a frontal solver as fast or faster than ILU preconditioned UCG methods. This is because the cross section of pipe contains few elements relative to the length of pipe resulting in a small bandwidth.

2 Flow through a tight 90° bend

The computational model was applied to the flow of water through a tight 90° bend of circular cross section at two laminar Reynolds numbers: 500, 1093, and at a turbulent Reynolds number: 43000. Experimental measurements were available for validation of the numerical simulations. This was data from Imperial College [5], tabulated laser anemometry measurements of mean velocity, ie. streamwise velocity u scaled by bulk velocity U_B or $\frac{u}{U_B}$. The first station for measurements was located 0.58 pipe diameters upstream of the bend. This was taken as the inlet flow plane for the computations. The pipe diameter was $48.0mm$ and the radius of curvature of the bend was $134.4mm$ or 2.8 times its diameter. The curvature of this bend was too tight for the analytical solutions of Dean [4] to be valid.

2.1 Setting up the numerical model

Inlet velocity boundary conditions were taken from the experiment. These were partially developed flows on the circular plane. A least squares fit was applied to the flow data on a plane 0.58 pipe diameters on the straight pipe or 'tangent' upstream of the bend. 'Best fit' polynomials of the form $u = f(d^o, d, d^2, ..., d^n)U_B$ were used to interpolate the experimental values, where $f(d^o, d, d^2, ..., d^n)$ is a polynomial defined in table 2 with $d = 1 - \frac{y}{R}$, when y is radial distance from the wall, and R the pipe radius. These

Table 2: Least squares fit to experimental inlet velocity data

$U_B(Re500) = 0.0105m/s$, $U_B(Re1093) = 0.0230m/s$, $U_B(Re43000) = 0.92m/s$							
Re=	domain	d^0	d^1	d^2	d^3	d^4	d^5
500	all	1.675	0.147	1.272	-3.051	0.7104	
1093	$d \leq 0.4$	1.485					
	else	1.501101	-0.416229	2.26953	-3.354248		
43000	$d < 0.7$	1.1375					
	else	1086.58	-7071.454	18306.0	-23547.85	15058.11	-3831.365

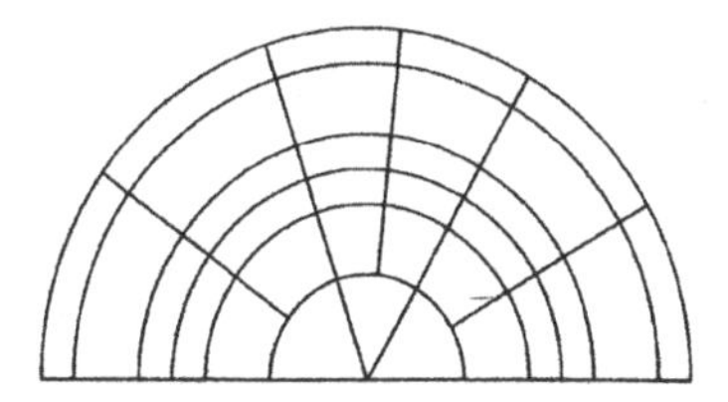

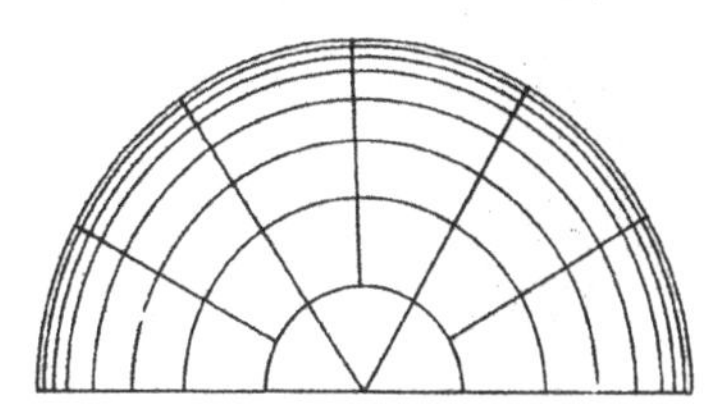

Figure 2: Cross section of finite element mesh for the Re=500 and Re=1093 computations.

polynomials gave a radially symmetric velocity profile at the flow inlet plane. The flow was assumed radially symmetric for simplicity. In reality it is slightly asymmetric on this plane because the strong curvature of the flow from the 90° bend feeds some of this information upstream.

The finite element meshes took into account vertical symmetry so that only half of the circular cross section, a semi-circular section, was discretised. For the Re=500 computations the mesh consisted of 429 elements arranged into 13 segments of 33 elements each. The distribution of these 33 elements can be appreciated from Fig. 2 which shows a cross section of pipe. At higher Reynolds numbers a more refined mesh was deemed necessary in the cross section in order to capture the stronger secondary flow pattern. The mesh for the Re=1093 computations consisted of 585 elements or 13 segments of 45 elements each. Fig. 2 illustrates the distribution of these 45 elements.

The number of velocity nodes using 27 noded elements was 4077 and 5481 for Re=500 and Re=1093 meshes respectively. The minimum maximum front width was 784 (with continuous tri-linear pressure) for the Re=1093 mesh. The distribution of the element segments was the same for both meshes. Downstream of the bend, the computational domain extended to a section of straight tubing (tangent) just 1.0 pipe diameters in length. The tangent upstream of the bend was modelled by two segments of elements, the tangent downstream from the bend by three so as to accurately model the iterative traction boundary condition on the outlet plane. The mesh for the Re=43000 case was much finer than the laminar ones. It had 1375 elements, or 25 segments of 55 elements each, with 18 of these segments placed at regular 5° intervals. The arrangement is shown in Fig. 3. The number of segments on the downstream tangent was four. The number of velocity nodes was 12393 and the minimum maximum front width was 1211. The Re=43000 case was discretised with continuous tri-linear pressure only.

2.2 Laminar Flow results

Three types of discretisation were run on a given mesh for each laminar Reynolds number: 500, 1093, investigated. These discretisations varied only in the treatment for the

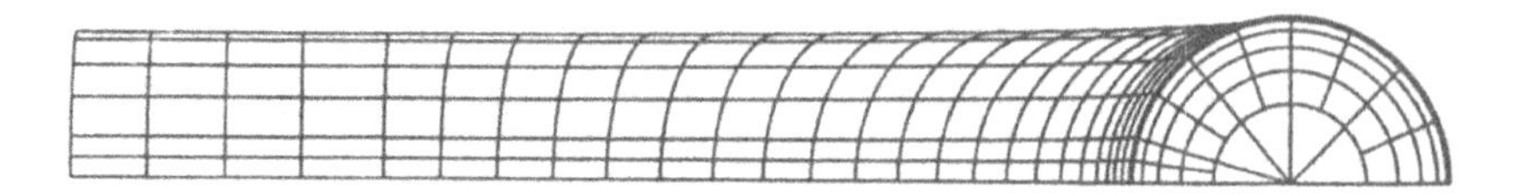

Figure 3: Fine grid arrangement for the Re=43000 computations.

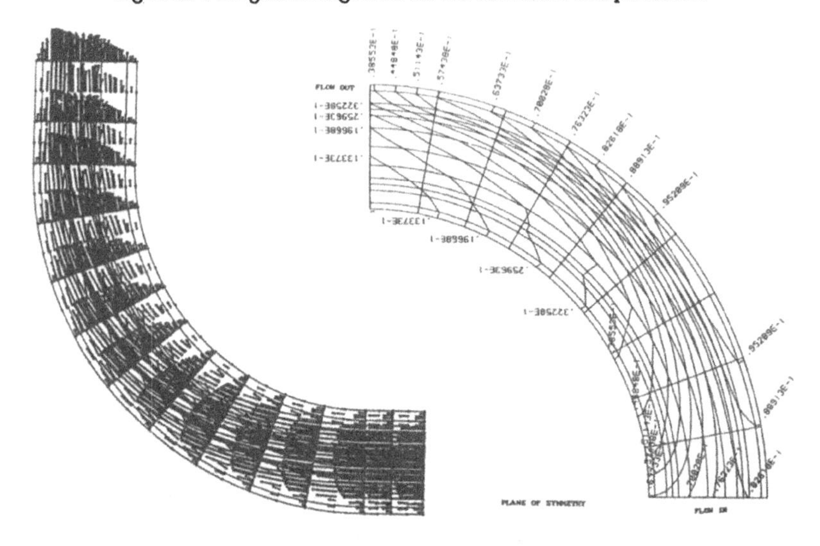

Figure 4: Re=500 computation: velocity vectors and pressure contours.

pressure.

Fig. 4 illustrates a 'transparent' 3D vector plot for a Re=500 computation. It clearly illustrates a developed secondary flow pattern superimposed on the primary developed flow pattern at the outlet. Physically, this developed secondary flow pattern does not decay until many pipe diameters downstream of the bend. The update traction boundary conditions are therefore quite handy since the computational domain was truncated only one pipe diameter after the bend. Results for $Re = 500$ were superior to those for $Re = 1093$ on this outlet cross section when compared with experimental measurements. Another observation is that streamlines turn prior to the 90° bend. This is is because of solution information travelling upstream.

Computations at Re=500 with Galerkin weighting showed fast convergence in the non-linearity and no wiggles or instabilities in the velocity solution. Agreement with experimental results was pretty good. Fig. 5 compares streamwise velocity values, obtained from PD4, PD8 and PC8 discretisations, to the experimental measurements (the solid line) along a given laser anemometry ray. There is more agreement between the discontinuous approximations PD4 and PD8 than between PD4 and PC8, which seems to indicate that the nature of the pressure approximation is more important than the number of pressure variables and associated incompressibility constraints involved in the approximation.

Galerkin computations for Re=1093 showed a deteriorating non-linear convergence when compared to Galerkin computations for Re=500. Indeed, all three discretisations took more than twice as many Picard iterations to converge than at Re=500. Par-

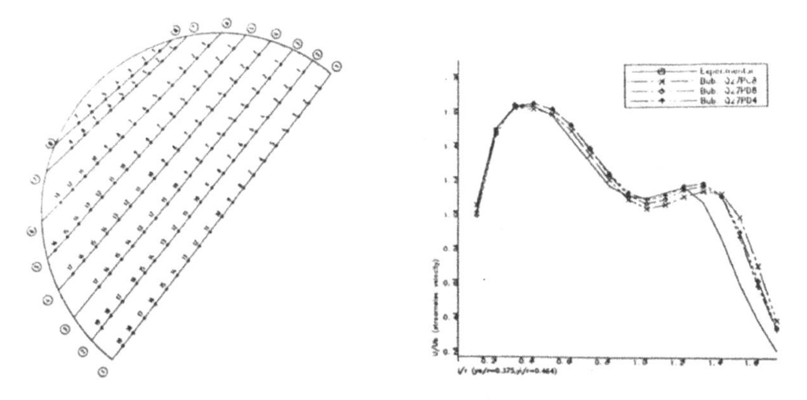

Figure 5: Laser anemometry rays and comparisons along line 4 on 60° section, Re=500.

ticularly bad was the Q27PC8 which displayed some wiggles in the velocity solution in a small region just upstream from the bend. However, both Q27PD4 and Q27PD8 remained stable in this and other flow regions.

The absence of 'wiggles' in the velocity solution using discontinuous pressure elements when the continuous pressure approximation gave wiggles in velocity, for the same mesh and Reynolds number, remains to be fully understood. The conjecture is that the pressure drops inside the bend, eg. as illustrated in Fig. 4 for the Re=500 case, are better modelled by a discontinuous rather than by a continuous polynomial for pressure. It would seem that a 'multi-dimensional cell Reynolds number', if it could be properly defined, is not enough to predict the stability of the Galerkin method for the u-P incompressible formulations studied. The pressure discretisation should also be taken into consideration.

Interestingly enough, the incompressibility estimator (21) was useful to detect an instability for the velocity approximation using continuous pressure elements (PC8) at Re=500. Consider Fig. 6, which shows graphs with finite element number on the horizontal scale and finite element incompressibility estimator (21) on the vertical scale. Approximations using PD4, PD8 and PC8 are shown for the Re=500 case and represent 'incompressibility signatures' for each of the numerical approximations used. The element numbering is optimised for the frontal Gauss elimination so that the numbering is orderly — starting from flow inlet and finishing at the flow outlet and sequentially numbered in each finite element layer representing a pipe cross section. The peaks reflect regions of poor incompressibility satisfaction which tend to occur at regular intervals in the inside wall of the bend. Both PD8 and PD4 (the discontinuous pressure approximations) show similar incompressibility patterns which are different from those pertaining to the PC8 approximation. The graphs are scaled but it is easy to see an unusually large peak, poor incompressibility, in a region upstream of the 90° bend for the PC8 approximation (circled). Comparison to similar graphs for the Re=1093 case, Fig. 6, show this same peak magnified to a very poor level of incompressibility satisfaction (circled). Indeed, for PC8 at Re=500 the instability reflects only in the velocity gradients, while at Re=1093, this instability can also be detected as 'wiggles' in the velocity solution.

The incompressibility estimator (21) was also used to determine the effect of the magnitude of the penalty number on the approximation. This was to change the scale of graphs in Fig. 6 but leave the pattern of the graphs unchanged. PD4 and PD8 approximations of Lagrange multiplier type, solving for discontinuous pressures explicitly using (16), confirmed this behaviour. It seems as if small penalty numbers have little or no effect on the quality of solution other than to change the element incompressibility uniformly everywhere. For the Re=500 computations, the penalty parameter could not

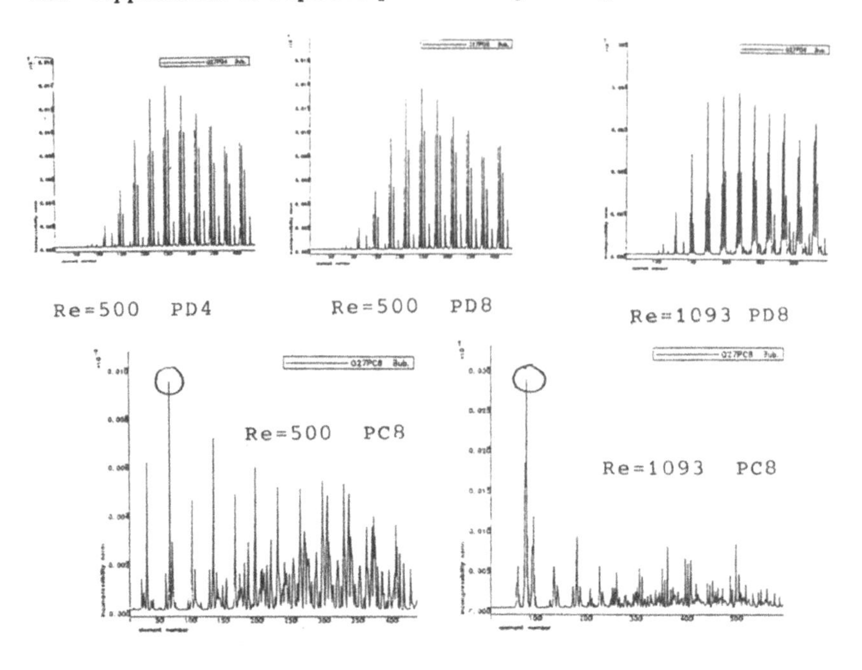

Figure 6: Graphs of finite element incompressibility for the converged solution.

be chosen smaller than:

$$\Upsilon = \frac{1.0 \times 10^8}{\rho} \tag{31}$$

in equation (17). The presence of ρ in (31) follows from the fact that the Navier–Stokes equations as computed were divided by density. In the case of the Re=1093 computations the penalty number was selected equal to or greater than:

$$\Upsilon = \frac{4.0 \times 10^8}{\rho} \tag{32}$$

These values of the penalty number must be divided by the fluid viscosity if one is to get a true feel for the absolute penalty number used. It would appear that the penalty number required for these computations was orders of magnitude greater than the corresponding penalty numbers employed for two dimensional simulations. However, the high accuracy of arithmetic of the Cray-XMP/48 made these large penalty numbers viable. The penalty numbers in (31) and (32) were increased by orders of magnitude with no apparent difficulties.

The inconsistent SUPG method described in section 1.4 was applied to the Q27PC8 approximation for Re=1093. Various levels of γ were tested. Only a very small γ was required to dramatically improve the non-linear convergence; this was enough, with higher levels of γ making little difference in total number of Picard iterations required for convergence. The accuracy of the computation was severely affected unless a very small value for $\gamma = 1/6$ was chosen. Higher γ values and in particular $\gamma = 1.0$ which represented 'optimal' upwinding, overdamped the solution. Fig. 7 illustrates just that.

It becomes difficult to assess precisely why the value for γ required for good accuracy and stability was so small. The SUPG method was not consistently applied to all

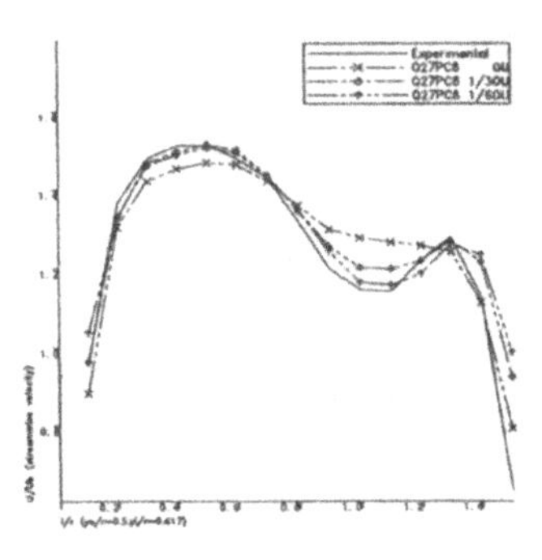

Figure 7: The effect of various SUPG levels, line 5 on 60° section.

terms. Indeed, the pressure gradient term was missed out by the SUPG formulation. In addition, ∇P^{o} should have been incorporated into the SUPG weighting function. Finally, the term 'optimal' stems from analysis of a 1D two point boundary value linear homogeneous convection-diffusion problem with linear trial functions ! This 'optimal' SUPG level is unlikely to correspond to a 3D non-linear vector equation system using tri-quadratic trial and test functions.

2.3 Turbulent Flow results

Details of the continuation method for solution of these equations can be obtained from [7].

The turbulent flow through the tight 90° bend had been previously modelled with some success by Iacovides using a plane by plane finite volume method [10]. Unlike flow through a bend of square cross section, secondary flows in the bend of circular cross section are promoted almost exclusively by the inertia terms and not by turbulence. Hence, a $k - \varepsilon$ model, which embodies a great number of isotropy assumptions, gave acceptable results in the study by Iacovides and was therefore used in this study.

However, the algorithm for prescribing near wall boundary conditions in the $k - \varepsilon$ model proved extremely important. For example Fig. 8 illustrates the secondary flow patterns obtained by two near wall logarithmic law models at a cross section 20° into the bend (top), and at a cross section 60° into the bend (bottom). The pattern with a rather damped recirculation near the walls was obtained by assuming a streamwise unidimensional Law of the Wall, ie. there is only one calculation for a streamwise shear stress (pictures on the left). An intermediate near wall treatment allowed for both a streamwise and a circumferential or 'slip' shear stress (not shown). This resulted in the evaluation of two independent shears and of two boundary condition values: one along the streamwise direction, and one along the slip direction. These two boundary conditions were decomposed into u, v and w boundary conditions on the near wall surface. Finally, the pattern with the most recirculation (pictures on the right) is for a wall treatment which allows the two shear stresses to interact. This made matters worse as far as obtaining non-linear convergence was concerned, but gave a much stronger and more realistic secondary flow field near the walls, and better agreement with experimental results for mean velocity.

Fig. 9 compares between experimental and numerical results which used the most 'slippery' wall boundary conditions. The streamwise velocity contours were drawn from experimental measurements. The agreement is only qualitative. It seems that a wall treatment promoting stronger secondary flow near the walls would help to obtain better agreement with the experiments.

All turbulence computations were obtained under Galerkin weighting and with a

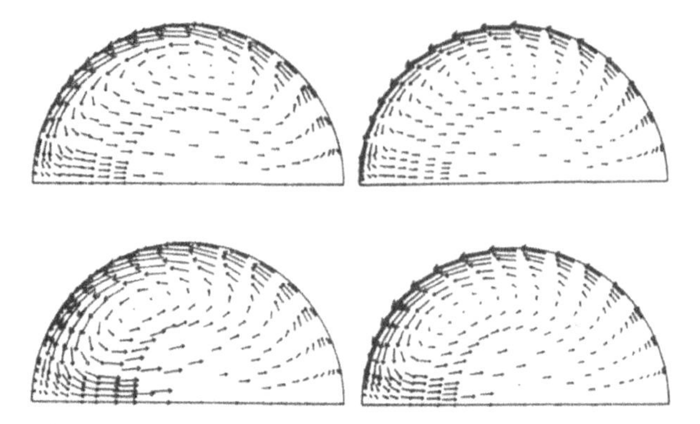

Figure 8: Near wall modelling effect on secondary flow intensity.

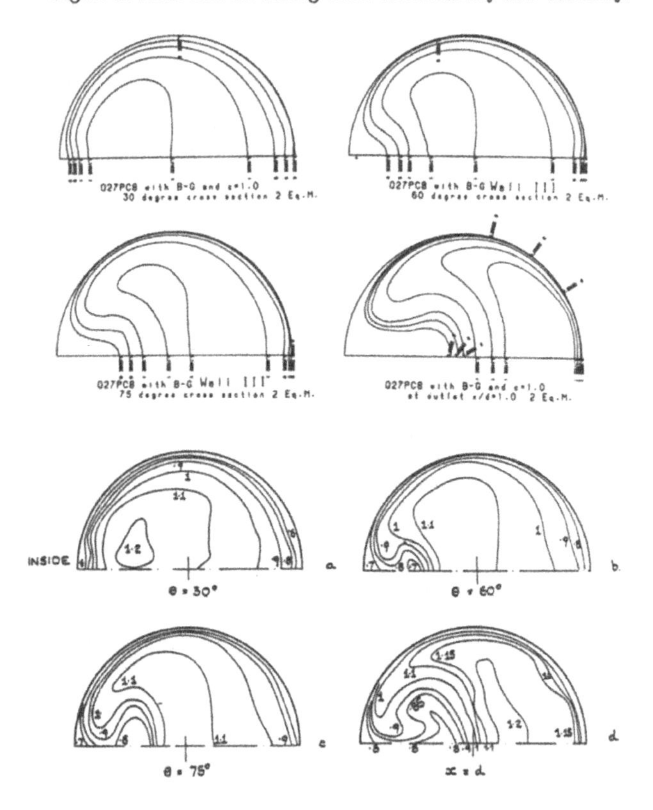

Figure 9: Comparison of $k-\varepsilon$ model to experiments at 30°, 60°, 75°, and $1.0D$ after the bend. Top: computations. Bottom: experiments.

Table 3: Cray-XMP/48 CPU times, Re=500 runs.

dof = 12231 - 2065 = 10166 (Penalty methods)		
dof = 12833 - 2065 = 10768 (Continuous pressure)		
Element type	**Weighting**	**CPU(secs)/Pic.it.**
Q27PC8	Galerkin	76
Q27PD8	Galerkin	64
Q27PD4	Galerkin	65
Q27PC8	SUPG opt.	79
Q27PD8	SUPG opt.	67
Q27PD4	SUPG opt.	67

Table 4: Cray-XMP/48 CPU times, Re=1093 runs.

constrained dof = 2429		
* pressures are solved for, ie: $\Upsilon = \infty$		
Element type	**Weighting**	**CPU(secs)/Pic.it.**
Q27PC8	Galerkin	158
Q27PC8	SUPG 1/6	158
Q27PD8	Galerkin	131
Q27PD4	Galerkin	130
Q27PD4F*	Galerkin	190

continuous tri-linear pressure. The biggest difficulty was the non-linear convergence of the scheme. More 'slippery' boundary conditions near the walls would be of advantage as far as the accuracy of the computations is concerned but these result in poorer non-linear convergence.

A discontinuous pressure approximation should be very advantageous for good non-linear convergence rates since the pressure gradients in the flow are much sharper than for the laminar flow cases studied. The difficulty lies in the implementation of a penalty method which is a very localised number with turbulence modelling. Explicit computation of the discontinuous pressures on the other hand would be prohibitive. The implementation of a penalty method in turbulence flow is not impossible and would be of interest for reasons associated with good non-linear convergence, and stability of the numerical solutions obtained with the Galerkin finite element method.

2.4 CPU times for laminar flow

Table 3 and table 4 give an idea of the CPU times per Picard iteration required for the solution of (16) and (17) on the two meshes used in laminar flow, under the various methods of approximation.

References

[1] J. Boussinesq. *Essai sur la theorie des eaux courantes.* Mem. pres. Acad. Sci. XXIII, vol. 4, no. 1, pg. 1-680, Paris 1877.

[2] C.A. Brebbia. *The unification of finite elements and boundary elements* in *Unification of finite element methods.* Ed. H. Kardestuncer. North-Holland Mathematics Studies 94, Elsevier Science Publishers B.V., 1984.

[3] A.N. Brooks and T.J.R. Hughes. *SUPG formulations for convection dominated flows with particular emphasis on the incompressible Navier-Stokes equations.* Comp. Meth. Appl. Mech. Engng., vol. 32, pg. 199-259 , 1982.

[4] W.R. Dean. *The streamline motion of a fluid in a curved pipe.* Phil. Mag., vol 7, no. 4, pg 208, and no. 5, pg. 673, 1928.

[5] M.M. Enayet, M.M. Gibson, A.M.K.P. Taylor and M. Yianneskis. *Laser doppler measurements of laminar and turbulent flow in a pipe bend.* NASA contract report 3551. Contract NASW-3258, May 1982.

[6] P. Hood. *Frontal solution program for unsymmetric matrices.* Int. J. Num. Meth. Engng., vol. 10, pg. 379-399.

[7] D. Howard. *Numerical Techniques for the Simulation of Three Dimensional Swirling Flow.* Ph.D. thesis, University College of Swansea, 1988.

[8] D. Howard, W.M. Connolley and J.S. Rollett. *Unsymmetric conjugate gradient methods and sparse direct methods in finite element flow simulation.* Int. J. Num. Meth. in Fluids, vol. 10, pg. 925-45, 1990.

[9] T.J.R. Hughes. *The finite element method: linear static and dynamic finite element analysis.* Prentice-Hall International Inc., 1987.

[10] H. Iacovides. *Momentum and heat transfer in flow through* 180° *bends of circular cross section.* PhD thesis. UMIST 1986.

[11] B.E. Launder and D.B. Spalding. *The numerical computation of turbulent flows.* Comp. Meth. Appl. Mech. Engng., vol. 3, pg. 269-289, 1974.

[12] D.S. Malkus and T.J.R. Hughes. *Mixed FEM - Reduced and selective integration techniques: A unification of concepts.* Comp. Meth. Appl. Mech. Engng., vol. 15, pg. 63-81, 1978.

[13] G.L. Mellor and H.J. Herring. *A survey of the mean turbulent field closure models.* AIAA Journal, vol. 11, no. 5, pg. 590-599, May 1973.

[14] S. Nakazawa. *Finite element analysis applied to polymer processing.* PhD thesis, Chemical Engineering, University College of Swansea, 1982.

[15] C. Taylor, J. Rance and J.O. Medwell. *A note on the imposition of traction bondary conditions when using the FEM for solving incompressible flow problems.* Proc. Int. Conf. Num. Meth. Lam. and Turb. Flow, pg. 345-352, 1985.

[16] C. Taylor, C.E. Thomas, and K. Morgan. *Analysis of turbulent flow with separation using the FEM.* in chap. 10 of *Computational techniques in transient and turbulent flow.* Ed: C. Taylor and K. Morgan, Pineridge Press, Swansea 1981.

[17] C.E. Thomas. *Analysis of confined turbulent flows.* PhD thesis, Civil Engineering, University College of Swansea, 1982.

[18] C.-C. Yu and J.C. Heinrich. *Petrov–Galerkin method for multidimensional, time-dependent, convective–diffusion equations.* Int. J. Numer. Meth. Engng., vol. 24, 2201-2215, 1987.

[19] O.C. Zienkiewicz, J. P.-Villotte and S. Toyoshima. *Iterative method for constrained and mixed approximation. An inexpensive improvement of FEM performance.* Internal report C/R/489/84, Inst. for Num. Meth. in Engng., Univ. College of Swansea, 1984.

Preconditioning Techniques for the Numerical Solution of the Stokes Problem

J. Atanga, D. Silvester

Department of Mathematics, University of Manchester Institute of Science and Technology, P.O. Box 88, Manchester M60 1QD, U.K.

Abstract

Mixed finite-element discretisation of the Stokes equations gives rise to an indefinite linear system of algebraic equations of the form

$$\begin{pmatrix} A & B^t \\ B & O \end{pmatrix} \begin{pmatrix} u \\ p \end{pmatrix} = \begin{pmatrix} f \\ g \end{pmatrix}$$

for the velocity and pressure fields. Applying block Gaussian elimination this indefinite problem can be transformed into a symmetric positive definite system

$$BA^{-1}B^t p = BA^{-1}f - g,$$

for the pressure unknowns. A standard conjugate gradient method can then be applied. Each iterative stage requires the solution of two discrete Poisson problems, which can also be solved using a standard conjugate gradient method, so the complete algorithm is a two-level iterative process. In some of our recent work [1], this idea is extended to the case when the underlying mixed finite element approximation is not stable in a Babuška-Brezzi sense, and in this paper we build on this work. In particular, we discuss different preconditioning techniques for both levels of iteration, and present results of some associated numerical experiments.

Key words and phrases. Stabilised finite element, Stokes equation.

1. Introduction.

The Preconditioned Conjugate Gradient (PCG) method is an efficient iterative procedure for solving linear systems of equations with a symmetric positive definite matrix, as arising from finite element discretisations of elliptic self-adjoint partial differential equations. Recent effort has centred on the development of efficient preconditioners for such problems so as to ensure rapid convergence. The use of diagonal (scaling) and incomplete factorisation preconditioners is standard practice (see for example, Kershaw [2], and Meijerink and Van der Vorst [3]). These preconditioners require knowledge only of the linear system to be solved and not of the underlying differential equation. When applied to a typical discretised differential operator like the Laplacian, the effectiveness of these preconditioners deteriorates as the grid size gets smaller. In recent years more sophisticated preconditioners have been developed. These ensure that the spectral condition number of the preconditioned system remains bounded independent of the grid size, or grows only logarithmically. One such preconditioner, which we consider here, results from the use of a hierarchical finite element basis instead of the more standard Lagrangian basis.

In this paper, we aim to solve the linear system of equations

$$\begin{pmatrix} A & B^t \\ B & O \end{pmatrix} \begin{pmatrix} u \\ p \end{pmatrix} = \begin{pmatrix} f \\ g \end{pmatrix}, \qquad (1.1)$$

originating from a mixed finite element approximation of the Stokes equation. Our discussion here will be based on a two-dimensional discretisation, although all the algorithms we discuss extend quite easily to the three-dimensional case. In the above, A is a symmetric positive definite $N_v \times N_v$ matrix, B is $N_p \times N_v$ matrix corresponding to the coupling term and is of full rank if a stable mixed method is used, O is the $N_p \times N_p$ zero matrix, f and g are vectors resulting from inhomogenous data, and u and p denote vectors of length N_v and N_p corresponding to nodal velocity and pressure values respectively. For a full description, see for example Thomasset [4].

In recent years, techniques of regularisation have been developed which allow the use of low-order conforming mixed approximations of incompressible flows, and provide reasonable results for both velocity and pressure fields (cf. Brezzi and Pitkaranta [5], Hughes and Franca [6], Brezzi and Douglas [7]). One such technique, which we pursue here, is the symmetric regularisation of Silvester and Kechkar [8], which gives rise to the *stabilised* linear system

$$\begin{pmatrix} A & B^t \\ B & -\beta C \end{pmatrix} \begin{pmatrix} u \\ p \end{pmatrix} = \begin{pmatrix} f \\ g \end{pmatrix}, \tag{1.2}$$

where C is a symmetric positive semi-definite matrix such that the 'stabilisation condition'

$$B^t p = 0 \quad \Rightarrow \quad p^t C p \neq 0 \tag{1.3}$$

is satisfied, and $\beta > 0$ is the stabilisation parameter.

A standard way of devising iterative solution schemes for the indefinite linear system (1.2) is to apply block Gaussian elimination to reduce equation (1.2) to

$$(\beta C + BA^{-1}B^t)p = BA^{-1}f - g, \tag{1.4}$$

uncoupling the discrete pressure field. The discrete velocity field is then given via

$$Au = f - B^t p. \tag{1.5}$$

The stabilised pressure matrix $K_\beta \equiv \beta C + BA^{-1}B^t$ is symmetric and positive definite. When scaled by the pressure mass matrix M, it gives rise to a matrix $M^{-1/2}(\beta C + BA^{-1}B^t)M^{-1/2}$ which has a spectral condition number independent of the grid parameter h (for a proof see Atanga [9]). Consequently, equation (1.4) can be solved efficiently using a preconditioned conjugate gradient method, giving rise to an 'outer iteration'. The computation of a (velocity) search direction ${s_v}^i$, at the ith outer iteration, is then associated with a linear system of the form

$$A{s_v}^i = B^t {s_p}^i \tag{1.6}$$

where ${s_p}^i$ is the current (pressure) search direction, and matrices A and B^t are as defined above. Note that, for a standard two-dimensional Stokes formulation, the solution of (1.6) amounts to solving two discrete Poisson problems, one for each component of velocity. If a preconditioned conjugate gradient algorithm is then applied to (1.6), an 'inner iteration' is introduced, so the overall algorithm is obviously a two-level iterative process.

The aim of this paper is to discuss the use of different preconditioners for both levels of iteration. In Section 2 the preconditioning of the outer iteration is considered. A hierarchical basis preconditioner for solving the discrete Poisson problems associated with the inner iteration is then developed in Section 3. The results of numerical experiments are presented in Section 4.

2. Preconditioning the Outer Iteration.

For a quasi-uniform sequence of grids the system matrix K_β in (1.4) can be well-conditioned without scaling by M the pressure mass matrix. In such a case the potential gains arising from preconditioning the outer iteration will be limited. However, for a non-quasi-uniform succession of grids the spectral condition number of the stabilised pressure matrix does not tend to a constant, and so the number of iterations required to satisfy some tolerance, increases with each refinement of the grid. The use of a preconditioner for the outer iteration becomes neccessary in this case, especially bearing in mind that at each outer iteration a number of inner iterations have to be performed. In the standard case of a stable mixed method, several preconditioners have been proposed to this end, (see for example, Cahouet and Chabard [10]). The value of this is not clear however, if the preconditioner is either difficult to construct, or else is expensive to apply. In [11], Vincent and Boyer discuss the preconditioning of the outer iteration in the context of the local stabilisation method [8]. They present results which demonstrate the efficiency of a preconditioner which is both simple to construct and fully exploits the block diagonal structure of the stabilisation matrix arising in [8]. Our experience suggests that the need for such a preconditioner is only really apparent in the case of a non-quasi-uniform refinement strategy, in which case it appears to be crucial for the success of the two-level approach.

In the light of these remarks, further illustrated by our results in Section 4, it is evident that the critical component of the overall algorithm is the inner iteration, the preconditioning of which is now addressed.

3. A Hierarchical Basis Preconditioner for the Inner Iteration.

As discussed above, in the solution of (1.2) the matrix A is the finite element approximation of the (vector–)Laplacian, thus its condition progressively gets worse as the grid is refined, irrespective of the uniformity of the grid. Indeed, it is well known (see for example, Johnson [12]) that the spectral condition number of a uniform grid discretisation of the Laplacian is $O(h^{-2})$ where h is the grid parameter. Preconditioning the inner iteration is therefore of crucial importance to the success of the overall scheme.

A relatively recent notion in finite element methodology is that of replacing the standard basis functions by 'hierarchical' basis functions (cf. Zienkiewicz et al. [13], Axelsson and Gustafsson [14], Yserentant [15,16]). Using this approach the condition number of the discretised two-dimensional Laplacian can be reduced from $O(h^{-2})$ to $O(\log h^{-1})^2$, so that the performance of any iterative solution algorithm will be greatly improved.

As an illustrative example, we focus here on a mixed finite element approximation of the Stokes equations using a piecewise bilinear velocity and a piecewise constant pressure (the Q_1–P_0 mixed method). The preconditioning of the inner iteration will be achieved by simply replacing the standard basis functions for the velocity field by hierarchical basis functions, while maintaining a standard basis for the pressure field.

To describe the 'hierarchical preconditioner' in a general setting, we assume that the domain Ω is a bounded open region, and that we can construct some initial subdivision consisting of rectangular elements. The generalisation to a quadrilateral subdivision is straightforward. The stiffness matrix $A^{(0)}$ then corresponds to a standard piecewise bilinear set V_0 of basis functions defined on the initial mesh Ω_0, with entries

$$(A^{(0)})_{i,j} = a(\phi_i{}^{(0)}, \phi_j{}^{(0)}), \tag{3.1}$$

where $V_0 = \{\phi_i{}^{(0)}; i = 1, 2 \ldots, N_0\}$. $a(\cdot,\cdot)$ is a symmetric bilinear form associated with a vector or scalar Laplacian.

Each rectangle of the initial mesh is now divided into four. A set of new mesh nodes $N_1{}^{(1)}$ can then be associated with the set of basis functions $V_1{}^{(1)} = \{\phi_i{}^{(1)}; i = 1, 2 \ldots, N_1 - N_0\}$. This set of basis functions $V_1{}^{(1)}$ has support on the four rectangles (or fewer for boundary nodes) which have the corresponding node as a vertex. Such a basis function is bilinear on each of these rectangles, continuous on the edges, and zero outside the union of them. The total set of basis functions on the second level is $V_1 = V_1^{(1)} \cup V_0$. If we number the nodes such that the new points come first then the overall stiffness matrix has the following form,

$$A^{(1)} = \begin{pmatrix} A_{11}{}^{(1)} & A_{12}{}^{(1)} \\ A_{21}{}^{(1)} & A^{(0)} \end{pmatrix}. \tag{3.2}$$

Here

$$(A_{11}{}^{(1)})_{i,j} = a(\phi_i{}^{(1)}, \phi_j{}^{(1)}) \qquad i,j = 1, \ldots, N_1{}^{(1)}$$

$$(A_{12}{}^{(1)})_{j,i} = a(\phi_i{}^{(0)}, \phi_j{}^{(1)}) \qquad i = 1, \ldots, N_0 \qquad j = 1, \ldots, N_1{}^{(1)}$$

$$A_{21}{}^{(1)} = A_{12}{}^{(1)^t}$$

with A_0 being the stiffness matrix on the first level.

This procedure can be generalised to k levels, when we have the set of node points $N_k = N_k{}^{(1)} \cup N_{k-1}$. In this case $N_k{}^{(1)}$ is the set of new node points on level k, and the corresponding set of basis functions is defined by $V_k{}^{(1)} = \{\phi_i{}^{(k)}; i = 1, 2, \ldots, N_k{}^{(1)}\}$. The total set of basis functions on this level is $V_k = V_k{}^{(1)} \cup V_{k-1}$ The stiffness matrix on level k is then of the form

$$A^{(k)} = \begin{pmatrix} A_{11}{}^{(k)} & A_{12}{}^{(k)} \\ A_{21}{}^{(k)} & A^{(k-1)} \end{pmatrix}. \tag{3.3}$$

with

$$(A_{11}{}^{(k)})_{i,j} = a(\phi_i{}^{(k)}, \phi_j{}^{(k)}) \qquad i,j = 1, \ldots, N_k{}^{(1)}$$

$$(A_{12}{}^{(k)})_{j,i} = a(\phi_i{}^{(k-1)}, \phi_j{}^{(k)}) \qquad i = 1, \ldots, N_{k-1} \qquad j = 1, \ldots, N_k{}^{(1)}$$

$$A_{21}{}^{(k)} = A_{12}{}^{(k)^t}$$

Although for an 'optimal' method a general k-level scheme is required, the macro-element construction of the local stabilisation process [8] lends itself to a two-level scheme in a very natural way. Hence we restrict attention to a simple two-level scheme here. Our aim is then limited to that of showing the degree of improvement possible with only a minor change in the data structure of the stabilised mixed method. To this end, working in a locally stabilised Stokes framework using a two-level hierarchical basis for the velocity, and a standard basis for the pressure, equation (1.2) is modified to

$$\begin{pmatrix} \tilde{A} & \tilde{B}^t \\ \tilde{B} & -\beta C \end{pmatrix} \begin{pmatrix} \tilde{u} \\ p \end{pmatrix} = \begin{pmatrix} \tilde{f} \\ \tilde{g} \end{pmatrix}, \tag{3.4}$$

Furthermore it is easily shown that

$$\tilde{A} = S^t A S$$

$$\tilde{B} = BT$$

where S and T are transformation matrices representing the change in the definition of the piecewise bilinear basis defining the discrete velocity field. Note that the solution to (3.4) above gives the nodal values of the pressure and velocities defined at the coarse grid (macro-element) vertices, but that the components in the solution vector corresponding to the additional fine grid nodes are to be interpreted as corrections to the interpolated values of velocity on the coarse grid.

Reformulating equation (3.4) as in Section 1, we obtain the hierarchical basis stabilised pressure equation,

$$(\beta C + \tilde{B}\tilde{A}^{-1}\tilde{B}^t)p = \tilde{B}\tilde{A}^{-1}\tilde{f} - \tilde{g}. \tag{3.5}$$

A 'preconditioning' of the inner iteration is then achieved by applying the basic two-level algorithm of Section 1 to equation (3.5) instead of to the original system (1.4). The point being that the coefficient matrix for the inner iteration is then $\tilde{A}$, whose condition is much better than that of A. This is illustrated by the numerical results in the next section.

4. Numerical Results.

In this section some numerical results are presented to show the effectiveness of the hierarchical preconditioner described above. As a test example we solved the lid-driven cavity problem. By exploiting the symmetry of the Stokes solution only half the domain was modelled. The discretisation was by uniformly refined grids of locally stabilised Q_1–P_0 square elements. As pointed out above, a feature of the local stabilisation [8] is that the grid must be constructed from appropriate macroelements to ensure stability, hence the mixed method is ideally suited to a two-level hierarchical basis construction.

The iteration counts we present correspond to the case of standard CG for the outer iteration. The tolerance for convergence was a reduction of 10^{-6} in the L_2–norm of the residual at both levels of iteration. The computations were done on an Amdahl VP1100 vector supercomputer.

	$\beta = 10^{-2}$		$\beta = 1$		$\beta = 10^{2}$	
Grid	Outer	Inner	Outer	Inner	Outer	Inner
2×4	9	54	8	48	8	46
4×8	18	252	17	238	22	307
8×16	25	722	20	577	27	789
16×32	26	1484	21	1193	27	1581
32×64	24	2686	20	2234	27	3099
64×128	22	4843	19	4141	24	5400

Table 4.1.

Number of iterations with standard basis functions

The first table (4.1) shows the number of iterations for some different values of β using standard basis functions, (i.e. without preconditioning the inner iteration). Clearly the number of outer iterations tends to a constant as $h \to 0$, while the number of inner iterations grows like $O(h)$, in agreement with standard CG convergence theory, (see [12] for details). The need for a preconditioner for the inner iteration is obvious.

	$\beta = 10^{-2}$		$\beta = 1$		$\beta = 10^2$	
Grid	Outer	Inner	Outer	Inner	Outer	Inner
2×4	9	54	8	48	8	48
4×8	18	250	17	237	22	302
8×16	25	525	20	424	27	570
16×32	26	926	21	750	27	971
32×64	24	1607	20	1344	27	1839
64×128	22	2876	19	2468	24	3178

Table 4.2

Number of iterations with hierarchical basis functions

The second table (4.2) gives iteration counts for the same problem except using the two-level hierarchical basis for the velocity space as discussed above. As expected the condition of the matrix $\tilde{A}$ is much improved and this is reflected by the inner iteration count. The point to note is that the number of inner iterations on a given grid in Table 4.2 is comparable with that of the inner iteration count in Table 4.1 corresponding to the same value of β on the corresponding coarse grid, (consistent with the hierarchical basis theory). The outer iteration counts in the two tables are identical, showing that the condition of the stabilised pressure matrix is unaffected by the change in basis of the velocity approximation subspace.

The stabilisation parameter β does not seem to affect the number of iterations much, but the best value appears to lie in the interval $(10^{-2}, 1)$. A full discussion on the way that the choice of β affects solution accuracy is given in [8]. A value of $\beta = 0.1$ is 'optimal' in both respects.

5. Concluding Remarks.

For the particular finite element spaces used here, the application of a two-level hierarchical basis preconditioner with a two-level iteration scheme has demonstrated some satisfactory results. The extension of this into a multi-level hierarchical basis preconditioner could provide very efficient iterative solvers for incompressible flow problems.

References.

1. J. Atanga and D. Silvester, "Iterative methods for stabilised mixed velocity-pressure finite elements", to appear in *Int. J. Numer. Methods in Fluids.*

2. D. Kershaw, "The incomplete Cholesky conjugate gradient method for the iterative solution of linear equations", *J. Comput. Phys.*, v. 26, 1978, pp. 43–65.

3. J. Meijernik and M. Van der Vorst, "An iterative solution method for linear systems of which the systems matrix in a symmetric M-matrix", *Math. Comp.*, v. 31, 1977, pp. 148–162.

4. F. Thomasset, *Implementation of finite element methods for Navier-Stokes equations*, Springer Series in Computational Physics, Springer-Verlag, Berlin, 1981.

5. F. Brezzi and J. Pitkäranta, "On the stabilisation of finite element approximations of the Stokes problem," *Efficient Solutions of Elliptic Systems*, Notes on Numerical Fluid Mechanics Vol.10 (W. Hackbusch, ed.), Vieweg, Braunschweig, 1984, pp. 11–19.

6. T.J.R. Hughes and L.P. Franca, "A new finite element formulation for CFD: VII. The Stokes problem with various well-posed bound-

ary conditions: Symmetric formulations that converge for all velocity/pressure spaces," *Comput. Methods Appl. Mech. Engrg.*, v. 65, 1987, pp. 85–96.

7. F. Brezzi and J. Douglas, "Stabilized mixed methods for the Stokes problem," *Numerische Mathematik*, v. 53, 1988, pp. 225–235.

8. D.J. Silvester and N. Kechkar, "Stabilised bilinear-constant velocity-pressure finite elements for the conjugate gradient solution of the Stokes problem", *Comput. Methods Appl. Mech. Engrg.*, v. 79, 1990, pp. 71–86.

9. J.N. Atanga. Ph.D. Thesis. In preparation.

10. J. Cahouet and J.P. Chabard, "Some fast 3D finite element solvers for the generalised Stokes problem", *Int. J. Numer. Methods in Fluids*, v. 8, 1988, pp. 869–895.

11. C. Vincent and R. Boyer, "A preconditioned conjugate gradient Uzawa-type method for the solution of the Stokes problem by mixed Q1–P0 stabilized finite elements", to appear in *Int. J. Numer. Methods in Fluids*.

12. C. Johnson, *Numerical solution of partial differential equations by the finite element method*, Cambridge University Press, Cambridge, 1990.

13. O.C. Zienkiewicz, D.W. Kelly, J. Gago, and I. Babuska, "Hierarchical finite element approaches, error estimates and adaptive refinement," in *The Mathematics of Finite Elements and Applications IV*, (J.R. Whiteman, ed.), Mafelap 1981, London, 1982.

14. O. Axelsson and I. Gustafsson, "Preconditioning and two-level multigrid methods of arbitrary degree of approximation", *Math Comp.*, v. 40, 1983, pp. 219–242.

15. H. Yserentant, "On the multi-level splitting of finite element spaces", *Numerische Mathematik*, v. 49, 1986, pp. 379–412.

16. H. Yserentant, "On the multi-level splitting of finite element spaces for indefinite elliptic boundary value problems", *SIAM J. Numer. Anal.*, v. 23, 1986, pp. 581–595.

Prediction of Swirling Flow in a Corrugated Channel

F.S. Henry (*), M.W. Collins (*), M. Ciofalo (**)
(*) Thermo Fluids Engineering, Research Centre, City University, London, ECIV OHB, U.K.
(**) Department of Nuclear Engineering, University of Palermo, 90128 Palermo, Italy

ABSTRACT

Laminar and turbulent flow in a corrugated channel is simulated using HARWELL-FLOW3D. The channel represents a typical cell within a proposed design for the heat-transfer element of a rotary regenerator. Velocity vectors in planes perpendicular and parallel to the axis of a corrugation are visualised using Harwell's graphics package, OUTPROC. Both programs were run on Harwell's Cray 2. The velocity vector plots show clear evidence of swirl, which is thought to be the mechanism responsible for the high rates of heat transfer in this type of heat-exchanger geometry. The swirl strength is shown to be a function of the channel geometry.

INTRODUCTION

Regenerators are used to increase the thermal efficiency of power plants. This is achieved by recovering some of the heat that would otherwise be exhausted to the atmosphere. While there are a variety of designs currently in use, this paper is restricted to consideration of the details of the flow in the heat-exchanger of a type known as the rotary regenerator. This type of regenerator is used, for example, to preheat the combustion air supply of fossil-fired power plants. Discussion of the use of rotary regenerators as combustion-air preheaters can be found in Chojnowski and Chew [1], and Chew [2], among others. The heat-exchanger of a rotary regenerator is a porous metal drum, or cylinder. The cylinder is composed of a matrix of closely packed corrugated steel plates, and is typically 10 metres in diameter and two metres deep. The main flow direction is aligned with the axis of the cylinder. The cylinder rotates slowly, and is exposed alternately to hot exhaust gases and cool incoming air. This results in a transfer of heat from the exhaust gases to the incoming air.

Inherent in the basic design of rotary regenerators are a number of interesting problems. These include: minimizing leakage between the hot and cool gas streams; accommodating distortion of the drum due to thermal expansion; reducing carry over of hot gases into the incoming air stream; and keeping fouling to a minimum. Obviously, there is a connection between fouling and the geometrical design of the heat-transfer element. Also, as might be expected, some geometrical designs have higher heat-transfer rates than others. The mechanism thought responsible for the enhanced rates of heat transfer is the subject of this paper. It should be noted that the flow geometry under consideration is found in various other types of heat-transfer equipment, and hence, the results discussed are not restricted solely to rotatory regenerators.

The predictions to be discussed are part of a comprehensive experimental and numerical study of the flow and heat transfer in crossed-corrugated channels. The work is funded by PowerGen, U.K. Other aspects of this work have been reported in Ciofalo, Collins and Perrone [3], and Stasiek, Collins, and Chew [4]. The flows to be described were calculated using

HARWELL-FLOW3D. Of particular interest was the detection of any swirl in the flow, as it was thought that swirl might play a significant role in producing the measured high heat-transfer rates. The geometry of the problem makes it exceedingly difficult to detect swirl experimentally; however, a numerical simulation is well suited for this task. The predicted flow fields were of such complexity that it was necessary to employ a graphics package to search for the swirl, and Harwell's OUTPROC was used for this purpose. Both programs were run on Harwell's Cray 2 using the IBM frontend. While, both FLOW3D and OUTPROC can be run on machines other than a Cray, the complex three-dimensional nature of the present application, made it essential to use a supercomputer in order to achieve realistic turn-around times.

FLOW GEOMETRY

Various corrugation designs have been developed, largely on an empirical basis. The design used for this study, shown in Figure 1, was chosen largely on the basis that it was relatively easy to model numerically. However, the design chosen does resemble some of the proposed high-efficiency configurations discussed by Chojnowski and Chew [1], and Chew [2]. The design considered is formed by placing one sinusoidally corrugated plate on top of another. The plates are arranged so that the corrugations of one plate are at an angle, θ, to those in the other. Fluid flows through the two sets of parallel channels thus formed, and the two flow streams share a common interface. For the purposes of this paper, the corrugated sheets are assumed to be infinite in extent. This allows a unitary cell, shown in Figure 2, to be considered. This cell has two inlets and two outlets. The flow in the cell is assumed to be fully developed and thus doubly periodic; i.e., in directions parallel to the corrugations.

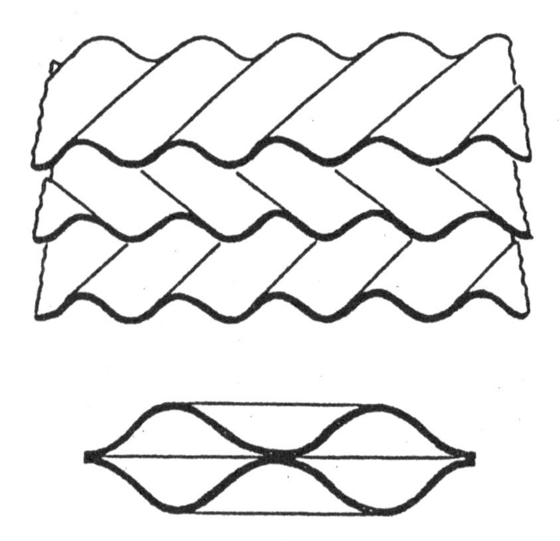

Figure 1. Perspective view (upper) and section normal to main flow (lower) of the crossed corrugated heat transfer elements.

NUMERICAL SOLUTION OF FLOW EQUATIONS

Discrete versions of the Navier-Stokes equations were solved using HARWELL-FLOW3D (Release 2.1), which is a general, three-dimensional, flow modelling code developed by AEA Technology Harwell. The code solves for both laminar and turbulent flows, and for heat transfer. In the turbulent case, the Reynolds stresses are estimated using the k–ε model, and boundary conditions on solid walls are satisfied using standard wall functions. FLOW3D uses a control volume approach on a non-staggered grid. The code offers the option of non-orthogonal,

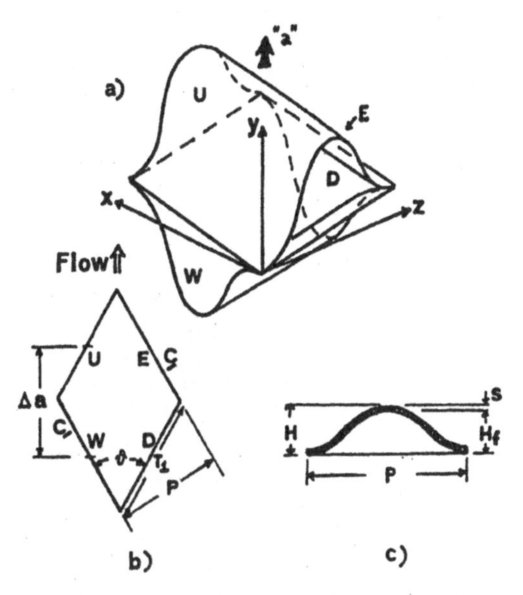

Figure 2. Perspective view (a); section at y=0 (b); and section C-C, normal to top corrugation, (c) of the unitary cell.

body-fitted coordinates which were necessary for the flow geometry under consideration. Details of the development of FLOW3D with body-fitted coordinates have been given by Burns et al. [5]. General details of FLOW3D have been given by Jones et al. [6].

The inlet/outlet pairs D/U and W/E, see Figure 2, were designated to be planes of periodicity. As the pressure is inherently non-periodic in the main flow direction, it was necessary to modify the pressure terms in the x- and z-momentum equations. In both equations, the pressure term was decomposed into a periodic component, p', and a linearly varying component; i.e., for the x-momentum equation, the pressure gradient can be written

$$-\frac{\partial p}{\partial x}=-\frac{\partial p'}{\partial x}+\frac{\Delta p}{\sqrt{2}\Delta a}.$$

where Δp is the mean pressure drop between inlet and outlet. The distance Δa is the mean distance between inlet and outlet (see Figure 2). A similar expression can be written for the pressure gradient in the z-direction. The above decomposition was achieved by adding the term $\Delta p/\sqrt{2}\Delta a$ to the momentum equations in the form of a source.

The equations were solved for an imposed mass flow rate. This was specified in terms of a required Reynolds number. Thus Δp had to be found as part of the iterative solution. At each iteration, Δp was adjusted by a factor proportional to the ratio of the current Reynolds number to the required value. The adjustment was underrelaxed to reduce oscillation in the solution. To further reduce oscillation, the modification outlined in Henry and Collins [7] was included. This entailed keeping a running average of Δp and reinitialising the current value with the average every fifty iterations.

GRIDDING

Only brief details of the grid construction will be presented here, as full details have been given elsewhere. See, for instance, Fodemski and Collins [8]. Essentially, construction of the non-orthogonal grid is divided into two parts. First, an auxiliary, two-dimensional, grid is defined on the left-hand half of the inlet face of the top channel (the face marked 'D' in Figure 2). A sample auxiliary grid is shown in Figure 3. The full three-dimensional grid is then constructed from reflections, translations and rotations of the auxiliary grid. A typical result is shown in Figure 4. Also shown, is the solution domain in computational space. In the turbulent case, the control volumes adjacent to the solid surfaces are set to a predetermined size to ensure that the first computational points are above the viscous sublayer. This latter is an essential restriction of the turbulence model as used.

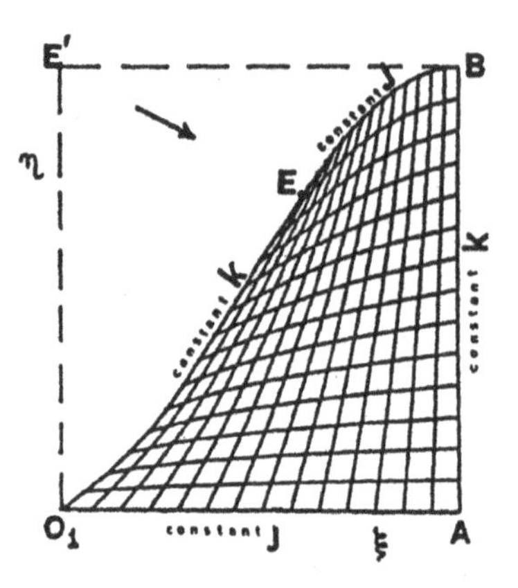

Figure 3. Sample auxiliary grid.

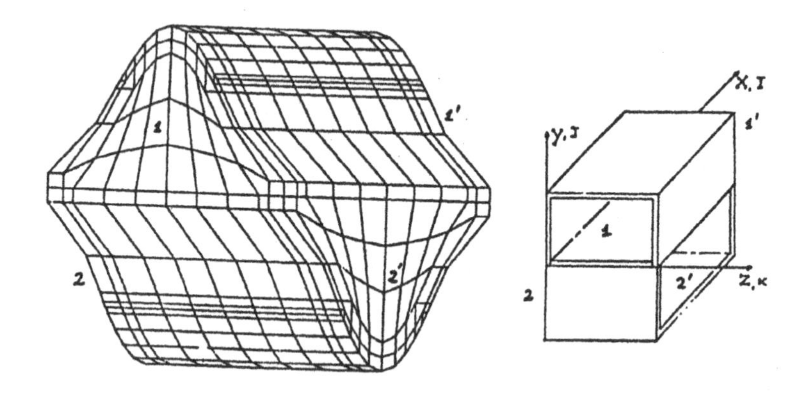

Figure 4. Sample three dimensional grid (left); and channel in computational space (right).

The grid shown in Figure 4 includes the corrugated plate. While this is unnecessary for the present calculations, it was included for future computations of the heat transfer through the plate. In this case, the number of cells across the plate thickness would be increased from the default value of one. The grid shown also includes dummy control volumes which are used by FLOW3D to implement boundary conditions.

GRAPHICS

Velocity vector plots were constructed from the output of FLOW3D using Harwell's post-processing package OUTPROC. In particular, velocity vectors in two planes were considered. The first was a plane aligned with the axis of the top corrugation. This corresponded to one of the planes for which measurements were available. The second was a plane that was normal to the top corrugation axis (plane C-C in Figure 2). It was assumed that it would be in this plane that the maximum swirl would occur.

OUTPROC uses an input data structure that is similar to that used in the Command Language frontend of FLOW3D. The user defines the required picture using a series of commands, subcommands, and keywords in a data file. In the following, only those aspects of OUTPROC that are essential to the discussion will be outlined. Full details of OUTPROC can be found in Jackson and Winters [9] and Winters and Jackson [10]. OUTPROC allows the user to define a plane at any orientation to the computational grid. The plane is defined by specifying two orthogonal lines with a common origin. If the required plane happens to lie on a grid plane, it is only necessary to specify that plane in terms of the grid plane (integer) number. This was the case for plots of velocity vectors in the plane aligned with the top corrugation axis. If, as is the case with plane C-C, the plane does not lie along a grid plane, OUTPROC treats the picture as three-dimensional, and it is necessary to define the perspective view. This is accomplished by defining the position of the eye and the direction in which the eye is looking.

It is also necessary to define the projections of the total velocity vectors onto the plane of interest. This is done in the Fortran subroutine VECUSR. By the nature of the problem, the y-components of velocity are always in the plane of interest. The x- and z-components of the projected velocities can be calculated by considering a rotation of axes. That is, the plane in question can be assumed to align with a rotated z-axis; the z'-axis, say. Hence, denoting the x-component of the projected velocity as u' and the z-component as w', the required components can be calculated using:

$$u' = (w \cos \alpha + u \sin \alpha) \sin \alpha$$

and

$$w' = (w \cos \alpha + u \sin \alpha) \cos \alpha$$

where u and w are the x- and z-components, respectively, of the original velocity vector, and α is the angle between the plane and the z-axis. In the case of plane C-C, $\alpha = \theta/2 - \pi/4$, and in the case of the plane aligned with the corrugation axis, $\alpha = \pi/4 - \theta/2$.

RESULTS AND DISCUSSION

Predicted velocity vectors in plane C-C are shown in Figure 5 for turbulent flow in a corrugation geometry that corresponds to the experimental rig. The velocity vectors show clear evidence of swirl of the flow in the upper corrugation. Due to the symmetry of the problem, it can be expected that the flow in the lower corrugation has a corresponding swirl. It should be noted that the main flow direction in the upper corrugation is normal to the paper, while in the lower corrugation is at an angle, $\pi/2 - \theta$, to the paper. It should also be noted that the double outline of the corrugation is due to the perspective nature of the picture. That is, one outline is in front of plane C-C and the other behind.

The corresponding view of the velocity vectors in the plane aligned with the top corrugation axis is shown in Figure 6. Note that the vectors are drawn to a different scale to that in Figure 5. In the upper part of this picture, the direction of the flow is in the plane of the paper and its magnitude is approximately equal to the mean velocity (5 m/s). However, in the lower

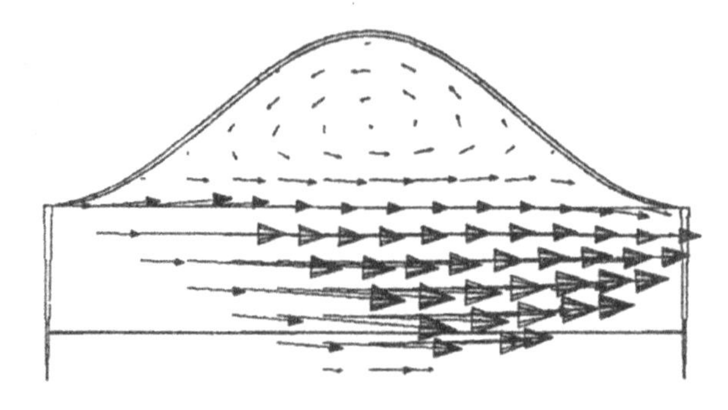

Figure 5. Velocity vectors in plane C-C. Turbulent flow

part of the picture the flow is at an angle, θ, to the paper. Evidence of the effect of swirl can also be seen in the lower part of this figure; i.e., the vectors to the left are deflected up and the vectors to the right are deflected down. The predicted plots are in broad agreement with the PIV measurements of Shand [11]. Detailed comparison is, however, difficult due to the sparseness of the measured velocity field.

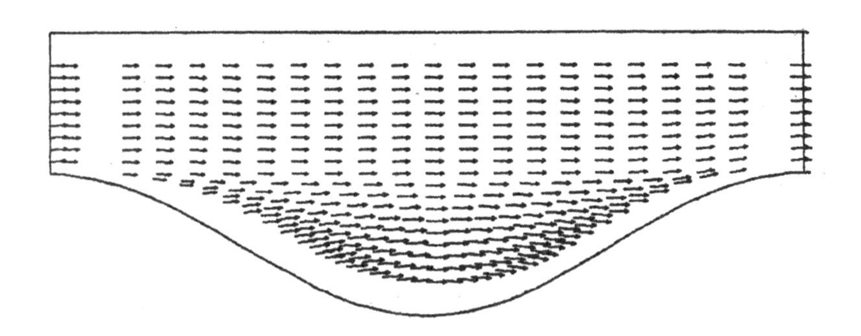

Figure 6. Velocity vectors in a plane aligned with the axis of the top corrugation. Turbulent flow.

The corresponding velocity vector plots for laminar flow are shown in Figures 7 and 8. It should be noted that it was necessary to increase the scale of both pictures from those used for the turbulent plots shown in Figures 5 and 6. Also, the grid used for the laminar calculation was coarser than that used for the turbulent case. In the laminar case, the grid had a total of 16 control volumes in each direction, including the dummy cells; while in the turbulent case, 24 control volumes were used in each direction. It can be seen in Figure 7 that while swirl can again be seen in plane C-C, its strength relative to the main flow appears reduced. In this case, the mean velocity is approximately 1 m/s. Again, the flow in the upper part of Figure 8 is aligned with the paper, and hence, the velocity vectors in this area have a magnitude approximately equal to the mean. Evidence of the reduced strength of the swirl can be seen in the lower part of Figure 8; i.e., it is hard to discern any deflection of the vectors from the horizontal.

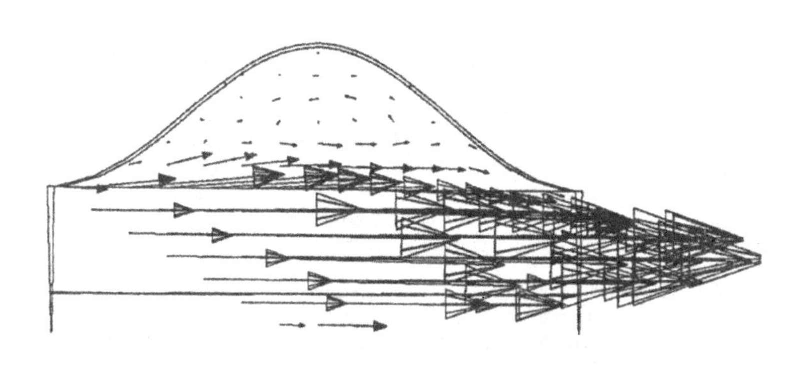

Figure 7. Velocity vectors in plane C-C. Laminar flow

The influence of the corrugation angle, θ, on the swirl strength is shown in the cross plane plots of Figure 9. Except for the angle θ, both channels have the same corrugation geometry. The pitch-to-height ratio is somewhat less than one half that of the experimental rig. The flow is turbulent, and the nominal mass flow rate is the same in both plots. Note that the seemingly flat profile in the lower corrugation of the lower plot is simply due to OUTPROC cropping the picture; i.e., some of the vectors were too large to fit in the defined drawing area. It can be seen that the swirl strength increases as the corrugation angle increases. An explanation for this increase can be seen in the lower portions of the plots. As the angle increases, the flow in the lower part of the plane becomes relatively stronger. That is, the plane rotates towards being completely aligned with the main flow direction of the lower corrugation. Ciofalo et al. [3] have found that heat-transfer rates and overall pressure drop increase with increasing corrugation angle. The plots of Figure 9 would suggest that these increases are related to an increase in swirl.

The influence of the corrugation pitch-to-height ratio on the swirl strength is shown in the cross plane plots of Figure 10. Both channels have the same corrugation geometry except for the value of pitch-to-height ratio. In fact, both plots have the same height; i.e., the pitch-to-height ratio was changed by changing the pitch. Again, the flow is turbulent, and the nominal mass flow rate is the same in both plots. While the effect is not as pronounced as that found for a

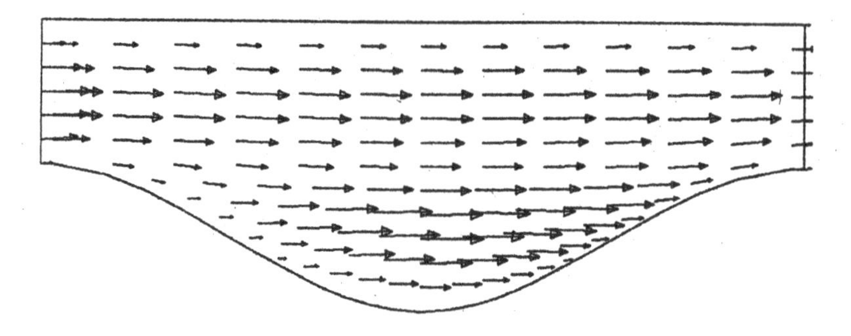

Figure 8. Velocity vectors in a plane aligned with the axis of the top corrugation. Laminar flow.

change in θ, it can be seen that the upper of the two plots of Figure 10 has a slightly stronger swirl. Hence, an increase in pitch-to-height ratio would appear to produce a decrease in swirl. Ciofalo et al. [3] have found that heat-transfer rates and overall pressure drop decrease with increasing pitch-to-height ratio. Hence, the plots of Figure 10 offer further support to the hypothesis that flow swirl is the mechanism responsible for the high heat-transfer rates associated with this type of heat-exchanger geometry.

With one exception, all predictions discussed were calculated on 16×16×16 grids. As mentioned previously, the exception was the calculation for the turbulent case shown in Figures 5 and 6. This was produced using a 24×24×24 grid. The justification for using these rather sparse grids is that only gross features of the flow were of interest, and it was necessary to keep the cost of each run to a minimum. Typically, for the turbulent runs on a 16^3 grid, convergence was achieved in under 1000 iterations. Ciofalo et al. [3] have discussed grid dependence, and other numerical details for similar calculations on an IBM 3090 (without a vector facility). They found grid independence was reached on a 32^3 grid. On the IBM, each iteration of a turbulent run on a 24^3 grid took approximately 15 CPU seconds; whereas, on the Cray 2, a similar job took 1.12 CPU seconds per iteration.

While it has been predicted that the flow is swirling as it passes through the corrugated channel, the turbulent predictions, at least, should be viewed with some caution. The reasons for this are the following. The k–ε model is known to model poorly highly swirling flows. The problem is compounded by the use of wall functions at the solid walls, which in their standard form do not account for swirl. FLOW3D does have an optional Reynolds stress model, which can be expected to model swirling flows more accurately. This may be used in future predictions. Also, the Reynolds numbers were relatively low (reflecting actual conditions in industry) which meant that a significant portion of the flow could not be modelled. That is, the control volumes adjacent to the wall had to be made rather large to ensure the first computational points were in the fully-turbulent part of the flow. To overcome this problem, a Low-Reynolds-number turbulence model, which will not required wall functions, is currently being developed.

A further source of error, which applies equally to the laminar predictions, is that the grid

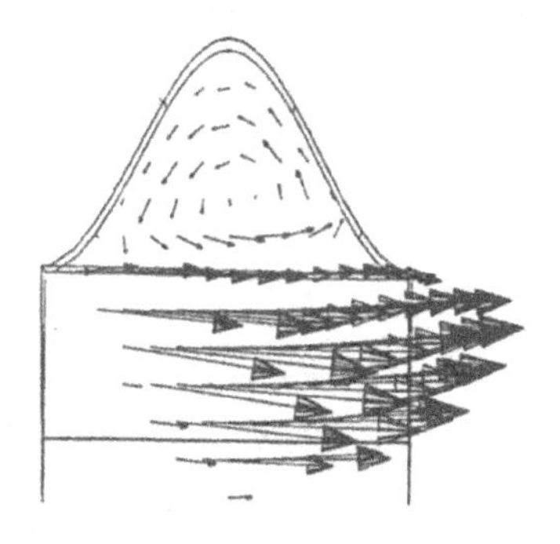

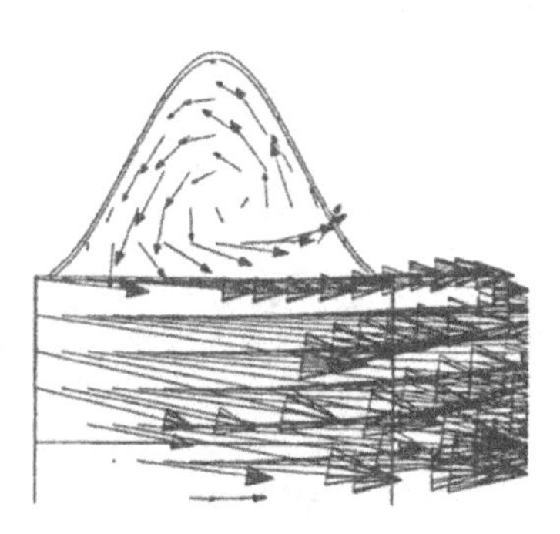

Figure 9. Effect of corrugation angle on velocity vectors in plane C-C. Upper plot: $\theta = 30°$. Lower plot: $\theta = 60°$.

is highly skewed, and significantly non-orthogonal in certain areas on the solid wall. Both conditions are known to cause error. However, calculations for pipe flow using a similar grid revealed that the error in wall values of stress and heat transfer were confined to the area of maximum non-orthogonality, and that the mean values were close to those expected. It should be mentioned that Release 3 of FLOW3D, released after the grid had been designed, includes the capability of a multi-block grid which would probably allow the construction of a more nearly-orthogonal and less skewed grid.

CONCLUSION

Plots of predicted velocity vectors in two planes within the crossed-corrugated channel have been produced using the OUTPROC post-processing package and data generated using FLOW3D. The plots of velocity vectors in planes normal to a corrugation axis show clear evidence of swirl. The swirl appears to be stronger in turbulent flow than in the laminar case. Plots of velocity vectors in a plane aligned with a corrugation axis broadly agree with experimental data. There would appear to be a correlation between an increase in heat-transfer

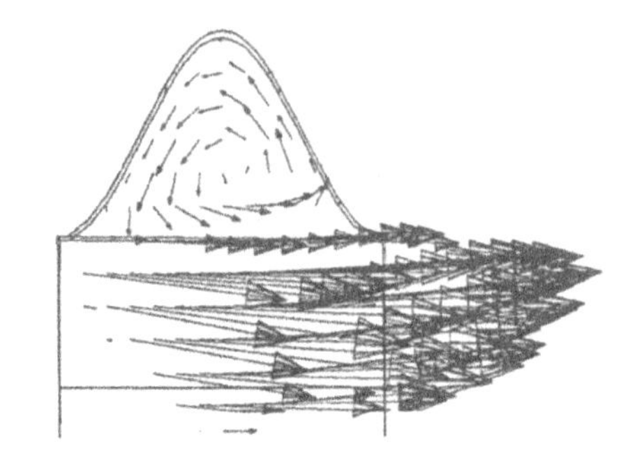

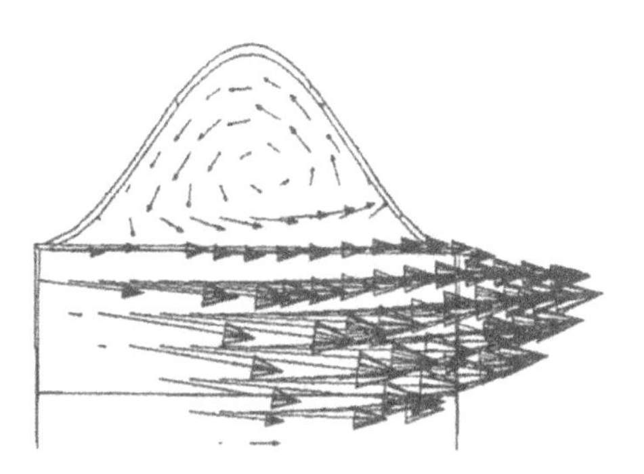

Figure 10. Effect of pitch-to-height ratio on velocity vectors in plane C-C. Upper plot: pitch-to-height ratio = 1.6. Lower plot: pitch-to-height ratio = 2.1.

rate and an increase in swirl. Work is under way to improve and extend the predictions discussed in this paper. In particular, a Low-Reynolds-number turbulence model is being developed.

ACKNOWLEDGMENT

This work was supported by a PowerGen contract number 23965/DEA. The authors would like to thank Mr. Peter Chew at Ratcliffe Technology Centre, and Dr. I.P. Jones and his colleagues at AEA Technology Harwell for many helpful discussions. FLOW3D is, of course, a Harwell propriety code.

REFERENCES

1. Chojnowski, B., and Chew, P.E. Getting the Best out of Rotary Air Heaters, CEGB Research Journal, pp. 14-21, May 1978.
2. Chew, P.E. Rotary Air Preheaters on Power Station Boilers, Proceedings of the

Institute of Energy Symposium on Waste Heat Recovery and Utilisation, Porstmouth, U.K., September, 1985.

3. Ciofalo, M., Collins, M.W., and Perrone, G. Turbulent Flow and Heat Transfer predictions for Cross-corrugated Rotary Regenerators, to appear in the Proc. Eurotech Direct '91 - Thermofluids Engineering, Birmingham, U.K., July, 1991.

4. Stasiek, J., Collins, M.W., and Chew, P. Liquid Crystal Mapping of Local Heat Transfer in Crossed-corrugated Geometrical Elements for Air Heat Exchangers, to appear in the Proc. Eurotech Direct '91 - Thermofluids Engineering, Birmingham, U.K., July, 1991.

5. Burns, A.D., Wilkes, N.S., Jones, I.P., and Kightley J.R. FLOW3D: body-fitted coordinates, Harwell Report AERE-R-12262, 1986.

6. Jones, I.P., Kightley J.R., Thompson, C.P., and Wilkes, N.S. FLOW3D, A Computer Code for the Prediction of Laminar and Turbulent Heat Transfer: Release 1, Harwell Report AERE-R-11825, 1985.

7. Henry, F.S. and Collins, M.W. Prediction of Flow over Helically Ribbed Surfaces, accepted for publication in the International Journal for Numerical Methods in Fluids.

8. Fodemski, T.R., and Collins, M.W. Computer Simulation of Flow and Heat Transfer in a Corrugated Geometry Using the Code FLOW3D, Release 2. Thermo-fluids Engineering Research Centre, City University, Final Report. 1988

9. Jackson, C.P., and Winters, K.H. A Guide to Post-processing Using the TGIN Language, AERE Report, 1987

10. Winters, K.H., and Jackson, C.P. TGIN Language Reference Manual: OUTPUT DATA Commands, AERE Report, 1987.

11. Shand, A.M. Particle Image Velocimetry in a Crossed Corrugated Acrylic Section, AEA Technology Harwell Report AEA-EE-0088, 1990.

On the Numerical Simulation of Shock Waves in an Annular Crevice and its Implementation on the IBM ES/3090 with Vector Facility

R.J. Gathmann (*), F.K. Hebeker (**),
S. Schöffel (***)

(*) Institut de Mécanique de Grenoble, F-38041 Grenoble, France

(**) IBM Scientific Center, D-6900 Heidelberg, Germany

(***) Daimler-Benz AG, D-7000 Stuttgart 80, Germany

ABSTRACT

We report on a new FORTRAN code called PICUS to simulate shock wave propagation in complex geometry, with special emphasis on its implementation on modern supercomputers (here the IBM ES/3090 with vector facility). The L-shaped geometry is decomposed into rectangles in order to get a favourable data structure and to apply locally one-dimensional numerical schemes. We use vanLeer's second-order shock-capturing scheme including Roe's approximate Riemann solver, and we describe how to optimize the resulting code. CPU time measurements recommend PICUS as particularly well fitted for processing on vector (and parallel) processors. The long-time behaviour of PICUS has been checked by computation of the von Kármán vortex street.

INTRODUCTION

Due to its significance to quite a lot of problems of engineering application, the problem of simulating shock waves (propagating and interacting mutually with each other and with walls) by means of numerical tools and computers has found much interest over recent years (see e.g. [3] , [9] and the references cited there). In the present report we will apply some recent numerical methods of CFD, namely the shock capturing methods introduced by vanLeer [8] , Roe [5] et al., with respect to some specific points of our interest. First, we treat an L-shaped domain by means of a domain-decomposition approach. This proves particularly useful in view of our main focus, i.e. how to realize shock-capturing schemes in an efficient way on modern supercomputers, namely on the IBM ES/3090, with respect to vector (and parallel) processing (cf. [7] , too). The new code, called PICUS, has been successfully applied to an important problem of modeling flow under knock conditions in internal combustion engines, a subject to be treated in more detail in forthcoming papers (cf. [6]).

Our numerical approach is summarized as follows. The mathematical model is introduced in Sect. 2, namely the Navier Stokes equations in thin-layer approximation, with appropriate boundary conditions. Then the L-shaped domain is decomposed and, alternating between both subregions, operator and dimensional splitting is used to reduce the numerical effort. The resulting system of conservation laws (including source terms) is treated using a second-order vanLeer scheme including Roe's approximate Riemann solver. This is the content of Sect. 3, and in Sect. 4 we describe briefly the new FORTRAN code PICUS based on the present numerical algorithm. Sect. 5 is devoted to vectorizing and optimizing this code for the IBM ES/3090. Finally, in Sect. 6, as a physical example we present some numerical results of shock waves interfering in the channel and forming a von Kármán vortex street.

The code PICUS has been developed to treat some more sophisticated effects including threedimensional geometry, non-equilibrium chemistry, and coupling with external heat conducting material. But in this short report we restrict ourselves to considering plane, (chemically) inert flow in an L-shaped geometry. But even in this simplified model, the need of supercomputers here is attributed to the requirement of very fine computational grids for resolving the intended physical phenomena with sufficient accuracy.

2. GOVERNING DIFFERENTIAL EQUATIONS

The underlying mathematical model takes advantage of the nonstationary Thin-Layer Navier Stokes equations which read in divergence form

$$\frac{\partial}{\partial t} U + \operatorname{div} F(U) = S(U) \tag{1}$$

where $U = (\rho, \rho u, \rho v, \rho e)^T$,

$$F = \begin{pmatrix} \rho u & \rho v \\ \rho u^2 + p & \rho u v \\ \rho u v & \rho u^2 + p \\ (\rho e + p)u & (\rho e + p)v \end{pmatrix} , \quad S = \begin{pmatrix} 0 \\ \mu u_{yy} \\ \mu v_{yy} \\ \lambda \Theta_{yy} + \mu(u_y^2 + v_y^2) \end{pmatrix}$$

and

$$p = (\kappa - 1)\left(\rho e - \frac{1}{2}\rho(u^2 + v^2)\right) \ , \quad \Theta = \frac{p}{\rho R} \ . \tag{2}$$

Here ρ, u, v, e, p, Θ denote density, velocity in x –, resp. y –direction, specific total energy, pressure, temperature in the flow field, and κ, λ, μ the isentropic exponent, thermal conductivity, and dynamic viscosity of the fluid (all assumed as constants).

The underlying L-shaped domain G for the flow problem studied here is shown in Fig.1 below. This domain is split into two subdomains G_1 and G_2. We assume Euler flow in G_1:

$$\lambda = \mu = 0 \ \ in \ G_1 \tag{3}$$

so that we assume for all flow quantities as boundary conditions:

$$\frac{\partial}{\partial n} = 0 \ \ on \ \Gamma_1 \tag{4}$$

with the exception of vanishing normal velocity:

$$u = 0 \ \ on \ \Gamma_1 \ . \tag{5}$$

On the coupling boundary Γ_2 all corresponding variable are continuously aligned when alternating between both subregions. The open boundary condition on Γ_3 is chosen in a way to allow for smooth outflow of the reflected waves (absorbing boundary condition).

In G_2 we assume the Thin-Layer equations to hold with

$$\lambda \, , \ \mu > 0 \ \ in \ G_2 \ .$$

Consequently, we choose the same boundary condition on Γ_1 as before, but require no-slip conditions on Γ_4:

$$u = v = 0 \ \ on \ \Gamma_4 \, . \tag{6}$$

For the temperature we set

$$\Theta = \Theta_w \ \ on \ \Gamma_4 \tag{7}$$

with given Θ_w. Following Hirschel and Groh [2], we obtain an additional condition for ρ from the continuity equation:

$$\rho_t + \rho v_y = 0 \ \ on \ \Gamma_4 \, . \tag{8}$$

3. THE NUMERICAL APPROACH

Each of the subdomains is decomposed into a total of $(NXj - 3) \times (NYj - 3)$, $j = 1{,}2$ equally sized interior cells, and a lot of further cells serve for smooth data flow between the subdomains as well as for proper modeling of the boundary conditions. Both subregions are treated alternatingly once per time step, so that the following numerical procedure needs to be described for one subdomain only.

For any time step we use dimensional splitting in order to update the flow quantities alternatingly, first in $x-$ then in y –direction. This means that for any time step, for instance in x –direction, the system

$$\frac{\partial}{\partial t} U + \frac{\partial}{\partial x} F^1(U) = S(U) \tag{9}$$

is solved (F^1 denoting the first column of F), subjected to the given boundary conditions.

The diffusive source terms are added explicitly to the conservative variables in a predictor-corrector fashion. This strategy, valid for small diffusive coefficients, is called "rapid solver" algorithm by MacCormack [4] . Consequently, our algorithm is composed as follows. Assume U^n to be given on the time level t_n , compute U^{n+1} from this strategy (where V summarizes the primitive variables ρ, u, v, p corresponding to U, and Δx = cell width, Δt = time step).

Algorithm:

1. Compute

$$(V^n_{i\pm 1/2})^{\mp} = V^n_i \pm \frac{1}{2} \delta V^n_i \tag{10}$$

where

$$\delta V_i = ave(V_{i+1} - V_i ,\ V_i - V_{i-1}) \tag{11}$$

with vanAlbada's slope limiter

$$ave(a,b) = \frac{a+b}{2}\left(1 - \frac{(a-b)^2}{a^2+b^2+\sigma}\right) \tag{12}$$

($\sigma > 0$ a small bias of order $O(\Delta x^2)$).

2. For an intermediate time step $t_{n+1/2}$ compute

$$U^{n+1/2}_i = U^n_i - \frac{\Delta t}{2\Delta x}\{F(V^n_{i+1/2})^- - F(V^n_{i-1/2})^+\} + \frac{\Delta t}{2} S(U^n_i)$$

then

$$\left(V^{n+1/2}_{i\pm 1/2}\right)^{\mp} = V^{n+1/2}_i \pm \frac{1}{2} \delta V^n_i \tag{13}$$

and finally

$$F_{i+1/2} = F_{Roe}\left(\left(V^{n+1/2}_{i+1/2}\right)^-, \left(V^{n+1/2}_{i+1-1/2}\right)^+\right) \tag{14}$$

(see [5] for the averaged fluxes F_{Roe}).

3. Update

$$U_i^{n+1} = U_i^n - \frac{\Delta t}{\Delta x}(F_{i+1/2} - F_{i-1/2}) + \Delta t\, S\left(U_i^{n+1/2}\right). \quad (15)$$

Since the diffusive coefficients are only small, the well-known restrictions for the time step are of the same order of magnitude for treating both the diffusive and the convective parts of the differential equations. This is the reason that, for simplicity, the diffusive terms are treated by means of an explicit scheme. In fact, for maximum time step size Δt_{n+1} we allow

$$\Delta t_{n+1} = 0.9 \min(\Delta t^{\times},\ 2\Delta t^{\times} - \Delta t_n) \quad (16)$$

where

$$\Delta t^{\times} = \left(\frac{1}{\Delta t_D} + \frac{1}{\Delta t_C}\right) \quad (17)$$

(Δt_D , Δt_C denote the maximum time steps due to explicit treatment of the diffusive or convective parts of the differential equations, resp.).

4. THE CODE PICUS

Based on the present algorithm a new FORTRAN code called PICUS has been created (which, nevertheless, is able to treat essentially more involved problems than just the present model problem: see the remarks above). We discuss the efficient implementation of the (two-dimensional) version labeled PICUS-2 (or its basic version PICUS-200) on the IBM ES/3090 with vector facility.

Both subdomains (see Fig.1) are served by a total of six large arrays, *STATEj*, *PLINj*, and *FLUXj* (j=1,2), each of which has the dimension

$$(NXj + 1) \times (NYj + 1) \times 4\ .$$

In the following we restrict ourselves to consider only the domain G_2 only, since for G_1 similar terms hold. In practice

$$NX2 \approx 1000\ ,\quad NY2 \approx 100$$

(underscoring the need of supercomputers).

The solution is stored in *STATE2*, alternatingly in primitive or conservative variables. The array *PLIN2* contains the (limited) slopes as entries of the approximate Riemann solver. And the array *FLUX2* serves as a workspace for both the primitive variables and the fluxes. Consequently, the specified large arrays are exploited as far as possible.

For achieving high performance rates with the vector facility it is crucial that the first argument of these arrays is that of the largest size (*NX*2), leading to long vectors.

This code is structured as follows. The main program contains preprocessing (initialization, domain decomposition, ...) and the time stepping loop. For any time step the twin subroutines *ADVANCEX* and *ADVANCEY*, serving to update the solution by dimension splitting, are called once for each of the subregions G_1 and G_2.

The principal subroutines then are *PIECELIN** and *SOLVEROE** (where * stands for *x* or *y*), called once by *ADVANCE** Here the routine *PIECELIN** serves to compute the flow quantities for an intermediate time step by means of a (slope-limited) finite-volume method. After that the second routine *SOLVEROE** is called, where Roe's scheme to compute the fluxes (on the intermediate time level) followed by conservatively updating the solution is used. Both routines *PIECELIN** call a function *AVE* where the slope-limiter is evaluated.

Fig.2 sheds light upon the merits of the new code PICUS-2 as compared with PLM2DTL, an academic code (of the RWTH Aachen) with related objectives in mind (for reasons of comparison, both codes were run including nonequilibrium chemistry). Here the parameters were $NX2 = 128$, $NY2 = 16$, with 10 time steps run (the domain G_1 treated by PICUS only is modeled by a coarse grid so that its contribution to the total CPU time is low). It turns out that the basic version PICUS-200 reduces the elapsed time by factor 10 and its tuned version PICUS-230 even by factor 20 (all runs carried out in vector mode). The subsequent section shows how to achieve this improvement.

5. VECTORIZING AND OPTIMIZING

Taking full advantage of the IBM ES/3090 vector facility requires some knowledge of its special structure. But as it will show up from the present section, only little specialists skill of the VF is required to attain an essential saving of CPU time. The key is just structuring the data and the data flow favourably (arrays *STATEj*, *PLINj*, *FLUXj*, and its handling !).

The result is shown in Fig.3. It shows as a function of *NX*2 (with $NY2 = 0.1 \times NX2$) the elapsed times (in seconds) of the

- basic version PICUS-200 run in scalar mode (upper graph)
- basic version PICUS-200 run in vector mode (midst graph)
- optimized version PICUS-230 run in vector mode (lower graph).

Consequently, even the basic version enjoys a slight vector-to-scalar speedup of about 1.2 which, for the IBM ES/3090 with its excellent scalar processor as the basis of comparison, is by no means self-evident but rather shows an already favourable data flow and structure of the algorithm. In the following we will explain shortly how to tune this code to attain a vector-to-scalar speedup of about 2.5 (see the smoothed graph of Fig.4).

The following strategy leads to reducing the CPU time in an efficient way:

1. A "hot-spot analysis" (a tool provided by the IBM VS FORTRAN Interactive Debug) yields a run-time statistics of the quota of the total CPU time used by the subroutines. This points out those subroutines tunable with best efficiency.

2. The compiler vector report shows the DO-loops run in vector or scalar mode, indicating where to start best vectorizing the DO-loops.

3. In some cases it is hardly possible to vectorize a DO-loop (due to some recurrence $a_i = a_{i-1}$, for instance). In this case exchanging all the subroutine or this part of the algorithm is required.

Concentrating ourselves to the most essential features of the code we summarize the tools and ways to optimize. The main tools are as follows:

- Removing calls of external subroutines in DO-loops by generating in-line code (eventually by use of the precompiler VAST-2),

- Removing recurrences by splitting DO-loops and eventually modifying the source code,

- Resorting to the IBM Engineering and Scientific Subroutine Library (ESSL) which consists of currently 288 optimized numerical subroutines,

- Avoiding multiple computation of quantities sent to memory, and improving further the data flow.

This leads to the final version, called PICUS-230, with a CPU time reduced by factor 2.65 as compared with the original version PICUS-200 (both run in vector mode, with NX2 = 500, see Fig.3). This means that, by efficient use of the vector facility, a saving of more than 60 percent of the overall CPU time has been achieved as compared with the basic version (in vector mode, both). As compared with the basic version run in scalar mode, even a saving of about 70 percent results. See Fig.3.
The final version PICUS-230 enjoys a vector-to-scalar speedup of about 2.5 (see Fig.4). This shows excellent cost/performance ratio, particularly

if the low additional costs of the vector facility and the moderate effort are taken into account.

6. PHYSICAL EXAMPLE

The physical character of the present analysis is illuminated by the following example. A lot of further test computations have been carried out and will be treated in forthcoming papers.

Let two shock waves enter successively the channel, namely in a way that the second wave enters in just that moment the first one is reflected at the end of the channel. Both waves are then interacting, and they produce a wave train (a "pseudo-shock wave") and one shock wave turning again towards the end of the channel. Ultimately, at a later stage, a von Kármán vortex street has fully developed (see Fig.5).

Typical values of dynamic viscosity μ for hydrocarbon air mixtures are chosen. Thermal conductivity is calculated from dynamic viscosity employing the Eucken relation. The propagation Mach number Ma of the shock wave entering G_2 coming from G_1 is taken as 1.3 in correspondence with typical measurements of pressure increase due to knock events in engines.

This computation, with about 64,000 internal cells of G_2, consumed about 24 CPU hours on the IBM ES/3090 .

CONCLUSIONS

We described our test computations and results on vectorizing the new FORTRAN code PICUS to simulate shock waves in an L-shaped domain, with special emphasis of the IBM ES/3090 with vector facility. The algorithm is based on vanLeer's second-order shock-capturing scheme using Roe's approximate Riemann solver. With its final version, it shows up faster by factor 20 than a previous (academic) code with related objectives in mind. The new code results in a vector-to-scalar speedup of about 2.5. The long-time behaviour of PICUS has been checked by computation of the von Kármán vortex street.

REFERENCES AND FIGURES

1. Gentzsch, W. and Glückert, S. The Processor IBM 3090 with Vector Facility (in German), Praxis der Informationsverarbeitung, Vol. 10, pp. 24-30, 1987.

2. Hirschel, E.H. and Groh, A. Wall-compatibility Conditions for the Solution of the Navier Stokes Equations, Journal of Computational Physics, Vol. 53, pp. 346-350, 1984.

3. LeVeque,R.J. Numerical Methods for Conservation Laws, Birkhäuser, Basel, 1990.

4. MacCormack, R.W. A Rapid Solver for Hyperbolic Systems of Equations, Lecture Notes in Physics, Vol. 59, pp. 307-317.

5. Roe, P.L. Characteristic-based Schemes for the Euler equations, Annual Review of Fluid Mechanics, Vol. 18, pp. 337-365, 1986.

6. Schöffel, S., Hebeker, F.K. and Gathmann R.J. The Code PICUS for Analyzing Wall Loading and Damage Mechanism under Knock Conditions in Internal Combustion Engines, Internal Report, IBM Heidelberg Scientific Center, 99 pp., 1991.

7. Thiel, U. and Reuter, R. Vectorization of an Upwind Difference Scheme for the Two-dimensional Euler Equations, Technical Report 90.03.003, IBM Heidelberg Scientific Center, 17 pp., 1990.

8. VanLeer, B. Towards the Ultimate Conservative Difference Scheme, Part 5, Journal of Computational Physics, Vol. 32, pp. 101-136, 1979.

9. Yee, H.C. A Class of High-resolution Explicit and Implicit Shock-capturing Methods, VKI Lecture Series 1989-04, 216 pp., 1989.

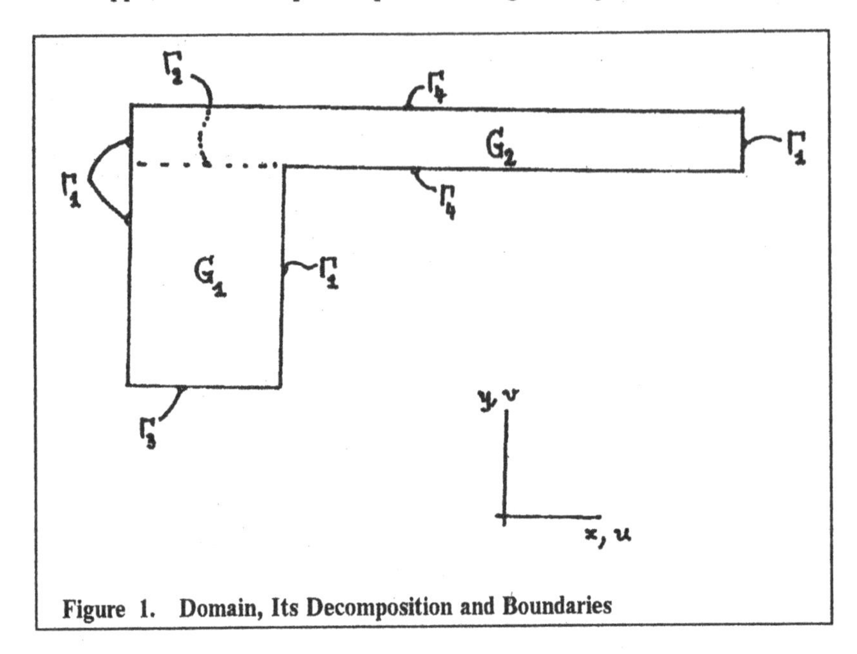

Figure 1. Domain, Its Decomposition and Boundaries

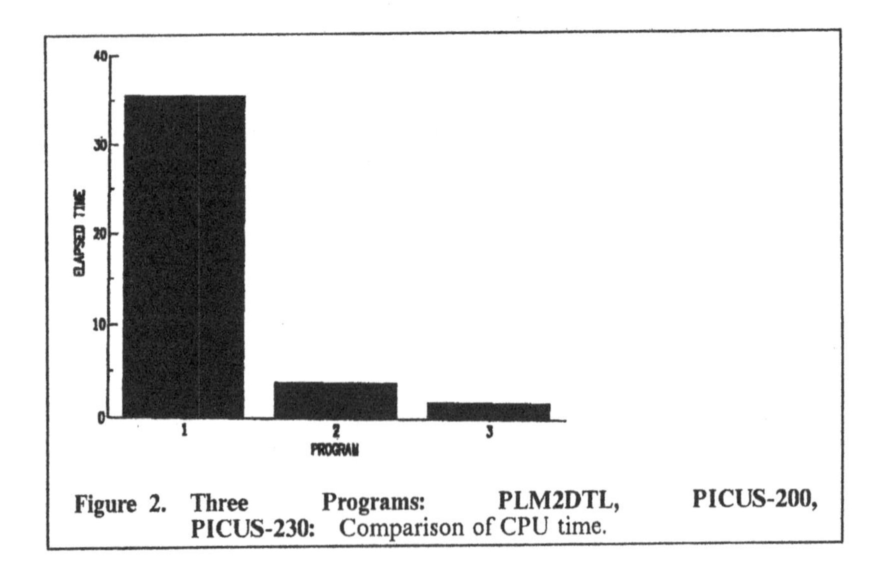

Figure 2. Three Programs: PLM2DTL, PICUS-200, PICUS-230: Comparison of CPU time.

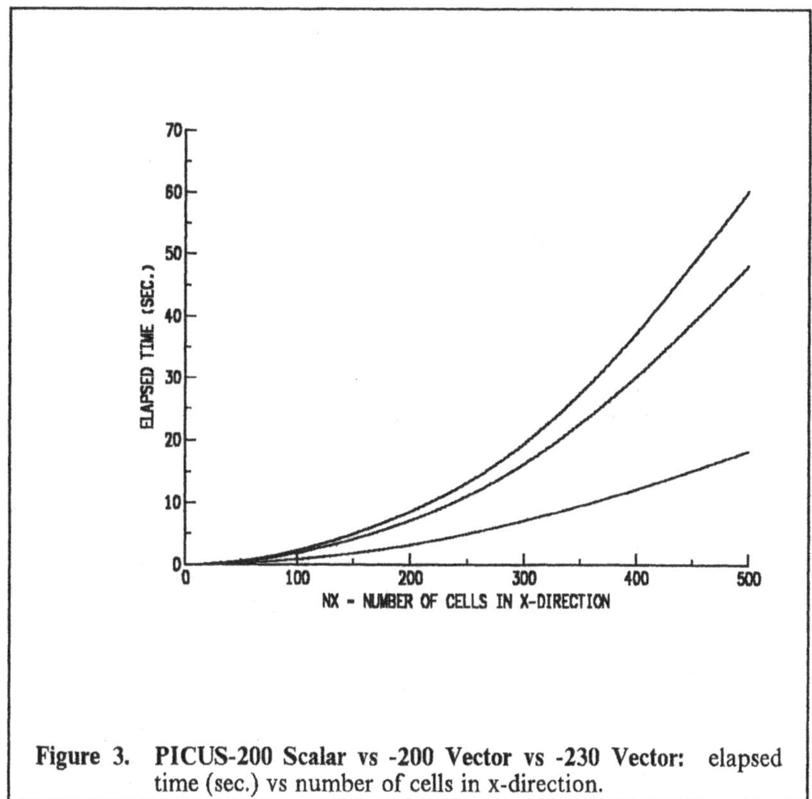

Figure 3. PICUS-200 Scalar vs -200 Vector vs -230 Vector: elapsed time (sec.) vs number of cells in x-direction.

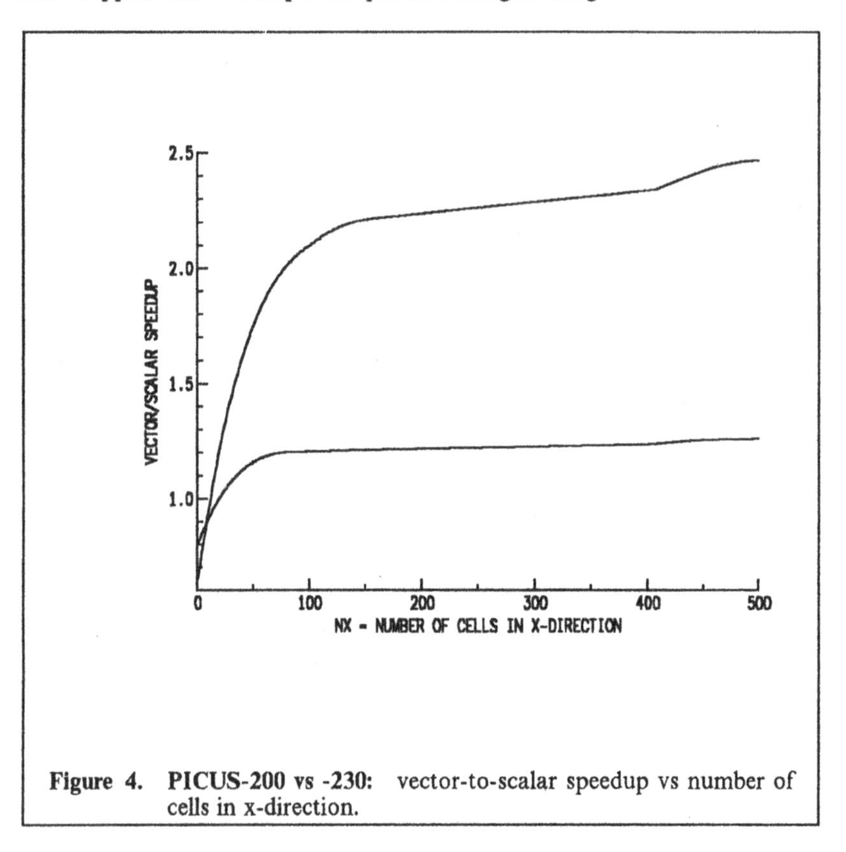

Figure 4. PICUS-200 vs -230: vector-to-scalar speedup vs number of cells in x-direction.

Figure 5. Wave Train and von Kármán Vortex Street

High Speed Laminar Flows Over Cavities

D.R. Emerson (*), D.I.A. Poll (**)

(*) S.E.R.C., Daresbury Laboratory, Warrington WA4 4AD, U.K.

(**) Department of Aeronautical Engineering, University of Manchester, Manchester M13 9LP, U.K.

ABSTRACT

The explicit MacCormack finite-volume method, with a Total Variation Diminishing (TVD) scheme for stability, has been used to solve the Navier-Stokes equations to simulate the flow over rectangular cavities. The computations indicate that the oscillatory frequencies are essentially independent of wall temperature, stagnation temperature, boundary layer thickness at separation and unit Reynolds number. However, the magnitude of the pressure fluctuations depend strongly on the wall temperature and boundary layer thickness at separation.

INTRODUCTION

Cavities occur frequently in aeronautical engineering. As a design aid they can be used to house internal stores, landing gear, optical sensing equipment and wrap-around fins for missiles. Cavities can also be produced by accidental damage, for example micro-meteorite impacts on re-entry vehicles, expansion or contraction of dissimilar materials in contact, and fabrication misalignment.

The presence of a cavity causes the boundary layer to separate and subsequently reattach at some point downstream. This produces significant aerodynamic effects, such as fluctuating pressures, noise, heat transfer and increased drag. These may lead to damage to internal stores, impaired efficiency or structural fatigue. For example, Gibson[1] identified wheel-wells as the primary source of noise on the Lockheed C-5 Galaxy in the landing/take-off configuration whilst Cantani et al[2] demonstrated that cavity drag can seriously affect missile range. Internal stores must pass through the separated shear-layer during deployment. This may adversely affect the trajectory or, in extreme situations, cause structural failure to the store itself. Nestler et al[3] investigated the effects of cavities on heat transfer in hypersonic flows. Re-entry vehicles are subjected to very high rates of heat transfer which can cause erosion of recessed windows and problems may arise from surface geometry changes resulting from differential expansion of dissimilar materials. This may lead to the formation of steps and cavities which are a potential source of premature transition. Buffeting is the result of oscillation of the separated shear-layers which can produce intense pressure fluctuations. Under certain conditions, the spectrum associated with these oscillations can have discrete components. Rossiter[4] found that cavities with length-to-depth ratios (L/D) > 4 had, in general, a broadband spectrum whilst cavities with L/D < 4 had discrete frequencies present i.e. the cavity was resonating. Charwat et al[5] classified shallow cavity flow (L/D > 1) as "open" or "closed". Open flow occurs when the cavity free shear-layer bridges the cavity and reattaches on the downstream face. Resonance effects occur only in open cavity flows. Closed flow occurs when L/D is greater than a critical value and the cavity free shear-layer reattaches to the cavity floor before impinging on the downstream face.

The primary objective of the research was to develop a numerical method that would simulate high speed flow over a rectangular cavity. Prior to the code development, a comprehensive review of the numerical methods applied to cavity flows was undertaken by Emerson[6]. Based on this information, an explicit MacCormack TVD scheme was developed to solve the unsteady 2-D Navier-Stokes equations. From the review, it was clear that the computations would be computer intensive and the code was therefore developed to run on the AMDAHL VP1100 Vector Processor at the Manchester Computing Centre. A highly vectorised code was developed by using a vectorisation tool called SAMPLER. Limited computational resources prevented an in-depth investigation into cavity flows and the results obtained employ a relatively coarse grid to resolve the attached boundary-layer and the cavity shear-layer. However, the results indicate good agreement with available prediction methods and qualitative agreement with experimental observations. Recently, the code has been modified to take advantage of the parallel hardware available at Daresbury Laboratory. By employing a domain decomposition strategy and embedding the code in FORTNET, a portable message-passing environment developed by Allen et al[7], it has been possible to test the code on the Intel iPSC/860 32-node hypercube. Some preliminary results of this work are also presented.

THE GOVERNING EQUATIONS

Unsteady flows involving separation and recirculation require the solution of the Navier-Stokes equations. For 2-D flows, they can be written as follows:

$$\frac{\partial Q}{\partial t} + \frac{\partial F}{\partial x} + \frac{\partial G}{\partial y} = 0 \tag{1}$$

where

$$Q = (\rho, \rho u, \rho v, E)^{\dagger} \tag{2}$$

$$F = F_i - F_v \tag{3}$$

$$G = G_i - G_v \tag{4}$$

and the superscript † represents the transpose of the column vector and the subscripts i and v represent the inviscid and viscous terms, respectively. These are given by:

$$F_i = \left(\rho u, P + \rho u^2, \rho uv, u\{E + P\}\right)^{\dagger} \tag{5}$$

$$F_v = (0, \tau_{xx}, \tau_{xy}, u\tau_{xx} + v\tau_{xy} + k\partial T/\partial x)^{\dagger} \tag{6}$$

$$G_i = \left(\rho v, \rho uv, P + \rho v^2, v\{E + P\}\right)^{\dagger} \tag{7}$$

$$G_v = (0, \tau_{yx}, \tau_{yy}, u\tau_{yx} + v\tau_{yy} + k\partial T/\partial y)^{\dagger} \tag{8}$$

where the symbols T, P, ρ and E represent the temperature, pressure, density and total energy per unit volume, respectively, and the x and y velocity components are represented by u and v. Assuming that Stokes' hypothesis for the bulk viscosity is valid, the shear stress terms can be expressed as:

$$\tau_{xx} = \mu\left\{\frac{4}{3}\frac{\partial u}{\partial x} - \frac{2}{3}\frac{\partial v}{\partial y}\right\} \tag{9}$$

$$\tau_{yy} = \mu\left\{\frac{4}{3}\frac{\partial v}{\partial y} - \frac{2}{3}\frac{\partial u}{\partial x}\right\} \tag{10}$$

$$\tau_{xy} = \tau_{yx} = \mu\left\{\frac{\partial u}{\partial y} - \frac{\partial v}{\partial x}\right\} \tag{11}$$

where μ and k represent the molecular viscosity and thermal conductivity, respectively. The pressure, P, is related to the density and temperature by an equation of state which for a thermally perfect gas, is given by:

$$P = (\gamma - 1)\left\{E - \frac{1}{2}\rho[u^2 + v^2]\right\} \tag{12}$$

and the ratio of specific heats, γ, is constant and for air has the value 1.4. The molecular viscosity was obtained from Sutherland's law:

$$\frac{\mu}{\mu_{ref}} = \left\{\frac{T}{T_{ref}}\right\}^{1/2} \frac{1 + S_o/T_{ref}}{1 + (S_o/T_{ref})(T_{ref}/T)} \tag{13}$$

where $S_o = 110.4K$ is Sutherland's constant and the subscript ref represents a reference condition. These were taken to be the conditions at the edge of the boundary layer.

THE NUMERICAL METHOD

The solution procedure adopted was that suggested by MacCormack and Paullay[8] and employs a finite-volume, operator-split approach. The Navier-Stokes equations are integrated over a control volume as follows:

$$\frac{\partial}{\partial t}\int\int\int_{\Omega} Q dV + \int\int\int_{\Omega}\left\{\frac{\partial F}{\partial x} + \frac{\partial G}{\partial y}\right\}dV = 0 \tag{14}$$

$$\frac{\partial}{\partial t}\int\int\int_{\Omega} Q dV + \int\int\int_{\Omega} \underline{\nabla}.\underline{H} dV = 0 \tag{15}$$

where

$$\underline{H} = \{\rho\underline{V}, P\underline{\hat{e}}_x + \rho u\underline{V}, P\underline{\hat{e}}_y + \rho v\underline{V}, (E + P)\underline{V}\}^\dagger \tag{16}$$

and $\underline{\hat{e}}_x$ and $\underline{\hat{e}}_y$ are unit vectors in the x and y directions, respectively, and $\underline{V} = u\underline{\hat{e}}_x + v\underline{\hat{e}}_y$. The divergence theorem can now be applied to the flux terms to give:

$$\frac{\partial}{\partial t}\int\int\int_{\Omega} Q dV + \int\int_{\Gamma} \underline{H}.\underline{\hat{n}} dS = 0 \tag{17}$$

where Γ = the cell boundary and $\underline{\hat{n}}$ = the outward unit normal. In 2–D, the "volume" is actually the cell area. The computational grid for a generic cavity consists entirely of rectangular cells, which simplifies the numerical algorithm considerably. The operator-split scheme can be written:

$$Q^{n+1} = L_x(\Delta t)L_y(\Delta t)Q^n \tag{18}$$

and is first order accurate in time and second order accurate in space. The $L_y(\Delta t)$ operator is given by:

$$\overline{Q^{n+1/2}_{ic,jc}} = Q^n_{ic,jc} - \frac{\Delta t}{\Delta y}\{G_{ic,jc+1} - G_{ic,jc}\} \tag{19}$$

$$Q^{n+1/2}_{ic,jc} = \left\{\frac{Q^n_{ic,jc} + \overline{Q^{n+1/2}_{ic,jc}}}{2}\right\} - \frac{\Delta t}{2\Delta y}\left\{\overline{G^{n+1/2}_{ic,jc}} - \overline{G^{n+1/2}_{ic,jc-1}}\right\} \tag{20}$$

where the overbar symbols represent intermediate values at time level n and the subscripts ic and jc represent values stored at cell centres. The $L_x(\Delta t)$ operator is defined in a similar manner. The viscous stress and heat transfer terms were solved with a weighted central-differencing to allow for the non-uniform grid employed.

INITIAL AND BOUNDARY CONDITIONS

Based upon experimental evidence, the upstream influence of the cavity extends only a few cavity depths and, consequently, the inflow boundary was located 3D upstream of the cavity leading edge. The influence of the cavity downstream of the trailing edge has not been extensively studied but, from experimental observations, depends upon external conditions and geometry. However, from numerical tests, locating the downstream boundary 6D beyond the trailing edge was a suitable choice. The location of the upper computational boundary was also determined from numerical experimentation and, for the conditions under investigation, this boundary was placed 4D above the flat plate.

The flowfield in and around a cavity is altered dramatically if an oscillatory flow develops. Therefore, to initialise the flowfield, relatively simple conditions were specified. However, one of the important parameters is the boundary layer thickness at separation, δ_s, and it was decided that this would be an input parameter. From laminar boundary-layer theory, δ varies as the square root of the distance, and this relation was used to determine the boundary layer thickness at all x-stations along the flat plate. The u-velocity profile was then determined from a one-half power law approximation:

$$\frac{u}{u_e} = \left\{\frac{y}{\delta}\right\}^{1/2} \qquad y < \delta \tag{21}$$

$$\frac{u}{u_e} = 1 \qquad y \geq \delta \tag{22}$$

and the v-component of velocity was set to zero. The temperature profile was then determined from Crocco's relation by assuming a zero pressure-gradient ($\partial P/\partial y = 0$) and unit Prandtl number ($Pr = 1$). The density was then determined from the equation of state. After initialisation, the Prandtl number was set to 0.72. Within the cavity itself, all parameters were set to their wall values.

At the inflow boundary, all parameters were held constant, whilst at the downstream boundary, zeroth order extrapolation of conserved quantities was employed. To avoid reflections at the upper computational boundary, the parameters were set after checking the flow direction. When the flow was outgoing, zeroth order extrapolation was used, and for incoming flow, the pressure was extrapolated from the computational domain and the density obtained from the isentropic relation ($P/\rho^\gamma = constant$). The u-component of velocity was then set to its freestream value and v was set to zero. Along all solid boundaries, the no-slip condition was specified and the wall pressure obtained from the cell-centre normal to the boundary ($\partial P/\partial n = 0$). The wall temperature was either constant (isothermal wall) or obtained from the adiabatic condition ($\partial T/\partial n = 0$). The density was then obtained from the equation of state.

The computational grid consisted of 120 $\times$ 100 node points with 50 $\times$ 30 points located within the cavity. An algebraic grid generation technique was employed with stretching applied to areas involving separation and re-attachment. However, to reduce the computation times involved, the boundary layer was divided up into equi-spaced intervals. Although this does not adequately resolve the boundary layer, it was considered sufficiently accurate for the code to be tested. A typical grid is illustrated in Figure 1.

THE MacCORMACK TVD SCHEME

Total Variation Diminishing (TVD) schemes were introduced by Harten[9] and they have very desirable properties. They are monotonicity preserving and the dissipation term is parameter free. Following the criteria established by Harten[9], Davis[10] modified the MacCormack predictor-corrector scheme to be TVD by appending a flux limiter to the corrector stage. To illustrate this, consider the Euler equations in one spatial dimension:

$$\frac{\partial Q}{\partial t} + \frac{\partial F}{\partial x} = 0 \tag{23}$$

where

$$Q = (\rho, \rho u, E)^t \tag{24}$$

and

$$F = \left(\rho u, P + \rho u^2, u\{E + P\}\right)^t \tag{25}$$

The MacCormack TVD scheme may then be written as:

$$\overline{Q_i^{n+1}} = Q^n - \lambda(F_{i+1}^n - F_i^n) \tag{26}$$

$$Q_i^{n+1} = \left(\frac{\overline{Q_i^{n+1}} + Q_i^n}{2}\right) - \frac{\lambda}{2}\left(\overline{F_i^{n+1}} - \overline{F_{i-1}^{n+1}}\right) + F_{lim}^n \tag{27}$$

where $\lambda = \Delta t/\Delta x$ and

$$F_{lim}^n = (F_i^+ + F_{i+1}^-)\Delta Q_{i+1/2}^n - (F_{i-1}^+ + F_i^-)\Delta Q_{i-1/2}^n \tag{28}$$

The terms $\Delta Q_{i+1/2}^n = Q_{i+1}^n - Q_i^n$ etc. and

$$F_i^\pm = F(r_i^\pm) = 0.5C(\nu)\{1 - \phi(r_i^\pm)\} \tag{29}$$

with

$$C(\nu) = \nu(1-\nu) \qquad \nu \leq 0.5 \tag{30}$$

$$C(\nu) = 0.25 \qquad \nu > 0.5 \tag{31}$$

where $\nu = \lambda(|u| + c)$ and c is the local speed of sound. The remaining terms are evaluated from

$$r_i^\pm = \frac{\left[\Delta Q_{i-1/2}^n, \Delta Q_{i+1/2}^n\right]}{\left[\Delta Q_{i\pm1/2}^n, \Delta Q_{i\pm1/2}^n\right]} \tag{32}$$

where [...,...] denotes the inner dot product of the column vector Q and

$$\phi(r) = max(0, min\{2r, 1\}) \tag{33}$$

To avoid any division by zero, a small amount (10^{-10}) was added to the numerator and denominator in Equation (32).

RESULTS AND DISCUSSION

The first set of results to be presented are for an adiabatic wall and the initial conditions and geometry are shown in Table 1. The computational results are, where possible, compared to the experimental data of Zhang[11], which was for a turbulent approaching boundary-layer. However, it was felt that all the qualitative feaures (e.g. the oscillatory flowfield, location of primary vortex etc.) should be captured.

Table 1 Initial Conditions For Adiabatic Wall

$M_e = 1.5$	L/D = 3
$Re = 3\times10^7$/metre	D = 0.015m
$\delta_s = 0.3D$	$T_o = 290K$

The computational data were collected between $100 \le t \le 300$ (where t is the non-dimensional time). Although oscillatory flow was established before 100t, it was found that the magnitudes of the pressure maxima and minima had not always reached their extreme values. The CPU time required for the computation was 4566 seconds. Figure 2 shows the pressure history along the trailing edge at y/D = 0.972 (with y/D = 0 along the cavity floor). It is clear that the flow has entered a self-sustained oscillatory process. To determine the oscillatory frequency, the pressure history was converted to the frequency domain by employing a Fast Fourier Transform (FFT) containing 4096 data samples. The Sound Pressure Level (SPL) was obtained from:

$$SPL = 20log_{10}(P/q_e) \tag{34}$$

where $q_e = 1/2\rho_e u_e^2$ is the dynamic pressure. The results from the transformed data are illustrated in Figure 3 in terms of a non-dimensional frequency, or Strouhal number, where $Str = fL/u_e$, and f is the frequency in Hertz. The dominant oscillatory frequency has the value 0.57. Rossiter[4] derived a semi-empirical relation to predict the resonant modes of oscillation. This equation was modified by Heller et al[12] for more accurate predictions at supersonic speeds and is given by:

$$S_H = (n - 0.25)/\left(\frac{M_e}{M^*} + \frac{1}{k_v}\right) \qquad n = 1,2,3... \tag{35}$$

where n is the mode number, the vortex convective velocity ratio $k_v = 0.57$ and $M^* = \left(1 + \{\frac{\gamma-1}{2}\}M_e^2\right)^{1/2}$. Using Equation (35) the first four predicted Strouhal numbers are 0.25, 0.58, 0.92 and 1.25. The dominant Strouhal number corresponds to the predicted second mode but there is clearly no first mode visible in the computational results. However, Krishnamurty[13] observed that for laminar separating flow, over a range of similar geometries and for Mach numbers up to and including $M_e = 1.5$, a fundamental frequency (f_o) and its first harmonic ($2f_o$) were produced. From the computational results, the second peak occurs at $Str = 1.13$ and this corresponds to a frequency doubling to within 1% ; i.e. the second peak is the first harmonic of the fundamental mode. Second and third harmonics (the third harmonic was calculated but is not shown) occur at $Str = 1.70$ and 2.25, respectively, and correspond to integer multiples of the fundamental frequency.

The time-averaged pressure (P/P_e) is shown in Figure 4. The pressure distribution along the cavity floor is typical of that for an open cavity and the location of the primary vortex is in good agreement with experimental data, although the numerical values differ. Knowing the time-averaged pressure within the cavity, it is possible to determine the form drag. This was determined from:

$$C_d = \int_0^1 \frac{P_{rf} - P_{ff}}{q_e} dy \tag{36}$$

where the subscripts ff and rf indicate distributions along the front face and rear face of the cavity, respectively. The computationally determined drag coefficient was found to be 0.064. Zhang[11] found that the experimental drag was 0.048, which is 25% lower than the computed value. This difference may be a result of difficulty in locating pressure sensing devices near regions like the trailing edge.

More information about the flow process can be obtained from density contours. Figure 5 shows a typical distribution at time t = 291. A high pressure region is beginning to build up on the rear face caused by the entrained flow stagnating near the trailing edge. Vortices, located beyond the trailing edge, are being convected towards the downstream boundary. The convective velocity of these vortices (determined from data output at regular intervals) was estimated to be 0.53 and is in reasonable agreement with the value obtained by Rossiter[4], which was $k_v = 0.57$.

As stated, δ_s could be varied and numerical tests were therefore carried out to assess its effect on cavity flow. The conditions stated in Table 1 were retained with the exception of δ_s. The flow features were essentially the same as those already described. However, the magnitude of the pressure oscillation varied considerably. Figure 6 shows a typical variation of pressure with time and the transformed pressure history is illustrated in Figure 7. Figure 8 indicates how the fundamental frequency and its first three harmonics vary with δ_s/D. The computations indicate that the fundamental frequency is essentially independent of the oncoming boundary layer thickness at separation although the higher harmonics show some small dependence. However, the differences may be caused by having too few samples for the FFT, too short a sample time or insufficient grid resolution. As the magnitude of the pressure depends strongly on δ_s, it would be reasonable to expect the cavity drag to depend on δ_s and this is indeed the case. Figure 9 shows how the cavity drag depends on the boundary layer thickness at separation. It illustrates that thick boundary layers produce low drag, which is in qualitative agreement with experiment. From Figure 9, the drag appears to decay in an exponential fashion and it is hypothesised that the maximum drag occurs when δ_s/D lies between 0.1 and 0.2 for the conditions under investigation. This would suggest that there is an "optimum" boundary layer thickness which would allow the maximum amplification of the instabilities in the cavity free shear-layer. This value of δ_s would produce the highest drag and pressure maxima and the most noise, and it should therefore be avoided at the design stage. The results for the drag and Strouhal numbers, and the CPU time in seconds, are summarised in Table 2 below, where the subscripts 0, 1, 2

Table 2 Results For Different Boundary Layer Thicknesses

δ_s/D	Str_0	Str_1	Str_2	Str_3	C_d	J_{bl}	CPU
0.1	0.60	1.19	1.79	2.39	0.114	12	14330
0.2	0.57	1.14	1.73	2.30	0.116	16	9713
0.3	0.57	1.13	1.70	2.25	0.064	16	4566
0.4	0.56	1.10	1.64	2.19	0.039	20	6139

and 3 represent the fundamental Strouhal number and first three harmonics, respectively, and J_{bl} is the number of grid points in the boundary layer at separation. It is evident that thin boundary layers are CPU intensive.

Typical results for an isothermal wall will now be presented and compared with the experimental work performed by Stallings and Wilcox[14]. The conditions are summarised in Table 3:

Table 3 Initial Conditions For Isothermal Wall

$M_e = 2.16$	L/D = 3
$Re = 6.56\times10^6$/metre	D = 0.0254m
$\delta_s = 0.2D$	$T_o = 325K$

The ratio of wall-to-edge temperature, (T_w/T_e) was set to one for this case. However, other values of T_w/T_e were also considered. The data were stored between 100 < t < 225 and the CPU time required for the computation was 6138 seconds. The pressure history, shown in Figure 10, indicates that the oscillatory flow was not fully established until just after t = 150. However, the differences appear to be small and the results were considered to be representative for a qualitative description. More frequency components appear to be present and this is confirmed by the FFT results, shown in Figure 11. From Equation (35) the first four modes are predicted to be 0.23, 0.53, 0.83 and 1.13, respectively. The computed frequencies were found to be 0.22, 0.43 (dominant mode), 0.65 and 0.84 and correspond to a harmonic and two sub-harmonics of the fundamental frequency. This feature was always observed when $M_e > 2.0$. At Mach 2.5 and above, the harmonics produced were not, in general, simple integer or half-integer multiples of the dominant frequency.

The time-averaged pressure and Stanton number are shown in Figure 12 and Figure 13, respectively. The primary vortex can be clearly identified in the pressure distribution. The Stanton number, defined as:

$$St = \frac{-\dot{q}}{c_p \rho_e u_e (T_o - T_w)} \tag{37}$$

where $\dot{q} = -k_w(\partial T/\partial n)_w$ and c_p is the specific heat at constant pressure, shows a more erratic distribution but the qualitative appearance is good with the maximum heat transfer rate occurring at reattachment. Also shown in Figure 13 is the Stanton number for initialisation and a significant reduction is predicted. Very low heat transfer rates are predicted along the cavity floor and front face with some negative values apparent. There is no experimental evidence to support negative heat transfer rates within the cavity and it is believed that these may be caused by the grid being too coarse. Further numerical experiments with different T_w/T_e indicated that high wall temperatures produced a strong damping effect. For example, when $T_w/T_e = 5$, the pressure maxima were only slightly above freestream. However, low wall temperatures produced large variations in pressure maxima. These results are in qualitative agreement with the experimental work of Larson and Keating[15].

The average CPU time per iteration was 0.17 seconds. This includes the time to generate the grid, initialise the flow parameters, output data in to files (such as the pressure history required to produce frequency information) and plotting time. However, for the computational times involved, these represent only a small percentage of the total time. The most CPU intensive part of the code was in the determination of the TVD algorithm, which required approximately 55% of the CPU time.

PARALLEL DEVELOPMENT

As stated in the introduction, the code is currently being modified to run on the parallel hardware available at Daresbury. A domain decomposition strategy was employed to map the code onto the Intel iPSC/860. This code is typical of many encountered in CFD and has both global and local parameters. It is also necessary to provide each domain with halo data at the boundaries and the message passing was achieved by embedding the code in FORTNET. Only limited results are available at present: these are summarised in Table 4 and indicate the calculation and communication times (in seconds) for 20 iterations. A linear speedup is evident. However, further work is required to assess the optimum number of nodes before communication costs become too high.

Table 4 Preliminary Results From Parallel Implementation

Nodes	1	3
Calculation	51.42	17.55
Communication	0.0	2.46

CONCLUSIONS

The MacCormack TVD method has been used to solve the unsteady Navier-Stokes equations for high speed flow past a rectangular cavity. The results obtained are in good qualitative agreement with the available experimental data. The solution procedure is CPU intensive but the scheme can be vectorised efficiently. A simple domain decomposition strategy has also been employed to run the code on a distributed memory system. The message passing was accomplished by embedding the code in FORTNET and, from the limited information available, a linear speedup can be achieved.

ACKNOWLEDGEMENTS

The authors would like to thank the M.o.D., Royal Aircraft Establishment, Farnborough, for their help and support in this work.

REFERENCES

1. Gibson, J. S., "Non-Engine Aerodynamic Noise Investigation of a Large Aircraft," N.A.S.A. CR- 2378, 1974.
2. Cantani, U., Bertin, J. J., de Amicis, R., Massalo, S., and Bouslog, S. A., "Aerodynamic Characteristics for a Slender Missile with Wrap-Around Fins," Journal of Spacecraft and Rockets, Vol. 20, pp. 122–128, 1983.
3. Nestler, D. E., Saydah, A. R., and Auxer, W. L., "Heat Transfer to Steps and Cavities in Hypersonic Turbulent Flow," A.I.A.A. Paper 68-673, 1968.
4. Rossiter, J. E., "Wind Tunnel Experiments on the Flow Over Rectangular Cavities at Subsonic and Supersonic Speeds," R.A.E. R and M 3438, 1964.
5. Charwat, A. F., Dewey, C. F., Roos, J. N., and Hitz, J. A., "An Investigation of Separated Flows. Part 1: The Pressure Field," Journal of the Aeronautical Sciences, Vol. 28, pp. 457–470, 1961.
6. Emerson, D. R., Computation of Laminar Cavity Flows. PhD thesis, Department of Aeronautical Engineering, University of Manchester, U.K.,, 1991.
7. Allen, R. J., Heck, L., and Zurek, S., "Parallel Fortran in Scientific Computing: A New Occam Harness Called Fortnet," Computer Physics Communications, Vol. 59, pp. 325–344, 1990.
8. MacCormack, R. W. and Paullay, A. J., "Computational Efficiency Achieved by Time-Splitting of Finite Difference Operators," A.I.A.A. Paper 72-154, 1972.
9. Harten, A., "High Resolution Schemes for Hyperbolic Conservation Laws," J. Comp. Phys., Vol. 49, pp. 357–393, 1983.
10. Davis, S. F., "A Simplified TVD Finite Difference Scheme Via Artificial Viscosity," S.I.A.M.J. Sci. Stat. Comput., Vol. 18, pp. 1–19, 1987.
11. Zhang, X., An Experimental and Computational Investigation into Supersonic Shear Layer Driven Single and Multiple Cavity Flowfields. PhD thesis, Department of Engineering, University of Cambridge, 1987.
12. Heller, H. H., Holmes, D. G., and Covert, E. E., "Flow-Induced Pressure Oscillations in Shallow Cavities," Journal of Sound and Vibration, Vol. 18, pp. 545–553, 1971.
13. Krishnamurty, K., "Acoustic Radiation From Two-Dimensional Rectangular Cutouts in Aerodynamic Surfaces," N.A.C.A. TN- 3487, 1955.
14. Stallings, R. L. and Wilcox, F. J., "Experimental Cavity Pressure Distributions at Supersonic Speeds," N.A.S.A. TP- 2683, 1987.
15. Larson, H. K. and Keating, S. J., "Transition Reynolds Number of Separated Flows at Supersonic Speeds," N.A.S.A. TN- 349, 1960.

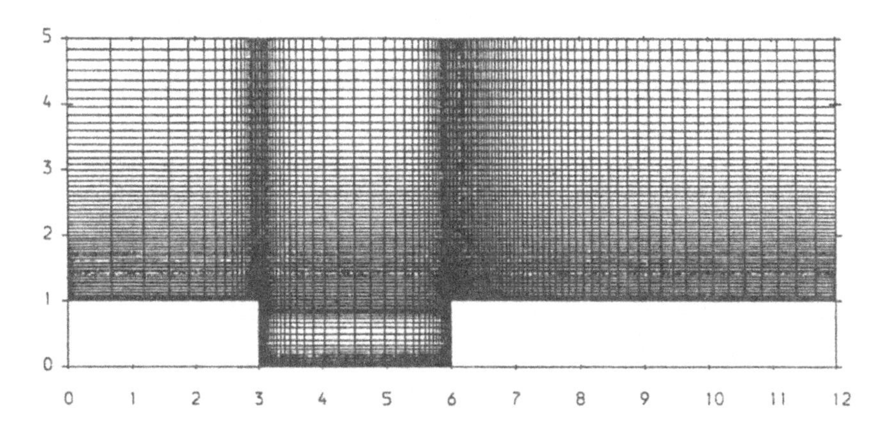

Figure 1: Typical Computational Grid

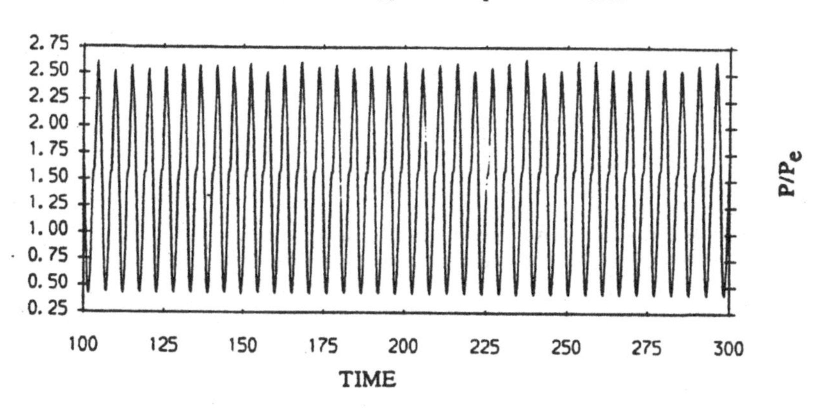

Figure 2: Pressure History On The Rear Face For Adiabatic Wall
y/D = 0.972 (y/D = 0 on cavity floor)
$M_e = 1.5, L/D = 3, Re = 3 \times 10^7/metre, \delta_s/D = 0.3$

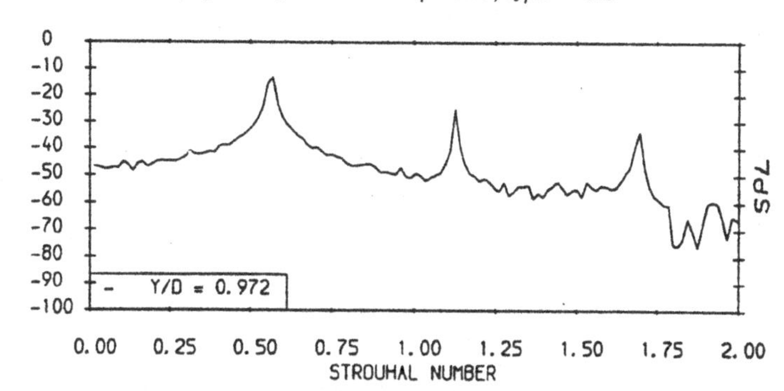

Figure 3: Sound Pressure Level
y/D = 0.972 (y/D = 0 on cavity floor)
$M_e = 1.5, L/D = 3, Re = 3 \times 10^7/metre, \delta_s/D = 0.3$

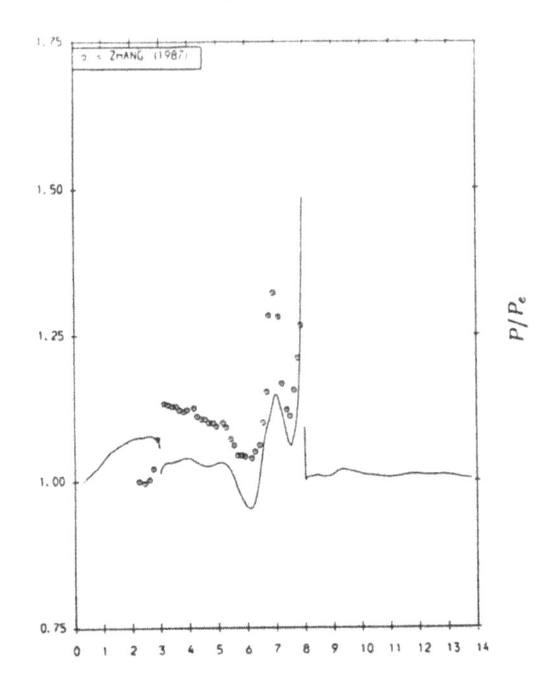

Figure 4: Time-Averaged Pressure Along Surface Perimeter (Inflow = 0)
Solid Line = computational results ; Symbols = data by Zhang (1987)
$M_e = 1.5, L/D = 3, Re = 3 \times 10^7/metre, \delta_s/D = 0.3$

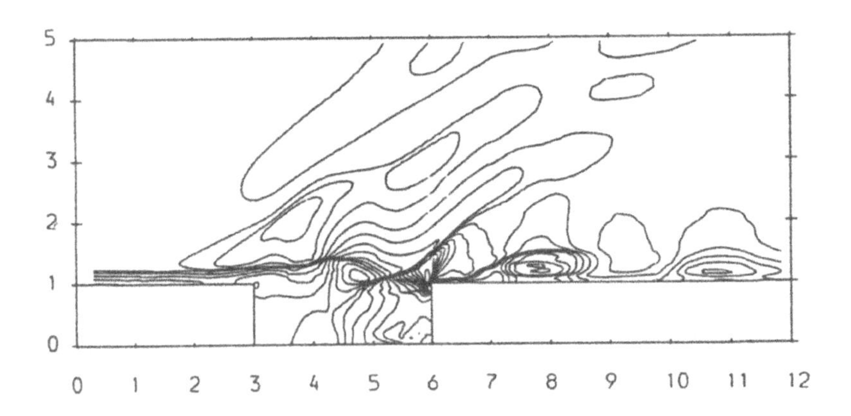

Figure 5: Instantaneous Density Contours At t = 291
30 Equi-spaced contour levels ; $p_{min} = 0.441, p_{max} = 1.689$
$Me = 1.5, L/D = 3, Re = 3 \times 10^7/metre, \delta_s/D = 0.3$

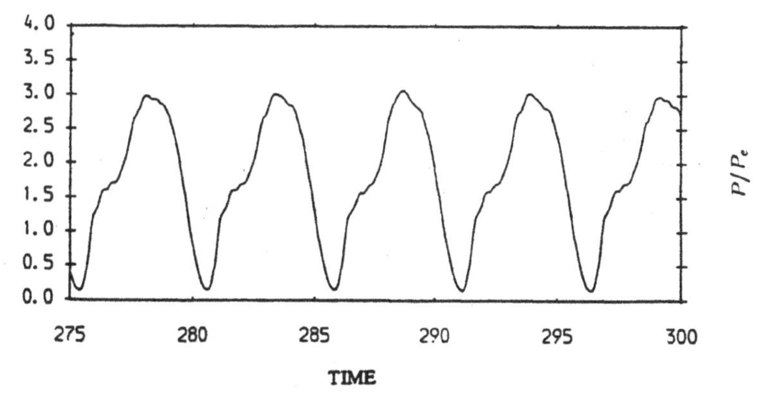

Figure 6: Pressure History For $\delta_s/D = 0.2$
y/D = 0.981 (y/D = 0 on cavity floor)
$M_e = 1.5, L/D = 3, Re = 3 \times 10^7/metre$

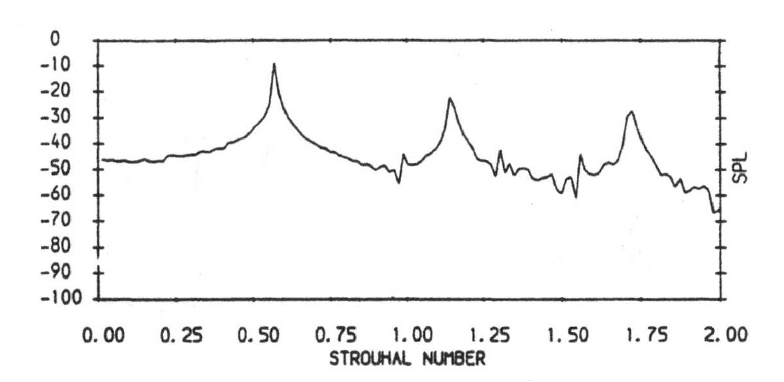

Figure 7: Sound Pressure Level
y/D = 0.981 (y/D = 0 on cavity floor)
$M_e = 1.5, L/D = 3, Re = 3 \times 10^7/metre, \delta_s/D = 0.2$

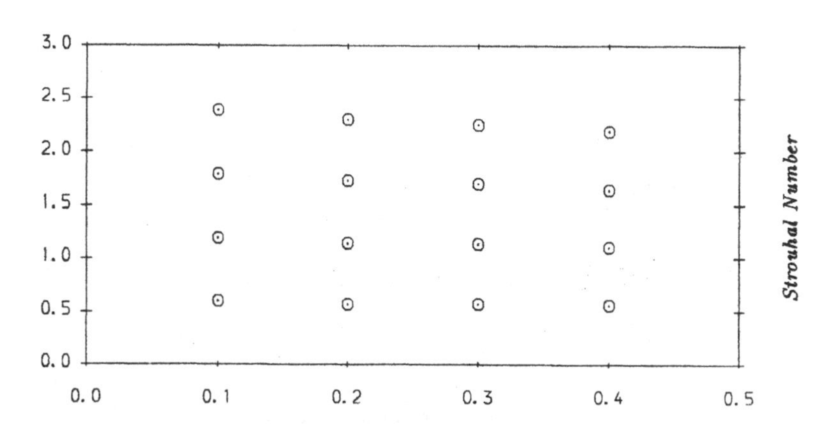

Figure 8: Variation Of Strouhal Number And Boundary Layer Thickness At Separation

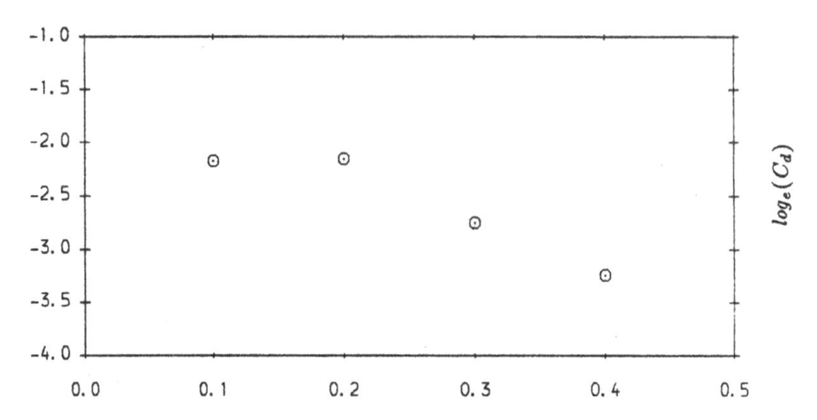

Figure 9: Variation Of Total Pressure Drag And Boundary Layer Thickness At Separation

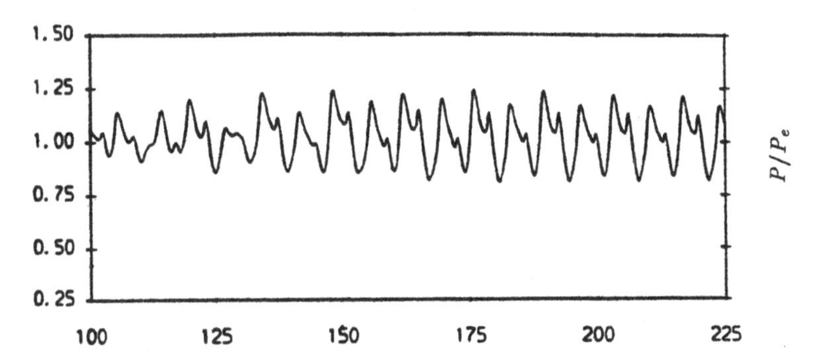

Figure 10: Pressure History Along Cavity Floor For Isothermal Wall
x/L = 0.262 (x/L = 0 on front face)
$M_e = 1.5, L/D = 3, Re = 6.56 \times 10^6/metre, \delta_s/D = 0.2$

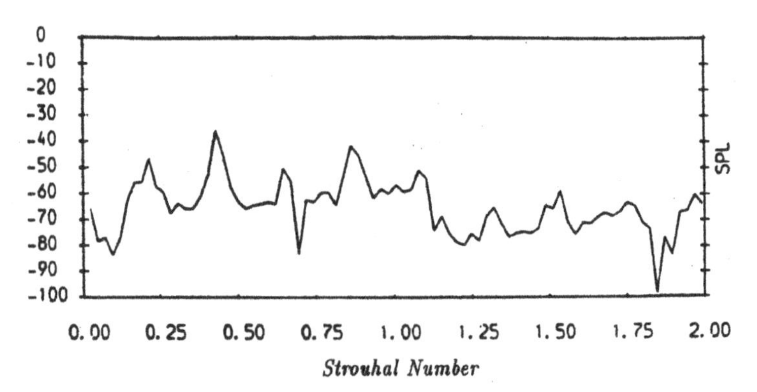

Figure 11: Sound Pressure Level
x/L = 0.262 (x/L = 0 on front face)
$M_e = 1.5, L/D = 3, Re = 6.56 \times 10^6/metre, \delta_s/D = 0.2$

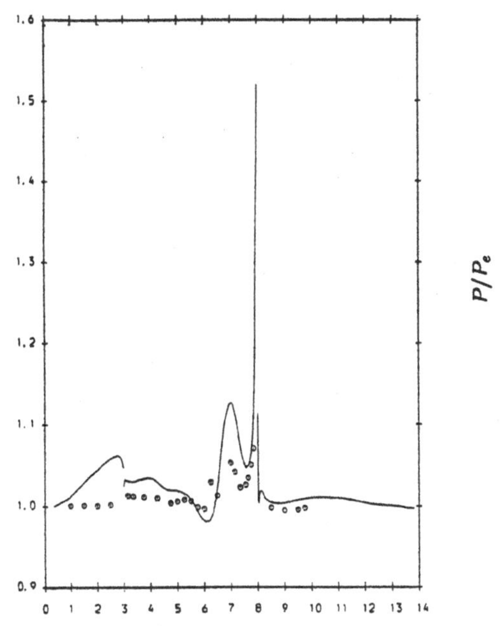

Figure 12: Time-Averaged Pressure Along Surface Perimeter (Inflow = 0)
Solid line = computational results ; Symbols = data by Stallings and Wilcox (1987)
$M_e = 1.5, L/D = 3, Re = 6.56 \times 10^6/metre, \delta_s/D = 0.2$

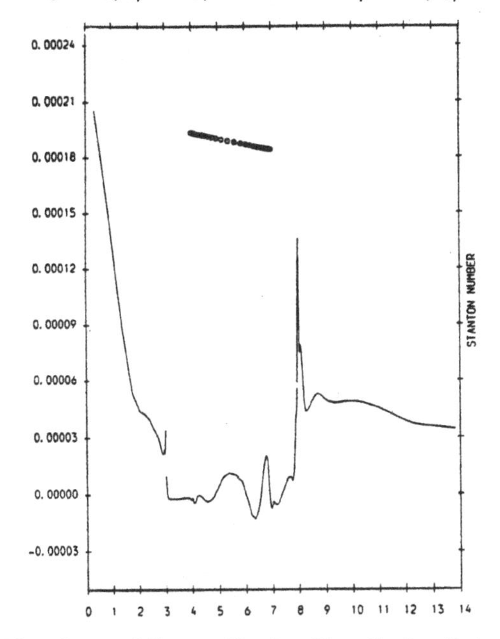

Figure 13: Time-Averaged Stanton Number Along Surface Perimeter (Inflow = 0)
Solid line = computational results ; Symbols = initial distribution
$M_e = 1.5, L/D = 3, Re = 6.56 \times 10^6/metre, \delta_s/D = 0.2$

Efficient Implementation of CG-type Algorithms for STAR-CD Flow Solver

M. Marrone (*), G. Radicati (*), F. Stella (**), M. Maggiore(*)

(*) IBM European Center for Scientific and Engineering Computing, Rome, Italy

(**) Department of Mechanical and Aeronautic Engineering, University of Rome 1, Rome, Italy

Abstract

The use of a general iterative solver for large sparse systems arising in a general fluid dynamics code, STAR-CD [1], is discussed.
The solution of the large systems accounts for about 75% of the floating point operations during the simulation on models of medium complexity. The solver we are testing treats problems with irregular sparsity patterns. Its implementation is based on a special internal data structure to vectorize the main operations, including the solution of sparse triangular systems that arise with ILU type preconditioners. With this solver we achieved a speed up of 2 over the course of the whole simulation over the original implementation.

[1] STAR-CD is a trademark of Computational Dynamics Ltd.

1 Introduction

In the last decade the numerical simulation of fluid dynamics analysis has became a very powerful tool in the solution of industrial and applied problems.

Many engineering and scientific programs have been developed using different numerical techniques:

- Finite element.
- Finite difference.
- Finite volume.

A crucial point common to all these formulations is the need of a fast solver for the several large sparse linear systems of equations occurring during the solution of the physical problem. Preconditioned conjugate gradient methods have proved very successful in solving these kind of linear systems.

Over the past years much effort has been spent in developing new CG type algorithms to solve non-symmetric problems, and many preconditioners have been considered, to increase the convergence rate of the basic algorithms. Much effort was also devoted to implementing CG type algorithms on vector supercomputers.

We have developed a general purpose solver with the following features:

1. Arbitrary sparsity pattern are treated.
2. A special internal data structure is used to vectorize the basic operations.
3. A Wavefront Technique is used to vectorize the solution of the sparse triangular system that arise by using ILU type preconditioners.

In this paper we discuss the use of the iterative solver we have developed within the STAR-CD general purpose Computational Fluid Dynamics package. Performance data and the relative speed-up are also commented on.

2 STAR-CD Description

STAR-CD is a suite of computer codes assembled by Computational Dynamics Ltd for the analysis of fluid flow, heat and mass transfer and chemical reaction in industrial and environmental circumstances. [2] In this paragraph we will briefly discuss the differential conservation equations and the numerical methodologies used in the core analysis code STAR for flow field simulation.

Although for the sake of simplicity the conservation equations will be presented in Cartesian tensor notation, it should be stated that in the development of finite volume numerical procedure embodied in STAR, the equations are transformed to a general curvilinear coordinate frame.

2.1 Mass and Momentum Conservation

The mass and momentum conservation for a general fluid flow, under steady condition, are written:

$$\frac{\partial \rho u_i}{\partial x_i} = 0 \tag{1}$$

$$\frac{\partial(\rho u_i u_j + \tau_{ij})}{\partial x_j} = \frac{\partial p}{\partial x_i} + s_i \tag{2}$$

where:

- $x_i \equiv$ Cartesian coordinate $i = 1, 2, 3$
- $u_i \equiv$ velocity in direction x_i
- $p \equiv$ pressure
- $\rho \equiv$ density
- $\tau_{ij} \equiv$ stress tensor

- $s_i \equiv$ momentum source

The stress tensor τ_{ij} is usually written for both laminar and turbulence flows in the following way:

- Newtonian Laminar Flow

$$\tau_{ij} = -\mu(s_{ij} + \frac{2}{3}\frac{\partial u_i}{\partial x_j}\delta_{ij}) \quad (3)$$

- Newtonian Turbulent Flow

$$\tau_{ij} = -\mu(s_{ij} + \frac{2}{3}\frac{\partial u_i}{\partial x_j}\delta_{ij}) + \overline{u_i' u_j'} \quad (4)$$

where s_{ij} is the rate of strain tensor given by:

$$s_{ij} = \frac{\partial u_i}{\partial x_j} + \frac{\partial u_i}{\partial x_i} \quad (5)$$

2.2 Heat transfer

Heat transfer is also provided for in STAR, using the following form of the energy conservation equations for a general fluid mixture at low Mach numbers:

$$\frac{\partial(\rho u_j h + F_{hj})}{\partial x_j} = S_h \quad (6)$$

Here:

$$h \equiv \bar{c}_p T - c_{po} T_o + \sum m_k H_k \quad (7)$$

- $T \equiv$ temperature
- $m_k \equiv$ mass fraction of mixture constituent k
- $H_k \equiv$ heat of formation of constituent k
- $\sum =$ summation over all mixture constituents
- $c_p \equiv$ mean specific heat at constant pressure referred to reference temperature T_o
- $c_{po} \equiv$ reference specific heat
- $F_{ij} \equiv$ diffusions enthalpy flux in direction x_j
- $S_h \equiv$ energy source

It should be noted that the static enthalpy h is defined here as the sum of the internal and chemical energies. For low speed flows of single component fluids with constant specific heat it reduces to the familiar temperature equation.

2.3 Turbulence Modelling

Mathematical models of turbulence are used to determine the Reynolds stresses and turbulence scalar fluxes. STAR currently contains three alternative turbulence representations. One comprises Differential Transport equations for the time averaged turbulence Kinetic Energy K and its dissipation rate ϵ. This is the widely used $k-\epsilon$ model. A second option dispenses with the ϵ equation and requires instead a user prescribed spatial distribution of the length scale l. This $k-l$ model is also well known. The third option allows the add viscosity to be prescribed directly by the user. Although it doesn't affect for the presented results, during the test cases we have used the standard $k-\epsilon$ model.

2.4 Discretization

The Differential Conservation equations are discretized by a Finite Volume procedure, whereby they are first integrated over the computational mesh cells, and then approximated in terms of the nodal values of the dependent variables. This approach has the merit, among others, of ensuring that the discretization form preserves the conservation properties of the parent differential equations. There is no reason to discuss here in details the discretization scheme used; we only want to point out a remarkable final point on discretization of the upwind scheme. STAR contains two different basic upwind schemes: a standard first order differencing ($\gamma = 0$) and a second order linear upwind differencing ($\gamma = 1$). The former is beneficial for numerical stability but can produce artificial "smearing" of step gradients of the function, while the latter is less diffusive but can give rise to overshoots and instabilities. Intermediate values for γ are also possible. We have used the first order differencing scheme in our test cases. This has no influence on our timing results.

2.5 Solution Technique

The STAR solver uses two different, well known, numerical procedures for the solution and time integration of field equations: SIMPLE and PISO (Pressure Implicit with Splitting of Operators). [3]

Both methods require the solution of very large linear systems for mass conservation and/or pressure integration. To solve these systems STAR uses the conjugate gradient method with Incomplete Cholesky preconditioning. [9] [8] The CG algorithm is a well known iterative method, especially suited for symmetric positive definite systems. It has been shown that this algorithm theoretically yields the exact answer after at most N steps, where N is the problem dimension. Given an initial guess x_0, the conjugate gradient method produces a sequence of approximations to the solution, x^*, of the form:

$$x_{i+1} = x_i + \alpha_i p_i \tag{8}$$

Here the vectors p_i are called the search direction. An important property of this algorithm is that the error vectors $x^* - x_i$ are minimized with respect to some suitable vector norm.

3 Vectorization

A large portion of the total CPU time for code of this type is devoted to the solutions of linear systems. In the our initial tests of STAR these costs amounted to around 75% of the global CPU time. Much can, however, be gained by vectorizing this part of the code.

From an implementation point of view, in fact, the performance of the preconditioned iterative method depends strongly on the optimization of two computational kernels:

- Solution of the triangular system.
- Matrix vector product.

Both kernels can be efficiently optimized for problems with an irregular sparsity pattern, provided a data structure appropriate to a vector architecture is adopted.

The vectorization of the triangular systems which arises from the preconditioning is non trivial. The required forward and backward substitution, in fact, are recursive in nature and therefore not well suited to vectorization well.

Our implementation makes use of the fact that when carrying out a sparse backward or forward substitution, not all of the $i-1$ previous components are required at any given i^{th} stage. The components can then be partitioned into the following level set, known as a wave-front: [10]

- Level 1: Those components independent of any other, given directly by the right hand side of the substitution. The first component lies in this class.
- Level 2: Those components depending only on the components of level 1.
- ....
- Level i: Those components depending only on the components of level $1, 2, ...i-1$

These sets form an ordered set, and the components belonging to each level set can be solved concurrently. By expanding the recursion in this way, vectorization can be achieved.

The data structure used to store the matrix and the preconditioner is an extension of the ITPACK scheme [1]. It is based on sorting the rows of the matrix by decreasing number of non zero elements. With this data structure a vectorization of the matrix-vector product can be achieved. To vectorize the iterative algorithm, we therefore need to perform some data transformation to the coefficient matrix before starting the iterative procedure.
This extra cost, which we include in evaluating the performance of the code, is offset by the performance gain due to the vectorization.

4 Numerical Tests

A number of preliminary tests were conducted on a test problem arising in the simulation of a cooling circuit using a mesh with 48606 points. Due to the geometrical complexity of the circuit, the resulting sparsity pattern of the matrix was irregular. This is shown in Figure 1.

In the following table we report the performance of the matrix vector product and solution of two sparse triangular linear systems, using the data structure described in Section 3. These results are compared with those obtained using a scalar code.

	Vector data format	Scalar data format
Matrix-vector	0.05 sec.	0.17 sec.
Triangular systems	0.08 sec.	0.20 sec.

Taking the CPU time spent in the solution of the linear system, a comparison of the STAR implementation and our implementation of the same algorithm CG was done. The STAR CG solver takes 99 sec. whereas our code took 33 sec. This gives a relative speed-up of 3 for the solution of linear system. The time spent for setting-up the new data structure was 4 seconds, negligible compared to the time spent in the iteration loop, that is 29 seconds. Fig. 2 shows residual trends as a function of the CPU time.

Replacing the original solver of STAR-CD with the vectorized solver we developed, but otherwise not changing anything else in the code, resulted in a speed-up of 2 over the whole simulation.

Although the matrices arising when using the discretization described in Paragraph 2 are positive definite, with more general fluid dynamics problems indefinite matrices can arise. In the Conjugate Gradient algorithm the definiteness of the matrix is a necessary condition to guarantee the properties of stability and convergence. The use of a general iterative solver, implementing more general iterative methods, allows us to treat this case without further work. In our code, in fact, using the same techniques described in Paragraph 3, the following methods have been implemented:

- Conjugate Gradient Squared (CGS) [6]
- Generalized Minimum Residual (GMRES) [4] [5]
- CGSTAB (More smoothly convergent version of CGS) [7]

These methods are well-known in the literature and are useful for indefinite matrices.

5 Physical case: Timings and Results

For a more complete test case, the question related to the optimization of the conversion efficiency of a catalytic converter has been studied. [11] [12] Catalytic converters seem in fact to provide the most viable solution to the stringent European standards imposed to pollutant emissions in the EC countries by 1992.

We focus on the influence exerted by a non uniform flow distribution at inlet of oxidizing catalytic converters, on the conversion efficiency evaluated channel by channel. The flow field in catalytic converter domain has been calculated using the STAR-CD version 2. The control volume is constituted by:

- the last part of exhaust manifold;
- the ellipse shaped duct connecting it with the converter shell;
- the honeycomb monolith itself and exhaust final duct.
- the ellipse shaped cylindrical shell containing the honeycomb monolith converter;
- the honeycomb monolith itself;
- the final exhaust duct constituted by the second and first above mentioned item.

Figure 3 depicts the computational mesh of the catalytic converter. The mesh is composed by 108,544 cells describing the fluid body and by 26,160 cells useful to describe the geometry of the ceramic monolith. Thanks to the geometrical and flow symmetries, the computational domain is the one shown. For this reason, besides wall surfaces, physical boundary conditions have been imposed only at the inlet, at the outlet and to the wall of calculation domain as follows:

- inlet section: constant gas velocity and assigned pressure;
- outlet section: zero gradients along flow streamlines, i.e. fully developed flow fields;
- axial boundary planes: zero normal velocity flow and zero normal gradients, i.e. symmetry condition;

- channels monolith internal surface: wall boundary condition, the channel wall is highly roughly to simulate the porous media.

The 3-D fluid dynamic model has been used to investigate the flow pattern at the converter inlet section, in order to provide the distribution of the exhaust gas rate entering the monolith passageways. By assuming such flow distribution, the reactive flow inside the single channels has been simulated by means of a 1-D model developed by the authors.

Final results were obtained in terms of mole fractions of pollutants at the end of the conversion process in the exhaust gas. The numerical analysis takes several hours of computing time on an IBM 3090/VF.

Two different versions of STAR-CD were compared for the final benchmark on the code:

1. Vector, optimized with STAR implementation of Conjugate Gradient: 1000 minutes $\simeq$ 17 hours.
2. Vector, optimized with our implementation of Conjugate Gradient: 550 minutes $\simeq$ 9 hours.

The overall speed-ups of 1.9 was obtained.

6 Conclusions

In summary we can conclude that:

- General iterative solvers offer a solution to many classes of linear algebraic problems and can be usefully used inside application programs. We have successfully used the same solver in some reservoir simulation programs. In particular, using a suitablely chosen implementation and matrix storage scheme, we obtained a factor 2 speed up over the code implemented in STAR-CD.
- Although the new solver perform the floating point operations in a different order with respect to the original STAR-CD solver, the results are equivalent up to round off errors.

References

[1] Paolini, G.V. and Radicati, G. (1989)
Data structure to Vectorize CG Algorithms for General Sparsity Patterns.
BIT, 29, 703-718.

[2] Computational Dynamics
STAR CD manuals.
Computational Dynamics Ltd., London

[3] R. I. Issa
Solution of the Implicit Discretized Fluid Flow Equations by Operator Splitting.
J. of Comp. Physics., 62, No. 1, 40-65.

[4] Saad, Y. and Schultz, M.H. (1986)
GMRES: a Generalized Minimal Residual Algorithm for Solving Nonsymmetric Linear Systems.
Siam J. Sci. Stat. Comput., 7, No. 3, 856-869.

[5] Wigton L. B., Yu N. J. and D. P. Young
GMRES: acceleration of fluid dynamics codes.hm for
Proceedings of the American Institute of Aeronautics and Astronautics 7th Computational Fluid Dynamics Conference, Cincinnati, OH, July 15-17, 7 15-17, 1985, pp. 67-74

[6] Sonneveld, P. (1989)
CGS, A Fast Lanczos Type Solver for Nonsymmetric Linear Systems.
SIAM J. Sci. Stat. Comput., 10, No. 1, 36-52.

[7] Van der Vorst, H. and Sonneveld, P. (1990)
CGSTAB: a More Smoothly Converging Variant of CGS.
Preprint of Delft University of Technology, May 21, 1990.

[8] Meijerink, J.A. and Van der Vorst, H. (1977)
An Iterative Solution Method For Linear Systems of Which the Coefficient Matrix is a Symmetric M-matrix.
Math. of Comp., 31, No. 137, 148-162.

[9] Melhem, R. (1987)
Toward Efficient Implementation of Preconditioned Conjugate Gradient Methods on Vector Supercomputers.
Int. Jour. of Supercomp. Applic., 1, No. 1, 70-88.

[10] Anderson, E. and Saad, Y. (1988)
Solving Sparse Triangular Linear System on Parallel Computers.
Center for Supercomputing Research and Development, Report No. 794, June 1988

[11] Bella G., Maggiore M., Rocco. V.
A Study of Inlet Flow Distorsion Effects on Automotive Catalytic

Converters
ASME paper 91 - ICE - 13.

[12] Bella G., Maggiore M., Rocco. V., Stella F., Succi F.
Automotive Catalytic Converters Performance Evaluation: a Computational Approach
ATA Journal, in press.

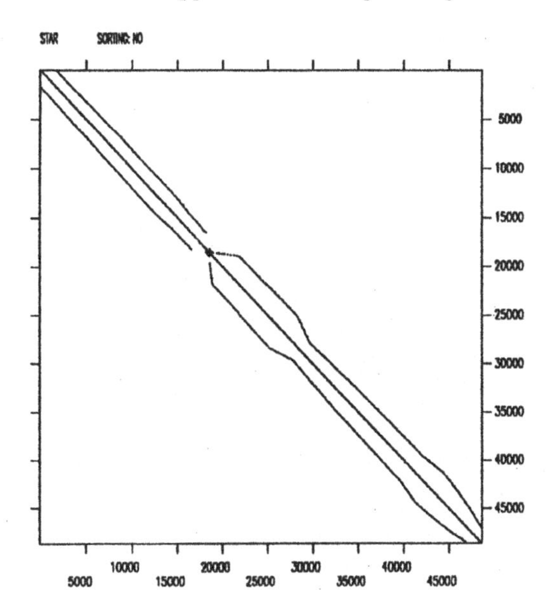

Figure 1: Sparsity pattern of STAR-CD matrix

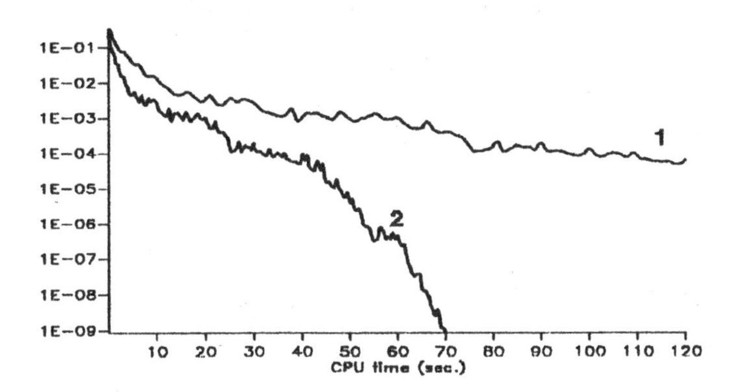

Figure 2: Residual trends for STAR-CG(1), our-CG(2)

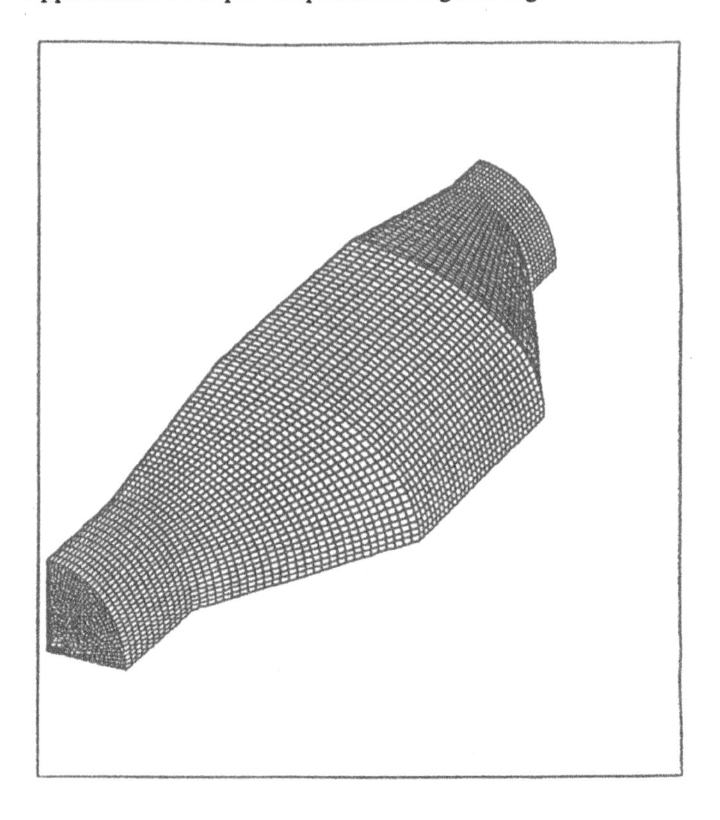

Figure 3: Catalytic converter computational mesh

Analysis of the Residual Optimization Potential and Runtime Analysis of the GW code on SNI Supercomputers

B. Romunde (*), H. Daniels (*), J. Neises (**), U. Proff (**)

(*) Institut für Wasserbau, Aachen University of Technology (RWTH), Mies-van-der-Rohe-Str. 1, 5100 Aachen, Germany

(**) Siemens Nixdorf Informationssysteme AG, Geschäftsstelle Supercomputer, Dreizehnmorgenweg 46, 5300 Bonn 2, Germany

1. SUMMARY

The GW code has been designed in 1987 for the computation of multi-aquifer groundwater flow problems on vectorcomputers as the Cyber 205 and the Cray X-MP. For a description of the code see i.g. Pelka, Peters (1986) and Peters (1988). Vectorcomputers have undergone further development since the early stages of the code implementation. In cooperation between the Aachen University of Technology (RWTH) and Siemens Nixdorf (SNI) the GW code was analyzed for residual optimization potential with respect to SNI supercomputers. Based on the existing code, vector compiler directives and recoding such as inlining were used to improve the performance. Most emphasis was put into the optimization of the solvers for the linear systems of equations resulting from the finite element formulation. The global matrix is random sparse, symmetric and positive definite. A direct skyline solver as well as iterative preconditioned conjugated gradient (PCG) solvers were tested. Especially the performance of the diagonally scaled PCG could be optimized substantially by recoding. The paper shows runtime results from two application data sets on SNI supercomputers and an IBM 3090 VF as a mainframe supercomputer with vector facility.

2. INTRODUCTION

In water resources management groundwater modelling becomes a tool of ever increasing importance. Some large scale groundwater management activities are being examined at the Institut für Wasserbau.

In a region close to Aachen, huge brown coal pits are operated. In open pit mining coal is gained from depths up to 500 m below the original ground surface. Though economically attractive, the brown coal mining activities cause severe ecological problems. The groundwater level has to be lowered below the bottom of the pits. Aquifer and aquitard formations which are underlying each other are influenced strongly by the groundwater depression. The water supply systems and the water levels in brooks and wetlands many kilometers away from the pits are subject to measurable changes.

As a tool to estimate the groundwater flow situation and groundwater balances in multi aquifer systems we developed and used the GW code during the last decade. As will be shown later, the size of the model problems easily reaches far beyond the capacity of todays supercomputers, unless model simplifications are accepted. But still, supercomputer power is required to obtain results in acceptable cycle times. Therefore, in cooperation between the SNI and the Institut für Wasserbau of the RWTH Aachen, the GW code was ported and optimized on SNI supercomputers.

3. THE GW CODE

Mathematical model

In principle, the GW code solves the well known threedimensional groundwater flow equation:

$$\frac{\partial}{\partial t}(n\rho) = \frac{\partial}{\partial x_i}\left[\frac{\rho\, k_{ij}}{\mu}\left[\frac{\partial p}{\partial x_j} + \rho\, g_j\right]\right] + Q \quad , i, j = 1, 2, 3 \quad . \tag{1}$$

Herein t denotes the time coordinate, n is the porosity, ρ the fluid density, x_i are the spatial coordinates, k_{ij} is the permeability tensor, μ is the dynamic fluid viscosity, p is the fluid pressure, g_j the vector of gravity and Q is a volume sink and source. Einstein's summation convection is used.

However, in practice it is impossible and beyond the scope of the research project to solve eq. (1) in three spatial dimensions for a multi aquifer system of a size of (L/H/B) = (50 km/ 400 m/ 30 km) on todays supercomputers. Therefore, the model problem is very much simplified:

1) The total geological formation is divided into a small number of discrete aquifers and aquitards.
2) Groundwater flow is assumed to be two-dimensional horizontal mainly in the aquifers. This is equivalent with the application of the Dupuit assumption in individual aquifers.
3) If aquifers are connected through aquitards, flow is assumed one-dimensional vertical in the aquitards.

The resulting flow equations are then for the aquifers

$$\frac{\partial(S\ h)}{\partial h}\frac{\partial h}{\partial t} = \frac{\partial}{\partial x_i} T(h) \frac{\partial h}{\partial x_i} + Q \quad , \qquad i = 1, 2 \tag{2}$$

and for the aquitards:

$$\frac{\partial(S\ h)}{\partial h}\frac{\partial h}{\partial t} = \frac{\partial}{\partial z} T(h) \frac{\partial h}{\partial z} + Q \quad . \tag{3}$$

Herein h denotes the potential or piezometric head, (Sh) is a dirac delta function in h for the storage coefficient:

$$(S\,h) = \begin{cases} h\,n & \text{for } h_b < h < h_t \\ h\,S_o & \text{for } h > h_t \\ 0 & \text{for } h < h_b \end{cases} \tag{4}$$

h_b and h_t denote the elevation of the bottom and the top of an individual aquifer or aquitard. S_o is the integrated value of the sum of the fluid and matrix compressibility in the aquifer or aquitard:

$$S_o = \int_{h_b}^{h_t} \frac{\partial}{\partial h}(n\,\rho)\,dh$$

T(h) is the nonlinear transmissivity function of an aquifer:

$$T(h) = \begin{cases} k_f\,(h - h_b) & \text{for } h_b < h < h_t \\ k_f\,(h_t - h_b) & \text{for } h > h_t \\ k_f \cdot 0{,}001 & \text{for } h < h_b \end{cases} \tag{5}$$

From eq. (4) it is clear that the nonlinear term $\partial S/\partial h$ requires very careful examination especially in time dependent calculations where h may easily jump over h_t or h_b in one time step Δt of the computation. As the nonlinearity in S also the nonlinearity in T requires an iterative solution.

Numerical model

Numerically, the flow equations for aquifers (eq. 2) and aquitards (eq. 3) are solved by the Galerkin finite element method (GFEM) and implicit finite differences for the approximation of the time dependent terms. With the GFEM method the GW code uses linear triangular finite elements for the aquifer spatial discretization and linear one dimensional elements for the aquitard spatial discretization. The discretizations of the aquifers are connected node by node by aquitard elements, if there is another aquifer to be connected to the one examined. Fig. 1 shows an aquifer and an aquitard finite element and fig. 2 a multi–aquifer subgrid.

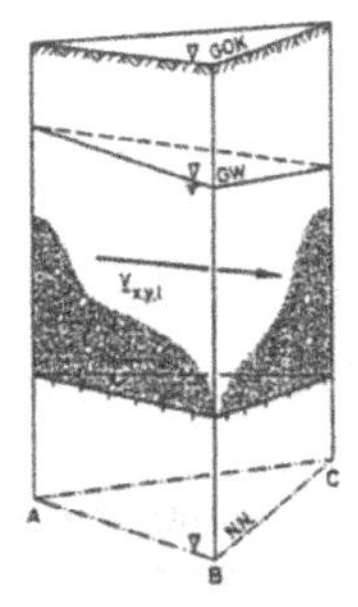

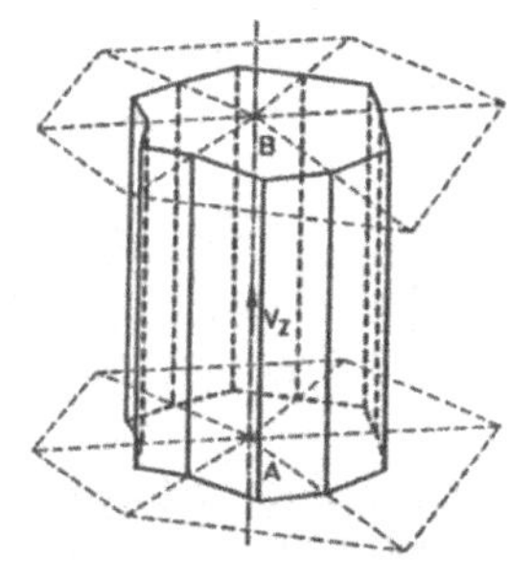

fig. 1: finite elements for a) aquifers, b) aquitards

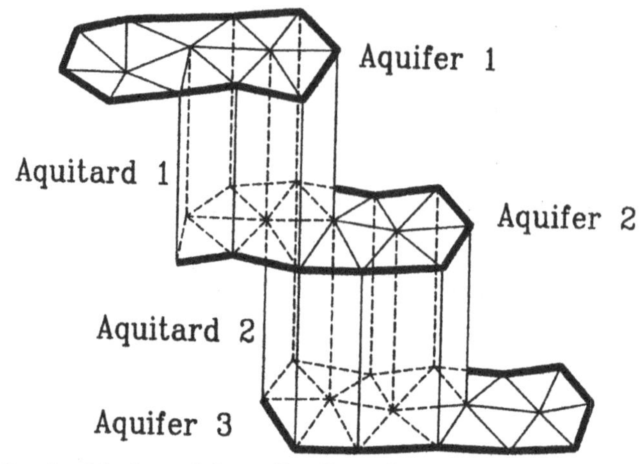

fig. 2: subgrid of a multi–aquifer discretization

The discrete algebraic variant of the equations (2) and (3) is with continuous linear C^0 approximation functions ϕ for the piezometric heads and piecewise discontinuous constant (0 or 1) C^{-1} approximation functions Ψ for storage coefficients and transmissivities:

$$(S + A)\, \underline{h}^{n+1} = S\, \underline{h}^{n} + \underline{f}^{n+1} \quad . \tag{6}$$

Herein $\underline{h}$ contains the values of the piezometric heads at the i=1, ..., N nodes of the finite element mesh. The number of triangles in the aquifer is M_n, the number of line elements in the aquitards is M_v and $M = M_n + M_v$. n and n+1 denote the new and the old time plane respectively. S is the storage matrix of

$$S_{mn} = \sum_{e=1}^{M} \frac{1}{\Delta t} \left[\frac{1}{\Delta \bar{h}} \int_{\bar{h}} (S \cdot \bar{h})\, d\bar{h} \right]^{e} \int_{\Omega^e} \phi_m \phi_n^T \, d\Omega^e \quad , n = 1, ..., N \, . \tag{7}$$

Herein Δt is the time step size, $\Delta \bar{h}$ is the change of the average element piezometric head $\bar{h}$ in the time step. Clearly $\Delta \bar{h}$ and the integral on $\bar{h}$ can only be evaluated in an iterative process within a time step.

A is the diffusion matrix of:

$$A_{mn} = \sum_{e=1}^{M} T(\bar{h})^e \int_{\Omega^e} \frac{\partial \phi_m}{\partial x_i} \frac{\partial \phi_n^T}{\partial x_i} \, d\Omega^e \quad , \qquad \begin{matrix} n = 1, ..., N \\ i = 1, 2, z \end{matrix} \tag{8}$$

$\underline{f}$ is the right hand side vector, which contains volume sinks and sources as well as a contribution from the application of Green's second identity on the diffusion operator:

$$\underline{f}_m = \sum_{e=1}^{M} Q^e \int_{\Omega^e} \phi_m \Psi_e \, d\Omega^e + T(\bar{h})^e \int_{\partial\Omega^e} \phi_m n_i \frac{\partial \phi_n^T}{\partial x_i} h_n^{n+1} \, \partial\Omega^e ,$$

$$n = 1, \cdots, N; \quad i = 1, 2, z \tag{9}$$

Herein n_i is the outward pointing normal vector on the boundary $\partial\Omega$ of the domain of computation $\Omega = \sum_e \Omega^e$. With linear approximation functions ϕ and piecewise continuous functions Ψ, $\underline{f}$ can be evaluated analytically:

$$\underline{f}_m = \underline{Q}_m \tag{9a}$$

where $\underline{Q}_m$ contains all the boundary and sink/source fluxes which correspond to the nodes in the finite element mesh. Similarly the integrals on ϕ in eq. (7) and (8) can be evaluated analytically for linear ϕ and triangles or line elements rather than by numerical Gauss point integration. The resulting integrated matrices are listed here for completeness:

	triangles	line elements
$\int_\Omega \phi_m \phi_n \, d\Omega :$	$\frac{A^e}{12} \begin{bmatrix} 2 & 1 & 1 \\ 1 & 2 & 1 \\ 1 & 1 & 2 \end{bmatrix}$	$\frac{L^e}{6} \begin{bmatrix} 2 & 1 \\ 1 & 2 \end{bmatrix}$
$\int_\Omega \frac{\partial \phi_m}{\partial x_i} \frac{\partial \phi_n^T}{\partial x_i} \, d\Omega :$	$\frac{1}{2A^e} \begin{bmatrix} b_2^1 & b_1 b_2 & b_1 b_3 \\ b_2 b_1 & b_2^2 & b_2 b_3 \\ b_3 b_1 & b_3 b_2 & b_3^2 \end{bmatrix}$	$\frac{1}{L^e} \begin{bmatrix} 1 & -1 \\ -1 & 1 \end{bmatrix}$

$$b_1 = y_2 - y_3 \quad , \quad b_2 = y_3 - y_1 \quad , \quad b_3 = y_1 - y_2$$

L^e is the length of a line element, A^e is the area of a triangle, y_i are the global coordinates of the 3 nodes of a triangle.

Implementation aspects (overview)

The vectorizable implementation of the element matrix evaluation and global matrix assembly was documented in Pelka and Peters (1986) for a stage of the GW code development, where

- Cholesky factorization was used to solve the linear symmetric positive definite global system of equations
- The available vectorcomputers were CDC Cyber 205 and Cray X–MP.

Peters (1988) and Peters, Romunde und Sartoretto (1988) documented the global matrix assembly and the vectorizable implementation of

- Preconditioned conjugated gradient solvers (PCG) combined with a compressed storage scheme to solve the global system of equations.
- Diagonal scaling, incomplete Cholesky and polynomial preconditioning was implemented.
- The available vector computers still were CDC Cyber 205 and Cray X–MP.

Runtime analysis showed that for any of the solvers most of the CPU time was spent on solving the large system of linear equations in each iteration step. This crucial part of the code therefore was subject to the main interest of the authors.

4. THE NEW SUPERCOMPUTER ENVIRONMENT

In 1989 the Aachen University of Technology (RWTH) bought an IBM 3090 600S VF and a SNI VP200–EX in April 1990. In March 1991 the VP200–EX was replaced by a S400/10. This new supercomputer environment raised the need to adapt the GW code to the SNI supercomputers.

The architecture of the Siemens Nixdorf S series

Similar to the previously installed Siemens Nixdorf VP200–EX supercomputers the S400/10 has a register to register vector unit and a scalar unit compatible to the IBM/370 architecture.

The scalar unit controls all operations. All scalar instructions are processed by the scalar unit itself with a sustained performance of 30 Mflops. Vector instructions are passed to the vector unit for execution. The vector unit consists of a vector execution control, the vector pipelines, the vector registers and the mask registers (fig. 3).

The S400/10 vector unit is build of

- 2 multifunctional pipelines, which execute vector multiplication, vector addition, logical vector operations, as well as linked triades (multiply and add).
- 1 divide pipeline
- 2 mask pipelines
- 2 load/store pipelines.

All pipelines are two–fold, that means they always operate on two operand pairs simultaneously. The performance of the vector unit is increased by parallel execution of two arithmetic, plus the two mask and the two load/store pipelines. With a cycle time of 4 ns, the S400/10 reaches a peak performance of 2Gflops (8 floating point operations per cycle). The vector pipelines operate on the vectors in the vector registers. With both load/store pipelines working in parallel a throughput of 8 Gbytes per second between memory and vector registers can be achieved.

The 64 Kbyte vector registers of the S400/10 can be configured dynamically between a maximum of 256 registers of 32 elements of 64 bit floating point numbers each and a minimum of 8 registers containing 1024 elements per vector. The configuration of the vector registers is done automatically by the compiler depending on loop length, the number of vectors and the number of vector operations in a loop. This optimization is done for each loop separately (fig. 4).

The operating system of the S400/10 at the RWTH Aachen is VSP/I and mostly the Fortran77/VP compiler is used. They will be replaced by VSP/S and Fortran77/EX very soon.

VSP/I and its successor VSP/S are similar to MVS with vector-processing enhancements. The main memory unit is split by the so called VP Line into scalar region and vector region. Scalar jobs are processed in virtual addressing mode with 4 Kbyte pages. Vector jobs use a logical addressing mode, with pages containing 1MB. Vector jobs have to be loaded in memory completely to be executed. They can be swapped in and out in units of vector pages.

Fortran77/VP and Fortran77/EX are standard ANSI Fortran77 compilers. The data formats are compatible to the IBM/370 architecture. Vectorization can be optimized by compiler directives.

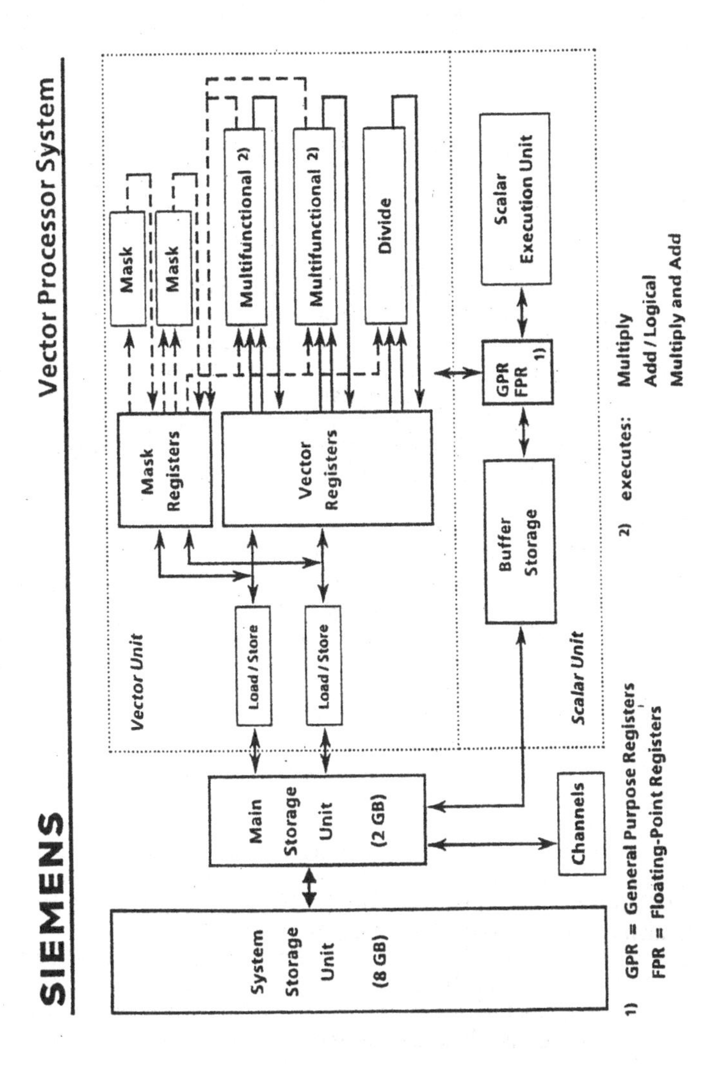

fig. 3: S Series: System Architecture

Number of Vector Registers	Length of Vector Register in Words of 64 Bit			
	S100	S200	S400	S600
256	16	16	32	64
128	32	32	64	128
64	64	64	128	256
32	128	128	256	512
16	256	256	512	1024
8	512	512	1024	2048

fig. 4: S Series: Vector Register Structure Patterns

5. THE TESTS

General remarks

The GW code tested here was originally coded on the CDC Cyber 205 and then ported to an IBM 3090. The simulation of Q8 vector instructions by subroutines for example realizing elementwise multiplication or gather/scatter operations is one of the program's most remarkable features. Thus there is a significant overhead in subroutine calls, that becomes obvious for small problems with short vectors. This additionally inhibits chaining of vector instructions and unnecessary load/store operations are needed. We measured cpu time required for

- I/0 and preprocessing
- matrix assembly and postprocessing
- solution of the system of linear equations.

We tested two problems as described below. The bulk of compution is done by solving the system of linear equations resulting from the finite element discretization. According to the problem the system has a very high sparsity. So we have focused our effort at this part. We tested the behaviour of 4 different solvers:

- DCG is a diagonally preconditioned CG–solver
- DCG–opt is the optimized version of DCG
- ICCG is a CG solver preconditioned by an incomplete Cholesky factorization (ILU).
- SKY is a skyline solver.

As SKY is about 20% faster and needs considerably less memory than a bandmatrix solver, we did not test any bandmatrix solver yet. Obviously ICCG is the most robust solver. It converges rapidly due to the good preconditioning. It is the fasted scalar solver.

The vectorization of ICCG is limited as the ILU factorization is recursive and the forward backward resubstitution must be executed in scalar mode. The good vectorization of the CG part is not effective because the ICCG method only needs a few iterations to converge.

DCG is highly vectorizable with a vector length of the number of nodes. It mainly consists of two inner products, three triades and a matrix vector multiplication. Its usage, however, is limited to well conditioned problems.

To eliminate the overhead of subroutine calls we optimized DCG by inlining the subroutines that realize the vector operations. After the inlining, isolated vector operations were combined to reduce the number of unnecessary load/store instructions and to achieve chaining effects. Thus a significant speed up was gained, in vector as well as in scalar mode.

SKY is a skyline solver. Vectorization is limited to single skylines because matrix factorization by the Cholesky method is recursive. In our examples, most of the skylines are shorter than 100 elements. As SKY needs much more floating point operations than the ILU and as it operates on short vectors compared to the CG method its performance is worse than the ICCG's.

The small problem "Büttelborn" (Rouvé et al., 1989)

The first test series was performed for a two–aquifer system in the vicinity of Frankfurt. Fig. 5 shows the surface mesh. Altogether 1896 nodes, 3686 triangular elements and 948 line elements were involved in the computation. For the test runs, the very first time step was calculated with 5 iterations of the nonlinearities. The global code performance in terms of absolute time required is shown in fig. 6 for the SNI S400/10 and as a benchmark backup also on one processor of the IBM 3090 600S VF. Always vector performance is plotted in the left half page and scalar performance in the right half page. Clearly DCG–opt outperforms DCG on the SNI systems. ICCG outperforms DCG, DCG–opt and SKY in the scalar mode, while in the vector mode DCG–opt is faster. Also on the IBM 3090 DCG–opt outperforms DCG in scalar mode.

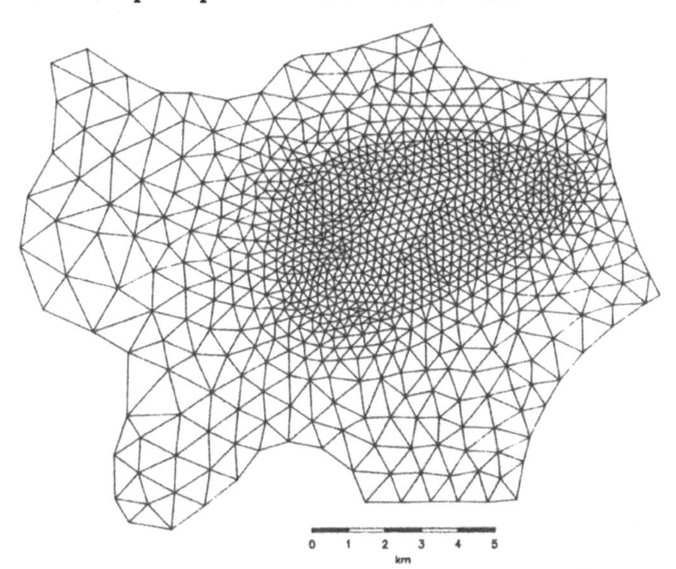

fig. 5: Surface discretization of "Büttelborn" problem

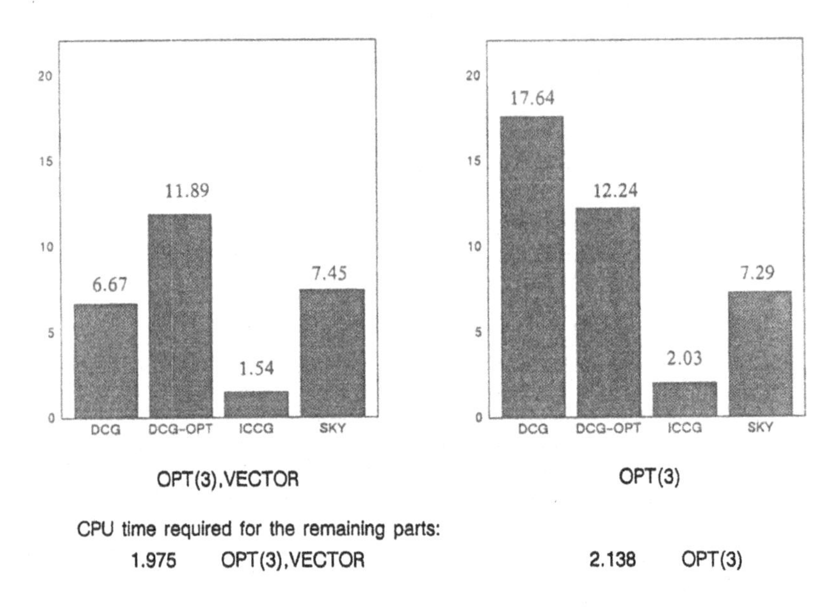

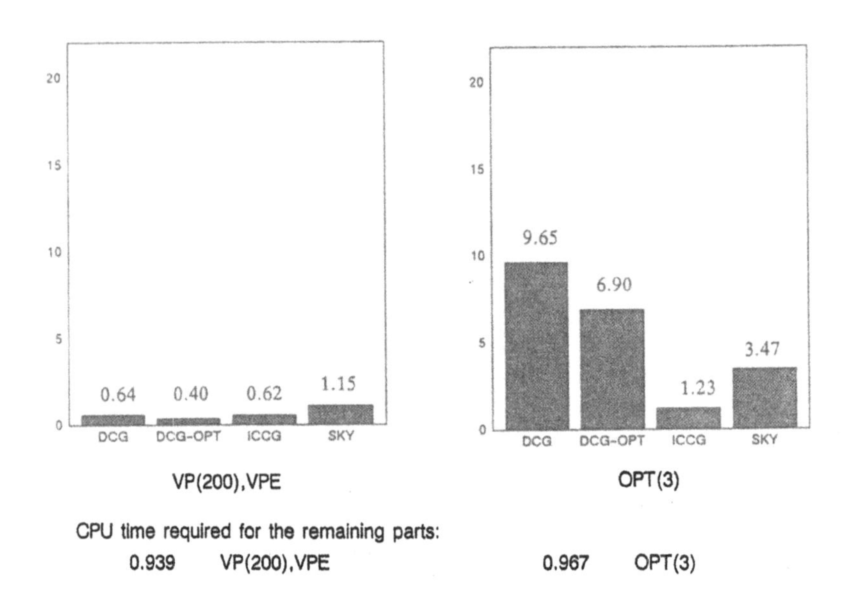

fig. 6: Absolute time required for "Büttelborn" problem

The large problem "Venloer Scholle" (Rouvé et al., 1990)
The second test series was performed for a multi–aquifer system in the brown coal region near Aachen. Five main aquifers are separated partly by four coal and clay/silt aquitards. Fig. 7 shows the mesh. Not all of the aquifers extend across the whole model area. Altogether 6760 nodes, 13341 triangular and 4879 line elements were used to describe the model area. For the test runs 12 time steps with 3 to 6 iterations of the nonlinearities were calculated. As a major result it was found that DCG (and DCG–opt of course) did not converge to a solution at all within a bounded number of CG iterations. In this case, clearly, ICCG out performs any other method on any machine by far (fig. 8).

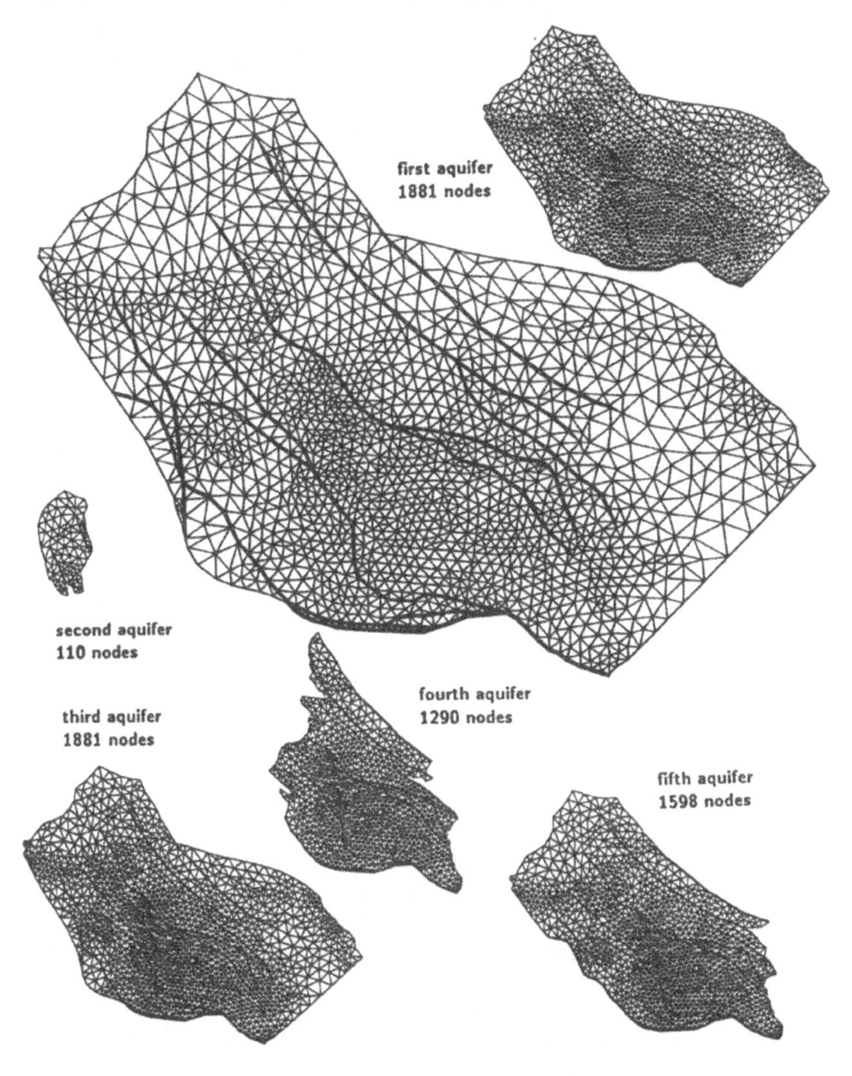

fig. 7: Discretization of "Venloer Scholle" problem

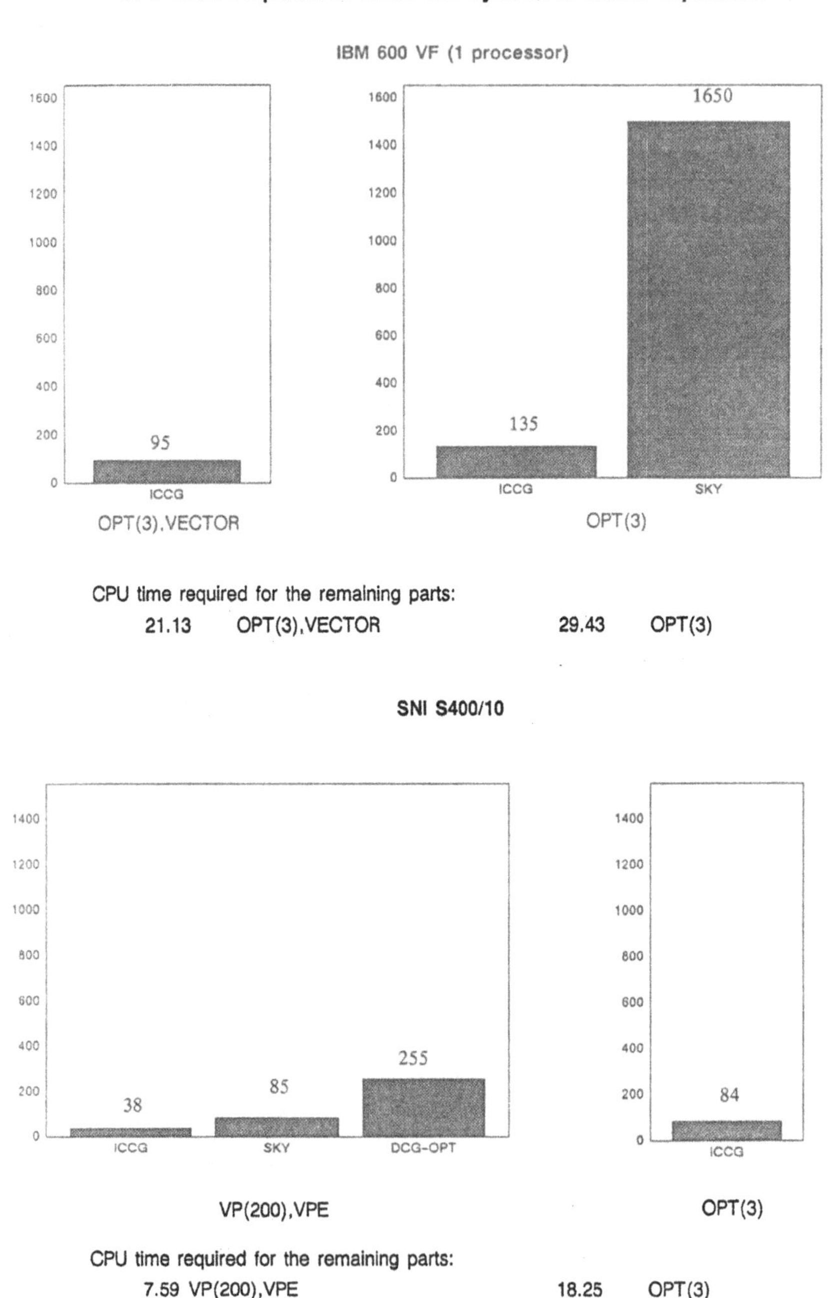

fig. 8: Absolute time required for "Venloer Scholle" problem

6. CONCLUSION

At first we should point out that there is a lot of further inlining and reorganization to be done on the GW code in order to get full optimization and best performance. The GW code's structure of small vector routines results in easy portable code but it inhibits optimal vectorization.

The central conclusion reads: The solver must correspond to the problem. A deeper insight into the problem's nature is essential to decide which solver should be used. The better performance of iterative methods in comparison to direct methods for medium and large sized problems – more than approximately 700 nodes – becomes obvious. Neither SKY, nor a bandmatrix solver performs, as well as the iterative methods. If an iterative solver is chosen, the decision on the most efficient preconditioning method has to be made. There are only a few criteria for a decision, especially the mesh's regularity and the eigenvalues distribution seem to have a great influence on the solver's convergence. If there should be a lack of insight into the problem, trial and error with short timesteps might help.

As preconditioning is the key to good convergence and optimal performance of an iterative solver, much further research is necessary in developing efficiently vectorizable preconditioners. There is a CG–type preconditioner which works well on eigenvalue problems solved by Rayleigh quotients. We will investigate on this.

7. REFERENCES

PETERS, A. (1988): Vectorized Programing Issues for FE Models; Advances in Water Resources, Vol. 11, No. 4, Dec. 1988, pp. 180 – 185.

PELKA, W.; A. PETERS (1986): Finite Element Ground–Water Models Implemented on Vector Computers; Int. J. Num. Meth. Fluids, John Wiley, Vol. 6, S. 913 – 925.

PETERS, A.; B. ROMUNDE and F. SARTORETTO (1988): Vectorized Implementation of Some MCG Codes for FE Solution of Large Groundwater Flow Problems; Proc. I Int. Conf. on Comp. Meth. in Flow Analysis, Okayama, Japan, Vol. 1.

ROUVE et al., Schlußbericht (1989): Simulation einer Grundwasserkontamination im Bereich der geplanten Hausmülldeponie des Kreises Groß–Gerau; unpublished report, Aachen.

ROUVE et al., Schlußbericht (1990): Grundwassermodell Venloer Scholle; unpublished report, Aachen.

A Comparison of Two Algorithms Vectorized to Solve Transient, Convective Heat Transfer on Supercomputers

S. Ganesh (*), W.C. Schreiber (*), C.H. Chuan (*), M.A.R. Sharif (**)

(*) The Mechanical Engineering Department, (**) The Engineering Mechanics Department, The University of Alabama, Tuscaloosa, Alabama, U.S.A.

SUMMARY

The efficiency with which a fluid flow algorithm can be computed on a supercomputer depends on two factors: economy and performance. In the present context, economy refers to the algorithm's ability to produce an accurate simulation in the least number of arithmetic calculations. The algorithm's performance, on the other hand, relates to the number of floating point operations per second (FLOPS) sustainable during the computation. As used in this sense, the two factors are not necessarily compatible; to exhibit maximum performance strictly in terms of FLOPS, an algorithm must usually sacrifice a degree of economy. The attainment of optimal algorithm efficiency requires a careful balance of both performance and economy.

In the present paper, two separate algorithms are compared for their efficiency and accuracy. The two codes differ markedly in their solution procedure. The Pressure Implicit with Splitting of Operators (PISO) [1,2] method has been demonstrated to be an exceptionally economical method to simulate transient, convective heat transfer in two or three dimensions. Our research has indicated that PISO can compute transient convection simulations two orders of magnitude faster than the Semi-Implicit Method for Pressure-Linked Equations, Revised (SIMPLER) method [3]. In order to gain optimal performance on a supercomputer, we have optimized the PISO algorithm for speed-up on a vector processor. The PISO method programmed for this test uses the power law [3] for calculating advection-diffusion at control volume interfaces. For the comparison case, an explicit method using a marker and cell technique (MAC) has been modified with a Bounded Directional Transportive Upwind Differencing Scheme (BDTUD) [4,5] in an attempt to give the advective transport a higher order of accuracy than that obtained using the power law. The MAC/BDTUD has also been vectorized to exploit pipelined architecture.

Driven cavity flow and natural convection in a box are solved to steady state on one processor of the Cray XMP to test each code for its efficiency. The solution accuracy of both codes is determined by comparing each against benchmark results for driven cavity flow.

DESCRIPTION OF TWO NUMERICAL METHODS

PISO method

References 1 and 2 describe in detail the theory and application of the PISO method. In reference 6, both of these methods are clearly summarized and compared for their respective abilities to solve steady state heat convection problems using a non-vectorized algorithm.

The convective transport equations for momentum, energy, and continuity in an incompressible fluid can be expressed as:

$$(\partial U/\partial t + U\nabla \cdot U) = -\nabla p + Re^{-1} \nabla \cdot (\nabla U) + Ra\,(Pr\,Re^2)^{-1}\,T \quad \text{momentum} \quad (1)$$

$$(\partial T/\partial t + U\nabla \cdot T) = Re^{-1}\,Pr^{-1}\,\nabla \cdot (\nabla T) \quad \text{energy} \quad (2)$$

$$\nabla \cdot U = 0 \quad \text{continuity} \quad (3)$$

where the variable **U** is the velocity vector which consists of velocities in the x, y, and z directions, and the body force is approximated by the Boussinesq expression for buoyancy. The dimensionless parameters important to the test cases are:

Reynolds number, $Re = U\,L\,/\,\nu$ Prandtl number, $Pr = \nu\,/\,\alpha$

Rayleigh number, $Ra = g\beta\,\Delta T\,L^3\,/\,\alpha\,\nu$

For purely forced convection, the Raleigh number is set to zero, and for purely natural convection, the Reynolds number is set to one. When the dimensionless numbers are used in this sense, the time scale is defined as $L\,/\,U_{CHAR}$ where L is the length scale and U_{CHAR} is defined differently for forced and natural convection. When the convection is buoyantly-driven, U_{CHAR} is defined as the characteristic length divided by the kinematic viscosity, $L\,/\,\nu$, and when forced convection is modeled, U_{CHAR} is equal to the lid's velocity.

These equations may be discretized using the finite volume method as explained in reference 7. The solution of the velocity field requires the simultaneous solution of the pressure field to which it is coupled. In compressible flows, the pressure field may be determined using a thermodynamic constitutive equation of state; however, for incompressible flows, the continuity equation must serve as the link between pressure and momentum.

The PISO procedure does not require cycle iteration for each time increment to solve the transient convective heat transfer problem. Instead, to arrive directly at the next time step, it computes an initial estimation of the velocity field and then follows this calculation with two correction steps for updating the pressure and velocity fields. An outline of the computational procedure used with the PISO method can be summarized as follows:

1. Using the pressure field calculated in the previous time step, the discretized momentum equations are solved to obtain an initial prediction for the next time step's velocity field.

2. Given this velocity field, a Poisson equation in terms of pressure is solved implicitly for an updated pressure field.

3. The pressure field resulting from step 2 is used in an explicit equation to update the velocity found in 1.

4. The two sets of velocity fields from steps 1 and 3 are used in a higher order Poisson equation in which the pressure variable is expressed implicitly. This equation is solved for an updated pressure field.

5. The velocity field is again updated using explicit equations.

6. The energy equation is solved implicitly.

Issa [1,2] describes the derivation of these equations in detail. The PISO as well as the SIMPLER methods are delineated and compared in a paper by Jang et al [6]. Issa has demonstrated with an error analysis that the PISO procedure's accuracy is limited only by the spatial discretization of the difference equations; therefore, iterating the PISO procedure is not warranted for improving the accuracy of the solution.

In stage 1, the velocity field is solved with an ADI (alternating direction implicit) solver which has been modified for vectorization. The ADI method was developed [8, 9, 10] for solving linear, transient, three-dimensional problems by dividing the time step, Δt, into three small time steps and solving the field of unknown variables over each small time step line-by-line using the Tridiagonal Matrix Algorithm, TDMA. A different sweep direction is used over each small time step to solve the variables implicitly in terms of the approximate solutions already determined for these variables. On a machine with serial processing, the sweep in the x direction for the ADI method may be performed using nested DO loops in the following sequence:

```
DO LOOP FOR THE Y DIRECTION
        DO LOOP FOR THE Z DIRECTION
                SOLVE TDMA FOR SCALAR VARIABLES IN
                THE X DIRECTION
        CLOSE DO LOOP IN Z DIRECTION
CLOSE DO LOOP IN Y DIRECTION
```

This algorithm is organized incorrectly for vectorization. Since the TDMA uses recurrence to solve for the variables in the x-direction implicitly, it can compute the variables in a serial fashion only. To vectorize the algorithm, the variables in each y-z plane are stored dynamically as a vector and the TDMA is modified to process an entire dynamically-stored vector rather than one separate scalar in the x-direction. In order to solve each y-z plane as a vector, the TDMA must effectively incorporate within itself the DO loops in the y and z directions. At each step within the vectorized TDMA scheme, the dynamically-stored vector of variables in the y-z plane is processed. With this modification, maximum vectorization is obtained and the pipe-line is kept full for a large majority of the time while the algorithm sweeps in each of the three different directions.

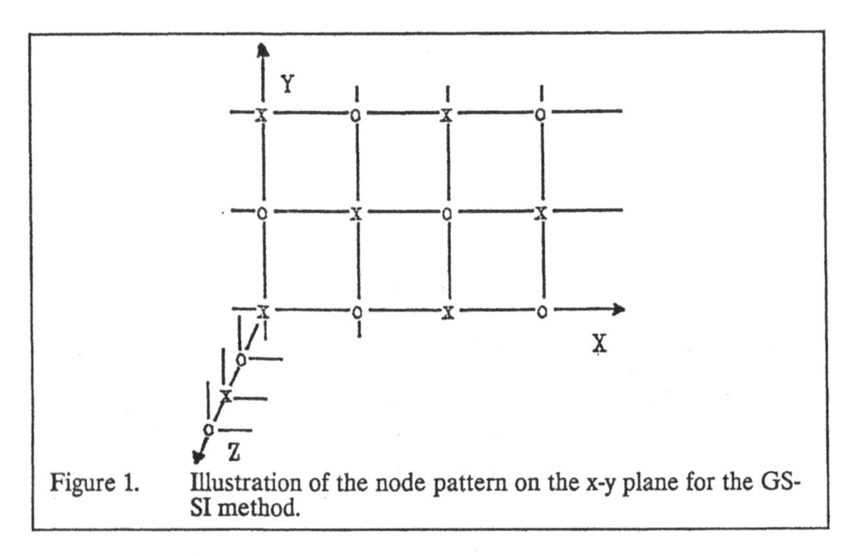

Figure 1. Illustration of the node pattern on the x-y plane for the GS-SI method.

Solving for the pressure in stages 2 and 4 presents a challenge different from that of solving the momentum equation. Boundary condition information for the pressure field is limited and prevents the use of the ADI method for solving the Poisson equation. Relative, and not absolute, values of the pressure are required for the determination of the velocity; hence, explicit boundary conditions for the pressure field are not necessary for a well-posed problem. If the ADI method is engaged for solving steps 2 and 4, the lack of apparent boundary conditions yields an indeterminate matrix problem. Point-by-point methods skirt this problem through the iteration they require for convergence of the solution. For solving the pressure field Successive Over-Relaxation, SOR [11], is a particularly effective point-by-point solver.

As a variant of the Gauss-Seidel method, however, the algorithm for SOR is serial and not suitable for vectorization; therefore, another Gauss-Seidel type technique must be substituted for SOR which allows both over-relaxation and vectorization. Young [12] describes the Gauss-

Seidel Semi-Iterative method (GS-SI) which was developed by Sheldon [13] and which applies the Gauss-Seidel method in a spatially leapfrog pattern. GS-SI may be over-relaxed to such a degree that it is faster than the SOR even as a serial solver. As an added bonus, the GS-SI is also vectorizable for speed-up increases on the pipe-lined supercomputer architecture. Figure 1 illustrates the x-y plane of a three-dimensional system of nodes illustrated as a pattern of alternate x's and o's. The GS-SI algorithm can calculate the pressure field by cycling between the two different points in the following way:

1. With an initial "guessed" value for all nodes, the values at the o points in terms of the x points are computed using the ordinary Gauss-Seidel method.

2. A relaxation factor is used to over-relax the newest values at all the o points.

3. Steps 1 and 2 are repeated for the x points.

4. The convergence criterion is checked, and the process is repeated until the solution converges.

For the purpose of vectorization, the x points and the o points could be dynamically stored as two separate vectors. For the present, however, the vectorization of the x and o points is only extended over the x direction. The rationale for this partial vectorization will be considered in the results section.

Explicit Marker and Cell Method with BDTUD

The Marker-and-Cell (MAC) technique was first introduced by Harlow and Welch [14] for solving incompressible fluid flow problems and later improved upon by Hirt et al [15] who named their revised version of the MAC code SOLA. This computational procedure solves directly for the primitive pressure and velocity variables. At each time step of the march, scalars are calculated explicitly from the scalar transport equation whereas the velocity approximations are calculated explicitly from the respective momentum equations. The velocity approximations are iteratively improved at every time step by adjusting the cell pressures and velocities until the mass conservation is sufficiently well satisfied for all the cells. Discussions about imposing the boundary conditions and the stability requirement for this procedure may be found elsewhere [5, 15].

The directional transportive upwind differencing (DTUD) scheme [16] for discretizing the convective terms in the convection-diffusion transport equations scheme has been shown to be effective for eliminating numerical diffusion in flows which are skewed with respect to computational grids. The BDTUD [4] is a bounded version of the DTUD scheme designed to eliminate the numerical oscillations associated with the DTUD.

Figure 2 illustrates the nine-node computational molecule used for the DTUD scheme. The convected value of ϕ to be used at the interface is determined by interpolating the nodal values of ϕ within the rectangle 'defgd' to a point located a distance β $\underline{V}$ Δt from the interface. The resultant velocity $\underline{V}$ is pictured at the left interface. The term, β, is a parameter which, when adjusted to 1/2, eliminates numerical diffusion and ensures that the accuracy of the scheme is second order.

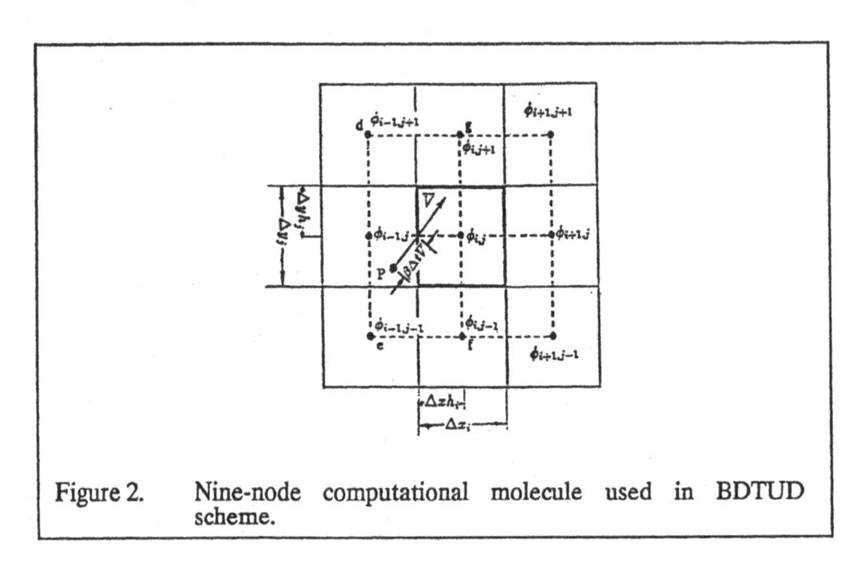

Figure 2. Nine-node computational molecule used in BDTUD scheme.

Bilinear interpolation and two-dimensional upwinding is used to determine the values of ϕ at the control volume interface. For the left interface, ϕ_L is calculated by the following scheme:

$$js = v_L / |v_L|; \quad ja = (1\text{-}js) / 2 \tag{4}$$

$$p_x = (\Delta x\, h_i + \beta u_L \Delta t) / \Delta x_{i-1}; \quad p_y = \beta\, |v_L|\, \Delta t / \Delta y_{j-1+ja} \tag{5}$$

$$\phi_L = (1\text{-}p_x)(1\text{-}p_y)\phi^n_{i,j} + p_x(1\text{-}p_y)\phi^n_{i-1,j} + (1\text{-}p_x)\, p_y \phi^n_{i,j-js} + p_x p_y \phi^n_{i-1,j-js} \tag{6}$$

where the superscript n denotes old time step.

For the bottom interface, ϕ_B is calculated by the following scheme:

$$is = u_B / |u_B|; \quad ia = (1\text{-}is) / 2 \tag{7}$$

$$p_y = (\Delta y\, h_j + b\, v_B \Delta t) / \Delta y_{j-1}; \quad p_x = \beta |u_B| \Delta t / \Delta x_{i-1+ia} \tag{8}$$

$$\phi_B = (1\text{-}p_x)(1\text{-}p_y)\phi^n_{i,j} + px(1\text{-}p_y)\phi^n_{i-is,j} + (1\text{-}p_x)\, p_y \phi^n_{i,j-1} + p_x p_y \phi^n_{i-is,j-1} \tag{9}$$

Approximations for ϕ_R and ϕ_T may be obtained similarly except that the subscripts L, i, B, and j are replaced by R, i+1, T, and j+1 respectively.

Using equations 4 - 9, the convection transport may be approximated as:

$$u\,\partial u/\partial x|_{i,j} \approx (u_R\phi_R - u_L\phi_L)\,/\,\Delta x_i \tag{10}$$

$$v\,\partial u/\partial y|_{i,j} \approx (v_T\phi_T - v_B\phi_B)\,/\,\Delta y_j \tag{11}$$

Because the DTUD scheme does not satisfy the discrete maximum principle which dictates that the solution to the transport equation should lie between the minimum and maximum of the bounding values, oscillations can arise in the solution. In order to ensure bounding, the BDTUD scheme calculates the interface fluxes predicted by both a higher order scheme, the DTUD, and by a lower order scheme, a full donor cell upwind. Using a flux correction algorithm [17], the two fluxes are weighted to yield a value of ϕ at the interface which has used the higher order scheme to greatest extent possible without introducing ripples.

Since it is an explicit procedure, the MAC method including BDTUD can be vectorized with no difficulty.

While it is limited to a time step several orders of magnitude smaller than the PISO method, the explicit method of computing fluid flow is competitive with the more implicit scheme for several reasons:

1. The explicit method is simple to program

2. The number of calculations required for each time step is relatively small.

3. Many realistic heat transfer simulations require the inclusion in equations 1-3 of source terms, defined by expressions or algorithms which must solved iteratively with the equations. Examples of such source terms arise in solving heat transfer problems using the enthalpy-porosity method for modeling phase change or using transformed equations in curvilinear coordinates for representing irregular domains. Even if an implicit algorithm is employed for solving equations 1-3, the presence of strongly-coupled source terms can make the overall solution procedure highly explicit.

4. Explicit methods are amenable to programming on computers that rely on pipelining or concurrent processing for high performance rates. When variables are updated in time explicitly in terms of the variables previously calculated at the old time step, special programming for parallel architectures is minimized.

RESULTS

Comparison of efficiency

The MAC/BDTUD and PISO algorithms were tested for their respective abilities on forced and natural convection simulations. The forced convection case consisted of a square cavity whose walls were maintained at T = 0 and whose lid is maintained at T = 1. At an initial time, the lid is impulsively started. For a Prandtl number of 1.0, the Reynolds numbers equaling 100 and 400 were used for the transient simulations. Steady state was considered to have been reached at dimensionless times equal to 30 and 60 for the case where the Reynolds number equaled 100 and 400, respectively. In order to examine the two methods' ability to predict natural convection, the transient flow of air (Pr = 0.733) was modeled in a square cavity whose top and bottom surfaces were insulated and whose vertical walls were maintained at constant temperatures of 0 and 1. The Rayleigh number of 14660 was used, and steady state was reached after a dimensionless time of 0.3. Both of the numerical methods for these three types of simulations were solved on grid sizes of 8x8, 30x30, and 60x60 using a wide range of time steps.

Driven Cavity

Reynolds Number	Grid Size	Performance, (MFLOPS) PISO	MAC/BDTUD
100	6 x 6	17-19	18-21
100	30 x 30	30-36	59
100	60 x 60	36-40	72
400	6 x 6	17-19	15-20
400	30 x 30	31-38	60
400	60 x 60	37-40	70

Natural Convection, Pr = 0.733, Ra = 14,600

Grid Size	Performance, (MFLOPS) PISO	MAC/BDTUD
6 x 6	17-19	13-15
30 x 30	31-36	57
60 x 60	37-40	71

Table 1. Performance sustained on the Cray XMP (one processor) by the cases run.

The convergence criteria for solving both the pressure and the velocity was 10^{-5} for both methods. In the PISO method, no relaxation factor was used with the ADI method and a relaxation factor of 1.6 was used to accelerate the SOR solver.

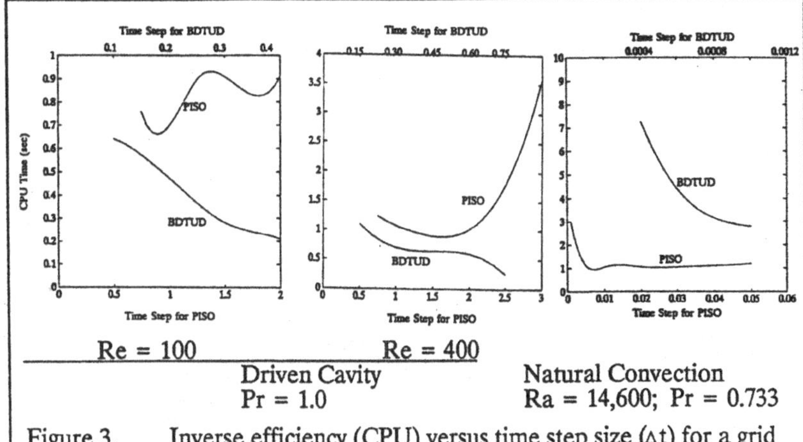

Figure 3. Inverse efficiency (CPU) versus time step size (Δt) for a grid on which $\Delta x = \Delta y = 0.1667$.

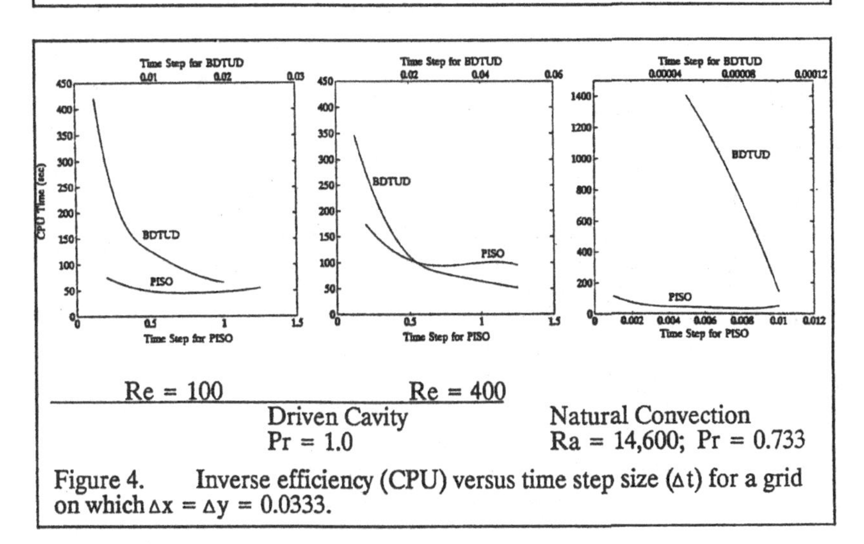

Figure 4. Inverse efficiency (CPU) versus time step size (Δt) for a grid on which $\Delta x = \Delta y = 0.0333$.

For the forced convection case with Reynolds number equal to 100, the performance for the MAC/BDTUD procedure was almost twice that of the PISO method. Table 1 lists the performance levels sustained by both methods under the different conditions. As would be suspected on a pipelined processor with 64 registers per vector, better performance is proportional to problem size. For the three problems posed, the type of problem did not seem to have much affect on performance. A performance range is given for some of the cases; generally, the higher performances were obtained at larger time steps.

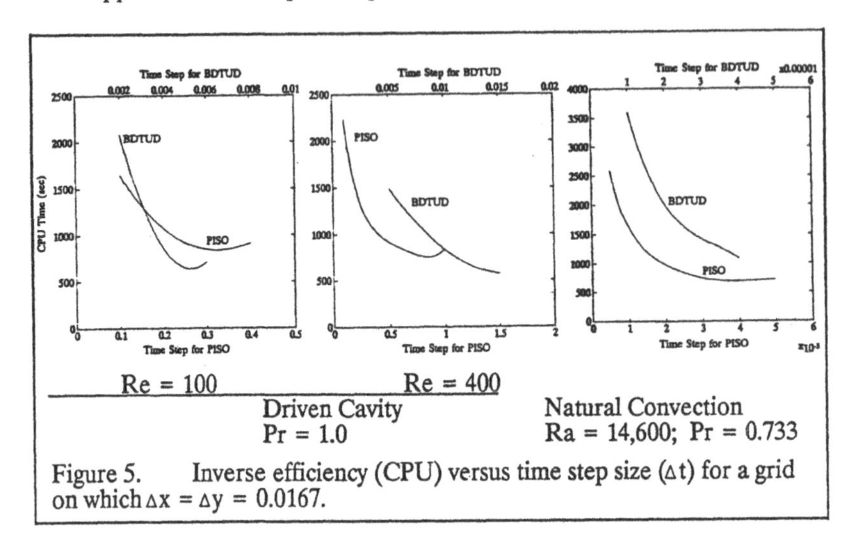

Figure 5. Inverse efficiency (CPU) versus time step size (Δt) for a grid on which $\Delta x = \Delta y = 0.0167$.

Performance, of course, is not the ultimate aim in a computer program. The economy must also be considered as an important factor in designing a computer program which will perform its function with the greatest efficiency.

Figures 3 through 5 are plots of CPU time versus time step size for the cases listed in table 1. As far as the greatest efficiency for the optimum time step, the two methods are competitive. The PISO method depends more on economy for its speed-up than the MAC/BDTUD, which relies more on performance. As indicated by figure 3, it was found that the MAC/BDTUD method was the most efficient for the coarse grid. For the 30 x 30 and 60 x 60 grids, the MAC/BDTUD method is competitive with PISO in terms of efficiency for optimum time step. The PISO method is more flexible than the MAC/BDTUD method in terms of the range of time steps for which it will compute efficiently. In figures 4 and 5, it is seen that the CPU time needed to run the simulation with the PISO method is fairly constant over a large range of time steps. The MAC/BDTUD method, however, is typically characterized by a steep curve which ends at an optimum time step value, beyond which the solution diverges. At it optimum time step, the CPU time needed for the explicit MAC/BDTUD is similar to that required by the implicit PISO method; however, the curves indicate that the determination of a near optimum time step is critical for the minimization of the CPU time needed by the MAC/BDTUD. Using dimensionless variables, the maximum stable time step for the explicit scheme can be estimated from the criteria,

$$\Delta t < (\Delta X)^2 \, Re / 2 \tag{12}$$

Using this estimate for the time step size can help in determining the optimum time step for computation, but experimentation is still required.

Comparison of accuracy

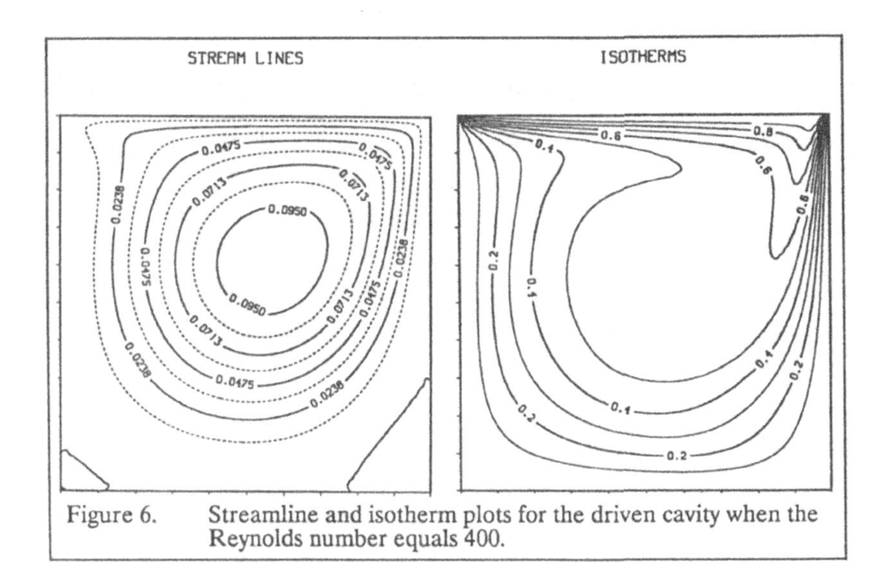

Figure 6. Streamline and isotherm plots for the driven cavity when the Reynolds number equals 400.

A study of the driven cavity flow run using very fine grids [18] was chosen as a benchmark in order to compare the two methods for their relative accuracy. Plots of streamlines and isotherms from results generated by both methods were basically indistinguishable from each other and from the benchmark results. Figure 6 illustrates the streamlines and temperature contours for the driven cavity flow when the Reynolds number is set to 400 and a 60x60 grid is used. Both numerical methods yield nearly identical plots and are able to pick up the secondary vortices in the lower corners. Velocity profiles are used to compare each method against the benchmark for more quantifiable results. For a Reynolds number of 1000, the u and v velocity profiles are determined at the vertical and horizontal centerlines of the domain. These results are compared to benchmark profiles using a formula to calculate relative error:

$$E = (100/N) \sum_{n=1}^{N} | \phi_e - \phi_c | \; / \; |\phi_{max}| \tag{13}$$

where: ϕ_e is the value of u or v computed by PISO or MAC/BDTUD

ϕ_c is the value of u or v found on figure 11 or 12 in reference 18 and ϕ_{max} is the value of maximum value of u or v in the profile

	MAC/BDTUD	PISO
u profile	16.74	20.46
v profile	43.24	42.74
	Grid Size = 0.167	
u profile	6.46	7.45
v profile	10.42	14.80
	Grid Size = 0.0333	
u profile	4.08	3.59
v profile	6.03	7.29
	Grid Size = 0.0167	

Table 2. Relative Error using equation 13 to compare velocity profiles with a benchmark [18] for Driven Cavity flow (Reynolds Number equals 1000).

The MAC/BDTUD method is shown to predict the velocity slightly more accurately than the PISO method. For the example chosen of viscous flow in a driven cavity, however, the difference seems insignificant in consideration of the extra computation needed for calculating the effect of skewed upwinding. For modeling a flow in which the viscosity is that of air, the power law [3] used on a fairly fine grid does not seem to exhibit noticeable numerical diffusion.

CONCLUSIONS

1. At their optimum time step values, PISO and MAC/BDTUD have comparable efficiencies when run on one processor of the Cray XMP

2. PISO exhibits superior efficiency over a much wider range of time steps than MAC/BDTUD.

3. Because of the ease with which it can be vectorized, the time-explicit MAC/BDTUD is probably the better candidate for parallel processing, especially on massively parallel machines using domain decomposition.

4. Use of BDTUD did not significantly enhance accuracy more than the use of the power law for the case of viscous flow in a driven cavity (Re = 1000).

ACKNOWLEDGEMENTS

The authors are indebted to the funding and services provided by:

1. Cray Research, Inc.
2. NSF/EPSCoR
3. Alabama Supercomputer Authority

REFERENCES

1. Issa, R. I.; "Solution of the Implicitly Discretized Fluid Flow Equations by Operator-Splitting"; J. Comp. Phys., vol. 62, pp. 40-65; 1985.

2. Issa, R. I., Gosman, A. D., and Watkins, A. P.; "The Computation of Compressible and Incompressible Recirculating Flows by a Non-Iterative Scheme"; J. Comp. Phys., vol. 62, pp. 66-82; 1985.

3. Patankar, S. V.; "A Calculation Procedure for Two-Dimensional Elliptic Situations", Num. Heat Transfer, vol. 2; 1979.

4. Sharif, M. A. R.; "A Bounded Directional Transportative upwind Differencing Scheme for Convection-Diffusion Problems"; submitted to J. Comp. Phys for review.

5. Sharif, M. A. R. and Busnaina, A. A.; "Assessment of Finite Difference Approximations for the Advection in the Simulation of Practical Flow Problems"; J. Comp. Phys., vol. 74, pp. 143-176; 1988.

6. Jang, D. S., Jetli, R., and Acharya, S.; "The Comparison of the PISO, SIMPLER, and the SIMPLEC algorithms for the treatment of the pressure-velocity coupling in steady flow problems"; Num. Heat Transfer, vol. 10, pp. 209-228; 1986.

7. Patankar, S. V.; Numerical Heat Transfer and Fluid Flow; Hemisphere Publ. Co.; 1980.

8. Peaceman, D. W. and Rachford, H. H.; " The Numerical Solution of Parabolic and Elliptic Difference Equations"; J. Soc. Ind. Appl. Math.; vol. 3, pp. 28-41; 1955.

9. Douglas, J.; "On the Numerical Integration of $\partial^2 u/\partial x^2 + \partial^2 u/\partial y^2 = \partial u/\partial t$ by Implicit Methods"; J. Soc. Ind. Appl. Math.; vol. 3, pp. 42-65; 1955.

10. Douglas, J. and Gunn, J. E.; "A general formulation of alternating direction methods - part I: Parabolic and Hyperbolic problems"; Numerische Mathematik, vol. 6, pp. 428-453; 1964.

11. Carnahan, B., Luther, H. A., and Wilkes, J. O.;"Applied Numerical Methods"; John Wiley Publ.; 1969

12. Young, D. M.; Iterative Solution of Large Linear Systems; Academic Press; 1971.

13. Sheldon, J. W.; "On the Spectral Norms of Several Iterative Processes"; J. Assoc. Comput. Mach., vol. 6, pp. 494-505; 1959.

14. Harlow, F. H. and Welch, J. E.; "Numerical Calculations of Time-Dependent Viscous Incompressible Flow of Fluid with Free Surface"; Phys. of Fluids, vol. 8, pp. 2182-2189; 1965.

15. Hirt, C. W., Nichols, B. D., and Romero, N. C.; "SOLA- A Numerical Solution Algorithm for Transient Fluid Flows"; Report LA-5852, Los Alamos Scientific Laboratory; 1975.

16. Eraslan, A., Lin, W., and Sharp, R.; "*FLOWER: A computer code for Simulating Three-Dimensional Flow, Temperature and Salinity Conditions in Rivers, Estuaries, and Coastal Regions*"; Oak Ridge National Laboratory Report No. ORNL/NUREG-8041, Oak Ridge TN; 1983.

17. Zalesak, S. T.; "Fully multidimensional Flux-Corrected Transport Algorithms for Fluids"; *J. Comp. Fluids*, vol. 31, pp. 248-283; 1979.

18. Schreiber, R., Keller, H.; "Driven Cavity Flow by Efficient Numerical Techniques"; *J. Comp. Phys.*, vol. 49, pp. 310-333; 1983.

Symmetric Versus Non-Symmetric Matrix Techniques: Implementation of Two Galerkin-FE Approaches for the Advection-Dispersion Equation

A. Peters (*), M. Eiermann (*), H. Daniels (**)
(*) IBM Scientific Center, D-6900 Heidelberg, Germany
(**) Institut für Wasserbau, RWTH Aachen, Germany

ABSTRACT

In this paper, we discuss the implementation on IBM 3090 VF and RISC 6000 of two solution approaches for the advection-dispersion equation. The first approach is based on the classical Crank-Nicholson scheme and leads to a non-symmetric system of linear equations which is iteratively solved by the conjugate gradient squared scheme and a variant of it. The second approach, introduced recently by Leismann and Frind (1989), places the advective component of the equation at the old time level and compensates for the resulting errors by introducing an artificial diffusion term. The scheme of Leismann and Frind yields a symmetric linear system that is solved by the most efficient variants of the conjugate gradient method.

INTRODUCTION

One of the most popular approaches for the solution of the advection-dispersion equation consists of a finite element approximation of the space derivatives and a weighted finite difference scheme for the integration of the time derivatives. This approach requires the assemblage and solution of a large but sparse system of linear equations in each iteration. Solving these systems is the most expensive step of the solution procedure.

Conjugate gradient (CG)-like methods are a popular choice among the existing algorithms for the solution of sparse systems of linear equations. They are easy to implement on a wide variety of computer architectures, require small computational effort per iteration and modest amounts of storage. However, in many cases the available CG-like methods are quite ineffective in the absence of preconditioning. Moreover, the selection of an appropriate CG-solver and preconditioner is a problem specific task that may require additional numerical experiments. We recall that solving a linear system of equations in the framework of approximating the solution of a partial differential equation is not a stand-alone problem. The convergence rate of any CG-like method employed in such a case depends primarily upon the choice of the discretization method.

In this work we develop two solution approaches for the solution of the advection-dispersion equation and the constituent algorithms are compared in terms of performance on different computers and accuracy:
- The first approach is based on an implicit weighted difference scheme for the approximation of time derivatives. It yields a non-symmetric system of linear equations. This system is solved by the recently developed conjugate gradient-squared (CGS) method (Sonneveld, 1989) and its variant CGSTAB (van der Vorst and Sonneveld, 1990).
- The second approach involves the semi-implicit time integration scheme of Leismann and Frind (1989) that yields a symmetric positive definite system of linear equations. This system is solved by the classical CG method (Hestenes and Stiefel, 1952).

We recall that the employed time integration schemes are compatible with a variety of discretization techniques for space derivatives, like e.g., finite differences, collocation or different types of finite elements. Experience has shown that linear finite elements provide a satisfactory approximation for the class of problems we are solving. Consequently, linear finite elements are used in both solution approaches.

First, we briefly review the employed discretization techniques and CG algorithms. Then we describe several programming techniques involved in the development of a computer code for computers with hierachical memory organization. We refer to the most time consuming stages of the solution. Finally, we present the results obtained with the proposed solution approaches for a case study: the simulation of a contamination from a landfill. The simulations are performed on a work station IBM RISC 6000 and on a main frame vector computer IBM 3090 VF.

DISCRETIZATION TECHNIQUES

The equation to be solved in this paper is the advection-dispersion equation

$$\frac{\partial c}{\partial t} + \frac{\partial}{\partial x_i}(v_i c) - \frac{\partial}{\partial x_i}\left(d_{ij}\frac{\partial c}{\partial x_j}\right) = 0\,, \qquad (1)$$

defined over a domain $G \subset \mathbb{R}^3$ with initial conditions (for $t = 0$) and Dirichlet and Cauchy conditions on the boundary ∂G. The unknown function c represents the solute concentration, t and x_i are the time and spatial coordinates, v_i is the velocity vector, and d_{ij} the dispersion tensor, defined as the sum of molecular diffusion and mechanical dispersion (Bear, 1979).

We apply the Galerkin weighted residual method, as described by Pinder and Gray (1977) to equation (1). The following matrix equation is obtained

$$S\frac{\partial \mathbf{c}}{\partial t} + (V + D + Q + P)\mathbf{c} = Q\mathbf{c}_R, \tag{2}$$

where vector $\mathbf{c}$ contains the unknown values c_n of the concentration at the N nodes of the finite element mesh. The coefficients of $\mathbf{c}_R$ are known concentration values of the fluid entering the domain through some parts of the boundary ∂G . The coefficients of the matrices S, V, D, Q and P are given by

$$S_{mn} = \int_G \phi_m \phi_n \, \mathrm{d}G\,, \quad V_{mn} = \int_G v_i \phi_m \frac{\partial \phi_n}{\partial x_i} \mathrm{d}G$$

$$D_{mn} = \int_G d_{ij} \frac{\partial \phi_m}{\partial x_j} \frac{\partial \phi_n}{\partial x_i} \mathrm{d}G\,, \quad P_{mn} = -\int_{\partial G} d_{ij} \frac{\partial \phi_m}{\partial x_j} \phi_n n_i \, \mathrm{d}\partial G\,,$$

$$Q_{mn} = \delta_{mn} \max\left[\int_{\partial G} n_i v_i \phi_n \, \mathrm{d}\partial G\,; 0\right] \quad (m,n = 1,\ldots,N), \tag{3}$$

where ϕ_n, ϕ_m are Chapeau basis functions. The fluid velocity v_i and the dispersion coefficient d_{ij} in the expressions (3) are assumed to be piecewise constant over the elements. While n_i in (3) are the components of the outward pointing normal on the boundary ∂G , δ_{ij} denotes the Kronecker delta function.

The "semi-discrete" equations (3), i.e. discrete in space and continuous in time, is integrated in time using the weighted finite difference scheme

$$S\frac{\mathbf{c}^{t+\Delta t} - \mathbf{c}^{t}}{\Delta t} + \theta_V V\mathbf{c}^{t+\Delta t} + (1-\theta_V)V\mathbf{c}^{t} + \theta_D D\mathbf{c}^{t+\Delta t} +$$

$$(1-\theta_D)D\mathbf{c}^{t} + \theta_Q Q\mathbf{c}^{t+\Delta t} + (1-\theta_Q)Q\mathbf{c}^{t} + \theta_P P\mathbf{c}^{t+\Delta t} + (1-\theta_P)P\mathbf{c}^{t}$$

$$= \theta_Q Q\mathbf{c}_R^{t+\Delta t} + (1-\theta_Q)Q\mathbf{c}_R^{t}, \tag{4}$$

where the top notations $t + \Delta t$ and t denote the new and old time levels, Δt is the time step and and θ_V, θ_D, θ_Q, θ_P are weighting factors.

Special cases of equation (4) are

- the fully implicit scheme ($\theta_V = \theta_D = \theta_Q = \theta_P = 1$),
- the Crank-Nicholson scheme ($\theta_V = \theta_D = \theta_Q = \theta_P = 0.5$),
- the explicit scheme ($\theta_V = \theta_D = \theta_Q = \theta_P = 0$).

While the coefficient matrices corresponding to the fully implicit scheme and Crank-Nicholson scheme are non-symmetric, the coefficient matrix of the explicit scheme is symmetric and positive definite. However, the explicit scheme is only conditionally stable, i.e. there are restrictive time and space discretization bounds above which the numerical solution is unstable. Consequently, many practitioners prefer to use the fully implicit scheme or the Crank-Nicholson scheme. These two schemes are unconditionally stable but not always very accurate.

Several authors (see e.g., Roache, 1982) have shown that the stability bounds of the explicit scheme can be relaxed partly by introducing an artificial diffusion term R into equation (4) to compensate for discretization errors:

$$R_{mn} = \frac{\Delta t}{2} \int_G v_i v_j \frac{\partial \phi_m}{\partial x_j} \frac{\partial \phi_n}{\partial x_i} \, dG \qquad (n = 1,\ldots,N). \tag{5}$$

Recently, Leismann and Frind (1989) have proposed a semi-implicit scheme that generates a symmetric, positive definite coefficient matrix. In this scheme, the advective term V (cf. (3.2)) is placed at the old time level and the matrix R (cf. (5)) is used again to compensate for the discretization errors. If R is center-weighted between the old and the new time level and with $\theta_V = 0$, $\theta_D = 1$, $\theta_Q = 0$, $\theta_P = 1$, the Leismann-Frind scheme leads to the matrix equation

$$\left[2\frac{S}{\Delta t} + 2D + 2P + R\right]\mathbf{c}^{t+\Delta t} =$$

$$\left[2\frac{S}{\Delta t} - 2V - 2Q - \mathrm{R}\right]\mathbf{c}^{t} + Q\mathbf{c}_R^{t}. \tag{6}$$

Leismann and Frind (1989) show that their scheme is unconditionally stable, and comparable in accuracy to the Crank-Nicholson scheme.

CG-LIKE METHODS

We describe briefly the CG-like algorithms employed in this paper.

- *The Preconditioned CG Method*

The linear system (9) has a coefficient matrix which is sparse, symmetric and positive definite. For ease of notation, we set

$$A \equiv \left[2\frac{S}{\Delta t} + 2D + 2P + R\right], \quad \mathbf{b} \equiv \left[2\frac{S}{\Delta t} - 2V - 2Q - R\right]\mathbf{c}^{t}$$

$+ Qc_R^k$, $\mathbf{c} \equiv \mathbf{c}^{t+\Delta t}$. Then (6) reads as $A\mathbf{c} = \mathbf{b}$ which we solve by the preconditioned CG method (see e.g., Golub and van Loan, 1989).

Given the initial guess $\mathbf{c}_0 \equiv \mathbf{c}^t$, we compute a sequence of approximants $\{\mathbf{c}_k\}_{k\geq 1}$ for the solution $\mathbf{c}$ as follows:

$$\mathbf{r}_0 = \mathbf{b} - A\mathbf{c}_0 , \tag{7.1}$$

$$\text{solve } M\mathbf{r}'_0 = \mathbf{r}_0 \text{ for } \mathbf{r}'_0 \text{ and set } \mathbf{p}_0 = \mathbf{r}'_0 . \tag{7.2}$$

Until the convergence criteria are fulfilled, compute for $k = 0,1, \ldots$

$$\alpha_k = (\mathbf{r}_k, \mathbf{r}'_k)/(\mathbf{p}_k, A\mathbf{p}_k) , \tag{7.3}$$

$$\mathbf{c}_{k+1} = \mathbf{c}_k + \alpha_k \mathbf{p}_k , \tag{7.4}$$

$$\mathbf{r}_{k+1} = \mathbf{r}_k - \alpha_k A\mathbf{p}_k . \tag{7.5}$$

$$\text{Solve } M\mathbf{r}'_{k+1} = \mathbf{r}_{k+1} \text{ for } \mathbf{r}'_{k+1} \text{ and compute} \tag{7.6}$$

$$\beta_k = (\mathbf{r}_{k+1}, \mathbf{r}'_{k+1})/(\mathbf{r}_k, \mathbf{r}'_k) , \tag{7.7}$$

$$\mathbf{p}_{k+1} = \mathbf{r}'_{k+1} + \beta_k \mathbf{p}_k . \tag{7.8}$$

Here, M is a suitably chosen preconditioning matrix.

- *The Preconditioned CGS Method*

The canonical generalization of the CG method to non-symmetric systems $A\mathbf{c} = \mathbf{b}$ is Lanczos' bi-conjugate gradient (BCG) method (Flechter, 1976). For symmetric systems, BCG reduces then to the classical CG method. In comparison to other CG-like algorithms for non-symmetric problems, like e.g. GMRES (Saad and Schulz, 1986), an important advantage of BCG stems from the fact that the residual vectors in each iteration are computed by a three-term recurrence relation. Like in the classical CG method, neither the work per iteration nor the amount of storage therefore increase during the iteration process. However, BCG does not construct optimal approximants for the solution in each iteration step as GMRES or CG do. Moreover, BCG is susceptible to breakdowns and near-breakdowns, i.e., division by 0 (or by very small numbers) may occur in the algorithm. A disadvantage of the BCG scheme is that it requires a multiplication by A^T, the transpose of the coefficient matrix, per iteration step. The CG-squared (CGS) method (Sonneveld, 1989) can be viewed as a variant of BCG which overcomes this difficulty. We apply the preconditioned CGS scheme to the linear system (4), where we chose $\theta_V = \theta_D = \theta_Q = \theta_R = 0.5$. Upon setting

$$\tilde{A} \equiv \left[2\frac{S}{\Delta t} + V + D + Q + P \right],$$

$$\tilde{\mathbf{b}} \equiv \left[2\frac{S}{\Delta t} - V - D - Q - P\right]\mathbf{c}^t + Q\mathbf{c}_R^{t+\Delta t} + Q\mathbf{c}_R^t$$

and $\mathbf{c} \equiv \mathbf{c}^{t+\Delta t}$, (4) is equivalent to $\tilde{A}\mathbf{c} = \tilde{\mathbf{b}}$. Given the initial guess $\mathbf{c}_0 \equiv \mathbf{c}^t$, and a suitable preconditioning matrix M, we compute a sequence of approximants $\{\mathbf{c}_k\}_{k\geq 1}$ to the solution $\mathbf{c}$ as follows:

$$\mathbf{r}_0 = \tilde{\mathbf{b}} - \tilde{A}\mathbf{c}_0 , \tag{8.1}$$

$$\text{solve } \tilde{M}\mathbf{r}'_0 = \mathbf{r}_0 \text{ for } \mathbf{r}'_0 \text{ and set}$$

$$\mathbf{q}_0 = \mathbf{p}_{-1} = \mathbf{0}, \quad \mathbf{r}_0 = \mathbf{r}'_0, \quad \beta_0 = 0 . \tag{8.2}$$

Until the convergence criteria are fulfilled, compute for $k = 0,1,...$

$$\mathbf{p}_k = \mathbf{r}_k + 2\beta_k\mathbf{q}_k + \beta_k^2\mathbf{p}_{k-1} . \tag{8.3}$$

$$\text{Solve } \tilde{M}\mathbf{t}_k = \tilde{A}\mathbf{p}_k \text{ for } \mathbf{t}_k \text{ and set} \tag{8.4}$$

$$\alpha_k = (\mathbf{r}'_0, \mathbf{r}_k)/(\mathbf{r}'_0, \mathbf{t}_k) , \tag{8.5}$$

$$\mathbf{s}_k = \mathbf{r}_k + \beta_k\mathbf{q}_k , \tag{8.6}$$

$$\mathbf{q}_{k+1} = \mathbf{s}_k - \alpha_k\mathbf{t}_k , \tag{8.7}$$

$$\mathbf{s}_k = \mathbf{s}_k + \mathbf{q}_k , \tag{8.8}$$

$$\mathbf{c}_{k+1} = \mathbf{c}_k + \alpha_k\mathbf{s}_k . \tag{8.9}$$

$$\text{Solve } \tilde{M}\mathbf{t}_k = \tilde{A}\mathbf{s}_k \text{ for } \mathbf{t}_k \text{ and set} \tag{8.10}$$

$$\mathbf{r}_{k+1} = \mathbf{r}_k - \alpha_k\mathbf{t}_k , \tag{8.11}$$

$$\beta_k = (\mathbf{r}'_0, \mathbf{r}_{k+1})/(\mathbf{r}'_0, \mathbf{r}_k) . \tag{8.12}$$

BCG and CGS have closely related convergence properties. It can be shown that CGS converges or diverges typically faster than BCG by a factor between one and two (Nachtigal et al., 1990).

- *The Preconditioned CGSTAB Method*

In many applications, it has been observed that the convergence history curve (i.e., $\|\mathbf{r}_k\|_2$ versus k) of the CGS method is very erratic. To avoid this, van der Vorst and Sonneveld (1990) recently developed the CGSTAB scheme which was announced to be "a more smoothly converging variant of CGS". We apply this algorithm to the system of linear equations (4).

Given the initial guess $\mathbf{c}_0 \equiv \mathbf{c}^t$, and a suitable preconditioning matrix $\tilde{M}$, we compute a sequence of approximants $\{\mathbf{c}_k\}_{k\geq 1}$ for the solution $\mathbf{c} \equiv \mathbf{c}^{t+\Delta t}$ as follows:

$$\mathbf{r}_0 = \tilde{\mathbf{b}} - \tilde{A}\mathbf{c}_0 , \tag{9.1}$$

$$\text{solve } \tilde{M}\mathbf{r}'_0 = \mathbf{r}_0 \text{ for } \mathbf{r}'_0 \text{ and set}$$

$$\mathbf{q}_0 = \mathbf{p}_0 = 0, \quad \mathbf{r}_0 = \mathbf{r}'_0, \quad \alpha_0 = \beta_0 = \omega_0 = 1 . \tag{9.2}$$

Until the convergence criteria are fulfilled, compute for $k = 1,2, \ldots$

$$\beta_{k+1} = (\mathbf{r}'_0, \mathbf{r}_k), \quad \omega_{k+1} = \frac{\beta_{k+1}}{\beta_k}\frac{\omega_k}{\alpha_k} , \tag{9.3}$$

$$\mathbf{q}_{k+1} = \mathbf{r}_k + \omega_{k+1}(\mathbf{q}_k - \alpha_k \mathbf{p}_k) . \tag{9.4}$$

$$\text{Solve } \tilde{M}\mathbf{p}_{k+1} = \tilde{A}\mathbf{q}_{k+1} \text{ for } \mathbf{p}_{k+1} \text{ and compute} \tag{9.5}$$

$$\omega_{k+1} = \beta_{k+1}/(\mathbf{r}'_0, \mathbf{p}_{k+1}) , \tag{9.6}$$

$$\mathbf{s}_k = \mathbf{r}_k - \omega_{k+1}\mathbf{p}_{k+1} . \tag{9.7}$$

$$\text{Solve } \tilde{M}\mathbf{t}_k = \tilde{A}\mathbf{s}_k \text{ for } \mathbf{t}_k \text{ and set} \tag{9.8}$$

$$\alpha_{k+1} = (\mathbf{t}_k, \mathbf{s}_k)/(\mathbf{t}_k, \mathbf{t}_k) , \tag{9.9}$$

$$\mathbf{c}_{k+1} = \mathbf{c}_k + \omega_{k+1}\mathbf{q}_{k+1} + \alpha_{k+1}\mathbf{s}_k , \tag{9.10}$$

$$\mathbf{r}_{k+1} = \mathbf{s}_k - \alpha_{k+1}\mathbf{t}_k . \tag{9.11}$$

During the implementation of the case study presented in the last Section, the classical CG method was preconditioned by the incomplete Cholesky factorization. This approach broke down during the matrix factorization due to the occurence of some negative square roots for some data sets. We then tried diagonal scaling which generated a converging scheme. The preconditioner is here given by

$$M = \text{diag}\,(A) . \tag{10}$$

In order to make a fair comparison between the Leismann-Frind scheme and the Crank-Nicholson scheme, the diagonal scaling technique was applied to the system of equations (4) as well. The preconditioning matrix used in conjunction with the CGS and CGSTAB schemes is

$$\tilde{M} = \text{diag}\,(\tilde{A}^T \tilde{A}) . \tag{11}$$

For a diagonal preconditioner M as e.g. defined by (20), no system of the form $M\mathbf{r}' = \mathbf{r}$ has to be solved during the iteration process. Instead, one scales the equations (4) by M^{-1} before starting the iteration.

We determine the operation counts per iteration for the algorithms presented above. The basic computations of the CGSTAB scheme (cf. (9.3) to (9.11)) are: two matrix vector multiplications, six saxpys (three vector multiply-and-add operations and and three vector multiply-and-subtract operations), four dot products and two times solving of the preconditioning system. The CGS scheme presented in equations (8.3) to (8.12) consists of two matrix vector multiplications, six saxpys (four vector multiply-and-add operations and two vector multiply-and-subtract operations), one vector addition, two dot products and the solution of two preconditioning systems. Each scheme requires one additional dot product if the residual norm $\|\mathbf{r}_k\|_2$ is employed in the stopping criterion.

The CGSTAB scheme evalutes two more dot products per iteration than the CGS scheme. Nevertheless, the additional work spent to perform these dot products is very small in comparison to the other computations. Virtually, any reduction in the number of iterations due to the better behaviour of the CGSTAB scheme would counterbalance this work.

The execution of one CG iteration (cf. (7.3) to (7.8)) requires one matrix vector multiplication, three saxpys, two dot products and the solution of one preconditioning system. This is half the work of the CGSTAB scheme. It is obvious that CGS or CGSTAB applied to the system (4) cannot outperform the CG method applied to (6), unless the former schemes converge at least twice as fast as the latter does. *In a complementary paper we have demonstrated that theoretically such situations are likely to occur for large grid Courant numbers and small Peclet numbers* (Eiermann et al. 1991).

IMPLEMENTATION ISSUES

The final goal of this work was to develop a computer program able to solve efficiently large groundwater contamination problems on computers with hierarchical memory organization. The designed program has been benchmarked on two computers of different classes that share this architectural feature: a main frame computer IBM 3090 with Vector Facility and a work station IBM RISC 6000.

Any attempt to achieve high performance on a computer should consider its architecture first. However, we abstain from a description of the architecture of IBM 3090 VF and RISC 6000 in this paper and refer only to some architecture dependent programming issues involved in the design of our code. Readers interested in presentations of the architecture of IBM 3090 VF and RISC 6000 are directed to the comprehensive collections of papers published in "IBM Systems Journal" (1986) and the manual "RISC System/6000 Technology" (1990). For anyone trying to optimize the performance of FORTRAN codes reading of the excellent articles by Liu and Strother (1988) and Bell (1990) is a must.

A FORTRAN program can achieve a resonable fraction of performance on IBM 3090 VF and RISC 6000 if it makes efficient use of the hierarchical storage system and takes advantage of compound instructions. Compound instructions, e.g. multiply-and-add, perform two floating-point operations at roughly the speed that would be required to perform each of the component operations.

The use of the following programming techniques warrants the efficient cache utilization for many applications:

- store frequently accessed data into contiguous memory locations and address them with stride one. The effective capacity of the cache increases when the cache lines have no gaps.
- segment data arrays of arbitrary length for the length of the cache and perform as many as possible of the computations involving the elements of the segment before they are flushed out by new data.
- buffer partial results into temporary data areas to avoid unnecessary store and load operations.

The following examples illustrate some approaches that makes possible the application of the mentioned techniques to the most time consuming parts of our code. Due to the variability in space and time of the velocity, dispersion coefficients and boundary conditions, the expressions (3) and (5) are recalculated and assembled into a large irregular sparse system of linear equations in each time step. The bulk of CPU time is consumed in the loops that compute the mass S , advection V, dispersion D and artificial dispersion R contributions to the linear system of linear equations (4) and (6) and solve these systems.

- *Assemblage of the Linear System*

The calculation of the mass, advection, dispersion and artificial dispersion contributions to the linear systems of equations (4) and (6) is based on a scheme proposed by Pelka and Peters (1986) for groundwater flow problems. Since the application of this scheme to the assemblage of the linear systems of equations (4) and (6) consists of similar steps, we illustrate only the assemblage of the mass, advection and dispersion components of coefficient matrix for the Crank-Nicholson scheme (eq. (4)).

Consider a finite element mesh consisting of N nodes and M triangles. The indices giving the relations between elements and nodes are placed into three vectors $\mathbf{k}_i$, $i = 1,2,3$ of length M (Fig.1), and the nodal coordinates in two vectors $\mathbf{x}_1$ and $\mathbf{x}_2$ of length N. The scheme consists of the following steps:

Step 1: Gather the vectors of nodal coordinate into six vectors of length M according to the index vectors $\mathbf{k}_i$

$$\mathbf{x}_{1i} = \mathbf{x}_1(\mathbf{k}_i), \quad \mathbf{x}_{2i} = \mathbf{x}_2(\mathbf{k}_i), \qquad (i = 1,2,3) \tag{12}$$

Step 2: Compute the vectors

$$\mathbf{f} = 0.5(\mathbf{x}_{11}{*}\,(\mathbf{x}_{22} - \mathbf{x}_{23}) + \mathbf{x}_{12}{*}\,(\mathbf{x}_{23} - \mathbf{x}_{21}) + \mathbf{x}_{13}{*}\,(\mathbf{x}_{21} - \mathbf{x}_{22})),$$

$$\mathbf{f}^{-1} = \mathbf{1}/\mathbf{f} \tag{13}$$

$$\mathbf{b}_i = 0.5\mathbf{f}^{-1}{*}(\mathbf{x}_{2j} - \mathbf{x}_{2k}), \quad \mathbf{c}_i = 0.5\mathbf{f}^{-1}{*}(\mathbf{x}_{1k} - \mathbf{x}_{1j}), \quad (i,j,k = 1,2,3)$$

where, " *" and "/" denote the componentwise multiplication and division operators, $\mathbf{1}$ is the unity vector and vector $\mathbf{f}$ contains the areas of the M finite elements. The preprocessing Steps 1 and 2 are executed once at the beginning of the time iteration.

Step 3: First, compute the components of the element mass matrices and store them into the vectors $\mathbf{S}_{ij}$ of length M

$$\mathbf{S}_{ii} = \frac{1}{6}\mathbf{f}, \quad (i = 1,2,3) \quad \mathbf{S}_{ij} = \frac{1}{12}\mathbf{f} \quad (i,j = 1,2,3 \;\; i \neq j) \tag{14}$$

Then, compute the components of the element advection matrices:

$$\mathbf{V}_{ij} = \frac{1}{3}\mathbf{f}{*}(\mathbf{b}_i{*}\mathbf{v}_1 + \mathbf{c}_i{*}\mathbf{v}_2) \quad (i,j = 1,2,3) \tag{15}$$

where, $\mathbf{v}_1$ and $\mathbf{v}_2$ are the element-wise constant components of the flow velocity.

Finally, compute the components of the element dispersion matrices:

$$\mathbf{D}_{ij} = \mathbf{f}{*}(\mathbf{b}_j{*}(\mathbf{b}_i{*}\mathbf{d}_{11} + \mathbf{c}_i{*}\mathbf{d}_{21}) + \mathbf{c}_j{*}(\mathbf{b}_i{*}\mathbf{d}_{12} + \mathbf{c}_i{*}\mathbf{d}_{22})) \tag{16}$$

where, the elements of the vectors $\mathbf{d}_{ij}$, $i,j = 1,2$ are element-wise constant dispersion coefficients. They are defined by the relations (Bear, 1979):

$$\mathbf{d}_{ij} = \delta_{ij}\mathbf{d}^{\star}_{ij} + \delta_{ij}\mathbf{a}_{II}{*}\mathbf{v} + (\mathbf{a}_I - \mathbf{a}_{II})\mathbf{v}_i{*}\mathbf{v}_j{*}\mathbf{v}^{-1} \quad (i,j = 1,2) \tag{17}$$

where, $\mathbf{v} = \sqrt{\mathbf{v}_1{*}\mathbf{v}_1 + \mathbf{v}_2{*}\mathbf{v}_2}$ and $\mathbf{v}^{-1} = \mathbf{1}/\mathbf{v}$. The vectors $\mathbf{d}^{\star}_{ij}$, and $\mathbf{a}_I$, $\mathbf{a}_{II}$ contain the elementwise constant coefficients of diffusion, longitudinal and transversal dispersivity, respectively.

We note that the vectors $\mathbf{S}_{ij}$, $\mathbf{V}_{ij}$ and $\mathbf{D}_{ij}$ store the mass, advection and dispersion components of the system of linear equations (4) elementwise while eq. (3) express the same components nodewise.

Step 4: Multiply the vectors $\mathbf{S}_{ij}$, $\mathbf{V}_{ij}$ and $\mathbf{D}_{ij}$ with the weighting factors θ_S, θ_V and θ_D, respectively, and accumulate the results into the vectors $\mathbf{A}_{ij}$ of length M according to the Crank-Nicholson scheme (eq. (4))

$$\mathbf{A}_{ij} = \frac{2}{\Delta t}\mathbf{S}_{ij} + \mathbf{V}_{ij} + \mathbf{D}_{ij} \quad (i,j = 1,2,3) \tag{18}$$

Step 5: Scatter the vectors $\mathbf{A}_{ij}$ $i,j = 1,2,3$ according to the index vectors $\mathbf{I}_{ij}$ into the coefficient matrix of the system of linear equations (4). The vectors $\mathbf{I}_{ij}$ indicate the locations of the element components in the coefficient matrix. The elements of the vectors $\mathbf{I}_{ij}$ which depend upon the node and element ordering of the mesh and the selected storage scheme for the coefficient matrix, are computed in a preprocessing step. If the non-zero entries of the coefficient matrix are stored into a vector $\mathbf{A}$, this step can be expressed as

$$\mathbf{A}(\mathbf{I}_{ij}) = \mathbf{A}(\mathbf{I}_{ij}) + \mathbf{A}_{ij} \quad (i,j = 1,2,3) \tag{19}$$

We note that the effort necessary to assemble the coefficient matrix consists of $9M$ additions and is invariant to the employed storage scheme.

Due to the variability in time of the velocity vectors, Steps 3 to 5 are repeated in each time step. The vectors in the expressions (14) to (18) are segmented automatically to the length of the cache and processed with stride one. Moreover, the compiler transforms most of the encountered constructs of type $fpa \times fpc \pm fpb$, where fpa , fpb and fpc are floating-point operands into compound instructions. The efficient reuse of the cache and registers is ensured if the expressions (14) to (18) are inlined and grouped in a small number of loops, such that each vector operand is reloaded fewer times and the partial results are available within the same iteration.

• *Solving the Linear System of Equations*

One comfortable way to solve the linear systems of equations (4) and (6) is to use the available subroutines from the Engineering and Scientific Subroutine Library (IBM ESSL, 1990). Another possibility is to implement the algorithms presented in the previous Section. In the absence of a sophisticated preconditioner, the only challenging programming issue associated with the implementation of the considered CG-like algorithms is the design of an efficient matrix vector multiplication subroutine. The importance of this issue in the context of the finite element application presented in the next Section is underlined by the results presented in Table 1: about 70% of the of the CPU time of a

CG-like iteration and more then 30% of the entire execution time of the program is consumed by the matrix vector multiplication subroutine.

Peters (1990) implemented and compared several sparse matrix vector multiplication techniques on an IBM 3090 VF. These techniques are based on the scalar and vector ITPACK storage schemes (Kincaid and Young, 1983) and on variants of them (Fernandes and Girdinio, 1989; Paolini and Radicati, 1989). The same techniques have been adapted for efficient use on RISC 6000 and applied to the CG-like algorithms discussed in the previous Section. We found that the scheme of Fernandes and Girdinio (1989) called ITPLUS give the best timing results for the problem presented in the next Section. To take advantage of the outer-loop vectorization capability of the VS FORTRAN compiler of IBM 3090 VF the compressed part of the matrix in the ITPLUS scheme was stored columnwise. The same compressed part of the matrix was stored rowwise for optimal processing on RISC 6000. The implemented matrix vector multiplication routines run run at a rate of about 16 Mflops on IBM 3090 VF (Model E) and about 8 Mflops on RISC 6000 (Model 530) for the matrix of the next Section.

A CASE STUDY

We apply the algorithms and procedures discussed in the previous Sections to a real life problem: the simulation of the groundwater contamination from the landfill Georgswerder in Northern Germany. The Georgswerder landfill is located south of the city of Hamburg on the island Wilhelmsburg in the river Elbe.

The detection of Seveso Dioxin 2,3,7,8 TCDD in the oil phase of the seepage fluid in 1983 initiated an extensive investigation and monitoring program for the landfill. At the request of the environmental protection agency of the city of Hamburg, (Amt für Gewässer- und Bodenschutz, Umweltbehörde Hamburg) the Institut für Wasserbau und Wasserwirtschaft of the Aachen University of Technology developed several groundwater contamination models as a part of the investigation program (Dorgarten and Daniels, 1989). The largest of these models is based on the algorithms and programming techniques discussed in this work. Figure 1 illustrates the discretization of the problem domain in $N = 1758$ nodes and $M = 3422$ elements.

Figure 2 shows the overlapped contour plots of the contamination plumes calculated with the Leismann-Frind scheme and the Crank-Nicholson scheme at the time twenty years from the begining of the contamination. The plotted contours overlap almost perfectly. Table 1 illustrates the results of the CPU-time measurements for the assemblage of the system of linear equations and its solution for the following combinations of algorithms:

- Leismann-Frind and CG preconditioned by diagonal scaling,

- Crank-Nicholson and CGS preconditioned by diagonal scaling,
- Crank-Nicholson and CGSTAB preconditioned by diagonal scaling.

All program variants have been implemented using the programming techniques described in the previous Section. As predicted by Eiermann et al. (1991) the Leismann-Frind/CG scheme outperforms the Crank-Nicholson/CGS scheme - not only that CGS and CGSTAB consume more CPU time than CG, but the time required to assemble the coefficient matrix for the Leismann-Frind method is shorter as well.

Table 1 - : Performance Comparison

Computer	Method	CPU Time (sec.) for 1000 Time Steps		
		Assemblage of the Linear System of Equations	Matrix Vector Multiplication	CG-like Algorithms
IBM 3090 VF	Leismann-Frind CG	66.	78.	114.
	Crank-Nicholson CGS	68.	95.	128.
	Crank-Nicholson CGSTAB	68.	99.	138.
RISC 6000	Leismann-Frind CG	100.	167.	275.
	Crank-Nicholson CGS	102.	204.	338.
	Crank-Nicholson CGSTAB	102.	220.	400.

REFERENCES

BEAR, J.: *Dynamics of Fluids in Porous Media.* Mc Graw Hill, New York, 1979.

BELL, R.: *IBM RISC System/6000 Performance Tuning for Numerically Intensive FORTRAN and C Programs.* GG24-3611, IBM Corporation 1990.

DORGARTEN, H. W. and H. DANIELS: Grundwasserstudie Deponie Georgswerder - Finite Elemente Modellierung bei der Planung eines sicheren und wirtschaftlichen Grundwasserschutzes. *Abfallwirtschaftsjournal* 6 (1989), pp. 69 - 74.

EIERMANN, M.; A. PETERS and H. DANIELS: Computational complexity analysis of two Galerkin-FE / CG-like approaches for the advection-dispersion equation, submitted to Seventh Int. Conf. Num. Methods in Laminar and Turbulent Flow, Stanford (1991)

FERNANDES, P. and P. GIRDINIO: A new storage scheme for an efficient implementation of the sparse matrix-vector product, *Parallel Comput.* 12 (1989), pp. 327 - 333.

FLETCHER, R.: Conjugate gradient methods for indefinite systems. In: G. A. WATSON, editor, *Lecture Notes in Math.* 506, pp. 73 - 89, Springer-Verlag, Berlin, 1976.

GOLUB, G. H. and C. F. VAN LOAN: *Matrix Computations,* 2nd Edition. The Johns Hopkins University Press, Baltimore, 1989.

HESTENES, M. R. and E. STIEFEL: Methods of conjugate gradients for solving linear systems. *J. Res. Nat. Bur. Standards* 49 (1952), pp. 409 - 436.

IBM ESSL, Engineering and Scientific Subroutine Library-Guide and Reference Release 4. SC23-0184-4, IBM Corporation 1990.

IBM VS Fortran Version 2 - Programming Guide Release 4. SC26-4222-04, IBM Corporation 1989.

IBM J. Res. Develop. Volume 30, No. 2, 1986.

IBM RISC System/6000 Technology. SA23-2619, IBM Corporation 1990.

KINCAID, D. R. and D. M. YOUNG: The ITPACK project: Past Present and Future, In: G. BIRKHOFF and A. SCHOERNSTADT, editors, *Eliptic Problem Solvers II.*, pp. 53 - 64, Academic Press, Orlando, 1983.

LEISMANN, H. M. and E. O. FRIND: A symmetric-matrix time integration scheme for the efficient solution of advection-dispersion problems. *Water Resources Research* 25 (1989), pp. 1133 - 1139.

LIU, B. and N. STROTHER: Programming in VS Fortran on the IBM 3090 for Maximum Vector Performance. *IEEE Computer* 21 (1988), pp. 65 - 76.

NACHTIGAL N.M., S.C. REDDY and L.N. TREFETHEN: How fast are nonsymmetric matrix iterations? *submitted to SIAM J. Sc. Stat. Comput..*

PAOLINI, G. V. and G. RADICATI DI BROZOLO: Data structures to vectorize CG algorithms for general sparsity patterns. *BIT* 29 (1989), pp. 703 - 719.

PELKA, W. and A. PETERS: FE Groundwater Models on Vector Computers. *Intern. J. Numer. Methods Fluids* 6 (1986), pp. 913 - 925.

PETERS, A.: Sparse Matrix Vector Multiplication Techniques on IBM 3090 VF. To appear in *Parallel Comput.* (1991).

PINDER, G. F. and W. G. GRAY: *Finite Element Simulation in Surface and Subsurface Hydrology.* Academic Press, New York, 1977.

ROACHE, P. J.: *Computational Fluid Dynamics.* Hermosa Publishers, Albuquerque, 1972.

SAAD, Y. and M. H. SCHULTZ: GMRES: A generalized minimal residual algorithm for solving nonsymmetric linear systems. *SIAM J. Sci. Stat. Comput.* 7 (1986), pp. 856 - 869.

SONNEVELD, P.: CGS, a fast Lanczos-type solver for nonsymmetric linear systems. *SIAM J. Sci. Stat. Comput.* 10 (1989), pp. 36 -52.

VAN DER VORST, H. A. and P. SONNEVELD: CGSTAB: A more smoothly converging variant of CG-S. IBM Europe Institute, Oberlech, 1990.

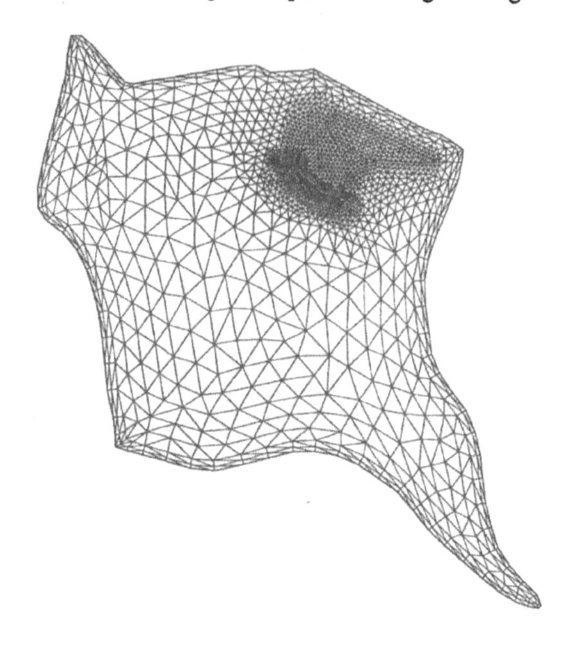

FIGURE 1: Finite element discretization of the project area.

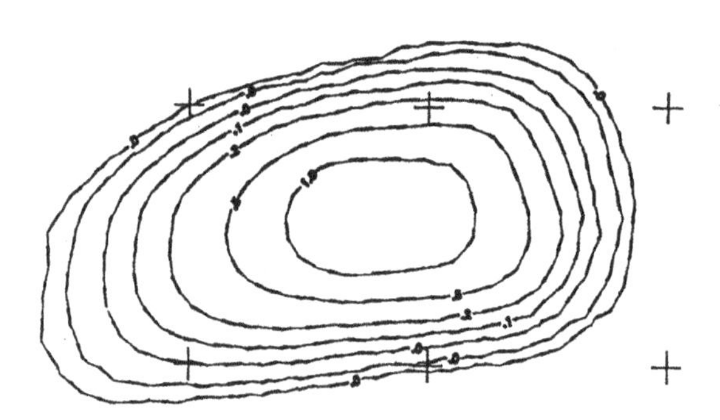

FIGURE 2: Contour plot of the contaminant plume for the Leismann-Frind scheme (-----) and the Crank-Nicholson scheme (- - -).

Numerical Simulation of Transient Response for Room Acoustic Design - Proposition of the Phase Element Integration Method (PEIM)

M. Tsuboi

Technical Research Institute, Obayashi Corporation, 4-640 Shimokiyoto, Kiyose-shi, Tokyo 204, Japan

ABSTRACT

Since the development of digital computers, the image source method and sound ray tracing method have been widely used for acoustic design of auditoriums. These methods, however, produce conspicuous errors when dealing with responses, including multiple reflections among boundaries of a room, especially in the low-frequency range. Therefore, it is very difficult to use these methods for room acoustic design of a concert hall where highly accurate prediction of a sound field covering the entire audio-frequency range is required.
The purpose of this paper is to propose an efficient method, the PEIM using supercomputer, for calculating a transient response which ranges over audio frequencies in a room. This method, based on the numerical integration of Kirchhoff's integral equation, carries out integration with technique for tracing identical phase elements like the Fresnel zone of a propagating wave-front. The effectiveness of this method is confirmed applying it in response calculations for multiple reflections at rigid planes and responses of an auditorium and comparing with results of calculations by the image source method and with measurement values.

1. INTRODUCTION

Acoustic design must create an attractive and comfortable acoustic space which meets a person's expectation. For this purpose, it is necessary to create the acoustic space aimed for by carrying out sound field prediction of high accuracy in designing and feeding back the results.

One technique for aiding such acoustic design often used is acoustic scaled model experiments with actual sounds emitted. However, for practical purposes, it is not an easy matter to match the acoustic characteristics of wall surface materials in accordance with design conditions such as wall surface materials and room configurations, and to conduct experiments on room configurations of a number of auditoriums.

Meanwhile, computer simulation by which it is possible to visually observe sound propagation naturally invisible on rapidly predicting the sound field at the planning stage is becoming an effective tool for designing. However, when the accuracy of sound field prediction is considered, it is not possible to calculate the transient response in the room for the entire audio-frequency range by the widely used image source method or the ray tracing method ignoring the wave motion properties of sound such as diffraction. Hence, a method of calculating sound fields of high accuracy is desired.

At present, the following have been presented as methods of calculating a sound field as accurately as possible based on Kirchhoff's integral equation equivalent to the wave equation.

An integral equation was formulated by Terai [1] with the velocity potential and normal differential at the boundary of the sound field as unknown functions and a rigorous solution method has been obtained. At present, the unknown functions on the boundary plane must be solved numerically, and it is difficult to intuitively find what the form and characteristics of the sound field are from the results.

For a more accurate prediction, Sakurai [2] used a curvilinear integral derived from Kirchhoff's boundary conditions and treated responses as superpositions of reflected sounds from rigid planes showing that calculation of early reflected sounds up to high frequencies is possible. However, this gradually becomes more difficult due to rapidly increasing image sources with calculations for higher-order reflections.

Sekiguchi et al. [3] tried calculating impulse responses employing "the method of finite sound ray integration" based on the technique of sound ray tracing. It appears to be difficult to accurately determine multiple reflected sounds, and the treatment of absorbent surfaces is not explained. Therefore, it is still difficult to use these methods for acoustic design of an actual auditorium.

In this paper, using a supercomputer which has a large capacity of the main memory and the high speed processing ability, the author proposes a practical method (PEIM) of applying Fresnel-Kirchhoff's integral equation for dealing with the transient response. Regarding the effectiveness of this method, it will be shown by comparative calculations and measurement values both of the multiple reflections and of auditorium responses described later.

2. SOUND FIELD CALCULATION TECHNIQUE WIDELY USED IN GENERAL FROM THE PAST

More than 20 years have gone by since computer simulations in the field of room acoustics began to be studied by Krokstad et al. in 1968. The techniques used most often at present are the image source method and the ray tracing method, because of both the establishment of algorithms and the convenience of being able to calculate macroscopic sound fields of basic room configurations even with personal computers. Sound is assumed to advance in a straight line like a laser beam, and moreover, be reflected with angle of incidence equal to angle of reflection as with reflection at an infinitely large rigid plane. Consequently, the techniques are employed limited to medium and high frequency ranges where wave lengths are amply short compared with the dimensions of the reflection plane in a room.

2.1 Ray Tracing Method

As shown in Fig. 1, this is a technique where sound rays to which acoustic energy has been allocated are radiated at equal solid angles in an indoor space and made to be mirror-reflected, and these are traced with elapse of time. This technique is suited to macroscopically observe sound fields of an auditorium such as service areas of loudspeakers and acoustic reflection plates, sound pressure contours, concentration phenomena of sound, etc.

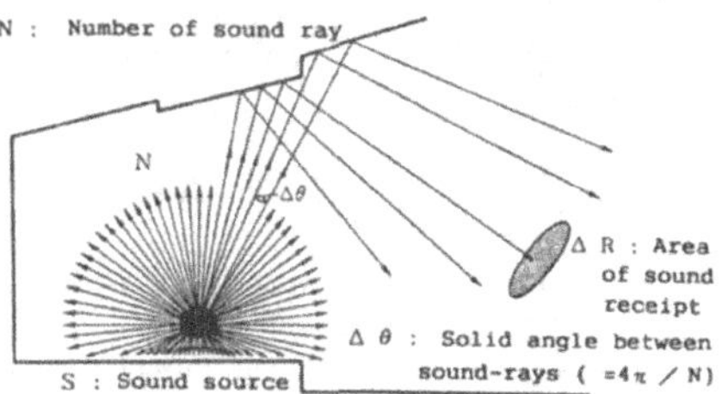

Fig. 1 Sound ray tracing method

2.2 Image Source Method

As shown in Fig. 2, this is a technique where the reflection paths to the observation point from all image sources are obtained and acoustic energy is calculated considering attenuation due to distance. This technique is suited to picking out observation points one by one. But, there is the drawback that if the number of wall surfaces of the model or the order of reflections is increased to calculate the reflected sound paths from all images, operation time will exponentially be increased.

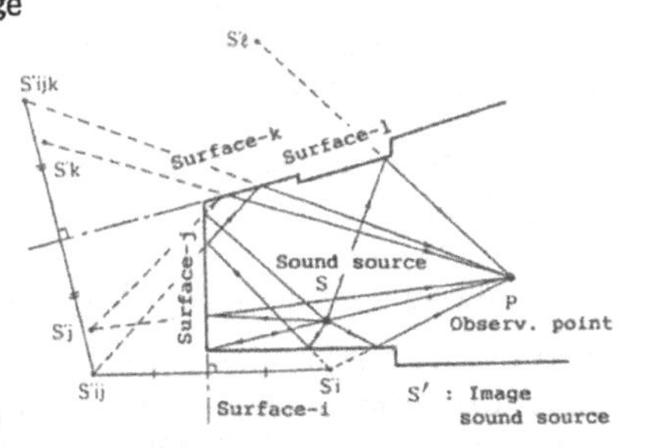

Fig. 2 Image sound source method

2.3 Problematic Points of the Techniques

These techniques are effective to an extent in examining acoustic troubles such as concentration phenomena of sound and echoes. To design an attractive and high-quality sound field which meets a person's expectation, it is necessary for a study to be carried out making predictions with good accuracy over the entire audio-frequency range including the low frequency range. However, with these calculation techniques, it is difficult to predict a low-frequency range of long wave length compared with the architectural dimensions and which wave motion properties cannot be ignored.

3. PEIM BASED ON FRESNEL-KIRCHHOFF'S INTEGRAL EQUATION

3.1 Direct Sound from Sound Source Point

Based on wave theory, the response from the source point Q to observation point P of Fig. 3 is considered. One of the wave-fronts at a certain instant of a sinusoidal spherical wave emitting sound from Q is a spherical surface Ω of radius r_0 with Q as the center. According to the principle of Huygens-Fresnel, the elements on this wave-front become new sound sources to produce secondary waves that are propagated. Therefore, the response at P is expressed by integration of the secondary

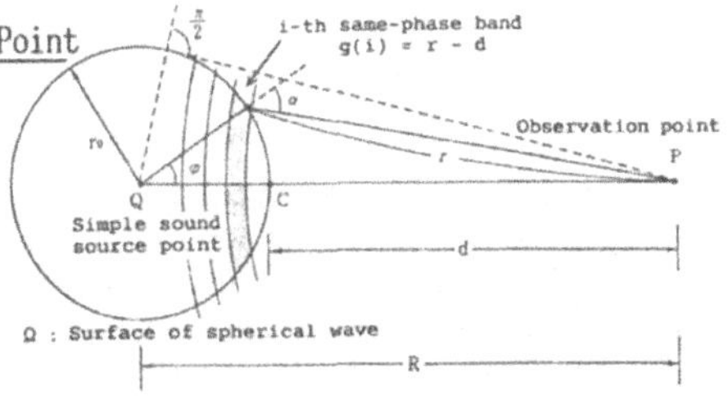

Fig. 3 Direct sound from the source

waves from the elements on this wave-front. Since elements on Ω are all of equal phases, the sound pressure U(P) of the observation point P omitting the time variation $\exp(-j\omega t)$ of the sound source is expressed by following equation :

$$U(P)=\frac{Ae^{jkr_0}}{r_0}\iint_\Omega \frac{e^{jkr}}{r}K(\alpha)dS \qquad \cdots\cdot(1)$$

where, $e^{-j\omega t} = \cos(\omega t) - j\sin(\omega t)$

j^2	=	-1,	k	=	$2\pi/\lambda$
ω	=	$2\pi f$: angular frequency,	A	:	amplitude
f	:	frequency (Hz),	λ	:	wave length
$K(\alpha)$	:	inclination factor,	Ω	:	spherical wave-front surface

Here, as a factor for discretization, the band-like part of same phases on the spherical surface comprising the potential at the time at which there is observation point response is focused on. If the area is dS, as is clearly shown in Fig. 3,

$$r^2=r_0^2+R^2-2r_0R\cos\varphi$$

Therefore,

$$dS=2\pi r_0\sin\varphi \cdot r_0 d\varphi=2\pi r_0/r dr$$

With the contributing sound pressure from the i-th same-phase band to the observation point as $U_i(P)$, and assuming inclination factor $K(\alpha) = K_i$ = constant on that band, $U_i(P)$ may be expressed by following equation :

$$U_i(P)=2\pi\frac{A}{R}e^{jkr_0}\cdot K_i\int_{d+g(i-1)}^{d+g(i)} e^{jkr}dr$$
$$=-j\cdot\lambda\cdot K_i\frac{Ae^{jkR}}{R}\{e^{jkg(i)}-e^{jkg(i-1)}\}$$

where, r_i : distance between the i-th same-phase band and observation point.
g(i): r_i-d (see Fig. 3)

Therefore, the sound pressure U(P) in case of the spherical surface Ω divided into N parts by the same-phase band, with the respective same-phase bands made into elements of division into M parts,

$$U(P) =\sum_{i=1}^{N}U_i(P)$$
$$=-j\lambda\frac{A}{R}e^{jkR}\sum_{i=1}^{N}\cdot\frac{K_i}{M}\sum_{m=1}^{M}\{e^{jkg(i)}-e^{jkg(i-1)}\} \qquad \cdots\cdot(2)$$

3.2 First Reflection of Rigid-plane Panel

In order to apply the approximate equation of diffraction obtained from Kirchhoff's integral equation to calculation of reflection characteristics, the reflection model is considered as the closed space of the diffraction model as shown in Fig. 4. In the reflection model, the spherical wave from the source Q is reflected at the rigid-plane panel surface Γ in the air.
The diffraction model passing the opening Γ possessing Kirchhoff's boundary conditions in a complementary relationship with the reflection panel Γ, from the image source Q' to the infinitely large cover panel Ω

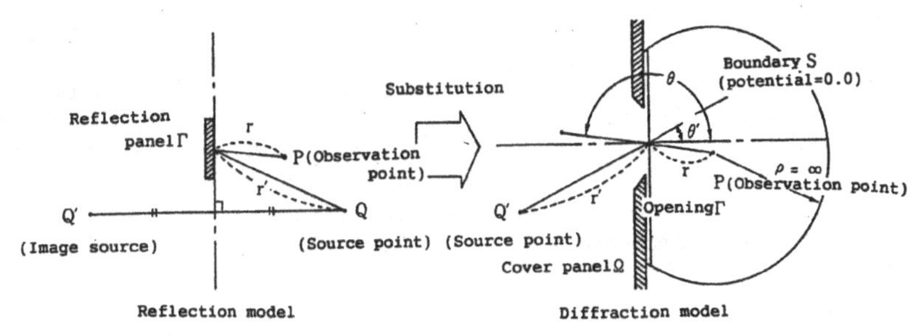

Fig. 4 Substitution diffraction model for reflection model

which includes the reflection panel Γ. In the diffraction model, the closed curved surface formed to both the rear of the cover panel and a part of the spherical surface of radius ρ enclosing P as shown in the figure are defined as the closed space S taking an integral. Considering $\rho \to \infty$ assuming that velocity potential of sound does not exist at the back of the panel, the contribution from the part on the spherical surface will be 0. Further, assuming the boundary conditions to be

Opening Γ of cover panel Ω : $u = \dfrac{Ae^{jkr'}}{r'}$ (spherical wave from Q')

$$\frac{\partial u}{\partial n} = \frac{Ae^{jkr'}}{r'}\left(jk - \frac{1}{r'}\right)\cos\theta'$$

Part of cover panel Ω other than opening : $u = 0$, $\dfrac{\partial u}{\partial n} = 0$

where, $\partial / \partial n$: partial differential coefficient in normal direction facing the inner side of S at various points on S

From Kirchhoff's integral equation, the time variation $\exp(-j\omega t)$ is omitted, and the sound pressure u(P) at the observation point P when a spherical wave is emitted from the sound source Q' may be expressed by the following equation :

$$\begin{aligned} u(P) &= \frac{1}{4\pi}\iint_s \left\{ u\frac{\partial}{\partial n}\left(\frac{e^{jkr}}{r}\right) - \frac{e^{jkr}}{r}\frac{\partial u}{\partial n} \right\} dS \\ &= \frac{1}{4\pi}\iint_s \frac{A}{r \cdot r'} e^{jk(r+r')} \cdot \left\{ \left(jk - \frac{1}{r}\right)\cos\theta - \left(jk - \frac{1}{r'}\right)\cos\theta' \right\} dS \\ &\fallingdotseq \frac{jk}{4\pi}\iint_\Gamma \frac{Ae^{jk(r+r')}}{r \cdot r'}(\cos\theta - \cos\theta')\, dS \qquad \cdots\cdot (3) \end{aligned}$$

This, however, is an approximate equation which assumes $\lambda \ll r$ and $\lambda \ll r'$ except for the vicinities both of the sound source point and of the observation point and ignores 1/r and 1/r'. Further, on substituting $k = 2\pi / \lambda$ in Eq. (3) and rearranging :

$$u(P) \fallingdotseq A\iint_\Gamma e^{jk(r+r')} \cdot j \cdot \frac{(\cos\theta - \cos\theta')}{2} \cdot \frac{1}{\lambda} \cdot \frac{1}{r \cdot r'} dS \qquad \cdots\cdot (4)$$

From Eq. (4), it can be seen that the influence on the observation point response per unit area ΔS of the reflection plane is in inverse proportion to the distance (r') from sound source point to reflection plane, distance (r) from reflection plane to observation point, and wave length (λ). As for the inclination factor ($\cos\theta$ - $\cos\theta'$), the reflected sound is indicated to become smaller with increasing the incidence angle of sound to the reflection plane, and the reflected sound becoming largest in case of perpendicular incidence. Here, Eq. (4) may be considered as being limited to elements on the spherical wave which causes the integration range of Eq. (1) to be incident at the reflection plane Γ. And if the two equations are considered as equal, the inclination factor $K(\alpha)$ in Eq. (1) will be expressed by

$$K(\alpha)=\frac{j}{2\lambda}(\cos\theta-\cos\theta')=-\frac{j}{2\lambda}(\cos\alpha+1)$$

With K_i = $K(\alpha_i)$ and substituting this in Eq. (2), the wave-front elements (i, m) incident to the reflection panel are integrated as weighting variables W(i, m) = 1.0, and others not integrated as being 0.0, Eq. (2) will be arranged and be expressed by

$$U(P)=-\frac{A}{R}e^{jkR}\sum_{i=1}^{N}\frac{(\cos\alpha_i+1)}{2}\frac{1}{M}\sum_{m=1}^{M}[\{e^{jkg(i)}-e^{jkg(i-1)}\}\times W(i,\ m)\cdots\cdots(5)$$

provided that α_i was taken to be the value on the wave-front at the middle between source point Q and observation point P, so that

$$\alpha_i=\cos^{-1}\left[\frac{R+\{g(i)+g(i-1)\}/2}{R}\right]$$

From this, it can be seen that in case of first reflection, it will suffice for numerical integration to be done by Eq. (5) for the spherical wave element incident to the reflection panel Γ. Further, in Eq. (5), the $1/(r\cdot r')$ and $1/\lambda$ in Eq. (4) are considered in terms of variation in size of element-area of same phase zone.

3.3 Multiple Reflections between Panels with Obstruction Panels

The transfer function H(f) (= response of observation point) for N reflections from the sound source point to the observation point including obstruction planes on the propagation path, if taken to consist of forth kinds of transfer functions, $H_m(f)$, $G_{01}(f)$, $G_{m(m+1)}(f)$, and $G_{n(n+1)}(f)$, can be expressed by the following equation :

$$\begin{aligned} H(f) = G_{01}(f) \times H_1(f) \times G_{12}(f) \times H_2(f) \times \cdots\cdots \times G_{(m-1)m}(f) \\ \times H_m(f) \times G_{m(m+1)} \times \cdots\cdots \times H_n(f) \times G_{n(n+1)}(f) \qquad \cdots\cdots (6) \end{aligned}$$

$H_m(f)$ are the transfer functions due to reflection at the M-th reflection plane ($1 \leq M \leq N$). $G_{01}(f)$ is the transfer function of the propagation path including obstruction planes between the sound source point and the first reflection plane. $G_{m(m+1)}(f)$ are the transfer functions of the propagation path including obstruction planes between the M-th reflection plane and the (M+1)-th reflection plane. $G_{n(n+1)}(f)$ are the transfer functions of the propagation path including obstruction planes between the N-th reflection plane and the observation point.

Eq. (6) indicates that as sound reflected at the first reflection plane is propagated to the second, third, $\cdots\cdots$, M-th reflection plane, the potential of the wave-front element which existed at the beginning is reflected only in part, and on striking against obstruction planes, gradually becomes lost, and the response of the observation point is determined by the potential of the wave-front element remaining in the end. therefore, the degrees that potentials of the individual wave-front elements of Eq. (5) are reflected and propagated, and the degrees of which they are lost over the propagation path from the source point of N-th reflection model to the observation point, were respectively traced with each part of the propagation path corresponding to $G_{m(m+1)}(f)$ and $H_m(f)$ of Eq. (6), and the potential energy of the wave-front element passing all of the propagation path was calculated. Similarly to the first reflection model, the response at the observation point of the N-th reflection model can be calculated by numerical integration through Eq. (5) of those potential energies as weighting factors W(i, m) of the same phase elements on the wave-front divided by d+g(i), (i = 1, 2, 3,····). The author has named this technique the "Phase Element Integration Method (PEIM)". The image of tracing of propagation of the wave-front elements for which wave motions can be considered over the entire propagation path is shown in Fig. 5. The third reflection elements of Q-P which cannot be detected in this geometrical technique can be numerically integrated by the PEIM.

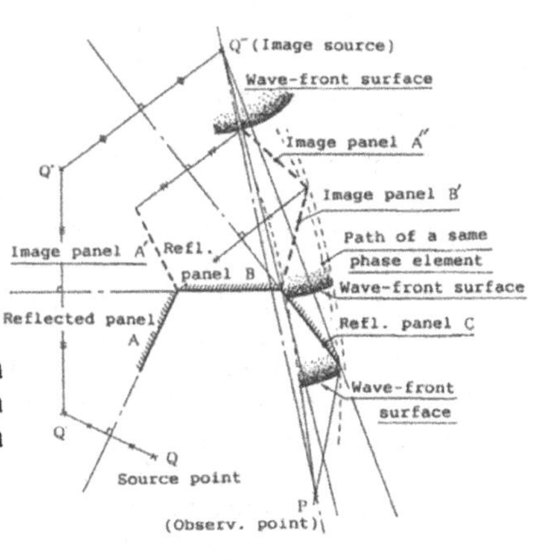

Fig. 5 Image of Tracing the same phase element

3.4 Calculation of Potential Energy of Same Phase Element

If the secondary wave-front elements of one half of the first Fresnel zone of frequencies corresponding to the potentials possessed by the divided same phase elements are propagated, then all of the potential of the same phase elements will be propagated. Therefore, by tracing the degree of existence of secondary wave-front elements from the sound source point to the wave--front (where there are same phase elements), and from the wave-front to the observation point, the potential energy of the same phase elements can be calculated.

3.5 Treatment of Absorbent Surface

The reflection coefficients of building materials are calculated using reverberation room absorption data generally published considering actual hall design. In case of calculating impulse response requiring absorption data for all frequencies, the data are supplemented with spline functions and transfer functions are calculated.

3.6 Techniques for Making Practical Use of PEIM

In case of application to an actual hall, to increase the range of integration and the number of divisions into elements of the wave-front will be effective for improving accuracy, but this will lead to an enormous increase in operation time and thus is not practical. And for examining variations in the sounds arriving at individual points of time over a long period from the emission of the sound, calculations would be made for

beyond the tenth reflection. But since the total number of propagation paths to be objects of calculation would be proportional to the power of the number of reflections at the number of planes of the model, calculations of the early reflected sounds of from the third to fourth reflections will be the practical limit.
Therefore, the author has devised the three techniques below for practicalization of PEIM :

① In order to efficiently select the path to be calculated, in application to an actual hall, sound rays are radiated in random directions as pre-processing to trace the path.
② To limit operation time, as shown in Fig. 6 in this paper, the limit to the integration range in calculations is made up to 2.5 times the wave length of 125 Hz with the discretized same phase elements at 1/8-wave-length intervals in the range of 2.5 times the wave lengths of the respective frequencies. This limitation of integration zone was determined by estimating the difference in the paths of the direction where reflected sound becomes 1/100 or under by the attenuation value of diffracted sound of Maekawa [4]. Outside of the integration range is ignored assuming the contribution to integration is 0.0 because the opposite phase (-) and same phase (+) are infinitely repeated to offset each other.
③ The potential energies of elements necessary for calculating impulse response, and moreover, elements which have been thinned out, are calculated on supplementation using spline functions.

3.7 Calculation of Impulse Response in PEIM

In calculation of impulse response, the sampling frequency was made 32 kHz dividing the wave length of 4kHz into eighths. The same phase elements to be calculated must correspond to this, but as stated in 3.6, the potential energies of these elements have been supplemented employing spline functions. The transfer functions at the respective propagation paths are obtained using PEIM, and the impulse responses at the propagation paths are calculated upon inverse Fourier transfer. The impulse response at the observation point is a synthesis of the impulse responses of all propagation paths.

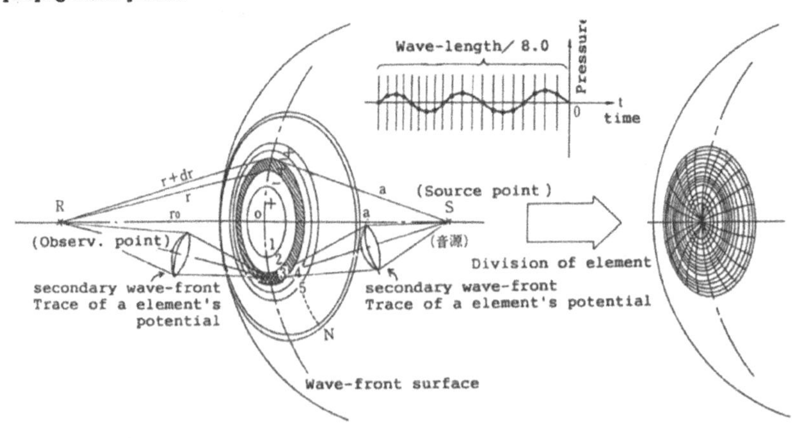

Fig. 6 Division of wave-front surface's element

3.8 Techniques for Vectorization of PEIM

In development of the PEIM program, software consisting of scientific and technological calculation libraries and numerical calculation libraries already prepared and made high speed were built into the supercomputer to increase the speed of the supercomputer by vectorization. Innovations were made such as to increase the length of the loop at the inner side so that there would be effect of vectorization.

4. COMPARISONS OF MEASUREMENTS AND CALCULATED RESULTS

4.1 First Reflection and Multiple Reflections of Rigid-plane Panels

Comparisons were made of calculated results and measurements of reflection properties to examine the appropriateness of PEIM. The measurements is using 1/8 scaled models. As shown in the following figures, frequencies are the values in the calculation model, and the frequencies in model experiments are eightfold high frequencies according to the law of similitude with wave length and dimensions. The reflection directivity characteristics and frequency characteristics of first reflection at the most fundamental single rigid-plane panel are shown in Figs. 7 and 8. The frequency characteristics of second reflection in the form of comparisons of calculated results and measurements under two kinds of element division conditions are shown in Fig. 9. These show that there is good agreement of calculation results of first reflection at a rigid-plane panel and measured values. The observation point of this second reflection model is aberrant from the reflective angle of the second reflection plane, but it was found it could be roughly approximated catching the diffracted wave. Element divisions were also compared with coarse meshes of 1026 elements considering application to an actual hall. Although, the level was slightly lower, it was learned that a rough approximation could be made. The reflection directivity characteristics and frequency characteristics for curved panels are shown in Figs. 10 and 11. This calculation model is approximated with six planes, but the small reflection plane (227 mm) divided and approximated in this way could not grasp the condition sufficiently, and it appears that the reflection characteristics have been calculated on the small side.

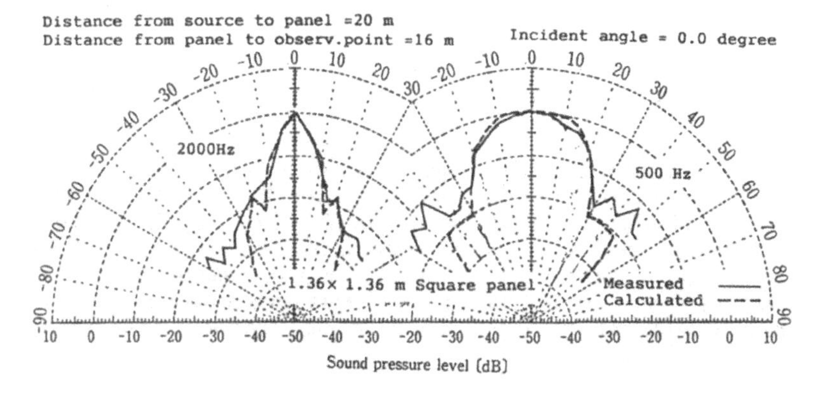

Fig. 7 Characteristic of directivity of first reflection

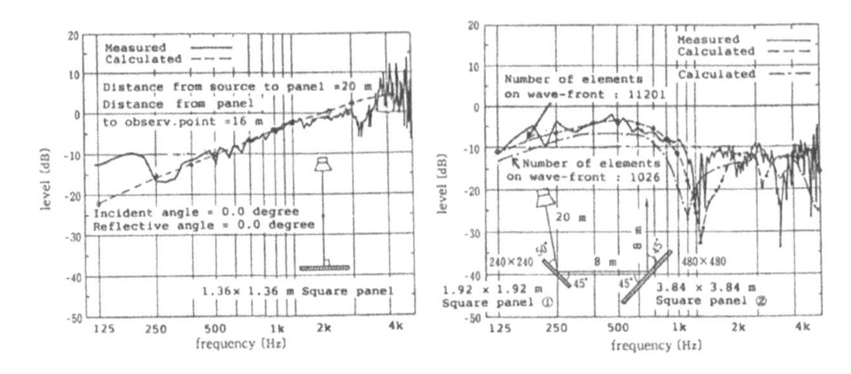

Fig. 8 Characteristic of frequency of first reflection

Fig. 9 Characteristic of frequency of multiple reflections

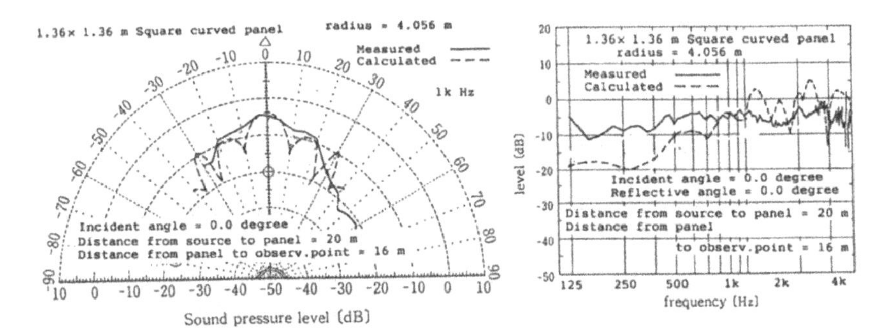

Fig. 10 Characteristic of directivity for curved panel

Fig. 11 Characteristic of frequency for curved panel

4.2 Applications to Actual Halls

The model of the Ashihara Memorial Hall (see Fig. 12) of Kounan Women's University in Kobe will be described here as an example of application to design. This hall has a volume of 16,200 m³ and a seating capacity of 1,784, and acoustic measurements were made after completion of construction. The analysis was performed with the ceiling and walls of three-dimensional curved surfaces approximated as a polyhedron, and modeling was done as a 524-plane model. The model configuration and the observation points at first-floor seats (42 points) are shown in Figs. 13 and 14.

As pre-processing, 65,500 sound rays were radiated in random directions, the propagation paths to be calculated were searched, and calculations were made by PEIM with only the paths as objects of integration. The measurements of stationary sound pressure level contours at 20 measuring points of 500 Hz and calculated values for 20 observation points corresponding to the measuring points are shown in Fig. 15. Simple comparisons cannot be made as the calculated values are the cumulative sound pressure levels from sound

Fig. 12 Photograph of Ashihara Memorial Hall

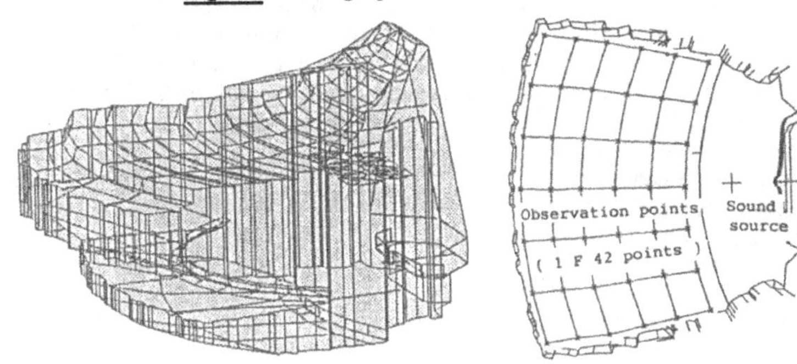

Fig. 13 Calculated model of the hall

Fig. 14 Observation points

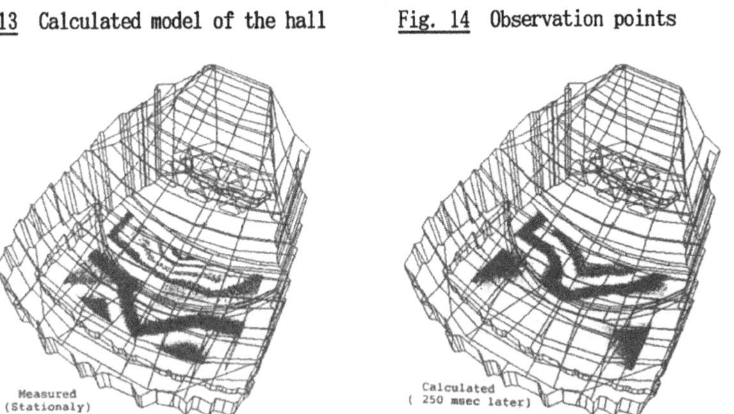

Fig. 15 Comparison with measured and calculated result of cumulative sound pressure level contours

emission to 250 msec, but they may be considered as transient distributions to a stationary state. The sound pressure levels according to arrival times are shown by a three-dimensional graph in Fig. 16 with the calculated values of transient response characteristics at the central observation point given by frequency in the order of 125 Hz, 250 Hz, 500 Hz, 1 kHz, 2 kHz, and 4 kHz. Comparisons of measurements and calculated values of echo-time patterns by frequencies are shown in Fig. 17. The sound sources differ with 1/1 octave band tone bursts of 10-msec width in measurements and pure tone bursts in calculated values, but there are good coincidences in transient responses at early reflections.

The time variations of cumulative level contours (42 observation points) of arrived sound pressures calculated for 125 Hz, 500 Hz, and 2 kHz are shown in Figs. 18 to 20. The operation time in this analysis example was approximately 13 hr (including 1 hr for pre-processing, 42 observation points) with SX-1EA on tracing the time-series data from sound emission to 250 msec. It appears that how accurately diffracted and diffused sounds are caught is the key to increased efficiency of calculations. The time variations of sound propagation are shown in Fig. 21.

The time variations of the cumulative level contours (42 observation points, 500Hz) by the image source method are shown in Fig. 22. In the image source method, the influences of small planes unrelated to frequency are the same as for a plane of infinitely large size and are taken to be excessively large. But in PEIM, it is smaller than the wave-length size and in actuality, the influence of a plane of small amount of reflection is small, and the difference due to frequency is calculated. Further, with regard to the number of propagation paths, a difference of more than 50:1500 was produced between the image source method and PEIM depending on whether or not there were diffraction paths.

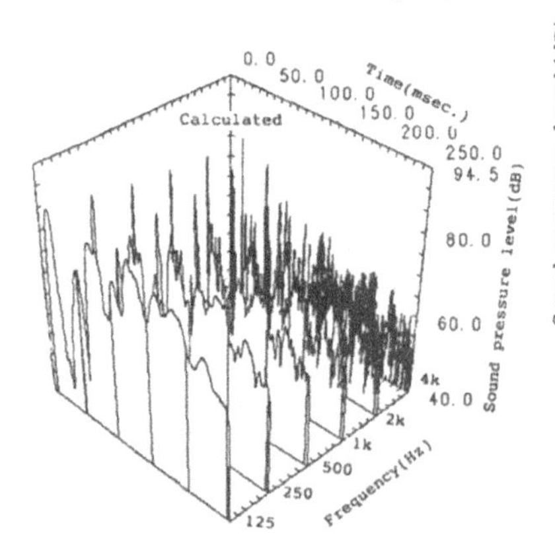

Fig. 16 Calculated transient responses

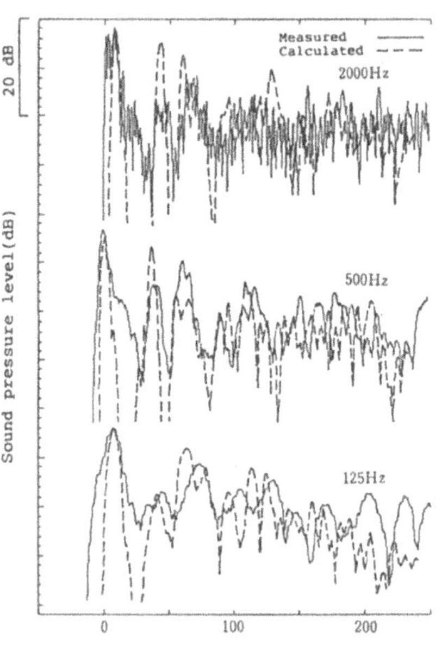

Fig. 17 Echo-time-patterns

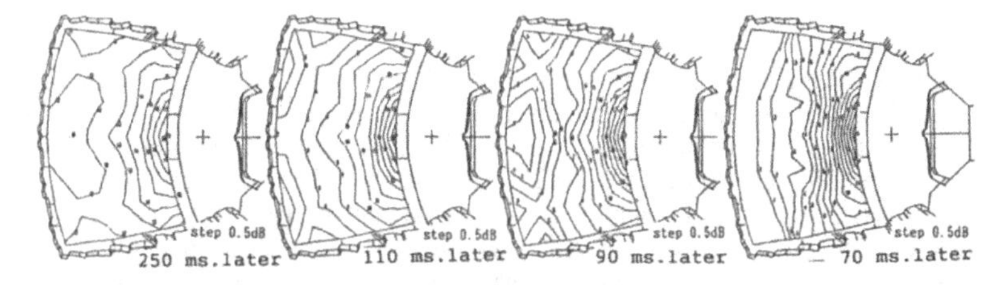

Fig. 18 Time variation of cumulative sound pressure level contours (125Hz)

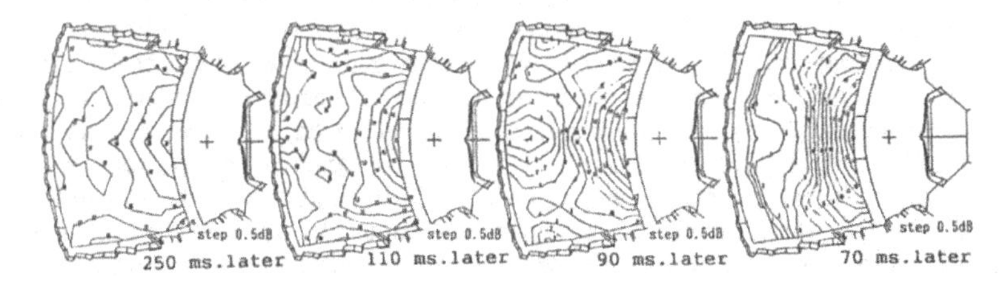

Fig. 19 Time variation of cumulative sound pressure level contours (500Hz)

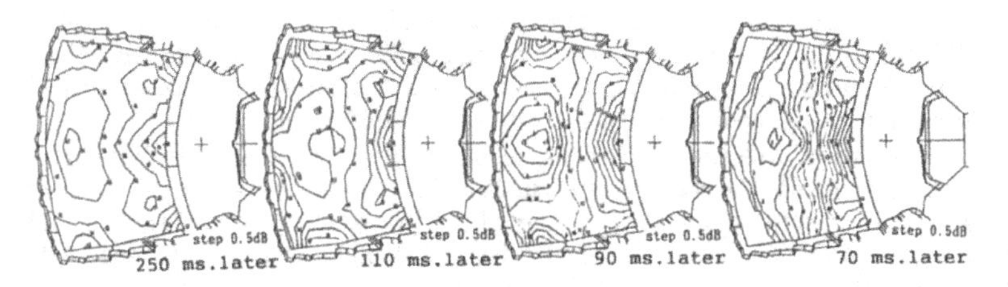

Fig. 20 Time variation of cumulative sound pressure level contours (2 kHz)

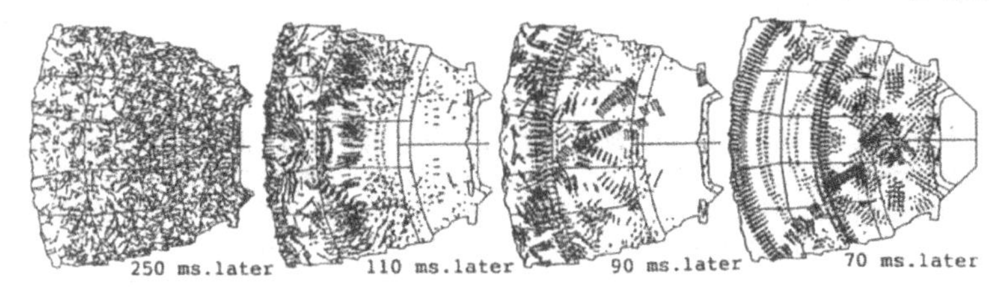

Fig. 21 Time variation of sound wave-front surface propagation

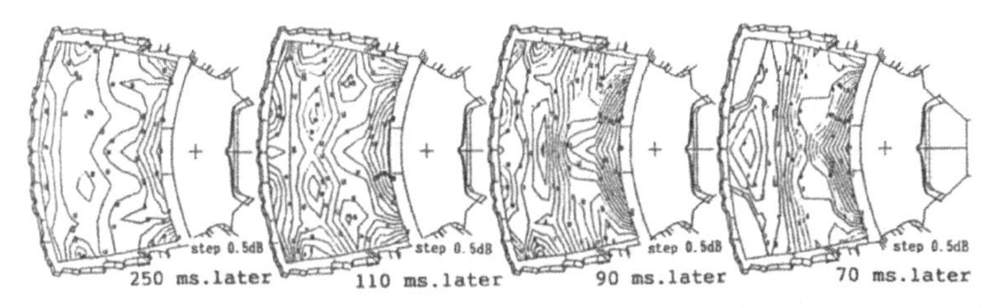

Fig. 22 The cumulative sound pressure level contours by image source method

5. CONCLUSION

The appropriateness of PEIM was confirmed from comparisons of the results of calculations using PEIM proposed in this paper and the results of measurements. In room acoustic design, PEIM using a supercomputer gives detailed spatial information in addition to the time-history, and it is possible for items of information to be visualized using a engineering work station. Further, by folding in dray-music (non-echoed music) to the impulse response (echo of auditorium) obtained by calculations, it is possible to listen to music after completion of construction at the design stage. In this way, it has become possible to provide more useful information for design of acoustic space called for by designers and owners. Hereafter, it is desirable for improvement in accuracy to be aimed for by carrying out an even greater number of comparison studies with measurement data for studies of details in analyses of PEIM such as the most efficient method of element division, integration range, and model preparation method. At the same time, it is intended to make comparison studies with other analysis methods and to build up a practical room acoustic field analysis system taking advantage of the respective features.

6. HARDWARE AND SOFTWARE USED

The supercomputer used in this analysis was an NEC SX-1EA of maximum speed of 330 MFLOPS. The engineering work station (EWS) used in three-dimensional graphic visualization of analysis was an SUN-4 (Sun Microsystems) and the software for pre/post-processing was an I-DEAS (SDRC Corporation).
As an example of imaging with the use of an EWS, the model configuration, the conditions of the wave-front surface 50 msec after emission of sound from a sound source on stage, and the time series variations in the cumulative sound pressure contours of 500 Hz at 100, 130, and 160 msec after sound emission are shown in Figs. 23, 24, and 25.
The author's company visualizes such analytical results which vary according to time through animation.

REFERENCES

1. T. Terai et al., J. Acoust. Soc. Jpn (E), vol.11, no.1, 1990, pp.1-10.
2. Y. Sakurai, J. Acoust. Soc. Jpn (E), vol.8, no.4, 1987, pp.127-138.
3. K. Sekiguchi et al., J. Acoust. Soc. Jpn (E), vol.6, 1985, pp.103-115
4. J. Maekawa, Architectural Acoustics, Kyoritsu Publishing, 1968, pp. 117 (in Japanese)

Fig. 23 Photograph of an auditorium

Fig. 24 Propagation of wave-front surface

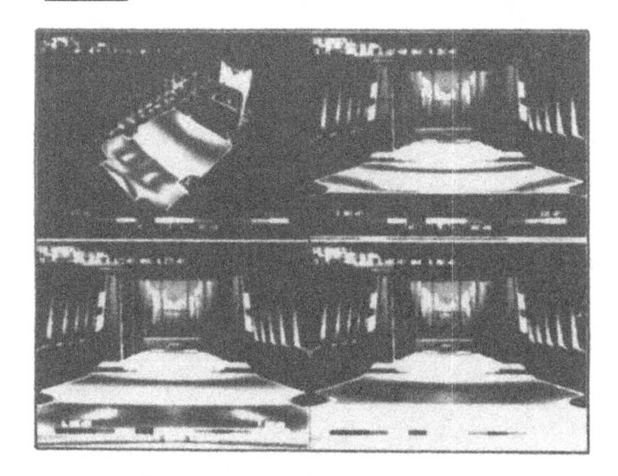

Fig. 25 Cumulative sound pressure level contour

Simulation of Nonlinear Contaminant Transport in Groundwater by a Higher Order Godunov-Mixed Finite Element Method

C. Dawson

Department of Mathematical Sciences, Rice University, Houston, Texas 77251, U.S.A.

Abstract: We consider the numerical simulation of contaminant transport in groundwater where the mathematical model includes a nonlinear adsorption term. The method we describe combines a higher order Godunov scheme for advection with a mixed finite element method for diffusion. The method is formulated in one space dimension, and numerical results for equilibrium and nonequilibrium adsorption are presented.

1 Introduction

As noted in two recent reports by the Environmental Protection Agency and the Department of Energy, the quality of groundwater at many locations in the United States is being threatened by the introduction of hazardous chemicals into the subsurface [1, 2]. Such chemicals include hydrocarbons, herbicides, and pesticides. In areas where groundwater is a major source of drinking water, such contamination poses a threat to the public health.

Modeling the flow of hazardous chemicals through the subsurface has seen increasing interest in recent years. The mathematical models which describe contaminant flow include terms which account for the aquifer characteristics (hydraulic conductivity, porosity, etc.), the presence of microorganisms capable of biodegrading certain compounds, and adsorption. Biodegradation is an important aspect of contaminant flow, as many compounds may be eliminated by natural biodegradation. Moreover, natural biodegradation processes may be enhanced to effectively remove contaminants [1]. Adsorption, which is a retardation/release reaction between the solute and the surface of the porous structure, is also a significant factor in contaminant movement. Adsorption can have the effects of segragating a hazardous compound from the groundwater, and slowing the overall movement of the compound.

In many cases contaminants are introduced to an aquifer through the unsaturated zone. A complete mathematical and numerical treatment of contaminant transport would include both the unsaturated and saturated zones. In this case, under certain simplifying assumptions, the mathematical description of the problem involves an elliptic/parabolic partial differential equation for pressure head, and a parabolic equation describing transport of contaminant. We will present a model which takes into account both the saturated and unsaturated regions. Numerically, we will focus on an algorithm for the transport equation only.

The author and M. F. Wheeler have developed and tested an algorithm for modeling multidimensional, multicomponent flow in the saturated zone which includes the effects of biodegradation and linear adsorption, see for example [3, 4]. In this code, a finite element modified method of characteristics models advection, and a mixed finite element method is used to calculate head and velocity. Nonlinear reaction terms modeling biodegradation are incorporated into the algorithm by time-splitting. In this paper, we consider the application of higher-order Godunov-mixed finite element methods to similar problems; however, we want to incorporate the effects of (possibly) nonlinear adsorption. Nonlinear adsorption complicates the model by the inclusion of a nonlinear time derivative and/or source term(s). In an earlier paper [5], the author and M. F. Wheeler described a first-order method for simulating these types of problems. Here, we describe a higher-order extension of this technique and use it to study nonlinear adsorption phenomena.

2 Problem Description

In saturated flow, the flow velocity $\mathbf{u}$ (L/T) of water satisfies Darcy's Law:

$$\mathbf{u} = -K\nabla h, \tag{1}$$

where K (L/T) is hydraulic conductivity and h (L) is the piezometric head. K is given by

$$K = \frac{k\gamma}{\mu}, \tag{2}$$

where k is the permeability of the medium and is possibly spatially dependent, γ is the specific weight of water, and μ is water viscosity.

In the unsaturated zone, where both air and water are present, if we assume air is stagnant and at a constant pressure, then

$$\mathbf{u} = -K(\Psi)\nabla h, \tag{3}$$

where in this case

$$K = \frac{kk_{rw}(\Psi)\gamma}{\mu}.$$

Here, $k_{rw}(\Psi)$ is the relative permeability of water, which is assumed to be a function of pressure head Ψ (L). The piezometric head is related to pressure head by

$$h = z + \Psi. \tag{4}$$

where z is the elevation of the water table. In the saturated zone, $\Psi \geq 0$, and in the unsaturated zone, $\Psi < 0$. If we extend the definition of $k_{rw}(\Psi)$ by $k_{rw}(\Psi) \equiv 1$ for $\Psi > 0$, then (3) is valid in both zones.

Let θ denote water content, which is a function of Ψ. In the saturated zone, $\theta(\Psi) \equiv \phi$, where ϕ is porosity. In the unsaturated zone, $0 < \theta < \phi$. Between these two zones lies the capillary fringe, or tension-saturated zone, where the functional relationship $\theta(\Psi)$ is soil-dependent. Water content and the flow velocity are related by

$$\theta_t + \nabla \cdot \mathbf{u} = q(x,t), \tag{5}$$

where $q(x,t)$ models source and sink terms. Thus, in the unsaturated zone, substituting (3) into (5), we obtain a nonlinear differential equation in Ψ, and in the saturated zone,

$$\nabla \cdot \mathbf{u} = -\nabla \cdot (K \nabla h) = q(x,t). \tag{6}$$

Thus, (5) is parabolic in the unsaturated zone, and elliptic in the saturated zone.

For more discussion on the saturated and unsaturated zones, and the capillary fringe, see [6] and [7].

Let c denote the concentration of some chemical species, which we will assume is soluble with water. Solute transport is described by [7]

$$(\theta c)_t + \rho A_t + \nabla \cdot (\mathbf{u}c - \theta D \nabla c) = r(x,c,t), \tag{7}$$

Here D (L^2/T) denotes the sum of the molecular diffusion and mechanical dispersion coefficients, and r is a source term, which could include point sources and sinks. The term ρA represents the amount of solute adsorbed, where ρ (M/V) is the bulk density.

In many cases of physical interest, flow is advection-dominated; that is, $\mathbf{u}$ is much "larger" than D, appropriately scaled. It is well known that standard finite element and finite difference techniques do not work well for these types of problems. However, the method we discuss here is capable of resolving sharp fronts without oscillation and with minimal numerical diffusion.

The term ρA is in general heterogeneous, depending on the adsorbent surfaces. The adsorption process can be divided into two classes, equilibrium and non-equilibrium. Adsorption is assumed to be in equilibrium when the reaction kinetics occur at a much faster rate than the rate of transport. A can be expressed as [8]

$$A = \lambda_1 \psi(c) + \lambda_2 s, \tag{8}$$

where $\lambda_i \geq 0$, $i = 1,2$, and $\lambda_1 + \lambda_2 = 1$; $\psi(c)$ describes adsorption at equilibrium sites, and s describes adsorption at non-equilibrium sites. The latter term is assumed to satisfy

$$s_t = k_a \Phi(c) - k_s s, \tag{9}$$

where k_a $(1/T)$ and k_s $(1/T)$ are nonnegative constants. The functions $\psi(c)$ and $\Phi(c)$ are adsorption isotherms. Two common isotherms are the Langmuir isotherm,

$$f_L(c) = \frac{K_d c}{1 + K_2 c}, \tag{10}$$

where K_d (V/M) and $K_2 > 0$, and the Freundlich isotherm,

$$f_F(c) = K_d c^p, \quad 0 < p \leq 1, \tag{11}$$

Note that $f_F(c)$ is not Lipschitz continuous at $c = 0$ for $0 < p < 1$.

3 A Numerical method for transport

For the purposes of discussing a numerical method for the transport equation (7), we assume water content θ and velocity $\mathbf{u}$ are given. For simplicity, we will assume saturated flow, thus $\theta \equiv \phi$. We are interested primarily in the effects of nonlinear adsorption. Thus, to ease the presentation, we reduce the problem to one space dimension, and assume no source terms are present in either (6) or (7). In this case, the flow velocity $\mathbf{u}$ becomes a scalar constant u.

First, consider the case of adsorption in equilibrium; i.e., $\lambda_1 = 1$ and $\lambda_2 = 0$ in (8). We assume $u > 0$ and prescribe an inflow concentration $c = c_0$ at $x = 0$. Equation (7) reduces to

$$c_t + \bar{\rho}\psi(c)_t + \bar{u}c_x - Dc_{xx} = 0, \quad x > 0, \ t > 0, \tag{12}$$

where $\bar{\rho} = \rho/\phi$, $\bar{u} = u/\phi$, and ψ is either the Freundlich or Langmuir isotherm. We assume initially that $c(x, 0) = c^0(x)$.

Let $\mu = c + \bar{\rho}\psi(c)$. Then $\mu(c)$ is a continuous, one-to-one, and onto mapping from $[0, \infty) \rightarrow [0, \infty)$. Thus, the inverse of μ, $\eta(\mu) = c$, exists, and (12) can be written as a parabolic equation (possibly degenerate) in μ:

$$\mu_t + \bar{u}\eta(\mu)_x - D\eta(\mu)_{xx} = 0, \quad x > 0, \ t > 0. \tag{13}$$

As mentioned earlier, if $\psi(c)$ is given by the Freundlich isotherm, with $0 < p < 1$, we encounter the difficulty that the nonlinearity c^p is non-Lipschitz at $c = 0$. Moreover, flow is generally advection-dominated, thus the solution exhibits sharp fronts. The effects of choosing $p < 1$ (as opposed to $p = 1$) are to make the fronts even sharper, and to further retard the flow of the chemical species. In fact, in the limit of zero diffusion with $p < 1$, solutions to (13) can exhibit shocks for smooth initial data.

For advection-dominated flow problems similar to (13), we have studied the application of higher-order Godunov-mixed methods (GMM). This class of methods was formulated and analyzed for nonlinear advection-diffusion equations in

[9]. A multidimensional extension of the method is described in [10]. We now describe the application of this algorithm to (13).

Assume the computational domain is truncated to a region $(0, \bar{x})$. At the point $\bar{x}$ assume the "outflow" condition

$$\mu_t + \bar{u}\eta(\mu)_x = 0, \quad \text{at } x = \bar{x}. \tag{14}$$

Let $0 = x_{1/2} < x_{3/2} < \ldots < x_{J+1/2} = \bar{x}$ be a partition of $[0, \bar{x}]$ into grid blocks $B_j = [x_{j-1/2}, x_{j+1/2}]$, and let x_j be the midpoint of B_j, $h_j = x_{j+1/2} - x_{j-1/2}$, and $h_{j+1/2} = (h_j + h_{j+1})/2$. Let $\Delta t > 0$ denote a time-stepping parameter, and let $t^n = n\Delta t$. For functions $g(x,t)$, let $g_j^n = g(x_j, t^n)$.

On each grid block B_j, approximate μ^n by a piecewise linear function w^n, where

$$w^n|_{B_j} = w_j^n + (x - x_j)\delta w_j^n, \quad j = 1, \ldots, J; \tag{15}$$

c^n is approximated by a piecewise constant function C^n, where

$$C^n|_{B_j} = C_j^n. \tag{16}$$

In (15),

$$w_j^n \equiv C_j^n + \bar{\rho}\psi(C_j^n). \tag{17}$$

Initially, set $C_j^0 = c^0(x_j)$, $j = 1, \ldots, J$. Assume for some time level t^n the approximate solution C^n is known. The GMM is based on splitting (13) into an advection equation:

$$\bar{\mu}_t + \bar{u}\eta(\bar{\mu})_x = 0, \quad x > 0, \quad t \in (t^n, t^{n+1}), \tag{18}$$

and a diffusion equation

$$\mu_t^* - D\eta(\mu^*)_{xx} = 0, \quad x > 0, \quad t \in (t^n, t^{n+1}). \tag{19}$$

Taking C^n as the initial condition, we apply a higher-order Godunov (HOG) scheme to (18). The solution generated from this step then serves as the initial condition for (19), where a mixed finite element method (MFEM) is applied. The result of these two steps is an approximation, C^{n+1}, to the solution at time t^{n+1}.

Let $\bar{w}^{n+1}$ be the HOG solution to (18) at time t^{n+1}. The HOG approach we will consider is similar to the MUSCL scheme proposed by van Leer [11]. Integrate (18) over the region $B_j \times [t^n, t^{n+1}]$ and apply the midpoint rule in time to obtain

$$\bar{w}_j^{n+1} = w_j^n - \frac{\Delta t}{h_j}\left[\eta(\bar{w}_{j+1/2}^{n+1/2}) - \eta(\bar{w}_{j-1/2}^{n+1/2})\right], \tag{20}$$

where $\bar{w}_j^{n+1} = \bar{w}^{n+1}|_{B_j}$.

The term $\eta(\bar{w}_{j+1/2}^{n+1/2})$ is an approximation to $\eta(\mu(x_{j+1/2}, t^{n+1/2}))$, and represents the advective flux across a grid block boundary. At the inflow boundary,

$$\eta(\bar{w}_{1/2}^{n+1/2}) = c_0^{n+1/2}.$$

At the other grid block boundaries, this term is approximated by characteristic tracing from the point $(x_{j+1/2}, t^{n+1/2})$ back to time t^n. Define the characteristics $x_L(t)$ and $x_R(t)$ satisfying

$$x_L{}'(t) = \max(0, \bar{u}\eta'(w_j^n)), \quad x_L(t^{n+1/2}) = x_{j+1/2}, \tag{21}$$

and

$$x_R{}'(t) = \min(0, \bar{u}\eta'(w_{j+1}^n)), \quad x_R(t^{n+1/2}) = x_{j+1/2}. \tag{22}$$

Let cfl be a specified parameter, $0 < cfl \le 1$. Assuming the CFL constraint

$$\lambda_j^n = \frac{\Delta t}{h_j}\bar{u}\eta'(w_j^n) \le cfl \le 1, \quad j = 1, \ldots, J, \tag{23}$$

then $x_L(t)$ crosses the $t = t^n$ axis at a point $x_{j,L}$, where $x_{j,L} \in B_j$, and $x_R(t)$ crosses at a point $x_{j,R} \in B_{j+1}$. In general, $\eta(\bar{w}_{j+1/2}^{n+1/2})$ is determined by upwinding; i.e., we evaluate η at $x_{j,L}$ or $x_{j,R}$, depending on the local direction of flow. Since we are assuming $\bar{u} > 0$, and since

$$\eta'(\mu(c)) = \frac{1}{\mu'(c)} \ge 0, \quad c \ge 0,$$

we find that

$$x_{j,L} = x_{j+1/2} - \frac{\Delta t}{2}\bar{u}\eta'(w_j^n), \tag{24}$$

and

$$\eta(\bar{w}_{j+1/2}^{n+1/2}) \equiv \eta(w^n(x_{j,L})). \tag{25}$$

The diffusion equation, (19), is handled using the MFEM with the lowest order Raviart-Thomas approximating spaces [12]. In [13], it is shown that this scheme, with the appropriate quadrature rule, is equivalent to block centered finite differences applied to (19). Let $\gamma(x,t)$ denote the diffusive flux,

$$\gamma(x,t) = -D\eta(\mu(x,t))_x = -Dc_x(x,t). \tag{26}$$

Using block centered finite differences in (26), $\gamma(x_{j+1/2}, t^n)$ is approximated by $\bar{\gamma}_{j+1/2}^n$, where

$$\bar{\gamma}_{j+1/2}^n = -D\frac{C_{j+1}^n - C_j^n}{h_{j+1/2}}, \quad j = 1, \ldots, J-1. \tag{27}$$

At the inflow boundary,

$$\bar{\gamma}_{1/2}^{n} = -2D\frac{C_1^n - c_0^n}{h_1}. \tag{28}$$

We discuss the handling of the outflow boundary condition (14) below.

Discretizing (19) using backward differencing in time and block centered differencing in space, and taking $\bar{w}^{n+1}$ as the initial condition, we obtain

$$\frac{w_j^{n+1} - \bar{w}_j^{n+1}}{\Delta t} + \frac{\bar{\gamma}_{j+1/2}^{n+1} - \bar{\gamma}_{j-1/2}^{n+1}}{h_j} = 0, \tag{29}$$

which holds for $j = 1, \ldots, J-1$. Substitute for $\bar{w}_j^{n+1}$ using (20) to obtain

$$\frac{w_j^{n+1} - w_j^n}{\Delta t} + \frac{\eta(\bar{w}_{j+1/2}^{n+1/2}) - \eta(\bar{w}_{j-1/2}^{n+1/2})}{h_j} + \frac{\bar{\gamma}_{j+1/2}^{n+1} - \bar{\gamma}_{j-1/2}^{n+1}}{h_j} = 0. \tag{30}$$

In block B_J, first compute $\bar{w}_J^{n+1}$ by

$$\frac{\bar{w}_J^{n+1} - w_J^n}{\Delta t} + \bar{u}\frac{\eta(\bar{w}_{J+1/2}^{n+1/2}) - \eta(\bar{w}_{J-1/2}^{n+1/2})}{h_J} = 0. \tag{31}$$

Set $C^{n+1}(\bar{x}) = \eta(\bar{w}_J^{n+1})$, and approximate the diffusive flux at $\bar{x}$ by

$$\bar{\gamma}_{J+1/2}^{n+1} = -2D\frac{C^{n+1}(\bar{x}) - C_J^{n+1}}{h_J}. \tag{32}$$

Then w_J^{n+1} is given by (30).

Substituting (17), (25), (27), (28), and (32) into (30), we obtain a tridiagonal system of equations for C_j^{n+1}, $j = 1, \ldots, J$, which is possibly nonlinear. Once C_j^{n+1} is determined, w_j^{n+1} is updated by (17).

The last step in the calculation at time t^{n+1} is the computing of the slopes, δw_j^{n+1}, $j = 1, \ldots, J$, which only appear in the characteristic tracing step. The slope in the last interval, δw_J^{n+1} is set to zero. In the remaining intervals, set

$$\delta w_j^{n+1} = \delta_{lim} w_j^{n+1} \cdot sign(w_{j+1}^{n+1} - w_{j-1}^{n+1}), \tag{33}$$

where

$$\delta_{lim} w_j^{n+1} = \begin{cases} \min(|\Delta_+ w_j^{n+1}|, |\Delta_- w_j^{n+1}|), & \text{if } \Delta_+ w_j^{n+1} \cdot \Delta_- w_j^{n+1} > 0, \\ 0, & \text{otherwise} . \end{cases} \tag{34}$$

Here $\Delta_+ w_j$ is the forward difference $(w_{j+1} - w_j)/h_{j+1/2}$. For $j = 2, \ldots, J-1$, $\Delta_- w_j = (w_j - w_{j-1})/h_{j-1/2}$, and $\Delta_- w_1 = 2(w_1 - c_0)/h_1$. The point of the procedure (33)-(34) is to compute a piecewise linear approximation without introducing new extrema into the approximate solution. Thus, in blocks where the solution already has a local extrema, the slope δw_j^{n+1} is set to zero.

Assuming

$$cfl \le \frac{1}{1+\frac{\alpha_l}{2}},$$

where

$$\alpha_l = \max_j \frac{h_j}{h_{j-1/2}},$$

the techniques given in [9] can be extended to prove that the approximate solution C^{n+1} defined above satisfies a maximum principle; that is,

$$\min\left(c_0, \min_j c^0(x_j)\right) \le C_j^{n+1} \le \max\left(c_0, \max_j c^0(x_j)\right). \tag{35}$$

Moreover, assuming linear, equilibrium adsorption (Freundlich isotherm, $p = 1$), and assuming a Dirichlet boundary condition is given at $\bar{x}$, then, for c sufficiently smooth, one can prove [9],

$$\max_n \left(\sum_j [c_j^n - C_j^n]^2 h_j\right)^{1/2} \le C(h + \Delta t), \tag{36}$$

where $h = \max_j h_j$. This pessimistic estimate says the scheme is at worst first order accurate. However, the accuracy depends on the number of local extrema in the approximate solution, that is, the number of times $\delta_{lim} w_j$ is set to zero in areas where $\mu_x \neq 0$. Heuristically, one expects this number to be small, in which case the error approaches $\mathcal{O}(h^2 + \Delta t)$. Numerical studies which verify this heuristic notion are given in [9]. In practice, we see much better performance from the scheme outlined above than from standard first-order schemes.

We can modify the definition of $\delta_{lim} w_j^{n+1}$ by

$$\delta_{lim} w_j^{n+1} = \min(|\Delta_+ w_j^{n+1}|, |\Delta_- w_j^{n+1}|). \tag{37}$$

In this case, one can prove that

$$\max_n \left(\sum_j [c_j^n - C_j^n]^2 h_j\right)^{1/2} \le C(h^2 + \Delta t); \tag{38}$$

however, (35) is no longer guaranteed to hold.

Remarks on simulating non-equilibrium adsorption. Assume $\lambda_1 = 0$ and $\lambda_2 = 1$, then we obtain the system of equations

$$c_t + \bar{u} c_x - D c_{xx} = -\rho(k_a \Phi(c) - k_s s) \equiv g(c, s), \tag{39}$$

$$s_t = k_a \Phi(c) - k_s s. \tag{40}$$

We discretize (39) as before, incorporating the source term $g(c, s)$ implicitly. We also apply an implicit discretization to (40). Thus at each time t^{n+1}, we have an implicit system of equations in $C^{n+1} \approx c^{n+1}$, and $S^{n+1} \approx s^{n+1}$. In this case, C^{n+1} is piecewise linear in space, and S^{n+1} is piecewise constant.

4 Numerical results

In this section, we present numerical results for both equilibrium and non-equilibrium adsorption assuming the adsorption kinetics is described by the Freundlich isotherm, (11). In particular, we will examine the effects of varying the expononent p.

We first consider the case of equilibrium adsorption; $\lambda_1 = 1$ and $\lambda_2 = 0$. When $p < 1$, we have a nonlinear tridiagonal system of equations to solve to determine C^{n+1}. We have solved this system numerically using a method of substitution. This approach converges slowly in some cases (especially at early time), but has the advantage that at each iteration, the approximate solution satisfies a maximum principle. Thus, we are guaranteed that the solution values stay nonnegative, which is crucial to the iteration, since c^p is undefined for negative c if $p < 1$.

In the simulations described below, we set $c_0 \equiv 1$, $\phi = .5$, $R \equiv \rho K_d/\phi = 1.5$, $u = 3\ cm/h$, $D = .20\ cm^2/h$, and $cfl = 1$. The computational domain is $0 < x \leq 100\ cm$, and the initial condition $c^0(x)$ is plotted in Figure 1.

We first consider the case $p = .8$. To test the convergence of the scheme, we compare the approximate solutions at $t = 20$ hours, generated using 25, 50, and 100 uniform grid blocks. As seen in Figure 1, the numerical solution appears to converge as the mesh and time step approach zero. This figure also shows the effect of choosing $p < 1$. Note that the initial condition has a fairly smooth front, while the solution at $t = 20$ hours has a much steeper front. This effect is not seen when $p = 1$.

In Figure 2, we compare solutions for $p = .5$, $.8$, and 1 at $t = 20$ hours. In these simulations, 100 uniform grid blocks are used. Figure 2 shows that decreasing p results in sharper fronts and substantial retardation of the solution. These results agree with the expected behavior of the solution, based on the mathematical model (12).

Next, we examine the case of non-equilibrium adsorption; $\lambda_1 = 0$ and $\lambda_2 = 1$. For these runs, we assume initially that $c^0(x) = 0$. The values of c_0, ϕ, u, and D are the same as those given above. We also set $\rho * K_d = 1.5$, and $k_a = k_s = .5$. In Figure 3, we compare the approximations to $c(x, t)$ at $t = 30$ hours, for $p = .5$, obtained using 25, 50, and 100 uniform grid blocks. This figure demonstrates that the approximate solution converges as the mesh and time step are refined.

In Figure 4, we compare solutions at 30 hours for $p = 1$, $.8$, and $.5$. The effects of reducing p in this case are more dramatic than in the equilibrium case.

Here, the solution for $p = .5$ has a much steeper front and lags substantially behind the $p = 1$ solution, while the $p = .8$ solution is intermediate. Comparing the equilibrium and non-equilibrium solutions, we see that their behavior is quite different. In particular, the solutions in the non-equilibrium case do not have the self-steepening quality that the equilibrium solutions possess (for $p < 1$).

5 Conclusions

In conclusion, the Godunov-mixed method described here gives solutions which agree with physical intuition. The method is convergent, and allows for stable and accurate approximations to contaminant transport problems.

The method has been applied to two case studies of equilibrium and nonequilibrium adsorption in one space dimension. Substantial differences in the numerical results were seen depending on the type of adsorption model chosen. Thus, the standard assumption of linear, equilibrium adsorption may not be appropriate for some contaminant species.

We have extended the numerical method discussed here to the modeling of two-dimensional, multicomponent saturated transport with biodegradation. At present, biodegradation is modeled by the Monod kinetics, and the method has been tested for three component flow involving a contaminant species, dissolved oxygen, and microorganisms. Some preliminary numerical results have been obtained, and will be presented at a later time.

The two-dimensional simulator uses a mixed finite element method to solve for flow velocity and head. We plan to extend this method to the more general case (5), which models unsaturated and saturated flow. Extending the transport simulator to include variable water content is straightforward. Thus, we hope in the near future to have a simulator capable of modeling contaminant transport through the unsaturated and saturated zones.

References

[1] J. M. Thomas, M. D. Lee, P. B. Bedient, R. C. Borden, L. W. Canter, and C. H. Ward, *Leaking underground storage tanks: remediation with emphasis on* in situ *biorestoration*, Environmental Protection Agency, 600/2-87,008, January, 1987.

[2] United States Department of Energy, *Site-directed subsurface environmental initiative, five year summary and plan for fundamental research in subsoils and in groundwater, FY1989-FY1993*, DOE/ER 034411, Office of Energy Research, April 1988.

[3] M. F. Wheeler and C. N. Dawson, *An operator-splitting method for advection-diffusion-reaction problems*, MAFELAP Proceedings VI, J. A. Whiteman, ed., Academic Press, pp. 463-482, 1988.

[4] C. Y. Chiang, C. N. Dawson, and M. F. Wheeler, *Modeling of* in-situ *biorestoration of organic compounds in groundwater*, to appear in Transport in Porous Media.

[5] C. N. Dawson and M. F. Wheeler, *Characteristic methods for modeling nonlinear adsorption in contaminant transport*, Proceedings, 8th International Conference on Computational Methods in Water Resources, Venice Italy, 1990, Computational Mechanics Publications, Southampton, U. K., pp. 305-314.

[6] J. Bear, Dynamics of Fluids in Porous Media, Dover Publications, New York, 1972.

[7] R. A. Freeze and J. A. Cherry, Groundwater, Prentice-Hall, Englewood Cliffs, New Jersey, 1979.

[8] C. J. van Duijn and P. Knabner, *Solute transport in porous media with equilibrium and non-equilibrium multiple-site adsorption: Travelling waves*, Institut für Mathematik, Universität Augsburg, Report No. 122, 1989.

[9] C. N. Dawson, *Godunov-mixed methods for advective flow problems in one space dimension*, to appear in SIAM J. Numer. Anal.

[10] C. N. Dawson, *Godunov-mixed methods for immiscible displacement*, International Journal for Numerical Methods in Fluids 11, pp. 835-847, 1990.

[11] B. van Leer, *Towards the ultimate conservative difference scheme via a second-order sequel to Godunov's method*, J. Comput. Phys. 32, pp. 101-136, 1979.

[12] P. A. Raviart and J. M. Thomas, *A mixed finite element method for 2nd order elliptic problems*, in Mathematical Aspects of the Finite Element Method, Rome 1975, Lecture Notes in Mathematics, Springer-Verlag, Berlin, 1977.

[13] T. F. Russell and M. F. Wheeler, *Finite element and finite difference methods for continuous flow problems*, in The Mathematics of Reservoir Simulation (R. E. Ewing, ed.), Frontiers in Science, SIAM, Philadelphia, 1983.

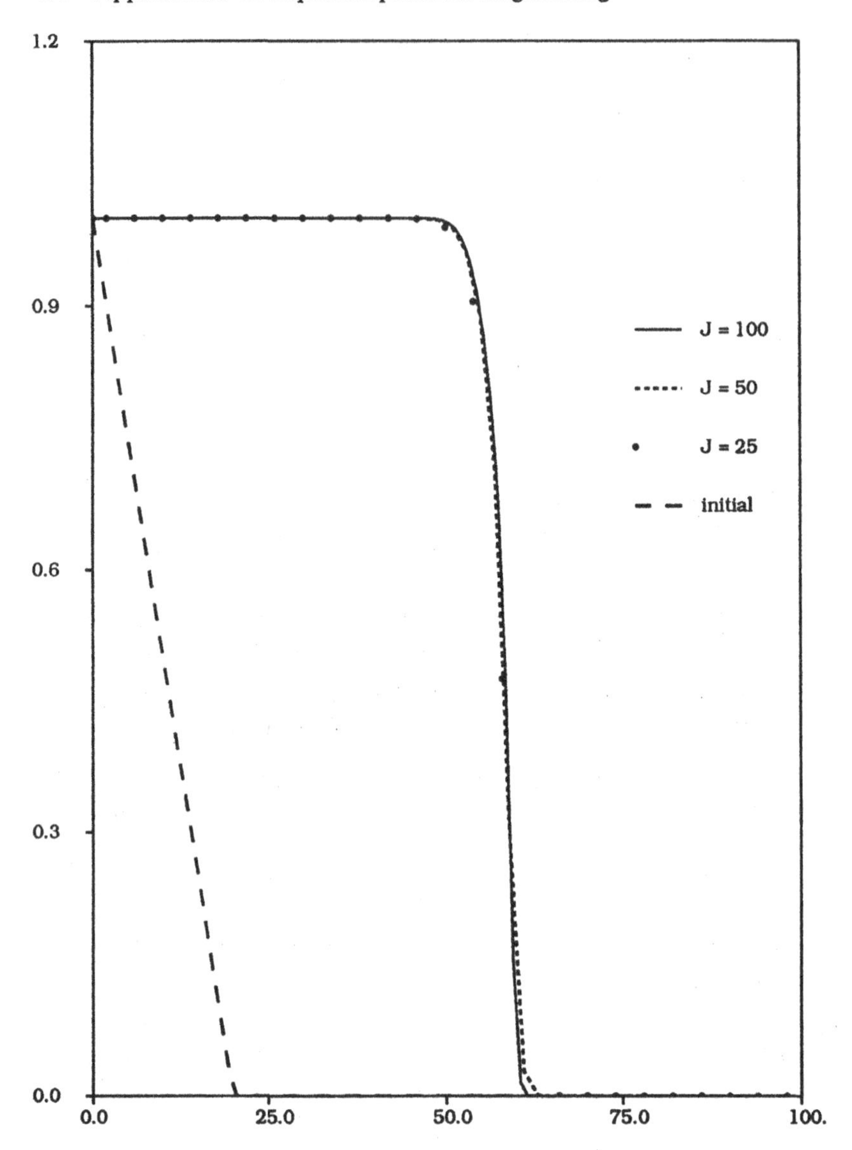

Figure 1: Test of convergence for $p = .8$.

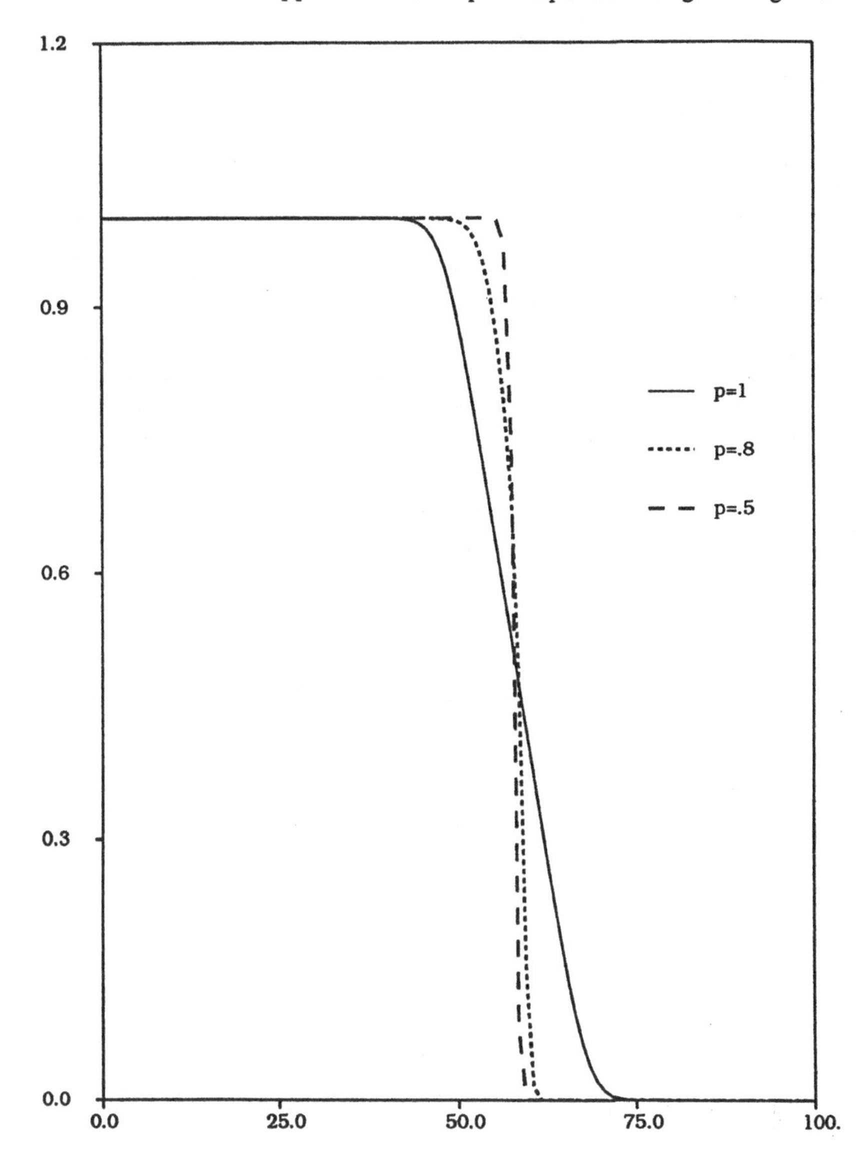

Figure 2: Comparison of $p = .5, .8$, and 1.

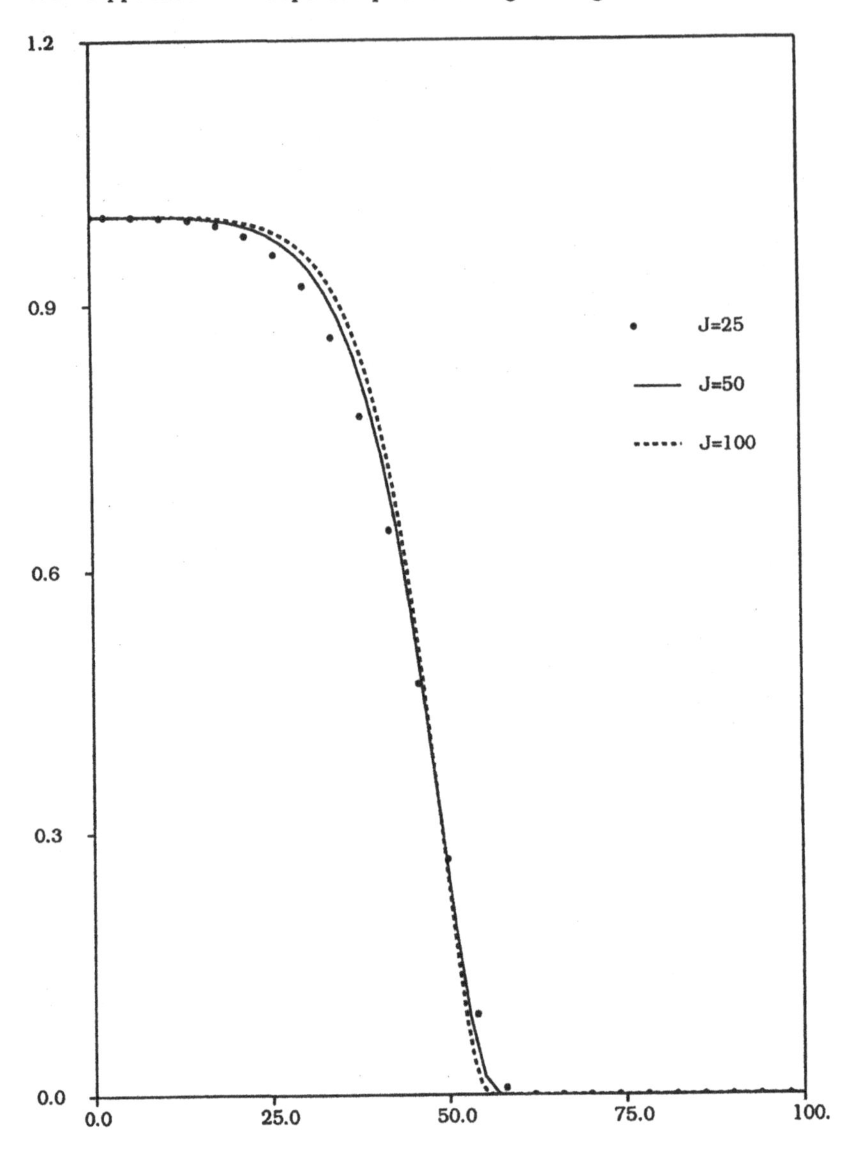

Figure 3: Test of convergence for non-equilibrium case, $p = .5$

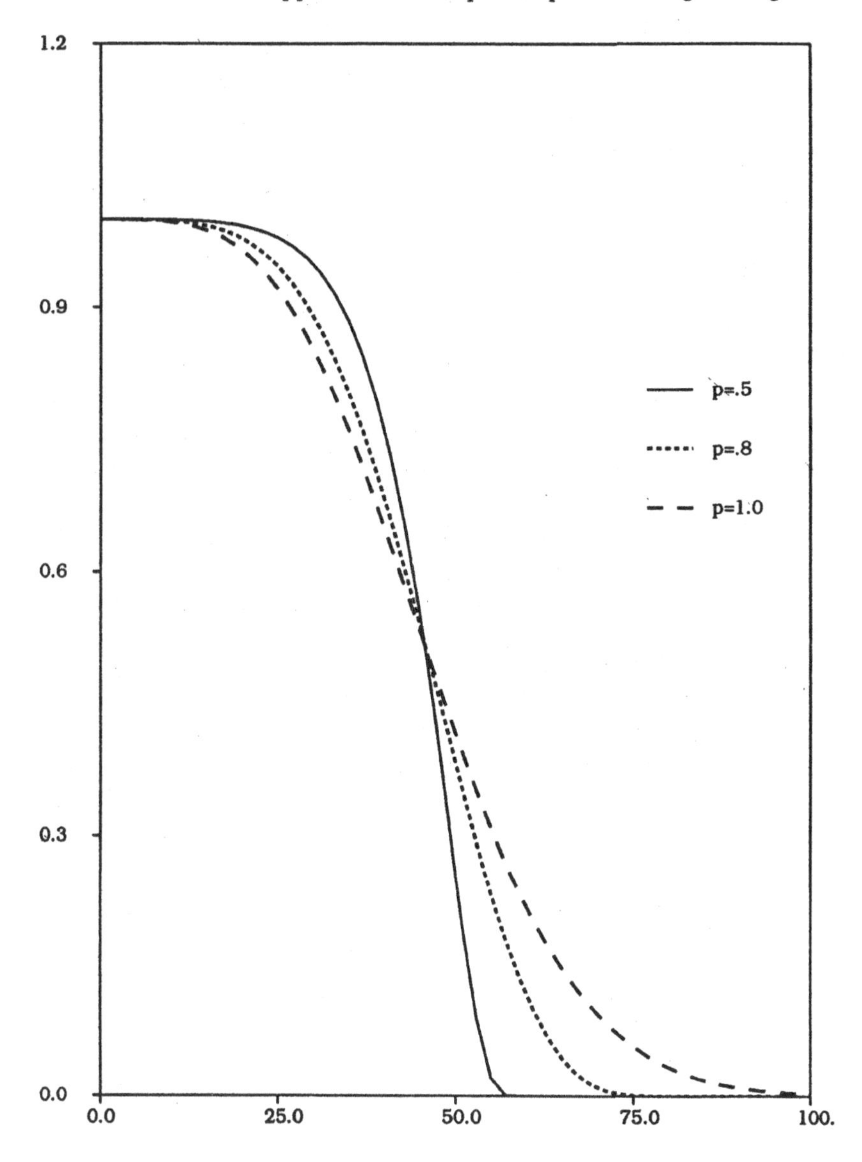

Figure 4: Comparison of $p = .5, .8$, and 1., nonequilibrium case

Time-Parallel Multigrid Solution of the Navier-Stokes Equations

G. Horton

Institut für Mathematische Maschinen und Datenverarbeitung III, Universität Erlangen-Nürnberg, D-8520 Erlangen, Germany

Abstract

We consider the problem of solving time-dependent partial differential equations on a MIMD computer. Conventional methods for the solution of this type of equation when discretized by an implicit method such as backward Euler proceed by solving a sequence of problems iteratively. It is shown that despite the sequential nature of this process, several processors may be employed to solve at several time-steps simultaneously, a technique which we shall refer to as *time-parallelism*. Details of a time-parallel multigrid method due to Hackbusch and of a smoothing technique applicable to the two-dimensional unsteady incompressible Navier-Stokes equations are given. Results obtained on a parallel Transputer system are presented.

1. Introduction

The solution of large non-linear systems of unsteady partial differential equations is one of the major classes of computationally intensive problems encountered in scientific and technical fields. For this reason, they have been studied intensively with a view to solution on parallel processors. Most approaches derive their parallelism from a subdivision of the computational domain in the space directions, yielding for example grid partitioning or domain decomposition methods. However, many unsteady problems require a number of grid points in the time direction which greatly exceeds that of the space directions. Thus it seems natural to ask whether parallelism can additionally be achieved along the time axis.

A new approach for parabolic equations based on the multigrid technique was introduced by Hackbusch in 1984 [5], where it was shown how the multigrid idea can be extended to apply to a sequence of time-steps simultaneously, as opposed to the standard solution procedure, whereby successive time-steps are treated in a sequential manner. In the following year, Burmeister [2] analysed the one-dimensional unsteady heat equation, obtaining convergence results which showed that the rate of convergence is independent of the number of time-steps to be solved in parallel. This encouraged further investigation of the problem, including a parallel implementation on the memory-coupled nearest-neighbour multiprocessor DIRMU by Bastian, Burmeister and Horton [1]. Subsequently Horton and Knirsch [8] have shown how the time-parallel approach can be used to enable the efficient parallel implementation of the extrapolation method to parabolic p.d.e.s, leading to a further increase in parallelism and a considerable increase in accuracy. Burmeister and Horton [3] have presented results for the time-parallel solution of the Navier-Stokes equations which will be extended in this paper to a larger number of processors and using a more efficient smoothing scheme.

We consider the two-dimensional incompressible Navier-Stokes equations :

$$\begin{aligned}\frac{\partial(\rho u)}{\partial t}+\frac{\partial}{\partial x}(\rho u^2-\mu\frac{\partial u}{\partial x})+\frac{\partial}{\partial y}(\rho uv-\mu\frac{\partial u}{\partial y}) &= -\frac{\partial p}{\partial x}+f^x\\ \frac{\partial(\rho v)}{\partial t}+\frac{\partial}{\partial x}(\rho uv-\mu\frac{\partial v}{\partial x})+\frac{\partial}{\partial y}(\rho v^2-\mu\frac{\partial v}{\partial y}) &= -\frac{\partial p}{\partial y}+f^y\\ \frac{\partial(\rho u)}{\partial x}+\frac{\partial(\rho v)}{\partial y} &= 0\end{aligned}$$

Note the particular form of the continuity equation due to the assumption of incompressibility $\rho = const.$ We further assume constant viscosity $\mu = const.$

In the following section, the non-linear time-parallel multigrid method is presented in a general form. The restriction and prolongation operators are seen to be similar to standard procedures, whereas the smoothing part has to be studied in more detail. The smoothing procedure is based on the well-known SIMPLE procedure of Patankar and Spalding [9], and is coupled with a pointwise ILU decomposition method due to Stone [10] for the solution of the resulting systems.

In section 3 the special time-parallel SIMPLE-like smoothing procedure is considered in detail, showing how the standard method can be extended to solve a set of Navier-Stokes problems at successive time-steps. It will be seen that all operations may be carried out in parallel across the time-steps.

Methods such as the time-parallel scheme presented here have been called "parallelization of the method", as opposed to the standard "parallelization of the data" type grid partitioning algorithms. One characteristic of such a method is that its numeric behaviour is dependent on the number of processors used. This motivates a more detailed analysis of the efficiency obtained by the method, which is outlined in section 5.

A test case based on the driven-cavity model problem is briefly described and results are presented for a variety of parameters. The results are discussed based on the concept of splitting the total efficiency obtained into numerically dependent and implementation-dependent components.

2. Time-Parallel Multi-Grid Method

Let a nonlinear partial differential equation for an unknown time and space dependent function $u = u(t, x, y)$ be given

$$\frac{\partial u}{\partial t}+\mathcal{L}(u)=q(t,x,y), \quad (x,y)\in\Omega\subset R^2, \quad 0<t\leq T\,.$$

The problem is assumed to have suitable boundary conditions and initial values for $t = 0$. Following the method of lines, the problem is discretized first in space by a

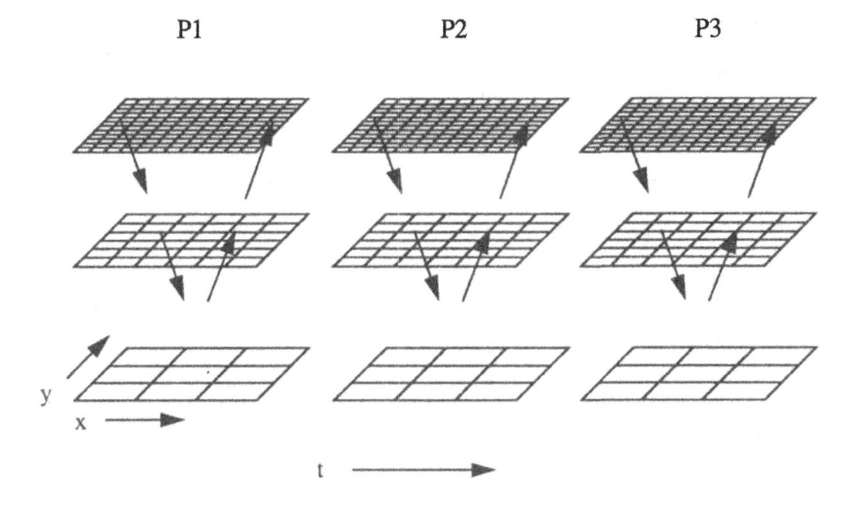

Figure 1: Time-Parallel Multigrid Method

discretization method like finite elements, finite differences or finite volumes. For this purpose a hierarchy of grids is used, indicated by a level $l \in \{0, \ldots, lmax\}$. The time integration is performed using the backward Euler formula. For ease of presentation, the time-step Δt is fixed.

The resulting discrete counterpart of the nonlinear problem for gridlevel l on a fixed time interval $[t_k, t_{k+m}]$ reads as follows. Using a given grid-function $u_{l,k}$, solve the set of discrete nonlinear problems

$$\tfrac{1}{\Delta t} u_{l,k+j} + \mathcal{L}_l \left(u_{l,k+j} \right) = q_{l,k+j} + \tfrac{1}{\Delta t} u_{l,k+j-1} \ , \ j = 1, \ldots, m \ .$$

Here $u_{l,k+j}$ denotes a grid function which is defined on a grid with mesh size h_l at time $t_{k+j} = (k+j)\Delta t$.

The usual solution procedure is to calculate the unknown grid functions $u_{l,k+1}, u_{l,k+2}, \ldots$ in a sequential manner. Since the discretization of the time derivative has been chosen to be implicit, this must be done in an iterative manner. Multigrid methods can, for example, be applied to the discrete nonlinear problem at each individual time-step, see [6]. Standard parallel techniques divide up the grid at each time-step among the processors in an attempt to perform each iteration in a shorter time. The sequential nature of the integration along the time axis is, however, retained.

Time-parallel methods may be derived by simply writing down the equations for several successive time-steps at once and applying an iterative solver to the extended system and by assigning one time-step to each processor. Speedup is achieved in this case not by accelerating the computation of each iteration, but by reducing the total number of iterations to be performed, each of which takes approximately the same time as in the serial case.

Following the ideas proposed in a paper by Hackbusch [5] and the implementation strategy described in [1], the set of nonlinear problems will be solved iteratively by means of multigrid techniques. In the following the nonlinear two-grid version will be discussed.

The i-th iterates $u_{l,k+j}^{(i)}$, $j = 1, \ldots, m$ may be given or already be calculated. The algorithm is multigrid in space only, that is, no coarsening is performed in the time direction. Figure 1 shows schematically the time-parallel multigrid algorithm for three processors.

The smoothing procedure, to be described for the Navier-Stokes case in the next section, is denoted by $\mathcal{S}_l$. The application of $\mathcal{S}_l$ ν-times will be denoted by $\mathcal{S}_l^{\nu}$.

Nonlinear Time-Parallel Multigrid Algorithm

For all time-steps $j = 1 \ldots m$ do :

- Pre-smoothing

$$u_{l,k+j}^{(i+\frac{1}{3})} = S_l^{\nu_1}\left(u_{l,k+j}^{(i)}, A_l(u_{l,k+j}^{(i)}), u_{l,k+j-1}^{(I)}\right)$$

- Compute defect

$$d_{l,k+j} = \left(D_l + A_l(u_{l,k+j}^{(i+\frac{1}{3})})\right) u_{l,k+j}^{(i+\frac{1}{3})} - q_{l,k+j} - D_l u_{l,k+j-1}^{(i+\frac{1}{3})}$$

- Restriction (with restriction operator r)

$$d_{l-1,k+j} = r d_{l,k+j}$$
$$u_{l-1,k+j} = r u_{l,k+j}^{(i+\frac{1}{3})}$$

- Solve coarse grid equation for $v_{l-1,k+j}$

$$\begin{aligned}(D_{l-1} + A_{l-1}(u_{l-1,k+j}))\, v_{l-1,k+j} &= d_{l-1,k+j} + D_{l-1} v_{l-1,k+j-1} \\ &\quad + (D_{l-1} + A_{l-1}(u_{l-1,k+j})) u_{l-1,k+j} \\ &\quad - D_{l-1} u_{l-1,k+j-1}\end{aligned}$$

- Prolongation and Correction (with prolongation operator p)

$$u_{l,k+j}^{(i+\frac{2}{3})} = u_{l,k+j}^{(i+\frac{1}{3})} - p(v_{l-1,k+j} - u_{l-1,k+j})$$

- Post-smoothing

$$u_{l,k+j}^{(i+1)} = S_l^{\nu_2}\left(u_{l,k+j}^{(i+\frac{2}{3})}, A_l(u_{l,k+j}^{(i+\frac{2}{3})}), u_{l,k+j-1}^{(I)}\right)$$

The algorithm can easily be extended to a multi-grid version by solving the coarse grid equation (a discrete nonlinear problem at level $l-1$) iteratively with starting guess $v_{l-1,k+j} := \tilde{u}_{k+j}$, $j = 1, \ldots, m$. For the choice $m = 1$ (one time-step to solve) the above algorithm reduces to the *full approximation storage* algorithm (FAS) . Main parts of the algorithm like defect calculation, restriction, prolongation and correction may be chosen in a standard way and moreover can be computed independently, and therefore in parallel in different time-steps.

The generalization of the above time-parallel multigrid method to nonlinear systems of equations is straightforward and can be done for the discrete counterpart of the incompressible Navier-Stokes equations

$$\begin{array}{rcl} D_l^x u_{l,k+j} + Q_l^x(u_{l,k+j}, v_{l,k+j}) + G_l^x p_{l,k+j} & = & f_{l,k+j}^x + D_l^x u_{l,k+j-1} \\ D_l^y v_{l,k+j} + Q_l^y(u_{l,k+j}, v_{l,k+j}) + G_l^y p_{l,k+j} & = & f_{l,k+j}^y + D_l^y v_{l,k+j-1} \\ -K_l^x u_{l,k+j} - K_l^y v_{l,k+j} & = & 0 \end{array}$$

$$, \; j = 1, \ldots, m$$

The nonlinear operators $Q_l^x(\cdot,\cdot)$ and $Q_l^y(\cdot,\cdot)$ include the nonlinearity of the Navier-Stokes equations, whilst the linear operators (matrices) G_l^x, G_l^y and K_l^x, K_l^y are discrete analogues of the gradient and the divergence operators. When finite differences are used, the diagonal matrices D_l^x, D_l^y become $\frac{1}{\Delta t} I_l$. In the case of finite volumes, the diagonal matrices may have varying coefficients. The discretization technique is the same as that described in [7],[8].

The particular smoothing procedure S_l applicable to the discrete Navier-Stokes-problem will be described in the following section, identifying in particular the time-parallelizability of the various components.

3. Time-Parallel SIMPLE Algorithm

The equations in this section are considered for a specific grid level l, which will be omitted henceforth for clarity. For ease of presentation let the set of unknowns denoted by $(u_j, v_j, p_j)^T$, $j = 1, \ldots, m$.

The SIMPLE-method (Semi-Implicit Method for Pressure-Linked Equations) was introduced by Patankar and Spalding [9] and is widely used as an iteration scheme for the solution of Navier-Stokes problems, both directly and as a smoother in multigrid solvers. The arising nonlinear problem for a fixed time-step t_j

$$\begin{array}{rcl} D^x u_j + Q^x(u_j, v_j) + G^x p_j & = & f_j^x + D^x u_{j-1} \\ D^y v_j + Q^y(u_j, v_j) + G^y p_j & = & f_j^y + D^y v_{j-1} \\ -K^x u_j - K^y v_j & = & 0 \end{array}$$

can be interpreted as a stationary Navier-Stokes problem enhanced with diagonal entries reflecting the time discretization.

One iteration step of the SIMPLE-method consists of a linearization of the nonlinear problem, a factorization of the system matrix and the sparse approximation of dense matrix inverses. The linearization is performed by a fixpoint iteration, where values from the previous iteration are used to determine the coefficients :

$$\begin{bmatrix} D^x + Q^x(u_j^{old}, v_j^{old}) & 0 & G^x \\ 0 & D^y + Q^y(u_j^{old}, v_j^{old}) & G^y \\ -K^x & -K^y & 0 \end{bmatrix} \begin{bmatrix} u_j^{new} \\ v_j^{new} \\ p_j \end{bmatrix} = \begin{bmatrix} f_j^x + D^x u_{j-1} \\ f_j^y + D^y v_{j-1} \\ 0 \end{bmatrix}$$

By grouping together the velocity grid-functions, the resulting system matrix is algebraically factorized

$$\left[\begin{array}{cc|c} Q_j^x & 0 & G^x \\ 0 & Q_j^y & G^y \\ \hline -K^x & -K^y & 0 \end{array}\right] =: \begin{bmatrix} Q_j & G \\ -K & 0 \end{bmatrix} = \begin{bmatrix} Q_j & 0 \\ -K & KQ_j^{-1}G \end{bmatrix} \begin{bmatrix} I & Q_j^{-1}G \\ 0 & I \end{bmatrix} \quad ;$$

whereby the abbreviations $Q_j^x := D^x + Q^x(u_j^{old}, v_j^{old})$ and $Q_j^y := D^y + Q^y(u_j^{old}, v_j^{old})$ are used.

Because Q_j^{-1} is a dense matrix, a sparse approximation R_j is used instead. The original SIMPLE-method is defined by using the diagonal part of Q_j

$$R_j^{-1} := \mathrm{diag}(Q_j)$$

yielding the approximate sparse factorization

$$\begin{bmatrix} Q_j & G \\ -K & 0 \end{bmatrix} \approx \begin{bmatrix} Q_j & 0 \\ -K & KR_jG \end{bmatrix} \begin{bmatrix} I & R_jG \\ 0 & I \end{bmatrix}$$

To obtain the time-parallel SIMPLE method, consider the extended system obtained by writing the discrete Navier-Stokes problem for m time-steps together:

$$\begin{bmatrix}
Q_1^x & 0 & G^x & & & & & & \\
0 & Q_1^y & G^y & & & & & & \\
-K^x & -K^y & 0 & & & & & & \\
-D^x & 0 & 0 & Q_2^x & 0 & G^x & & & \\
0 & -D^y & 0 & 0 & Q_2^y & G^y & & & \\
0 & 0 & 0 & -K^x & -K^y & 0 & & & \\
 & & & \ddots & & \ddots & & & \\
 & & & -D^x & 0 & 0 & Q_m^x & 0 & G^x \\
 & & & 0 & -D^y & 0 & 0 & Q_m^y & G^y \\
 & & & 0 & 0 & 0 & -K^x & -K^y & 0
\end{bmatrix}
\begin{bmatrix} u_1 \\ v_1 \\ p_1 \\ u_2 \\ v_2 \\ p_2 \\ \vdots \\ u_m \\ v_m \\ p_m \end{bmatrix} = \cdots$$

Analog to the standard SIMPLE method, an approximate factorization, which ignores some fill-in, is performed on the system to obtain the block matrices L and U:

$$L = \begin{bmatrix}
Q_1 & 0 & & & & \\
-K & KR_1G & & & & \\
-D & 0 & Q_2 & 0 & & \\
0 & 0 & -K & KR_2G & & \\
 & & \ddots & & \ddots & \\
 & & -D & 0 & Q_m & 0 \\
 & & 0 & 0 & -K & KR_mG
\end{bmatrix}$$

$$U = \begin{bmatrix}
I & R_1G & & & & \\
0 & I & & & & \\
 & & I & R_2G & & \\
 & & 0 & I & & \\
 & & & & \ddots & \\
 & & & & I & R_mG \\
 & & & & 0 & I
\end{bmatrix}$$

Note that owing to the structure of the matrix U, the systems at different time-steps in this equation are independent of each other and may therefore be solved in parallel at each time-step. The block L equation however contains dependencies between the time-steps via the diagonal matrices D, which must be considered more carefully with a view to parallelization.

The block L system can be rearranged, grouping the u, v and p equations together, and using the temporary storage vectors u^{temp}, v^{temp} and p^{temp} :

$$\begin{bmatrix} Q_1^x & & & \\ -D^x & Q_2^x & & \\ & \ddots & \ddots & \\ & & -D^x & Q_m^x \end{bmatrix} \begin{bmatrix} u_1^{temp} \\ u_2^{temp} \\ \vdots \\ u_m^{temp} \end{bmatrix} = \begin{bmatrix} f_1^x + D^x u_0 \\ f_2^x \\ \vdots \\ f_m^x \end{bmatrix}$$

$$\begin{bmatrix} Q_1^y & & & \\ -D^y & Q_2^y & & \\ & \ddots & \ddots & \\ & & -D^y & Q_m^y \end{bmatrix} \begin{bmatrix} v_1^{temp} \\ v_2^{temp} \\ \vdots \\ v_m^{temp} \end{bmatrix} = \begin{bmatrix} f_1^y + D^y v_0 \\ f_2^y \\ \vdots \\ f_m^y \end{bmatrix}$$

$$\begin{bmatrix} KR_1G & & & \\ & KR_2G & & \\ & & \ddots & \\ & & & KR_mG \end{bmatrix} \begin{bmatrix} p_1^{temp} \\ p_2^{temp} \\ \vdots \\ p_m^{temp} \end{bmatrix} = \begin{bmatrix} Ku_1^{temp} + Ku_1^{temp} \\ Ku_2^{temp} + Kv_2^{temp} \\ \vdots \\ Ku_m^{temp} + Kv_m^{temp} \end{bmatrix}$$

Since the matrix of the latter equation is block-diagonal, the p^{temp} equations at different time-steps are independent of each other and may thus be processed in parallel. However they depend on the new values of both u^{temp} and v^{temp} at the same time-step, and must therefore be processed after these two equations.

Note that the u^{temp} and v^{temp} equations are independent of each other and moreover that they have the same structure, both corresponding to unsteady linear scalar convection-diffusion type equations.

The problem of time-parallelization of the Navier-Stokes problem thus reduces to the parallel solution of unsteady linear scalar convection-diffusion type equations. By treating the block-off-diagonals D^x and D^y explicitly, i.e. using values of the preceding time-step from the preceding iteration, the matrices Q may be treated in the same way as for the standard, non-parallel case. In this work, the incomplete decomposition due to Stone [10] was used to invert the matrices Q. Note that the diagonal matrices D may also be included in the decomposition, which results however in a solution process in which grid point values must be communicated individually between neighbouring time-steps, see Burmeister and Horton [3]. This variant will, for the purposes of this paper, not be further considered.

5. Measurement of Efficiency

The time-parallel method presented in this paper is an example of an algorithm with computational redundancy. This means that the number of operations performed by the parallel algorithm may exceed that of the serial method. In the time-parallel case, this is due to the possible deterioration of the convergence rate as the system is enlarged.

Thus one may expect, in addition to the losses in efficiency incurred by synchronization and communication, a further loss owing to the additional work carried out. In order to identify these losses in efficiency more precisely, the total efficiency of the algorithm E_{tot} may be considered as a product of the *numeric* efficiency E_{num} and the *parallel* efficiency E_{par} defined as follows:

$$\begin{aligned} E_{par} &= \frac{T_{Iter}(1)}{T_{Iter}(p)} \\ E_{num} &= \frac{\#Iter(1)}{p * \#Iter(p)} \\ E_{tot} &= E_{num} * E_{par} \end{aligned}$$

where $T_{Iter}(p)$ represents the computation time for one iteration on p processors, and $\#Iter(p)$ the total number of iterations required by p processors to perform the integration.

Numeric efficiency describes the increase in the number of iterations needed by the parallel method to integrate a certain number of time-steps to a specified accuracy compared to the serial time-step-by-time-step approach. Parallel Efficiency is defined as the ratio of computation time for one iteration on one and on p processors. It describes losses due to communication overhead and synchronization.

The classically defined efficiency

$$E = \frac{T(1)}{p * T(p)}$$

where $T(p)$ represents the total computation time on p processors, was found to be almost equal to E_{tot}, differing only as a result of the small overhead incurred in the parallel algorithm between iterations. The results presented in the following section include both parallel and numeric efficiency, and a discussion on the factors that influence them.

6. Model problem and Results

The model problem used to test the algorithm is a variation on the standard Driven Cavity problem, where the lid of the cavity has velocity $u = \sin(t)$, instead of the constant velocity assumed for the steady case. The problem was discretized with a 64×64 finest grid using 5 grid levels and solved using V-cycles with $\nu_1 = \nu_2 = 3$. Tests were carried out for Reynolds numbers 1 and 1000 based on cavity width and the maximum lid velocity and for $\Delta t = \frac{\pi}{10}$ and $\Delta t = \frac{\pi}{1000}$. Table 1 gives the results obtained with a parallel Transputer System on up to 32 processors. T_{tot} is the total computation time in seconds, and T_{Iter} the computation time per iteration in seconds.

The parallel efficiency for a given number of processors is identical for each problem, since the arithmetic operations performed are the same in every case. It is seen to vary between 99% and 92% on 2 and 32 processors respectively. These efficiencies, which one may have expected to be considerably lower, given the number of grid points to be communicated in each iteration, reflect the fact that the communication fraction of the method is, in fact, relatively small. This is particularly due to the fact that, since an incompressible equation was solved, there is no dependency between time-steps in the pressure correction equation and thus a large number of arithmetic operations may be performed without the need to communicate. The high parallel efficiency is also due to the good ratio of communication bandwidth to floating point speed of the T800 Transputer.

Δt	Re	p	$\#Iter$	T_{Tot}	T_{Iter}	E_{num}	E_{par}	E_{tot}	E
$\frac{\pi}{10}$	1	1	1115	66395	59.2	1.00	1.00	1.00	1.00
		2	559	33672	60.1	1.00	0.99	0.99	0.99
		4	283	17432	60.8	0.98	0.97	0.93	0.95
		8	147	9027	61.1	0.95	0.97	0.95	0.92
		16	150	9321	61.7	0.93	0.96	0.89	0.89
		32	152	9736	64.2	0.92	0.92	0.85	0.85
	1000	1	911	54256	59.2	1.00	1.00	1.00	1.00
		2	459	27661	60.1	0.99	0.99	0.98	0.98
		4	235	14470	60.8	0.97	0.97	0.94	0.94
		8	123	7557	61.1	0.93	0.97	0.90	0.90
		16	127	7927	61.7	0.90	0.96	0.86	0.86
		32	129	8280	64.2	0.88	0.92	0.81	0.82
$\frac{\pi}{1000}$	1	1	896	53347	59.2	1.00	1.00	1.00	1.00
		2	448	27001	60.1	1.00	0.99	0.99	0.99
		4	226	13956	60.8	0.99	0.97	0.96	0.96
		8	118	7244	61.1	0.95	0.97	0.92	0.92
		16	122	7657	61.7	0.92	0.96	0.88	0.87
		32	125	8023	64.2	0.90	0.92	0.83	0.83
	1000	1	576	34295	59.2	1.00	1.00	1.00	1.00
		2	288	17365	60.1	1.00	0.99	0.99	0.99
		4	145	8849	60.8	0.99	0.97	0.96	0.97
		8	75	4619	61.1	0.96	0.97	0.93	0.93
		16	78	4875	61.7	0.92	0.96	0.88	0.88
		32	83	5365	64.2	0.87	0.92	0.80	0.80

Table 1: Efficiency results for the time-parallel method

The numeric efficiency is seen to be only weakly dependent on the values of the parameters Δt and Re that were chosen and is in all cases better than or equal to 87% in the 32 processor case.

The computed total efficiency E_{tot}, given in the penultimate column of the table, is seen to correspond closely to the classical efficiency E (final column). This shows that the costs incurred between iterations, in particular those associated with advancing the computation to a new set of time-steps, is comparatively small.

7. Discussion and Conclusions

In this paper a non-linear multigrid method based on the parabolic scheme of Hackbusch was used to solve the two-dimensional, incompressible, unsteady Navier-Stokes equations, which allows several time-steps to be processed simultaneously. Attention was focussed on the smoothing procedure, which is based on the well-known SIMPLE method. Experimental results show the behaviour of the method for various parameters and indicate high efficiencies ($\geq$ 80%) for up to 32 processors. The total efficiency was considered as a product of the parallel efficiency and the numeric efficiency, enabling a better understanding of losses incurred by communication and by a slight increase in the number of iterations needed in the parallel case.

The communication requirement of the time-parallel method is much higher than that of a standard grid partitioning approach. In the former, values of all grid points must be transferred to a neighbouring processor during an iteration, whilst in the latter case, only the values of edge points must be communicated. Thus the communicated data volume is considerably higher. The grids must be transferred in order to perform the smoothing and to calculate the defects which can be achieved with one transfer operation each. Each processor has only one send and one receive operation to perform during each communication phase, whilst in the standard grid partitioning case each subgrid has at least four neighbours, for each of which both send and receive operations must be carried out. Thus the time-parallel scheme, whilst having a higher communication data volume, requires a smaller number of messages, which is favourable for parallel machines with a high message latency or synchronization cost.

One additional advantage of the time-parallel approach over standard parallel methods is the retainment of larger vector lengths. Grid partitioning schemes suffer from the reduction of vector lengths incurred by the subdivision of the computational grid. The resulting loss in computing speed contradicts to an extent the object of the parallelization, namely the effective increase the available power of the machine. This is not true, however, for time-parallel methods, which assign a complete space grid to each processor, thus preserving the vector lengths of the original problem. In this respect time-parallel methods are evidently superior.

Several of the problems often associated with the implementation of space-parallel methods, such as load-balancing, restructuring of sequential code, and the mapping of the problem onto the parallel machine, are greatly simplified for the time-parallel case. Load balancing is obviously trivial, since each processor is assigned the complete problem at one time-step. Similarly, the time-parallel code is obtainable with little modification of an existing serial program. The topological requirements of the algorithm are again, minimal, since the discretization of the time derivative only requires communication between successive time-steps, which results in a linear processor configuration.

Although the time-parallel method presented here obtains high efficiencies, as shown in the previous section, it does not necessarily need to be considered solely as an alternative to standard space-parallel techniques. Since both methods may be combined, a substantial additional parallelism may be achieved at possibly higher efficiency than were either method to be used alone. Consider the case where the efficiency of a standard method drops sharply for a large number of processors because of unequal load, short vector lengths, or simply too high a communication-to-computation ratio. In such a case, a combination with time-parallelism may prove to be substantially more efficient.

Time parallelism is a general approach, which may in principle be applied to any unsteady partial differential equation, regardless of the space discretization technique used, as long as an implicit method such as Backward Euler or Crank-Nicholson is applied to the time derivative. It has been shown for the Navier-Stokes equations that high efficiencies are obtainable, and that the approach has a number of advantages over the standard space-parallel technique.

Acknowledgement

This work was supported by the Stiftung Volkswagenwerk within its program "Entwicklung von Berechnungsverfahren für Probleme der Strömungstechnik".

References

1. Bastian, P., Burmeister, J., Horton, G.: *Implementation of a parallel multigrid method for parabolic partial differential equations.*, in Hackbusch, W.(ed.): *Parallel algorithms for pdes*, Proceedings of the 6th GAMM-Seminar Kiel, January 19-21, 1990, Vieweg-Verlag, Wiesbaden, 1990.

2. Burmeister, J.: *Paralleles Lösen diskreter parabolischer Probleme mit Mehrgittertechniken.* Diplom thesis, Kiel, 1985.

3. Burmeister, J., Horton, G., *Time-Parallel Multigrid Solution of the Navier-Stokes Equations.*, in W. Hackbusch, U. Trottenberg (eds.) : Proceedings of the 3rd European Multigrid Conference, Bonn 1990, Birkhäuser Verlag, 1991.

4. Demirdzic, I., Peric, M., *Finite Volume Method for Prediction of Fluid Flow in Arbitrarily Shaped Domains with Moving Boundary.* Int. J. Num. Meth. Fluids, Vol. 10, pp. 771-790 (1990).

5. Hackbusch, W.: *Parabolic Multi-grid Methods.* In : Glowinski, R., Lions, J.-R.(eds.): *Computing Methods in Applied Sciences and Engineering, VI.* Proceedings of the 6th International Symposium on Computing Methods in Applied Sciences and Engineering. Versailles, France, December 12-16, 1983. North Holland 1984.

6. Hackbusch, W.: *Multi-grid Methods and Applications.* Heidelberg: Springer-Verlag 1985.

7. Hortmann, M., Peric, M., Scheuerer, G., *Finite Volume Multigrid Prediction of Laminar Natural Convection: Bench-Mark Solutions.*, Int. J. Num. Meth. Fluids, Vol. 11, pp. 189-207 (1990).

8. Horton G., Knirsch R., *Time-Parallel Multigrid in Extrapolation Method for Time-Dependent Partial Differential Equations.* Report 1/91, IMMD3, University of Erlangen-Nürnberg, 1991.

9. Patankar S., Spalding D.B., *A Calculation Procedure for Heat, Mass and Momentum Transfer in Three-Dimensional Parabolic Flows*, Int. J. Heat Mass Transfer, 15, 1972.

10. Stone, H. L., *Iterative Solution of Implicit Approximations of Multi-Dimensional Partial Differential Equations*, SIAM J. Num. Anal. 5, 1968.

A Finite Element Formulation for Nonlinear Wave Propagation with Application to Cavitation

G. Sandberg

Division of Structural Mechanics, Lund Institute of Technology, Lund University, P.O. Box 118, S-221 00 Lund, Sweden

SUMMARY

A new formulation for nonlinear, elastic wave propagation, based on the pressure and density, is introduced. The formulation accounts for a nonlinear relation between speed of sound and density. The pressure and density are interpolated by identical shape functions between the nodes. The equation of state is introduced in the discretized system, hence the nonlinearity is enforced point wise. This procedure concentrates on proper interpolation of the balance equations by relaxing the nonlinear constitutive relation between the nodes. A standard finite element procedure focuses on the constitutive equation and thereby produces a less consistent representation of the balance equation. In addition, the formulation accounts for fluid-structure interaction.

INTRODUCTION

We consider a fluid domain idealized as a nonlinear acoustic medium. The fluid interacts dynamically with a flexible structure. The objective is to describe a general method to account for a nonlinear equation of state.

An acoustic medium has only a limited interval in which the speed of sound can be considered as a constant in engineering applications. Beyond this interval the speed of sound is dependent on the density and also the rate of change of density. In particular the fluid can be considered incapable of transmitting tensile stresses. Such occurs when a pressure wave impinges on a flexible structure. The resulting scattered wave carries high negative pressure, which in fact can drop below zero absolute pressure, if the ambient hydrostatic pressure is sufficiently low. The result is the formation of a cavitated region, physically consisting of micro bubbles. See Cole[1] and Plesset[2].

Cavitating fluids due to underwater shocks were first given a finite element formulation by Newton[3–6] using the displacement potential, ψ , in the fluid. Later, Zienkiewicz et al[7] studied the effect on dams in connection with earthquakes. Felippa et al[8] reported studies on hull cavitations using staggered time integration routines and special boundary techniques.

Studies by Newton[3–6] indicated that the cavitating fluid might cause spurious pressure oscillations that generate pressurized zones within the cavitated region and vice versa. This phenomenon has been named frothing. The remedy is to introduce numerical damping, see Felippa et al[8]. The introduction of numerical damping calls for the calculation of, basically, $\partial^3\psi/\partial t^3$, which is not naturally present in the time stepping routine when using ψ as a primary variable. On the other hand, when ρ is used as primary variable, numerical damping is generated via $\partial\rho/\partial t$, which is naturally present in the calculation. Above all, ρ is a physical variable allowing more general constitutive laws, and boundary conditions that naturally occur, such as pressurized boundaries, need no special treatment.

BASIC FLUID EQUATIONS

With the nonlinear acoustic fluid hypothesis, the fluid is described as

$$\frac{\partial\rho}{\partial t} + \rho_s\nabla\cdot v = q \qquad \text{in } \Omega, \tag{1}$$

$$\rho_s\frac{\partial v}{\partial t} + \nabla p = \rho_s b \qquad \text{in } \Omega. \tag{2}$$

$$\nabla p\cdot\mathbf{n} = \frac{\partial v}{\partial t}\cdot\mathbf{n} = \ddot{u}_s\cdot\mathbf{n} \quad \text{on } \Gamma \tag{3}$$

$$p = \sigma(\rho,\dot{\rho}) \qquad \text{in } \Omega. \tag{4}$$

where ρ, p, $\boldsymbol{v}$ denotes the density, pressure and fluid velocity field. ρ_s is the static value of the density. Ω denotes the fluid domain and Γ is a flexible boundary. $\mathbf{n}$ is the outward normal vector along Γ and $\boldsymbol{u}_s$ is the displacement of structural members along Γ. $\boldsymbol{b}$ is the bodyforce and q is the added fluid mass per unit volume and time. Finally σ is an arbitrary, nonlinear function of ρ and $\partial\rho/\partial t$. By differentiating Equation (1) with respect to time and eliminating the velocity field we have

$$\frac{\partial^2\rho}{\partial t^2} - \nabla^2 p = -\rho_s\nabla\cdot b + \frac{\partial q}{\partial t}\,. \tag{5}$$

If a linear equation of state is used, i.e. $p = c^2\rho$, this is the standard acoustic wave equation.

Experiments show that a fluid has some slight ability to withstand tensile forces, see Cole[1], but there is no agreement as to the magnitude of the parameters that might be involved in σ. Still, a quite general constitutive law would be

$$p = \sigma(\rho, \frac{\partial\rho}{\partial t}) = \alpha(\rho)\rho + \beta(\rho)\frac{\partial\rho}{\partial t} \tag{6}$$

which, in the numerical example presented in this paper, is simplified to

$$\beta(\rho) \equiv 0, \tag{7}$$

$$\alpha(\rho) = c^2 \quad \text{when } \rho \geq -p_s/c^2, \tag{8a}$$

$$\alpha(\rho) = -p_s/\rho \quad \text{when } \rho \leq -p_s/c^2, \tag{8b}$$

although nothing in this formulation excludes more elaborate forms of constitutive equations. In fact, in the formulation σ is used without specifying its content. The consequence of Equations (7) - (8) is that, as long as the absolute pressure is above zero, there is a linear relation between the density and pressure and that, when the pressure drops to $-p_s$, this value is maintained during fluid expansion.

The weak form of Equation (5), and by employing Green's identity, yields

$$\int_\Omega \mathrm{w}\ddot{p}\ d\Omega + \int_\Omega (\nabla \mathrm{w}) \cdot (\nabla p)\ d\Omega = \int_\Gamma \mathrm{w}(\nabla p) \cdot \mathbf{n}\ d\Gamma - \rho_s \int_\Omega \mathrm{w}\nabla \cdot \boldsymbol{b}\ d\Omega + \int_\Omega \mathrm{w}\dot{q}\ d\Omega. \tag{9}$$

The surface integral in Equation (9) can be used to employ various boundary conditions. For instance, a simple transmitting boundary can be formulated as

$$\nabla p \cdot \mathbf{n} = -c\dot{\rho}.$$

In this paper, however, only the structural boundary is treated. Hence, the boundary condition in Equation (3) yields

$$\int_\Omega \mathrm{w}\ddot{p}\ d\Omega + \int_\Omega (\nabla \mathrm{w}) \cdot (\nabla p)\ d\Omega = \int_\Gamma \mathrm{w}\ddot{\boldsymbol{u}}_s \cdot \mathbf{n}\ d\Gamma - \rho_s \int_\Omega \mathrm{w}\nabla \cdot \boldsymbol{b}\ d\Omega + \int_\Omega \mathrm{w}\dot{q}\ d\Omega. \tag{10}$$

Substituting the constitutive law given in Equation (4) above into Equation (10) leads to

$$\int_\Omega \mathrm{w}\ddot{\rho}\ d\Omega + \int_\Omega \nabla \mathrm{w} \nabla \sigma\ d\Omega = -\rho_s \int_\Gamma \mathrm{w}\ddot{\boldsymbol{u}}_s \cdot \mathbf{n}\ d\Gamma + \rho_s \int_\Omega \nabla \mathrm{w} \cdot \boldsymbol{b}\ d\Omega + \int_\Omega \mathrm{w}\dot{q}\ d\Omega. \tag{11}$$

Because

$$\nabla \sigma = \frac{\partial \sigma}{\partial \dot{\rho}} \nabla \dot{\rho} + \frac{\partial \sigma}{\partial \rho} \nabla \rho \tag{12}$$

we get

$$\int_\Omega \mathrm{w}\ddot{\rho}\ d\Omega + \int_\Omega \nabla \mathrm{w} \frac{\partial \sigma}{\partial \dot{\rho}} \nabla \dot{\rho}\ d\Omega + \int_\Omega \nabla \mathrm{w} \frac{\partial \sigma}{\partial \rho} \nabla \rho\ d\Omega =$$

$$= -\rho_s \int_\Gamma \mathrm{w}\ddot{\boldsymbol{u}}_s \cdot \mathbf{n}\ d\Gamma + \rho_s \int_\Omega \nabla \mathrm{w} \cdot \boldsymbol{b}\ d\Omega + \int_\Omega \mathrm{w}\dot{q}\ d\Omega. \tag{13}$$

Equation (13) is used for the derivation of a conventional finite element scheme, and

Equation (10) for the proposed scheme.

BASIC STRUCTURAL EQUATION

The structural behaviour is governed by some differential equation

$$L(\boldsymbol{u}_s) = F_s(r,t) \tag{14}$$

with the boundary condition

$$F_f(r,t) = p(r,t) \cdot \mathbf{n}$$

due to the distributed pressure load along the fluid boundary.

A discretized formulation for the structure yields, in matrix notation,

$$M_s \ddot{U}_s + C_s \dot{U}_s + K_s U_s = L_{s,e} + L_f, \tag{15}$$

where M_s, C_s and K_s are the structural mass matrix, damping matrix and stiffness matrix, respectively. $L_{s,e}$ is the load vector due to the external structural loads and L_f is the load vector due to the fluid coupling effects.

Following Galerkin, the test functions and the trial functions in the structural domain are taken from the same set of functions. The functions are indicated by a subscript 's' i.e., N_s and the function set by $\{N_s\}$. The coupling vector, L_f, between the structural and fluid domains is thus

$$L_f = \int_\Gamma N_s \cdot \mathbf{n}\, p\, d\Gamma. \tag{16}$$

In the next Section two finite element schemes will be derived. The first of these is denoted 'Conventional scheme', and the second 'Proposed scheme'

FINITE ELEMENT DISCRETIZATION OF THE EQUATIONS

REMARK The two schemes derived below use the same notation for the same parts of the coupled fluid-structure matrices, though their content differs.

CONVENTIONAL SCHEME

Expanding ρ in terms of a set of shape functions, $\{N_f\}$,

$$\rho(r,t) = \sum_j N_f^j(r) \cdot \Pi_j(t), \tag{17}$$

where Π_j is the unknown value of ρ at node j. Further, if

$$w \in \{N_f\},$$

we get the discretized form of Equation (13)

$$M_f \ddot{\Pi} + C_f \dot{\Pi} + K_f \Pi = -L_s + L_b + L_q, \tag{18}$$

where

$$(M_f)_{ij} = \int_\Omega N_f^i N_f^j \, d\Omega,$$

$$(C_f)_{ij} = \int_\Omega (\nabla N_f^i) \cdot (\nabla N_f^j) \frac{\partial \sigma}{\partial \dot{\rho}} \, d\Omega,$$

$$(K_f)_{ij} = \int_\Omega (\nabla N_f^i) \cdot (\nabla N_f^j) \frac{\partial \sigma}{\partial \rho} \, d\Omega,$$

$$(L_s)_i = \rho_s \int_\Gamma N_f^i \, \ddot{\boldsymbol{u}}_s \, d\Gamma,$$

$$(L_b)_i = \rho_s \int_\Omega \nabla N_f^i \cdot \boldsymbol{b} \, d\Omega,$$

$$(L_q)_i = \int_\Omega N_f^i \, \dot{q} \, d\Omega,$$

(i = row index, j = column index) and Π is a column matrix for unknown nodal values of ρ.

Writing σ as in Equation (6)

$$\sigma(\rho,\dot{\rho}) = \alpha(\rho)\rho + \beta(\rho)\dot{\rho}$$

we get

$$\frac{\partial \sigma}{\partial \dot{\rho}} = \beta, \tag{19a}$$

$$\frac{\partial \sigma}{\partial \rho} = \rho \frac{d\alpha}{d\rho} + \alpha + \dot{\rho} \frac{d\beta}{d\rho}, \tag{19b}$$

and thus in Equation (18) we have

$$(C_f)_{ij} = \int_\Omega (\nabla N_f^i) \cdot (\nabla N_f^j) \beta \, d\Omega, \tag{20a}$$

$$(K_f)_{ij} = \int_\Omega (\nabla N_f^i) \cdot (\nabla N_f^j)(\rho \frac{d\alpha}{d\rho} + \alpha + \dot{\rho} \frac{d\beta}{d\rho}) \, d\Omega. \tag{20b}$$

As a practical approach, α and β might be chosen as piecewise constant functions. To each subdomain, i.e. to each element, we assign constant values α_e and β_e to α and β, respectively. The element contributions to the matrices in Equations (20) and (20b) are thus

$$(C_f)^e_{ij} = \beta_e \int_e (\nabla \overset{i}{N_f})(\nabla \overset{j}{N_f})\, dV, \tag{21a}$$

$$(K_f)^e_{ij} = \alpha_e \int_e (\nabla \overset{i}{N_f})(\nabla \overset{j}{N_f})\, dV. \tag{21b}$$

The constant values α_e and β_e should then be chosen on the basis of the nodal values of ρ at nodal points belonging to that subdomain.

Accounting for the fluid-structure coupling effects, the L_f matrix in Equation (14) and the L_s matrix in Equation (18) need to be reformulated. In Equation (15) we have

$$(L_f)_i = \int_\Gamma \overset{i}{N_s} \cdot \mathbf{n}\, p(r,t)\, d\Gamma. \tag{22}$$

A quasi-linearized form of Equation (22) is

$$(L_f)_i = \int_\Gamma \overset{i}{N_s} \cdot \mathbf{n}\, (\frac{\partial \sigma}{\partial \rho} \rho + \frac{\partial \sigma}{\partial \dot{\rho}} \dot{\rho})\, d\Gamma. \tag{23}$$

Expanding ρ in terms of the shape functions yields

$$L_f = C_c \dot{\Pi} + K_c \Pi, \tag{24}$$

where

$$(C_c)_{ij} = \int_\Gamma \overset{i}{N_s} \cdot \mathbf{n} \frac{\partial \sigma}{\partial \dot{\rho}} \overset{j}{N_f}\, d\Gamma, \tag{25a}$$

$$(K_c)_{ij} = \int_\Gamma \overset{i}{N_s} \cdot \mathbf{n} \frac{\partial \sigma}{\partial \rho} \overset{j}{N_f}\, d\Gamma. \tag{25b}$$

If the special form for σ in Equation (6) is used we get

$$(C_c)_{ij} = \int_\Gamma \overset{i}{N_s} \cdot \mathbf{n}\, \beta\, \overset{j}{N_f}\, d\Gamma, \tag{26a}$$

$$(K_c)_{ij} = \int_\Gamma \overset{i}{N_s} \cdot \mathbf{n}\, (\rho \frac{d\alpha}{d\rho} + \alpha + \dot{\rho} \frac{d\beta}{d\rho})\, \overset{j}{N_f}\, d\Gamma. \tag{26b}$$

Likewise, the coupling matrix in Equation (18) is rewritten using structural discretization

$$L_s = M_c\, \ddot{U}_s, \tag{27}$$

where

$$(M_c)_{ij} = \rho_s \int_\Gamma \overset{i}{N_f} \overset{j}{N_s} \cdot \mathbf{n}\, d\Gamma.$$

Assembling Equations (14) and (18) yields

$$\begin{bmatrix} M_s & 0 \\ M_c & M_f \end{bmatrix} \begin{bmatrix} \ddot{U}_s \\ \ddot{\Pi} \end{bmatrix} + \begin{bmatrix} C_s & -C_c \\ 0 & C_f \end{bmatrix} \begin{bmatrix} \dot{U}_s \\ \dot{\Pi} \end{bmatrix} + \begin{bmatrix} K_s & -K_c \\ 0 & K_f \end{bmatrix} \begin{bmatrix} U_s \\ \Pi \end{bmatrix} = \begin{bmatrix} L_{s,e} \\ L_b + L_q \end{bmatrix}. \tag{28}$$

This form is to be compared with the result of a proposed scheme in the next section.

PROPOSED SCHEME

An attractive approach as compared to the conventional finite element procedure is mentioned by Hughes et al[9] and references are made there to several authors working along this line. Christie et al[10] have termed it 'product approximation', Spradley et al[11] called it the 'general interpolants method'. Fletcher[12] and Fletcher et al[13] have introduced the 'group finite element formulation'. Their investigation clearly indicates a considerable gain in

* computer economy,
* accuracy, in particular for high order elements,

when treating a nonlinear problem. Their conclusions apply particularly with regard to compressible fluid flow. The explanation is that the scheme does not connect the nonlinear contributions between nodes to the same degree as the conventional procedure. This quality is central when computing a non-smooth solution, as is the case when the fluid is cavitating.

Both ρ and p are expanded in terms of the same set of trial functions

$$\rho(r,t) = \sum_j \overset{j}{N_f}(r) \cdot \Pi_j(t), \tag{29}$$

$$p(r,t) = \sum_j \overset{j}{N_f}(r) \cdot P_j(t), \tag{30}$$

where Π_j and P_j are the nodal values of ρ and p. Because

$$P_j(t) = \sigma(\Pi_j(t), \dot{\Pi}_j(t)) \tag{31}$$

we have

$$\nabla p(r,t) = \sum_j \nabla \overset{j}{N_f} \cdot \sigma(\Pi_j, \dot{\Pi}_j) \tag{32}$$

If $w \in \{N_f\}$, we get the discretized form of Equation (10)

$$M_f \ddot{\Pi} + K_f\, \sigma(\Pi, \dot{\Pi}) = -L_s + L_b + L_q, \tag{33}$$

where

$$(K_f)_{ij} = \int_\Omega (\nabla N_f^i) \cdot (\nabla N_f^j)\, d\Omega.$$

$\sigma(\Pi,\dot{\Pi})$ is a column matrix and at the i:th row we have

$$(\sigma(\Pi,\dot{\Pi}))_i = \sigma(\Pi_i,\dot{\Pi}_i).$$

The other matrices in Equation (33) are the same as in Equation (18). Writing σ as in Equation (6) we get

$$\sigma(\Pi_i,\dot{\Pi}_i) = \alpha(\Pi_i)\Pi_i + \beta(\Pi_i)\dot{\Pi}_i \tag{34}$$

and the second term in Equation (33) becomes

$$K_f\, \sigma(\Pi,\dot{\Pi}) = K_f\, A\, \Pi + K_f\, B\, \dot{\Pi}, \tag{35}$$

where A and B are diagonal matrices

$$A = \text{Diag}\ (\alpha(\Pi_i)),$$
$$B = \text{Diag}\ (\beta(\Pi_i)).$$

Following the same scheme, the coupling to the structural domain is rewritten as

$$L_f = K_c \sigma(\Pi,\dot{\Pi}), \tag{36}$$

where

$$(K_c)_{ij} = \int_\Gamma N_s^i \cdot \mathbf{n}\, N_f^j\, d\Gamma.$$

By using the constitutive law as expressed in Equation (6) we obtain

$$L_f = K_c\, A\, \Pi + K_c\, B\, \dot{\Pi}. \tag{37}$$

Assembling the Equations (14) and (33) yields

$$\begin{bmatrix} M_s & 0 \\ M_c & M_f \end{bmatrix} \begin{bmatrix} \ddot{U}_s \\ \ddot{\Pi} \end{bmatrix} + \begin{bmatrix} C_s & 0 \\ 0 & 0 \end{bmatrix} \begin{bmatrix} \dot{U}_s \\ \dot{\Pi} \end{bmatrix} + \begin{bmatrix} K_s & -K_c \\ 0 & K_f \end{bmatrix} \begin{bmatrix} U_s \\ \sigma(\Pi,\dot{\Pi}) \end{bmatrix} = \begin{bmatrix} L_{s,e} \\ L_b + L_q \end{bmatrix}. \tag{38}$$

or by using Equations (35) and (37)

$$\begin{bmatrix} M_s & 0 \\ M_c & M_f \end{bmatrix} \begin{bmatrix} \ddot{U}_s \\ \ddot{\Pi} \end{bmatrix} + \begin{bmatrix} C_s & -K_cB \\ 0 & K_fB \end{bmatrix} \begin{bmatrix} \dot{U}_s \\ \dot{\Pi} \end{bmatrix} + \begin{bmatrix} K_s & -K_cA \\ 0 & K_fA \end{bmatrix} \begin{bmatrix} U_s \\ \Pi \end{bmatrix} = \begin{bmatrix} L_{s,e} \\ L_b + L_q \end{bmatrix}. \quad (39)$$

The nonlinear part of the damping and stiffness matrices can be separated

$$\begin{bmatrix} C_s & -K_cB \\ 0 & K_fB \end{bmatrix} = \begin{bmatrix} C_s & -K_c \\ 0 & K_f \end{bmatrix} \begin{bmatrix} I & 0 \\ 0 & B \end{bmatrix}, \quad (40a)$$

$$\begin{bmatrix} K_s & -K_cA \\ 0 & K_fA \end{bmatrix} = \begin{bmatrix} K_s & -K_c \\ 0 & K_f \end{bmatrix} \begin{bmatrix} I & 0 \\ 0 & A \end{bmatrix}, \quad (40b)$$

where I is the identity matrix.

The benefit in the proposed scheme is obvious when Equation (28) is compared with Equation (38) and Equation (39). In Equations (38) and (39), the nonlinear contribution is kept outside the system matrices.

TIME STEPPING CONSIDERATIONS

Both the conventional and the proposed finite element scheme can be solved with various implicit and explicit methods. In this Section however, only the proposed scheme is discussed. Furthermore, the discussion is made with reference mainly to an implicit method used in the verification example presented.

The time stepping procedure needs the calculation of the tangential damping matrix, C_t, and the tangential stiffness matrix, K_t. If by

$$F_i = F_i(U_s, \dot{U}_s, \Pi, \dot{\Pi})$$

we mean the internal forces, then

$$F_i = \begin{bmatrix} C_s & 0 \\ 0 & 0 \end{bmatrix} \begin{bmatrix} \dot{U}_s \\ \dot{\Pi} \end{bmatrix} + \begin{bmatrix} K_s & -K_c \\ 0 & K_f \end{bmatrix} \begin{bmatrix} U_s \\ \sigma(\Pi, \dot{\Pi}) \end{bmatrix} \quad (41)$$

and

$$C_t = \frac{\partial F_i}{\partial(\dot{U}_s, \dot{\Pi})} = \begin{bmatrix} C_s & 0 \\ 0 & 0 \end{bmatrix} + \begin{bmatrix} K_s & -K_c \\ 0 & K_f \end{bmatrix} \begin{bmatrix} I & 0 \\ 0 & \dot{D} \end{bmatrix}, \quad (42a)$$

where $\dot{D} = \mathrm{Diag}\left[\frac{\partial\sigma(\Pi_j, \dot{\Pi}_j)}{\partial\dot{\Pi}_j}\right]$ and

$$K_t = \frac{\partial F_i}{\partial(U_s, \Pi)} = \begin{bmatrix} K_s & -K_c \\ 0 & K_f \end{bmatrix} \begin{bmatrix} I & 0 \\ 0 & D \end{bmatrix}, \quad (42b)$$

where $D = \text{Diag}\left[\frac{\partial\sigma(\Pi_j,\dot{\Pi}_j)}{\partial\Pi_j}\right]$.

If instead Equation (39) is used, i.e.,

$$F_i = \begin{bmatrix} C_s & -K_c \\ 0 & K_f \end{bmatrix} \begin{bmatrix} I & 0 \\ 0 & B \end{bmatrix} \begin{bmatrix} \ddot{U}_s \\ \ddot{\Pi} \end{bmatrix} + \begin{bmatrix} K_s & -K_c \\ 0 & K_f \end{bmatrix} \begin{bmatrix} I & 0 \\ 0 & A \end{bmatrix} \begin{bmatrix} \dot{U}_s \\ \dot{\Pi} \end{bmatrix}, \tag{43}$$

we get

$$C_t = \frac{\partial F_i}{\partial(\dot{U}_s,\dot{\Pi})} = \begin{bmatrix} C_s & -K_c \\ 0 & K_f \end{bmatrix} \begin{bmatrix} I & 0 \\ 0 & B \end{bmatrix}, \tag{44a}$$

$$K_t = \frac{\partial F_i}{\partial(U_s,\Pi)} = \begin{bmatrix} C_s & -K_c \\ 0 & K_f \end{bmatrix} \begin{bmatrix} I & 0 \\ 0 & B' \end{bmatrix} + \begin{bmatrix} K_s & -K_c \\ 0 & K_f \end{bmatrix} \begin{bmatrix} I & 0 \\ 0 & A'+A \end{bmatrix}, \tag{44b}$$

where

$$A' = \text{Diag}\left[\frac{\partial\alpha(\Pi_j)}{\partial\Pi_j} \cdot \Pi_j\right],$$

$$B' = \text{Diag}\left[\frac{\partial\beta(\Pi_j)}{\partial\Pi_j} \cdot \dot{\Pi}_j\right].$$

Finally, the constitutive law to be used in the numerical experiments and described in Equations (7)-(8) yields

$$C_t = \begin{bmatrix} C_c & 0 \\ 0 & 0 \end{bmatrix}, \tag{45a}$$

$$K_t = \begin{bmatrix} K_s & -K_c \\ 0 & K_f \end{bmatrix} \begin{bmatrix} I & 0 \\ 0 & A'+A \end{bmatrix}, \tag{45b}$$

and

$$A'+A = \text{Diag}\left[\frac{\partial\alpha(\Pi_j)}{\partial\Pi_j}\Pi_j + \alpha(\Pi_j)\right]$$

will have a zero diagonal element whenever the corresponding nodal point is cavitating i.e., when $\Pi_j \leq -p_o/c^2$, while those corresponding to a non-cavitating nodal point will take a constant value, c^2.

In order to prevent frothing (see the Introduction) artificial damping is introduced in the following manner. Let

$$\sigma(\Pi_j,\dot{\Pi}_j) = \eta\, \Delta t\, c^2\, \dot{\Pi}_j + \alpha(\Pi_j) \cdot \Pi_j, \tag{46}$$

where Δt is the time step used and η is a dimensionless damping coefficient. The matrix

B in Equation (43) becomes

$$B = \eta \cdot \Delta t\, c^2\, I. \tag{47}$$

This does not affect K_t because B' = 0 but C_t is changed according to Equation (44a).

Based on the expressions above an implicit time stepping routine has been employed for numerical verification in the next Section.

THE BLEICH-SANDLER EXAMPLE

The nonlinear fluid model presented has been verified in one test example. This example by Bleich-Sandler[14] is the only one in which an exact solution has been obtained. Originally, this is a one-dimensional problem, although the data for the numerical test correspond to the two-dimensional counterpart. The problem consists of a plate, initially at rest, on the surface of a semi-infinite space of fluid, see Figure 1. The surface mass is exposed to a plane pressure wave with a sudden rise and an exponential decay.

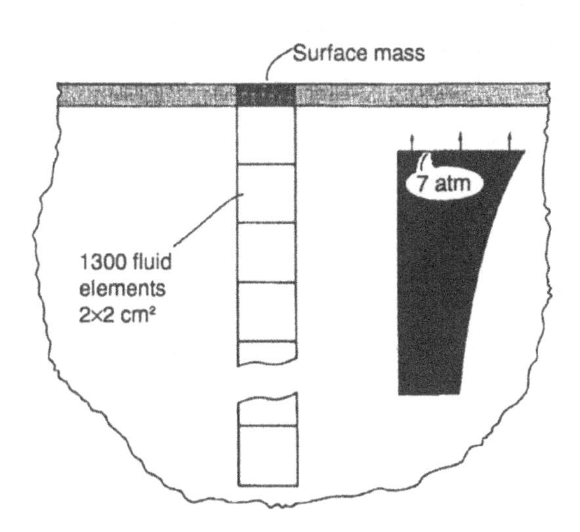

Figure 1. The Bleich-Sandler cavitating fluid example

The calculations were performed using 1300 fluid elements of 2 cm side length. This was enough to ensure that the boundary conditions at 'infinity' did not interfere with the interesting part of the solution. Physical data correspond to those used in Bleich-Sandler[14], although these are transformed into SI-units.

Speed of sound in the fluid	$c = 1423.4$ m/s,
Speed of sound in the cavitated region	$c_1 = 0$ m/s,
Density of the fluid	$\rho_s = 999.83$ kg/m^3,
Surface mass density	$m_s = 144.68$ kg/m^2,
Atmospheric pressure	$P_a = 1.0133 \cdot 10^5$ Pa,
Peak value of the incident wave	$P_p = 7.1016 \cdot 10^5$ Pa,
Decay length of the incident wave	$L = 1.445$ m,
Acceleration due to gravity	$g = 9.8146$ m/s^2.

The time integration procedure is described briefly in the previous Section. The check for convergence was made using the maximum norm, i.e.

$$\max_i |x_i| < \text{error}, \tag{48}$$

and the error was set at 10^{-3}. Normally it took two iterations per time station to satisfy this condition, and never more than three iterations.

At the surface of the fluid, the hydrostatic pressure P_h was equal to the atmospheric pressure and increased linearly into the fluid half space consistent with the acceleration due to gravity and the fluid density. The calculations were performed over 13 milliseconds using 1950 time stations, i.e., the length of the time step was $6.667 \cdot 10^{-6}$ second. At time zero the pressure wave was initialized as though it was starting at infinity, i.e.,

$$\rho(x,0) = \frac{P_p \exp(-x/L)}{c^2}, \tag{49a}$$

$$\dot{\rho}(x,0) = -\frac{c}{L}\rho(x,0), \tag{49b}$$

$$\ddot{\rho}(x,0) = \left(\frac{c}{L}\right)^2 \rho(x,0). \tag{49c}$$

At each time station, a check was made for cavitating nodes, that is, whether $\rho(x,t) < P_h/c^2$, or not.

In order to avoid numerical destruction of the wave front it was 'ramped' over 5 elements and the time was measured from the arrival of half the 'ramped' front at the surface mass. This had no impact on the timing of the velocity peak but made the pressure wave act as intended behind the front. Four sets of runs were performed with different artificial damping coefficients $\eta = 0.5, 0.25, 0.1$ and 0. In Figures 2, 3 and 5, the upward velocity of the surface mass for these cases is shown. The solid lines refer to the present analysis while the discrete symbols are taken from Bleich-Sandler[14]. (Note that the values in that paper are given in a non-dimensional form). The figures also show the upward velocity when no check for cavitation is made. As can be seen, the numerical result in the first cases is in excellent agreement with the analytical result, and the smoothing effect of the artificial damping has only slight impact on the timing of the solution. The cavitating zone opens after 0.36 ms, i.e., after the peak velocity has occurred. After 6 ms the noncavitating solution has a zero upward surface mass velocity while the cavitating fluid makes the surface mass velocity continue below zero. After 10.6 ms the cavitated zone closes and a secondary pressure wave reaches the surface mass after approximately 12 ms, which is clearly visible in Figures 2 and 3. Furthermore, there is a considerable difference in the peak displacement when the fluid is allowed to cavitate.

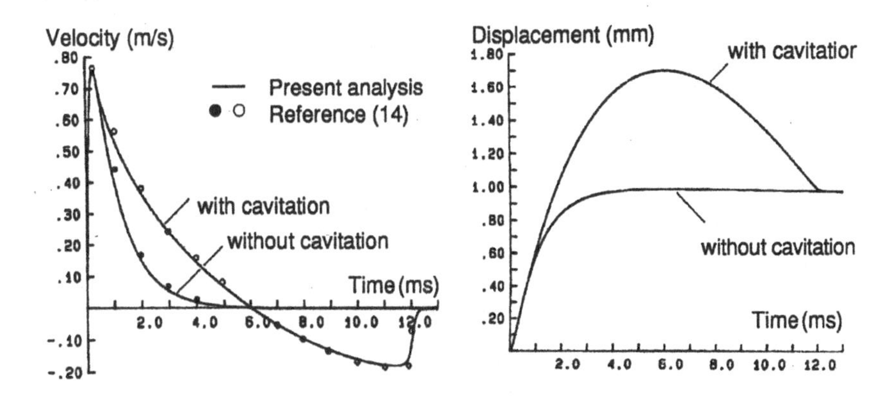

Figure 2. Upward velocity and displacement of the surface mass, $\eta = 0.5$

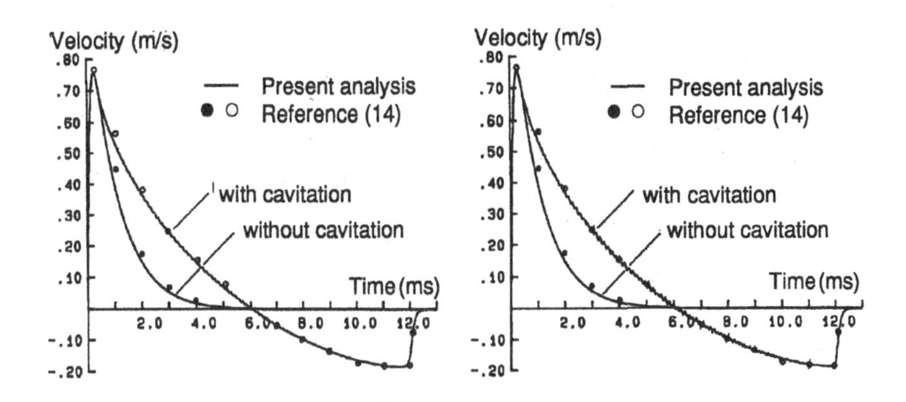

Figure 3. Upward velocity of the surface mass, $\eta = 0.25$ (left) and $\eta = 0.1$ (right)

In Figure 4 the cavitating zone is shown for $\eta = 0.5$ and $\eta = 0.1$. The interfaces between cavitated and non-cavitated regions are made up of horizontal lines because only cavitated nodes are displayed. The result agrees closely with that of Bleich-Sandler[14]. The cavitated region never reaches the surface and closes after 10.6 ms. For $\eta = 0.1$ the upper interface is 'fragmentized' due to the spurious pressure oscillations.

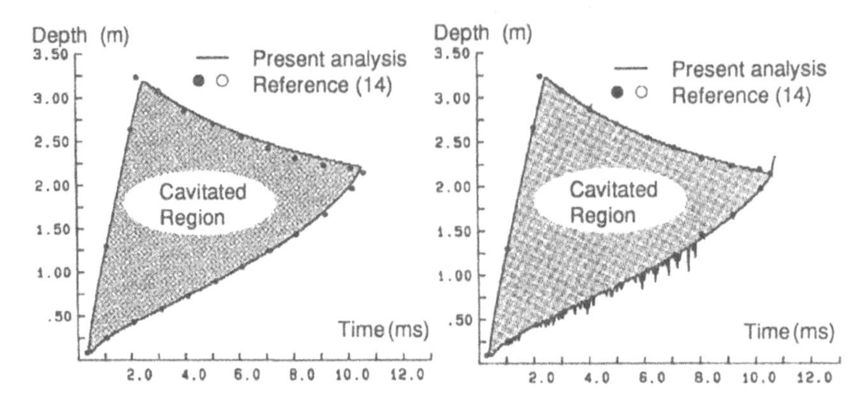

Figure 4. Time history of the cavitated region, η = 0.5 (left) and η = 0.1 (right)

These oscillations have a devastating effect on the solution when η = 0. Figure 5 shows the upward velocity of the surface plate for this case.

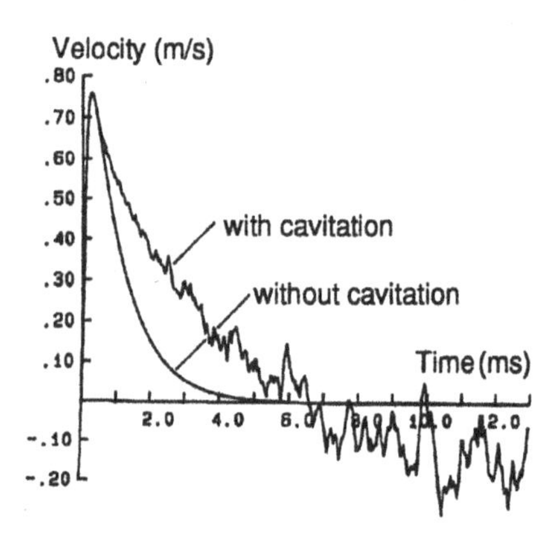

Figure 5. Upward velocity of the surface mass, η = 0.0

CONCLUSION

A numerical method to solve nonlinear wave propagation problems has been proposed. The formulation also accounts for fluid-structure interaction. The proposed scheme is verified against an analytical solution by Bleich-Sandler. The numerical result agrees very well with the analytical solution.

REFERENCES

1. COLE, R.H.: Underwater Explosions, Princeton University Press, 1948.

2. PLESSET, M.S.: Cavitation in Real Liquids, Bubble Dynamics, Elsevier Publishing, Amsterdam, 1964.

3. NEWTON, R.E.: Effects of Cavitation on Underwater Shock Loading - Part 1. NPS-69-78-013, Naval Postgraduate School, Monterey, California, 1978.

4. NEWTON, R.E.: Effects of Cavitation on Underwater Shock Loading - Plane Problem, Part 1. NPS-69-79-007PR, Naval Postgraduate School, Monterey, California, 1979.

5. NEWTON, R.E.: Effects of Cavitation on Underwater Shock Loading - Plane Problem, Part 2. NPS-69-80-001, Naval Postgraduate School, Monterey, California, 1980.

6. NEWTON, R.E.: Effects of Cavitation on Underwater Shock Loading - Plane Problem - Final Report. NPS-69-81-001A, Naval Postgraduate School, Monterey, California, 1981.

7. ZIENKIEWICZ, O.C., PAUL, D.K., HINTON, E.: 'Cavitation in Fluid-Structure Response (with particular reference to dams under earthquake loading)', Earthquake Engineering and Structural Dynamics, Vol. 11, pp. 463-481, 1983.

8. FELIPPA, C.A., DERUNTZ, J.A.: Finite Element Analysis of Shock-Induced Hull Cavitation. Computer Methods in Applied Mechanics and Engineering, Vol. 44, pp. 297-337, 1984.

9. HUGHES, T.J.R., TEZDUGAR, T.E.: Finite Element Methods for First-Order Hyperbolic Systems with Particular Emphasis on the Compressible Euler Equations. Computer Methods in Applied Mechanics and Engineering, Vol. 45, pp. 217-284, 1984.

10. CHRISTIE, I., GRIFFITHS, D.F., MITCHELL, A.R., SANZ-SERNA, J.M.: Product Approximation for Non-Linear Problems in the Finite Element Method, IMA Journal of Numerical Analysis, Vol. 1, pp. 253-266, 1981.

11. SPRADLEY, L.W., STALNAKER, J.F., RATLIFF, A.W.: Computation of three-dimensional Viscous Flows with the Navier-Stokes Equations, AIAA-80-1348, AIAA 13th Fluid and Plasma Dynamics Conference, Snowmass, Colorado, 1980.

12. FLETCHER, C.A.J.: The Group Finite Element Formulation, Computer Methods in Applied Mechanics and Engineering, Vol. 37, pp. 225-243, 1983.

13. FLETCHER, C.A.J., SRINIVAS, K.: On the Role of Mass Operators in the Group Finite Element Formulation, Computer Methods in Applied Mechanics and Engineering, Vol. 46, pp. 313-327, 1984.

14. BLEICH, H.H., SANDLER, I.S.: Interaction Between Structures and Bilinear Fluids, International Journal of Solids and Structures, Vol. 6, pp. 617-639, 1970.

SECTION 5: APPLICATIONS IN STRUCTURAL MECHANICS

Supercomputers in Materials Science

J.M. Hyde, A. Cerezo, M.G. Hetherington, G.D.W. Smith

Department of Materials, University of Oxford, Parks Road, Oxford, OX1 3PH, U.K.

ABSTRACT

A high performance graphics mini-supercomputer has been used for the visualisation and analysis of 3-dimensional atomic-scale chemical data from iron-chromium alloys obtained using the position-sensitive atom probe. This instrument allows us to follow the progress of phase transformation by spinodal decomposition which occurs in these alloys during thermal ageing. Experimental results are compared with the results of computer modelling of the atomic processes occuring during decomposition. Two types of modelling have been performed: Monte Carlo simulations and a numerical solution of the solid-state diffusion (Cahn-Hilliard) equation. It was found that both models produced similar microstructures to the experimental system, but the models differed in their later stages of development in a time regime where no experimental data was available.

INTRODUCTION

A fundamental aim of Materials Science is understanding what factors determine the macroscopic properties of a material, such as strength and fracture toughness, and how these can be controlled in the production of, for example, a metallic alloy. Computer simulations are widely used, for instance to verify theoretical models or to investigate systems where no accurate analytic theory exists. Knowledge of the microstructure of a material can be crucial in understanding how the particular properties arise, and as a result microscopy play a very significant role in Materials Science research. The microstructures which are observed are often produced by solid state diffusion which occurs during the heat treatments applied as part of the production route, or during thermal ageing in service. In modern materials, the scale of important microstructural features may be no more than a few atomic spacings, and microscopy at very high resolution is required. The highest resolution for chemical analysis of materials is provided by atom probe microanalysis, and in particular the position-sensitive atom probe (POSAP) [1,2] , which yields 3-dimensional compositional data from small volumes of material at sub-nanometre spatial resolution.

In this paper we consider thermal ageing of the iron–chromium system. The Fe–Cr spinodal system is a good experimental system to model because the small lattice mismatch and the lack of competing phase transformations (only an extremely sluggish sigma phase exists) makes it an ideal model system. This alloy is also a model for the ferrite phase in a class of industrial alloys known as duplex stainless steels which are used, for example, in the primary cooling circuits of pressurised water reactors. The ferrite phase in these alloys is thermodynamically unstable, which leads to embrittlement via spinodal decomposition [3]. As decomposition occurs the atoms in the ferrite cluster to produce iron-rich and chromium-rich regions, on a scale of 1–4nm, forming an interconnected structure similar to a sponge [4]. This structure is percolated, that is to say that it is possible to traverse the structure while remaining totally within either the chromium– or iron–rich region. Small cylindrical volumes of these materials have been analysed with the POSAP, in order to study how the spinodal decomposition process leads to the observed change in properties. Typically 100,000 atoms are collected with the POSAP and, with the aid of a graphics workstation, it is possible to reconstruct the atomic–scale chemistry and analyse the extent of decomposition. Qualitative information may be obtained by various graphical representations, from the simplest technique, where by each atom position is marked by an appropriately coloured dot, to drawing a three dimensional surface of constant composition to mark the interface between regions.

Understanding how spinodal decomposition leads to embrittlement requires a grasp of the atomic scale processes involved. The high computation performance available in top–end workstations now enables sufficiently large simulations of atomic scale process to yield realistic results. Two methods have been analysed, a Monte Carlo simulation and a numerical solution to the Cahn-Hilliard equation using a finite difference method. Comparison of experimental data and numerical simulations is leading to a better understanding of spinodal development and will enable predictions such as the effect of temperature on the lifetime of real alloys. Our work has been based using the Stardent™ ST1000 graphics mini–supercomputer whose architecture combines high computational performance with impressive graphics capabilities enabling both modelling and analysis (numerical and visual) of atomic scale processes in metals.

TECHNIQUES

The experimental and modelling techniques used to study spinodal decomposition in the iron-chromium system are briefly summarised in this section.

Position Sensitive Atom Probe

The POSAP is capable of measuring 3–dimensional composition variations within materials with sub–nanometre spatial resolution. The specimen is made in the form of a sharp needle with end radius of about 100nm. A high electrostatic field enables the controlled removal of surface atoms by field evaporation from the needle apex. The resulting ionized atoms are radially projected from the specimen, and strike a position–sensitive detector at a distance of several centimetres. This projection produces a magnification of about a million. A time–of–flight measurement allows the mass–to–charge ratio of the ion to be calculated, and the position of impact on the detector yields the original position on the specimen surface. The resolution for position determination is around 0.5nm. Continual removal of material yields depth profiling, and yields a full 3-dimensional analysis. Typically a cylindrical block 20nm in diameter and 15-20nm deep is analysed in a single experiment. POSAP analyses have been obtained from a series of Fe 45 at.% Cr alloys aged up to 500 hours at 500°C.

Monte Carlo Simulation

With the advent of cheaper and more powerful computing facilities modeling has become a powerful method for studying atomic scale processes. The Monte Carlo technique [5] implemented models atom scale diffusion by considering atoms 'jumping' between neighbouring lattice sites. A regular lattice of atoms of size 50x50x50 with periodic boundary conditions was used to model the atomic structure of an iron-chromium alloy. Each configuration of the lattice (X_{i+1}) is generated from the previous one (X_i) by considering the energy change ΔH when two nearest neighbour atoms are swapped. If the energy decreases the swap is accepted. If the energy increases the probability of acceptance W is equal to the metropolis function [6].

$$W(X_i \rightarrow X_{i+1}) \begin{cases} \exp(-\Delta H/kT) & \text{if } dH > 0 \\ 1 & \text{otherwise} \end{cases}$$

By locating the critical parameter $\Delta H/kT$, above which no order exists and comparing with critical temperature for the Fe-Cr system an absolute temperature scale was introduced into the calculations. There is good correlation between the miscibility gap generated by this Monte Carlo method and the experimentally determined miscibility gap from the Thermo-Calc databank system [7] shown in figure 1 despite the fact that the magnetic term has been ignored in the Monte Carlo Simulation.

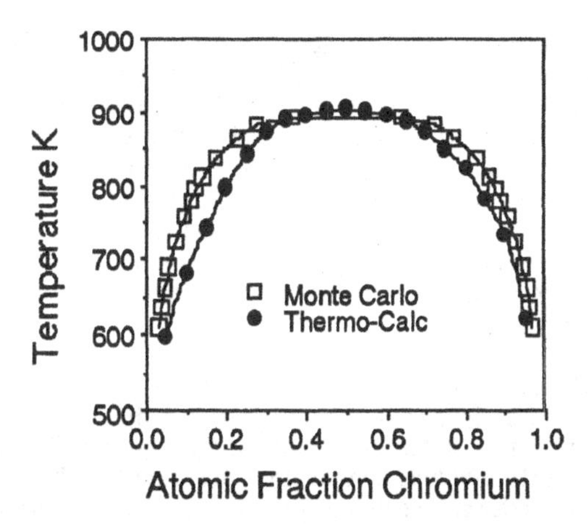

Figure 1. Miscibility gap for Monte Carlo model compare with the phase diagram calculated with the Thermo-Calc database.

Simulations of a 50 atomic percent binary alloy were run, quenching from an infinite temperature (random lattice) down to 600K, 700K and 800K and then aging up to 10000 Monte Carlo steps per lattice site (MCS).

Numerical Solution to the Non Linear Cahn-Hilliard Equation

Below a critical temperature the Ginzburg-Landau free energy ψ has a double well shape and the region where $\psi'' < 0$ is the spinodal. In this region the system is unstable in that infinitesimally small fluctuations in composition about the average composition lower the free energy. Cahn and Hilliard [8] proposed a continuum model for the time evolution of the concentration $u(\underline{x},t)$ for x in the spatial domain and $t > 0$ after the quench.

$$\frac{\partial u}{\partial t} = M \nabla^2 (\psi'(u) - \gamma \nabla^2 u)$$

where M is the mobility and γ a phenomenological constant modelling the interfacial energy. Due to the non-linearity the Cahn-Hilliard diffusion equation is not soluble analytically in greater than 1-D necessitating the need for a numerical calculation. Using a numerical model developed by Copetti and Elliott [9] based on the Galerkin finite element analysis, the phase decomposition on a regular three dimensional grid of size 64x64x64 has been followed. A quartic polynomial has been used for ψ, defined between -1 and 1, with the spinodal limits at $\pm 1/\sqrt{3}$. No temperature is implicit in this model, however an approximate value may be obtained by comparing the solid solubility limits with an experimental phase diagram. A comparison with the Fe-Cr phase diagram predicts a temperature of just above 800K. Since u is a continuous composition variable on the scale -1 to 1 a criteria was required to define points on the grid as A or B type atoms. In these simulations a positive composition was taken as A type and a negative composition as B type.

VISUALIZATION

Graphics supercomputers are now equipped with software designed for displaying 3D objects and have hardware configured for high speed rendering for real time animation. The simplest graphical representation of POSAP data possible is to use a colour coded pixel to mark each atom position, the colour indicating the atomic species. Since the pixel is the simplest graphics primitive, real time rendering is possible even with upwards of 100,000 atoms to display. This method together with a discussion on depth-cueing and Z-buffering has been considered by Cerezo and Hetherington [10] who found that this technique was particularly effective in the representation of isolated particles but confusing for interconnected structures.

However the field evaporation process does not permit true atomic mapping since small trajectory aberrations cause a small scatter in the calculated atomic positions, resulting in the loss of nearest neighbour information. Therefore the POSAP analysis primarily involves composition mapping which requires the data to be sampled. A regular compositional grid is generated, with resolution of the order of 1-2 atomic spacings to preserve detail at the finest level possible. At this resolution some positions in the data array will not contain any atom information from which a composition may be calculated (figure 2a). It is therefore necessary to introduce moving average sampling methods. Two techniques are used: simple smoothing and centre-weighted. In the first (figure 2b), all points within the smoothing volume contribute equally to the composition calculation. However, this will tend to smooth out fine detail present in the structure. The second method is heavily biassed towards atoms nearest to the centre of the smoothing volume (figure 2c) and thus is essentially a 'hole filling' algorithm. This results in a structure containing a higher level of fine detail. The result of this smoothing is a grid of compositional information which is used as the basis for volume rendering.

As previously stated spinodal decomposition in the Fe-Cr system leads to an interconnected structure somewhat similar to a sponge. The graphical techniques discussed in this section demonstrate some of the best techniques for visualising percolated or interconnected morphologies. The graphical output was generated using the Stardent™ Application Visualization System (AVS™).

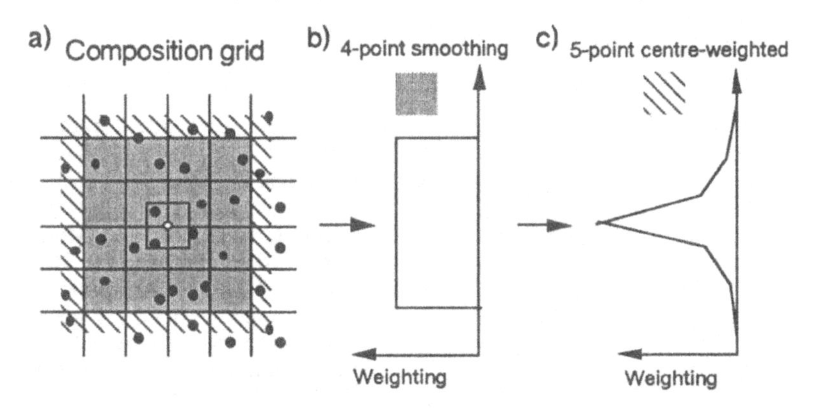

Figure 2. Sampling techniques.

Colour Coded Cross Sections

In the simplest visualization, a cross-section through the compositional grid (to reveal microstructure) may be drawn. Brightness may be used, with saturation set to zero, to generate a grey scale image, between black (minimum brightness), and white (maximum brightness) corresponding to the composition range. In figure 3 the white areas represent the chromium rich regions, and the the black areas represent the iron rich regions. The circular image (figure 3a) represents a cross-section through a cylinder of POSAP data from an Fe-Cr alloy aged for 500 hours at 500°C. Figures 3b and 3c show the results from a Monte Carlo simulation aged for 1000 MCS at 800 K and the numerical model aged for 0.3 time units respectively. The cross-sections show the range of composition variation and indicate interface diffuseness. In each case the interconnected structure can be clearly distinguished. In figures 3b & 3c the periodic boundary conditions are evident.

Binary Representation

With appropriate thresholding, the grid structure can be transformed into a binary grid of black and white points, where each voxel (volume element, or

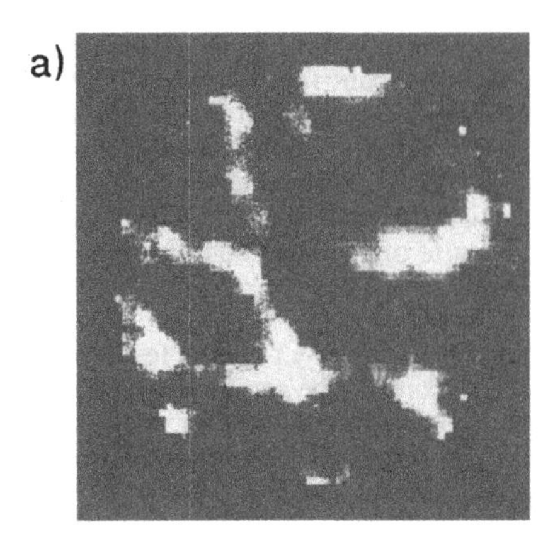

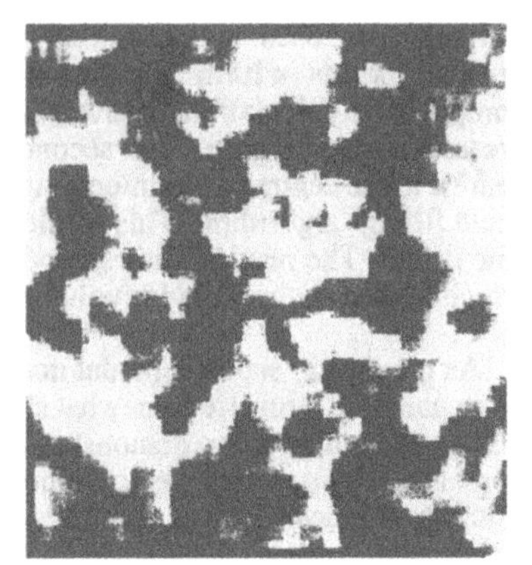

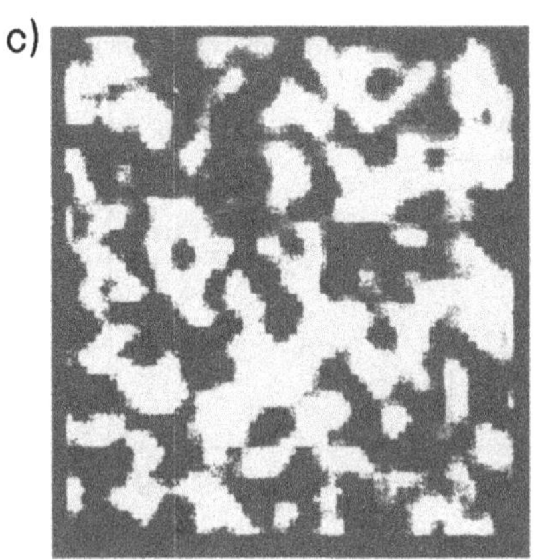

Figure 3. Grey scale images representing composition variations in a 2-D cross section through (a) An Fe-Cr alloy aged at 500°C for 500 h, (b) Monte Carlo Simulation aged for 1000 MCS at 800K and (c) Numerical model aged to 0.3 time steps. The white areas correspond to Cr-rich regions and the dark areas to Cr-depleted regions.

node in the grid) is assigned to one of the material's constituent phases. A cross-section through the microstructure yields a two dimensional black and white image where the boundary between black and white represents the interface between the phases. A scan through several adjacent cross-sections shows the details of the morphology. In particular it is possible to 'see' if the microstructure is percolated [11].

Isosurface

A logical progression of this two dimensional representation into three dimensions generates a surface of constant composition (isosurface) corresponding to the interface between regions. This is the three dimensional analogue of a contour line. In this representation one phase is shown coloured and the other transparent, a representation analogous to a view of a sponge where the sponge represents, for instance, the chromium rich phase and the air the iron rich phase. This has proved to be one of the best methods for visualising the complex morphology associated with percolating structures. By choosing the thresholding, or isosurface level, it is possible to select the interface composition of interest or examine, for instance, how diffuse the interface is. In order to improve the apparent three dimensionality of the data on a two dimensional screen it is useful to use shading techniques, such as gouraud or phong, to highlight the surface and generate contrast. Figure 4 shows an

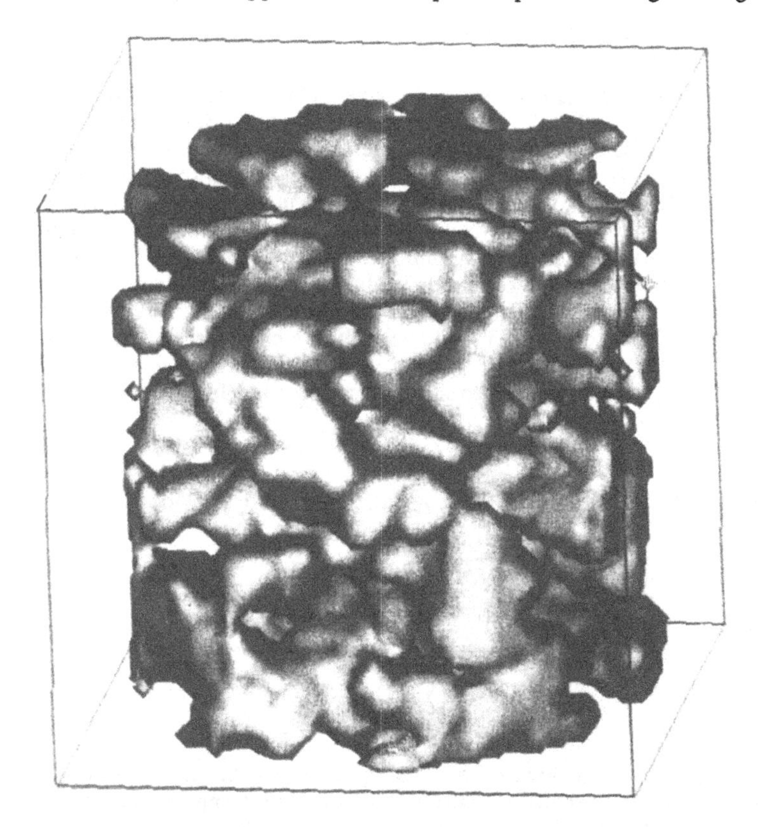

Figure 4 Isosurface drawn through all points in the grid with composition 50% Cr indicating the interface between the chromium rich phase (coloured) and iron-rich phase (transparent) for an Fe-Cr alloy aged for 500 h at 500°C.

example of an isosurface generated from experimental data in which the coloured phase represents the chromium rich regions. This visualization technique show that both phases are percolating.

Discussion

From visualization of both experimental and modelling data it is possible to make qualitative assessments and comparisons between the microstructures obtained. It was found that the Monte Carlo technique generated structures very similar to the experimental results. The Monte Carlo simulation at higher temperatures yielded a rougher interface for a given scale than at lower temperatures. The numerical model generated similar structures at low ageing times, however at longer annealing times the scale did not appear to increase while the interface roughness decreased.

CHARACTERIZATION OF STRUCTURE

The visualization techniques described enable us to analyse the microstructure qualitatively in terms of scale, amplitude and morphology. However if objective comparisons are to be made between different microstructures quantitative analysis is necessary. Scale and amplitude can be measured using other experimental techniques, for instance small angle neutron scattering experiments yield a structure function which is related to scale, and conventional 1-D atom probe analysis is used to measure amplitude. However the POSAP is unique in that full 3D morphological information is obtainable. New analysis techniques have been developed to characterise the morphology of the microstructures arising from spinodal decomposition. The analysis techniques have been applied to experimental results obtained from an Fe 45 at. % Cr alloy aged up to 500 hours at 500°C and compared to structures generated by the Monte Carlo simulation and numerical model.

Scale

The scale of a structure may be determined using an autocorrelation function R_k defined by :

$$R_k = \frac{N \sum_{1}^{N-K} (C_{(t)} - C_o)\ (C_{(t+k)} - C_o)}{(N-K) \sum_{1}^{N} (C_{(t)} - C_o)^2}$$

where $C_{(t)}$ is the composition of the tth sample, and C_o the mean composition. In 3D the samples are shells of uniform thickness with compositions $C_{(i)}$. The autocorrelation is analogous to the fourier transform of the structure function. It detects regular modulations in composition. For isolated particles the first minimum has been found to correspond to particle size and the first maximum to interparticle separation. For a spinodal structure the scale λ is related to the width of the domains, and taken to be twice the first minimum since when k is equal to $\lambda/2$, the numerator will be large and negative yielding a minimum in R_k.

The Monte Carlo results (figure 5a) show two interesting features. Firstly that scale is approximately independent of annealing temperature for a given number of Monte Carlo steps and secondly that the scale is proportional to t^k with a time exponent $k = 0.18\text{-}0.24$ in agreement with some previous atom probe and SANS results [12] but not in agreement with the Lifshitz-Slyosov [13] law which predicts an exponent of 1/3 for late stage growth. The linearized Cahn-Hilliard equation predicts that the early stages of growth should be exponential but this has only been seen in polymer mixtures [14]. In alloys the non-linear terms appear to dominate even at early ageing. The numerical results (figure 5b) show that once the initial random fluctuations have become large enough to generate a measurable scale ($t > 0.1$) a rapid increase in scale occurs ($0.1 < t < 1$) followed by a slower coarsening regime ($1 < t < 30$). An absolute value for scale has no meaning for this simulation since the lattice points do not correspond to atom positions but to composition sampling points. The experimental data showed a scale increasing to 2.2nm for the alloy aged for 500 hours at 500°C.

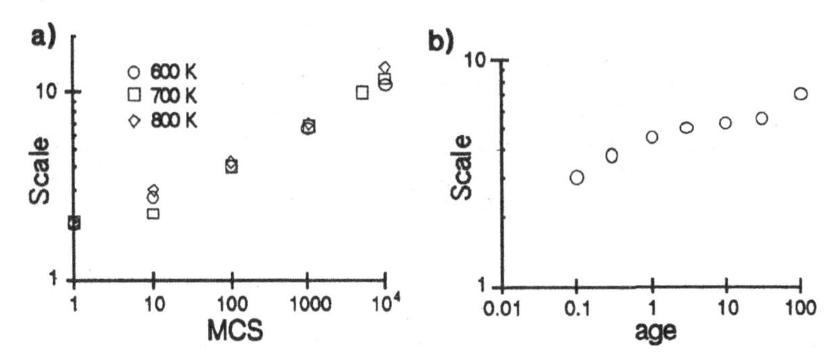

Figure 5. Development of scale for (a) Monte Carlo and (b) numerical models.

Composition Amplitude

In order to quantify the amplitude of the composition fluctuations, the atoms are collected into cubic blocks of N atoms and the frequency distribution obtained by plotting the observed number of blocks with a given number of A atoms against that number of atoms. The models developed to analyse this data describe the sample distribution of block compositions and thus only contain information on the amplitude of decay and not wavelength or spatial distribution. A random solid solution will generate a well defined binomial frequency distribution whereas in a system in which phase separation has occured it will lead to a broadening of this distribution. In the limiting case a very heavily decomposed sample will yield a distribution with two peaks corresponding to the development of two separate phases. Thus in a two phase microstructure, such as may occur from spinodal decomposition, the phase separation may be quantified by deconvoluting the two peaks. The choice of block size has been the subject of some discussion [10,15] since if too small a block size is chosen the result will be dominated by statistical noise, limiting the significance of the calculated amplitude. If too large a block size is chosen the composition fluctuations will be smoothed out. The details of calculation procedure have been published elsewhere [16,11]. The results in this paper use the two parameter model which assumes that the frequency distribution may be represented by the superposition of two Gaussian distributions from the non-linear theory developed by Langer, Bar-on and Miller [17]. The decomposition is represented by the separation of these peaks. A block size of 27 was chosen corresponding to a 3x3x3 cube of within which a composition was calculated. Fits to the model using a maximum likelihood estimator were significant at the 5% level.

The amplitude increases with ageing corresponding to the progress of the reaction and the generation of a two phase microstructure. At very late stages in the ageing the amplitude will tend to the difference in composition between the two phases. The amplitude for the experimental data (figure 6a) reaches a value of 0.35 - 0.4, whereas both the Monte Carlo and numerical models indicate greater amplitudes corresponding to longer ageing. The Monte Carlo simulation

(figure 6b) shows how temperature affects the amplitude of decomposition. Since the probability of an unfavourable swap is greater at higher temperatures, a cluster of like atoms will form more rapidly (in terms of MCS) at a lower temperature resulting in a greater amplitude. Since the Fe-Cr alloy was aged at 500°C the best fit for the Monte Carlo data should be at 800K. The results in figures 6a & 6b show fair quantitative correlation, with the most heavily aged alloy corresponding to a structure aged for 1000-10000 MCS at 800K. The results for the numerical simulation (figure 6c) again show two distinct regimes, a rapid increase in amplitude ($0.1 < t < 1$) and a plateau ($1 < t < 30$) corresponding to longer times than have been studied by either Monte Carlo or experiment. The best fit to the experimental data occurs for $0.1 < t < 0.3$.

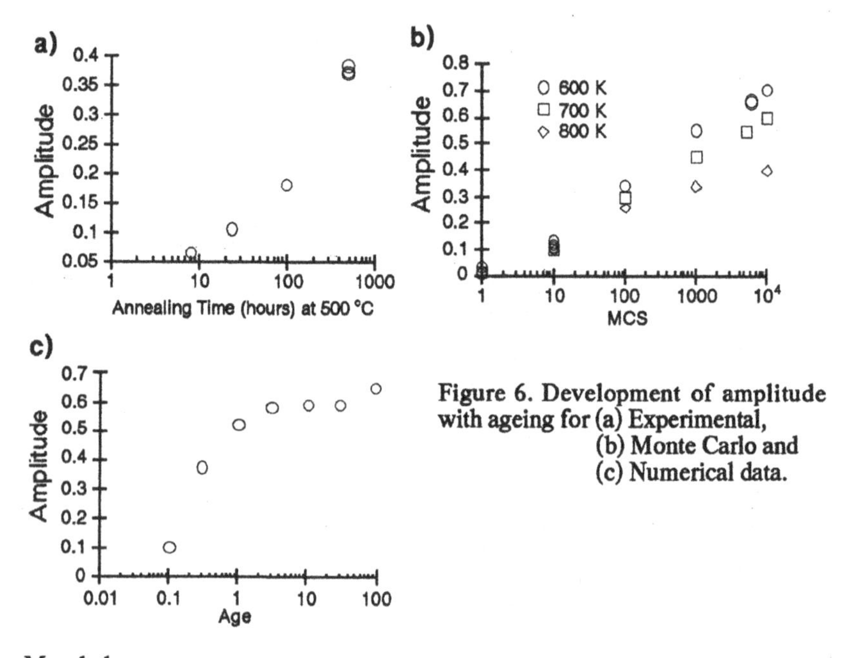

Figure 6. Development of amplitude with ageing for (a) Experimental, (b) Monte Carlo and (c) Numerical data.

Morphology

The spatial information obtained using the position sensitive atom probe enables studies of morphology that are impossible with any other technique. It has been speculated that the interconnected structure often associated with spinodal decomposition might have the properties of a fractal and thus a fractal dimension could be used to parameterize the structure. Recently the field of digital topology [18] has received much interest, mostly in two dimensions where many theoretical results are known and have been applied as a basis for image thinning, border following and object counting. Some of these results have been extended to three dimensions, yielding algorithms for image thinning and object counting. A complex morphology may be reduced to a basic framework to be analysed, for instance, in terms of the number of cavities and tunnels.

Fractal Analysis

Fractal objects are objects with an identical degree of irregularity on a wide range of length scales. The dimension is a measure of the fragmentation. It is possible to use the scaling behaviour of the binary grid used for image processing to define a fractal parameter for the microstructures obtained. The range of length scales is limited by the size of the analysed volume and the maximum resolution of the atom probe.

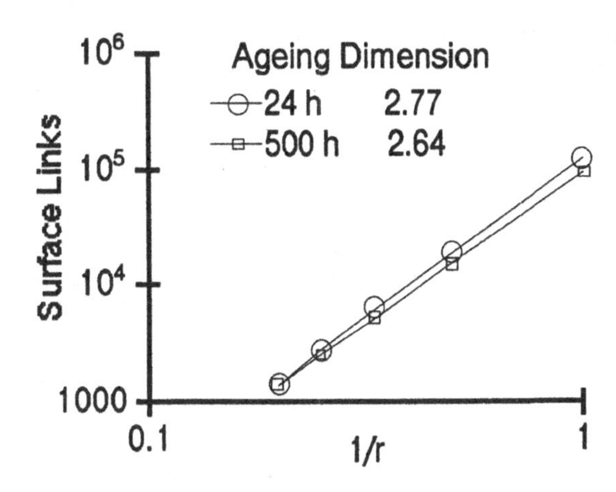

Figure 7. Calculating the fractal dimension for the morphology of an Fe 45% Cr alloy.

The binary grid structure used for the visualization may be used to define nodes (grid points inside B 'phase'), volume links (B→B) and surface links (A→B). A surface term may be defined by counting the number of surface links Sl, and by observing the scaling behaviour of this surface term as a function of grid resolution, we can define a fractal dimension for the surface [10]. The plot of ln(Sl) v. ln(1/r), where r is the length scale, has gradient equal to the dimension of the surface. Figure 7 shows this scaling behaviour for the iron-chromium alloy.

As the alloy ages the dimension decreases from 3 down towards 2 indicating that for a random percolating structure the surface is effectively volume filling, while after extended ageing the surface dimension tends towards the classical non-fractal limit of 2. Figure 8a shows the dimension decreasing with age for the experimental data. The Monte Carlo simulation (figure 8b) appears to show the dimension tending to a temperature dependent limit. This measure of dimension is sensitive to surface roughness, the higher the temperature the more diffuse the interface since at higher temperatures the probability of an energetically unfavourable swap increases. The numerical simulation (figure 8c) shows the dimension decreasing very rapidly down towards 2 indicating a much smoother interface region. Again the time period corresponding to the experimental results is 0.1 to 1 units. A visual assessment agrees with the above analysis. The experimental data shows similar behaviour to the Monte Carlo model aged up to 1000 MCS at 800 K.

Topological Characterization

In topological characterisation, two structures are considered identical if they can be transformed into one another without requiring any cuts. Thus a coffee mug and doughnut are therefore topologically identical since they both have only one hole or handle. Moving a dislocation through a connected microstructure, such as a sponge, will require cutting through the handles of the structure, and therefore the handle density of a percolated structure is the analogue of the particle density for separate particles.

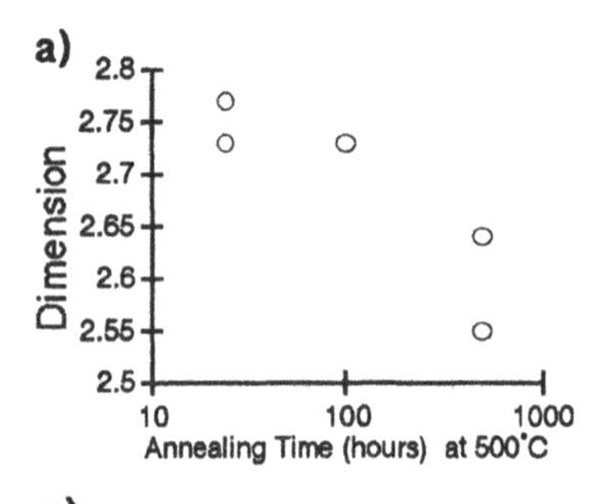

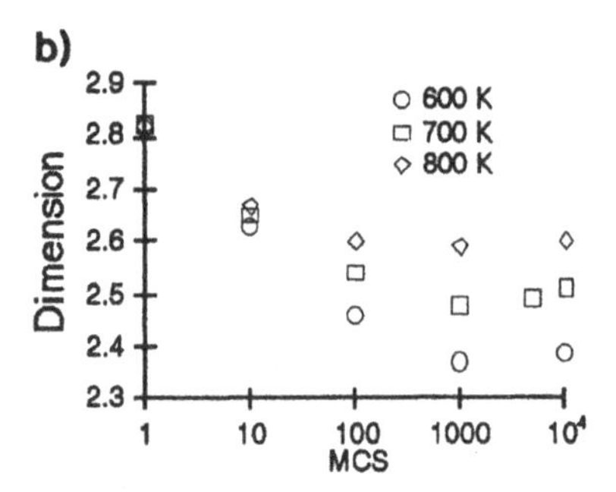

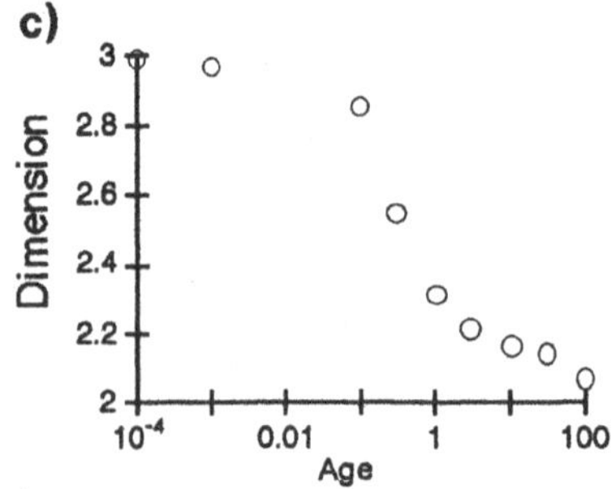

Figure 8. Fractal Dimension as a function of ageing time for (a) POSAP data, (b) Monte Carlo data and (c) Numerical simulation data.

Some suggestions on the quantification of interconnected microstructures were made by Camus et al. [19]. We have implemented an algorithm to calculate the number of handles in a 3D binary image derived from the compositional data [20]. This method counts only the total number of handles (and generates a density based on the number of atoms in the analysed volume) and makes no account of the size of loop. However by comparing the results for the smoothing techniques it is possible to measure both the total number of handles in the structure (centre-weighted smoothing) and also just the larger handles (simple smoothing).

As the structure decomposes and the scale increases the number of handles will decrease. The three techniques demonstrate that most of the handles are very small (figure 9). The Monte Carlo analysis (figures 9c & 9d) shows that interface properties are related to ageing temperature. Using simple smoothing, where all the small scale loops are smoothed out (ie loops around the interface region), shows that measuring the handle density is equivalent to scale analysis (approximately independent of annealing temperature), whereas analysis on the exact structure (which is known in these simulations) shows a marked dependence on annealing temperature (in agreement with the fractal analysis). The numerical method (figures 9e & 9f) shows the handle density decreasing very rapidly with age, similar to scale and fractal measurements.

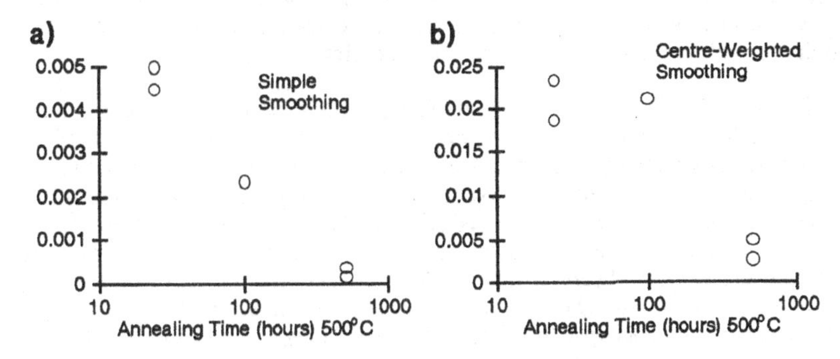

Figures 9a & b. Handle density as a function of ageing for the Fe 45 at. % Cr alloy aged at 500°C

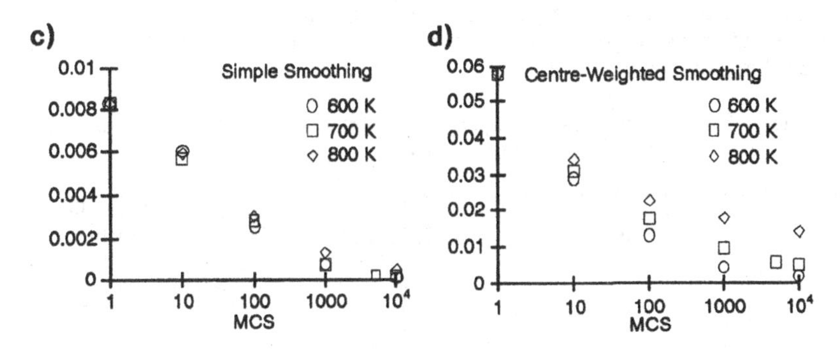

Figures 9c & d. Handle density as a function of ageing for the Monte Carlo Simulation

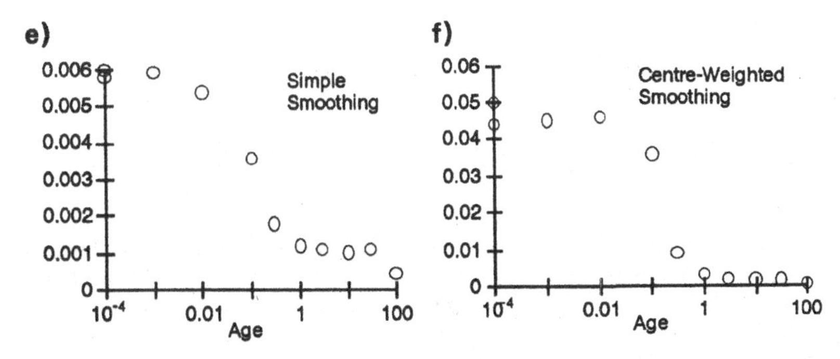

Figures 9e & f. Handle Density as a function of ageing for the numerical calculation.

CONCLUSIONS

None of the analysis techniques described would have been possible without the aid of computers. With the development of more powerful hardware, and improved software, visualization has become a vitally important scientific tool yielding qualitative information rapidly. The quantitative analysis techniques discussed enable the parametrization of atomic scale structures and also the ability to study the accuracy of models and theories.

The results indicate that the Monte Carlo method produces microstructures similar to the experimental Fe-Cr system. The very early stage growth of the numerical solution to the non linear Cahn-Hilliard equation ($t < 0.1$) could not be quantified with the techniques discussed since in this time regime the initial random fluctuations had not increased sufficiently to generate a structure. In the range 0.1 to 0.3 time units the numerical model produced structures similar to those experimentally observed, however at later ageings structures with a similar scale but smoother interfaces were observed. No experimental data has yet become available to analyse the effects of the very long term ageing equivalent to the most heavily aged numerical simulation. The table below (table 1) summarises the best fit structures generated by both the Monte Carlo method and numerical model to the iron-chromium alloy aged for 500 hours. Good fits have been obtained for scale, amplitude and fractal dimension. The handle densities for real data are somewhat lower than the models predict but still of the correct order of magnitude.

Table 1 : Comparison between experimental structures and those generated by Monte Carlo calculation and numerical model.

	Scale	Amplitude	Fractal Dimension	Handle Density Simple Smoothing	Handle Density Centre-Weight
Fe-Cr alloy 500h at 500°C	2.2 nm (c.10 lattice units)	0.38	2.6	0.0004	0.005
Monte Carlo 1000 MCS at 800 K	6.5	0.33	2.6	0.0013	0.018
Monte Carlo 10000 MCS at 800 K	10.25	0.40	2.6	0.0005	0.014
Numerical Model : t = 0.3	3.8	0.38	2.55	0.0018	0.009

The results in this paper demonstrate that both computer modelling techniques can accurately follow the spinodal decomposition although the exact correlation between real time and computer time has yet to be determined. Further experimental data is being compiled for a number of alloys with different Cr contents and annealing treatments, in order to extend the comparison with our computer models.

ACKNOWLEDGEMENTS

The authors wish to thank Professor Sir Peter Hirsch for provision of laboratory facilities, MGH and JMH would like to acknowledge the SERC for funding during this work and AC thanks The Royal Society for financial support and Wolfson College, Oxford for support in the form of a Research Fellowship. Thanks are due to Ms. M. Copetti and Professor C. Elliott for the numerical solution to the Cahn-Hilliard equation and to Dr M.K. Miller for supplying the Fe-Cr specimens.

---

1. A. Cerezo, T.J. Godfrey and G.D.W. Smith, Rev. Sci. Instrum., 59, 862 (1988)
2. M. K. Miller and G.D.W. Smith, Atom-Probe Microanalysis: Principles and Applications in Materials Science, (Materials Research Society, Pittsburg, 1989)
3. K. Binder, 'Alloy Phase Stability', ed. G.M. Stocks and A. Gonis Kluwer Academic Publishers, 233-262 (1989)
4 J.E. Brown, A. Cerezo, T.J. Godfrey, M.G. Hetherington and G.D.W. Smith, Mat. Sci. Technol., 6, 293 (1990).
5. K. Binder, D.W. Heermann 'Monte Carlo Methods in Statistical Physics' Springer-Verlag Berlin Heidelberg (1988)
6. N. Metropolis, A.W. Rosenbluth, A.H. Teller and E. Teller, J. Chem. Phys. 21, 1087 (1953)
7. B. Sundman, B. Janson and J. Andersson, Calphad, 9(2), 13 (1985)
8. J. W. Cahn and J. E. Hilliard J. Chem. Phys 28, 258 (1958)
9. M. Copetti and C. Elliott, Mater. Sci. and Technol., March (1990)
10. A. Cerezo and M.G. Hetherington, J. de Physique., C8-50 , 523 (1989)
11. A. Cerezo, T.J. Godfrey, C.R.M. Grovener, M.G. Hetherington, J.M. Hyde, J.A. Liddle, R.A.D. Mackenzie and G.D.W. Smith, EMSA Bulletin 20:2, 77-83 (1990)
12. M. Miller, L.L. Horton and S. Spooner, J. de Physique., C2-47, 409 (1986)
13. I.M. Lifshitz and V.V. Slyozov, J. Phys. Chem. Solids, 19, 35 (1961)
14. H.L. Snyder, P. Meakin and S. Reich, Macromolecules, 16, 757 (1983)
15. M.G. Hetherington and M.K. Miller, J. de Physique, C8-50, 535 (1989)
16. T.J. Godfrey, M.G. Hetherington, J.M. Sasson and G.D.W. Smith, J. de Physique, C6-49, 421 (1988)
17. J. S. Langer, M. Bar-on and H. D. Miller, Phys. Rev. A11, 1417 (1975)
18. T.Y. Kong and A. Rosenfeld, Comp. Vision, Graphics and Image Processing 48, 357-393 (1989)
19. P.P Camus, W.A. Soffa, S.S. Brenner and M.K. Miller, J. de Phys. 45-C9 (1984).
20. A. Cerezo, M.G. Hetherington, J.M. Hyde and M.K. Miller, in press Scripta Metall. et Mater., 25 (1991)

Supercomputer Applications in the Civil Engineering Field

T. Takeda, Y. Omote, H. Eto, A. Kawaguchi
Technical Research Institute, Obayashi Corporation, 4-640 Shimokiyoto, Kiyose-shi, Tokyo 204, Japan

ABSTRACT

This paper is divided into two parts. The first part introduces the state of the art of supercomputer application in the field of civil engineering in Japan. Some examples such as structural analyses, vibration analyses, soil mechanics, fluid analyses, acoustic analyses and marine analyses are introduced. The effective use of supercomputers and some requirements for performing computational engineering in the civil engineering field are also discussed. The second part presents specific examples of analytical research work which focuses on two on-going projects of the buiding design. A three dimensional and nonlinear response analysis of a reinforced concrete building under strong earthquake exitation is conducted for the first case. Then, both a two-dimensional and a three-dimensional fluid anlyses to the other project which is twin tower high-rise steel structure against the wind force are made.

INTRODUCTION

The use of supercomputers in the engineering field started in the early part of 1980 in Japan. There are reported to be about 190 units of supercomputers now working in Japan, with 25% of the units belonging to computer centers in the universities and 15% of the units being used in public research institutes primarily for basic research work. In addition, 60% of the 190 units of supercomputers are now being used in various research institutes of private companies, and the applications are expanding from basic research work to practical development and design.

The most encouraging users are particulary LSI or VLSI design in the electronic field, molecular design in chemistry and medicine and anti-impact design and/or fluid dynamics around a body in automobile production.
Utilization of supercomputers in the civil engineering field began actively in late 1980. Many kinds of structures are required to meet social needs, and these structures should be designed safely and economically to deal with natural external forces such as earthquakes, strong winds such as typhoons and tsunami. Moreover, pursuit of amenities for human living space is another important factor in building design and a high quality of design through

precise analytical techniques has been recently required.

This paper is divided into two parts. The first part introduces an idea of the effective use of supercomputers in civil engineering field and presents some analytical results against the practical design applications. It also demonstrates the present state of the art of technical level and discusses the possibility of future trends towards the performance of computational simulations instead of experimental tests. The second part presents a specific example of analytical research works, focusing on two kinds of on-going projects of building design with the use of a supercomputer.

1. The state of the art of supercomputer applications in the field of civil engineering in Japan

1.1 Effective Use of a Supercomputer

Although a supercomputer permits high-speed calculation and has a very large memory capacity, it mainly needs three additional things for effective use.
1) A specially developed fully vectorized software program is required. When a program is developed for general purposes on a commercial basis, there may be a little problem with vectorizing. If the program is developed on an in-house basis for a specific purpose, however, full attention should be paid to speeding the logic during the programing process. 2) Since a complicated and large sized analytical model is required, graphic treatment of the data for both input and output is necessary. Visualization techniques for pre and post data processing, save time and improve precision in judging the calculated results. 3) Animation techniques are necessary to judge the adequacy of computed results especially for dynamic analyses, and also to fully understand the dynamic characteristics.

1.2 Examples of Civil Engineering Applications

<u>Structural Analyses</u> These types of analyses are usually conducted by the finite element method(FEM).
Three dimensional stress-strain analysis of steel structures, reinforced concrete structures and steel-concrete conposite structures are the main focus. From the technical point of view, geometrical or material inelasticity, contact problem, velocity dependency of strain hardening, buckling phenomena and so on are the concerns. Figure 1.1 shows an impact reponse analysis made on a model of approximately 8,000 elements for a flying object colliding with a reinforced concrete slab.

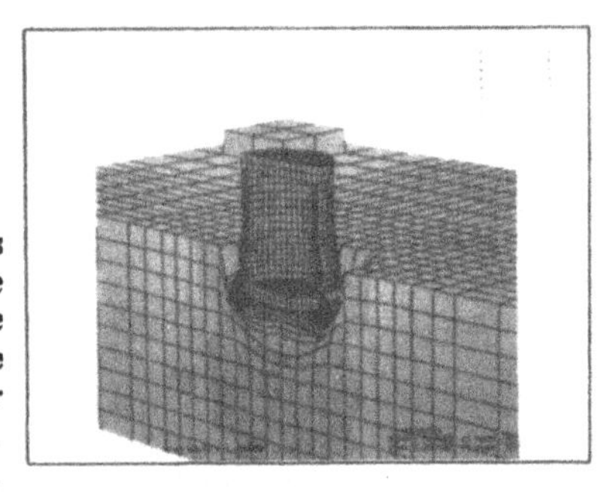

Fig.1.1 Impact analysis

<u>Vibration Analyses</u> An earthquake resistant design for a structure is very important in Japan and precise analytical research should be conducted beyond elasticity. A three dimensional frame structure, as shown in Fig.1.2, which consists of girders, columns and shear wall elements is modeled in

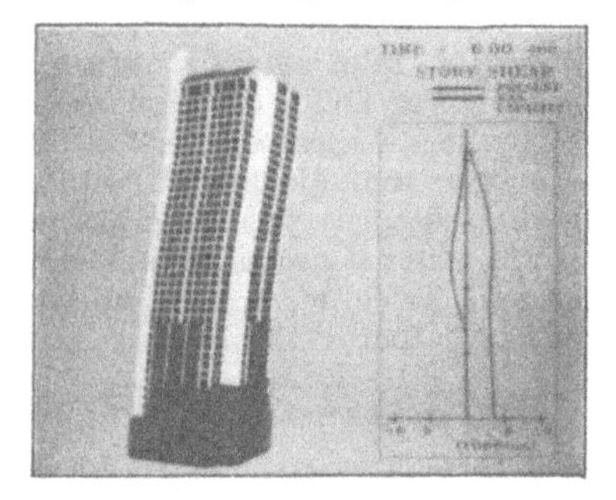

Fig.1.2 Vibration Analysis

consideration of the inelasticity of each element, and the mathematical model is calculated for horizontal and/or vertical earthquake excitations. Because element has six-degrees of freedom, the rigidity matrix is very large and the the step-by-step time history response requires a very long computation time for simulation of an earthquake. Other types of structures such as nuclear power plant structures are also an important object, and a soil-structure interaction analysisis required for the design of such a structure. Usually, FEM or BEM or a combination of these two is used to solve that kind of structure.

Soil Mechanics In order to construct a hard construction structure below ground level, such as a subway tunnel or underground fuel tank, it is very important to precisely anticipate the ground stress and strain relationship or settling behaviour of the structure during construction. A three dimensional analysis which takes into account the inelastic behaviour of soil is also necessary by FEM. Figure 1.3 shows an example of stress analysis for a deep tunnel during construction. Another important issue is to analyze the liquefaction mechanism such that observed in the Loma Prieta Earthquake in USA in 1989. Two dimensional earthquake response calculation is commonly used at present, because of the difficulty of defining the precise constitutive model of soil for a 3D stress analysis, in addition to the ability of a computer.

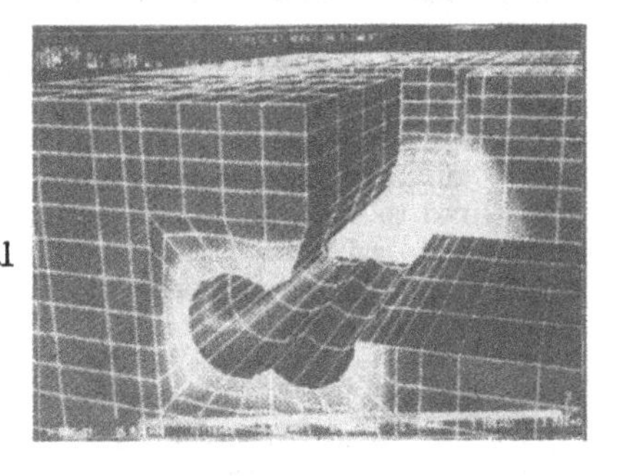

Fig.1.3 Soil Mechanics

Fluid Analyses In the civil engineering field, the need for fluid analyses is divided into two categories. One is to analyze the wind force on structures, especially tall buildings, membrane structures and suspension bridges, in order to create a wind resistant design. The other is for the purpose of air conditioning design, where air current and temperature distribution analyses are conducted inside the rooms. Mathematical calculations are based on a Navier-Stokes equation which can be solved by the differential method or by FEM. In order to calculate wind pressure distribution against a tall buiding, for example, the region for analysis should be taken as several times the size of the structure itself. Over 100,000 meshes are needed for the mathematical model and the number of steps for calculation is usually over 10,000.
A very high-speed supercomputer is required for 3-D analysis. Figure 1.4 shows an example of air flow simulation within and around a large space structure which has a retractable roof, using direct numerical simulation (DS) in the place of turbulence model.

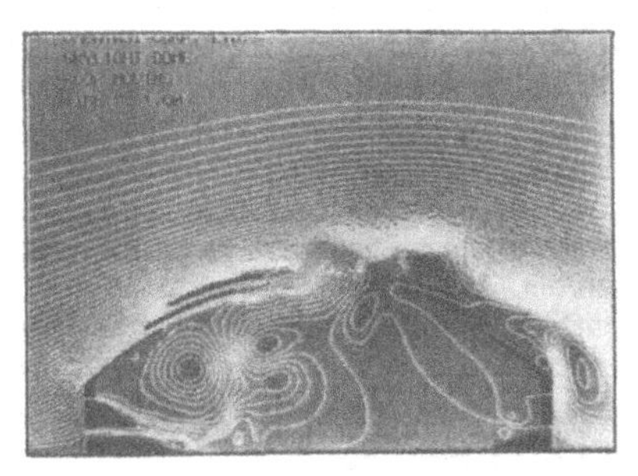

Fig.1.4 Fluid analysis

Acoustic Analyses An acoustic design has to create a very attractive and comfortable acoustical space which meets

a person's expectations. To make such an acoustic design, computer simulation is producing very useful support tools, because it permits us to promptly forecast an acoustic field and to visually produce the sound propagation with the use of an engineering work station(EWS). Figure 1.5 shows an example of an analysis of sound propagation through transient responses in an existing hall, based on Kirch-hoff's integral equation.
Because computer speed is limited, the number of sound sources on the stage is assumed to be one for this calculation. Multiple sound source analysis is required to simulate an orchestra performance and will necessitate the use of a more high-speed supercomputer than now exists.

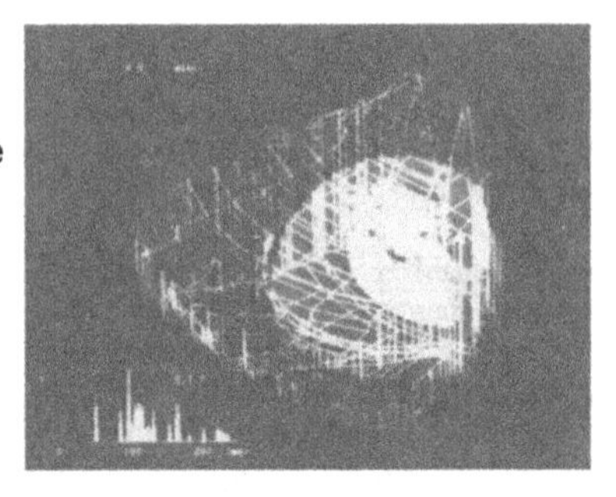

Fig.1.5 Acoustics Analysis

Marine Analyses In order to create a good marine environment, it is necessary to know how the environments surrounding the local sea waters may change, and affect a marine structure constructed there. Furthermore, from the point of view of disaster prevention, it is very important to numerically simulate and clarify the effects of a tsunami caused by near or distant earthquake upon water-front structures. Figure 1.6 shows an example of trance-oceanic wave propagation caused by the Chilean Tsunami, which took place in 1960. In this analysis, the linear long-wave theory on the polar coordinate system was applied, and the difference calculus grating was given intervals of 10 minutes, which made a total of 734,400 grids for calculation.

Fig.1.6 Tsunami Analysis

1.3 Computational Engineering with the Use of a Supercomputer

Conventionally, when a new design method or new construction technique is developed, the adequency of the design or technique is confirmed by experimental laboratory tests. However, when many tests are required or when it is difficult to do the tests with the use of a full scale structure and effects of natural phenomenon cannot be adequately tested, computer simulation provides a powerful approach to future design. To achive this computational engineering, there are some problems to be overcome. 1) The appropriateness of the application of the program to the object should be confirmed. 2) Fully tuned-up software which is suited for supercomputer is required. 3) Pre and post data processing should be prepared in order to reduce the total calculation procedure. 4) A data exchange system between a supercomputer and a EWS should be produced which makes it convenient for users to do scientific visualizations. 5) Data transmission speed between the supercomputer and the EWS must be as fast as possible to produce a real time visual simulation.

2. Three-dimensional Nonlinear Seismic Response Frame Analyses of Reinforced Concrete Highrise Buildings

2.1 Introduction

Today, social demands concerning usage, planning, design, etc. of buildings have become variegated and buildings of complex plan and elevation configurations have been increasing in number. Such buildings with eccentricity are liable to suffer local damage during a severe earthquake, which will hasten overall failure. But with such buildings, it is possible for safe and high-quality design to be made by grasping in detail the three-dimensional non-linear behaviors of buildings during severe earthquakes at the level of individual members.

Therefore, individual members were replaced with idealized member models, and a three-dimensional dynamic response frame analysis program, "DREAM-3D", was developed for supercomputer software.

An outline of this program is given here along with an example of application to seismic response analysis of a 23-story highrise reinforced concrete building with the use of SX-1EA.

2.2 Outline of DREAM-3D

2.2.1 Analysis Method

The building is modeled as a three-dimensional frame by models of members such as beams, columns, shear walls, and beam-column connections, and torsional deformation of the building as a whole has been considered.

The analysis uses incremental analysis method. Regarding earthquake input and wind input, it is possible for the waveforms of two directions horizontally and one direction vertically to be inputted simultaneously.

2.2.2 Member Model

The members are replaced by idealized member models so that it will be possible for the instantaneous stiffnesses of the members to be directly obtained from end forces and end displacements of members. Restoring force characteristics of members are defined independently for each kind of deformation component such as in the flexural, shearing and axial forces of the various components. The torsional deformations of the individual members are not considered.

<u>(1) Beam Member</u> For a beam member model, the degree of freedom in vertical plane only is considered regarding bending deformation, and a plane member model combining "Connected Two-cantilevers Model[1]" and "Member-end Rotational Springs Model[1]" is employed as shown in Fig. 2.1.

<u>(2) Column Member</u> During an earthquake, lateral forces in two directions and axial force act on a column and the intensities of these forces vary in a complex manner. The restoring force characteristics of bending moments in two directions (Mx, My) and axial force (N) of the column will be interacted, and it will be necessary for a column member model to be made capable of accurately expressing the triaxial yielding interactions of these forces.

A column member model considering steel construction uses "Divided Beam Model[2]", and expresses the triaxial yielding interactions by yielding curved surfaces which consists of two varieties of elliptical and parabolic.

In case of reinforced concrete construction, the hysteresis loops during earthquake will vary with complication due to cracking of concrete, yielding of reinforcing bars, etc., and it will not be appropriate to express the triaxial

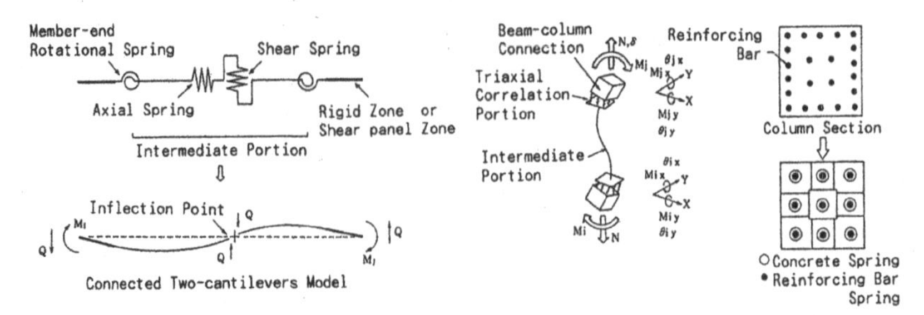

Fig. 2.1 Beam Member Model

Fig. 2.2 Reinforced Concrete Column Member Model

yielding interactions by yielding curved surfaces as with steel construction.

"Multi-springs Model[3]" dividing the cross section at member end into equivalent axial-direction springs expressing the axial characteristics of concrete and reinforcing bars, as shown in Fig. 2.2, is used for a reinforced concrete column member model in this program, and the triaxial yielding interactions can be expressed automatically and readily without using yielding curved surfaces.

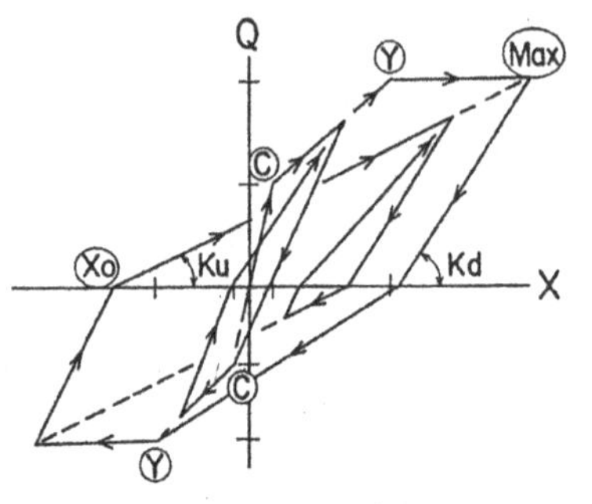

Fig. 2.3 Takeda's Model

(3) Shear Wall member For a shear wall member model, the model replacing the edge columns and wall panel portions with their respective elements is used. The edge columns are the same models as reinforced concrete column members, provided that the wall in-plane direction is pin-connected at both ends. The wall panel portion is expressed by the same model as a beam member, and is connected to edge columns by rigid beams at top and bottom.

2.2.3 Restoring Force Characteristics Model

As restoring force characteristics models, there are a great many available beginning with trilinear models represented by the Takeda's Model[4] used for reinforced concrete members, as shown in Fig. 2.3, and including bilinear models, multilinear models. Therefore, models suitable for analysis can be selected and used.

2.3 Seismic Response Analysis of 23-story Reinforced Concrete Highrise Building

2.3.1 Outline of Building

This building is a reinforced concrete apartment house building 23 stories above ground with one basement of open frame construction except that it has shear walls at the basement as shown in Fig. 2.4. It is featured by the plan and elevation configurations being irregular. Especially in the transverse direction which is Y direction, there are no beams at the part of midway up at frame ④.

The first natural periods in the individual directions are 1.37 sec (X direction), 1.51 sec (Y direction), 1.28 sec (θ direction).

2.3.2 Earthquake Resistant Design

In earthquake resistant design of this building with eccentricity, an ultimate strength design method is adopted so that input energy applied to the building during a severe earthquake is absorbed by the plasticization such as cracking of concrete, yielding of reinforcing bars, etc. of members. While moreover, the plasticization of members is not concentrated at limited parts of the building, but is evenly dispersed throughout the whole for safety and also economical design against severe earthquakes.

Accordingly, members are made to have restoring force characteristics of excellent ductility, besides which, at the final stage of design, nonlinear seismic response analyses should be performed in order to ascertain whether the designed building satisfy the design criteria. The design criteria for this building subjected to severe earthquakes whose maximum velocity is 50 cm/sec (kine) are as follows.

Collapse mechanism : a total collapse mechanism of beam-yielding type	
Story drift angle	$R \leqq 1/100$
Ductility factor of story	$\mu \leqq 2$
Ductility factor of member	$\mu \leqq 2$ for column
	$\mu \leqq 4$ for beam

In this building, as seismic response analyses are conducted by DREAM-3D, it is possible to ascertain at the level of individual members whether the design criteria are satisfied, and this enhances the quality of design in earthquake resistant design.

The ultimate lateral shear possessed in terms of base shear coefficient C_B of this building are 0.183 in the X direction and 0.166 in the Y direction.

2.3.3 Analysis Procedure

The input seismic wave in the Y direction of the building was assumed as the El Centro 1940 NS strong motion record for which response was greatest from the results of approximate analyses by shear-type lumped mass models with various strong earthquake records as input seismic waves. The maximum velocity assumed 50 kine (511 gal, 1.5 times the original record) considering a severe earthquake. The input seismic wave in the X direction assumed as the EL Centro 1940 EW strong motion record, and used amplifying the original record 1.5 times similarly to the Y direction.

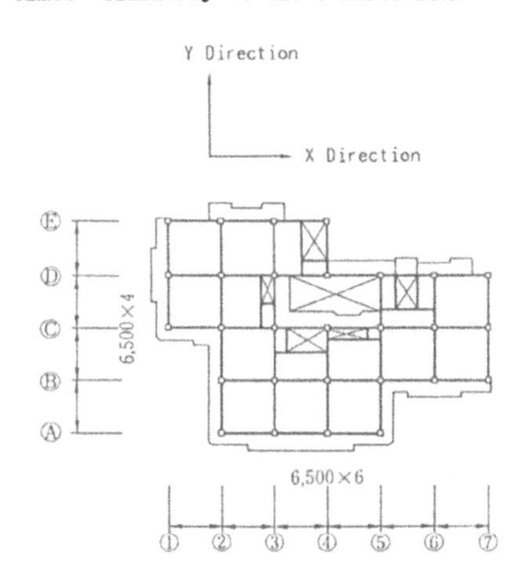

(a) Plan

(b) Elevation

Fig. 2.4 Analyzed Building

Both the one direction of "Y-Direction Input" and the two direction of "X·Y-Direction Input" simultaneously are assumed for calculations.

2.3.4 Analysis Results

An example of displacement diagrams is shown in Fig. 2.5、 where displacement amplified 15 times of actual value for the purpose of visual presentation.

(1) Maximum Response Value of Story Drift The distributions of maximum response values of story drift δ at the centers of gravity of stories of Y-Direction Input are shown in Fig. 2.6. The results of the three-dimensional frame analyses are roughly equal to the results of shear-type torsional response analyses.

Figure 2.7 shows that the maximum response values of story drifts for X·Y -Direction Input are larger than the analysis results for Y-Direction Input except for the top and bottom portion of the building.

The maximum response values of story drifts are in terms of deformation angle between story R = 1/125 (Y-Direction Input) and 1/118 (X·Y-Direction Input), ductility factor of story μ = 1.08 (Y-Direction Input) and 1.21 (X·Y-Direction Input). These values are within the design criteria and ascertained the safety of the building under a severe earthquake.

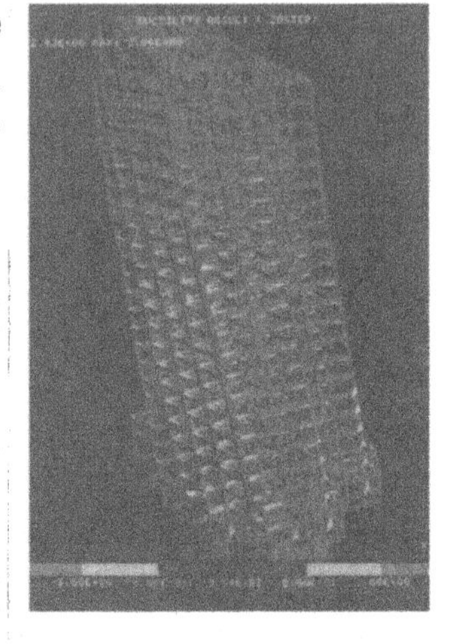

Fig. 2.5 Displacement Diagram

(2) Yielding Hinge Occurrence Locations and Ductility Factor of Beams and Columns

The yielding hinge occurrence locations of beams and columns of Frame ④ and ductility factors μ are shown in Fig. 2.8. The building is designed as beam-yielding type, but it has not reached the stage of collapse mechanism.

The results of analyses for X·Y-Direction

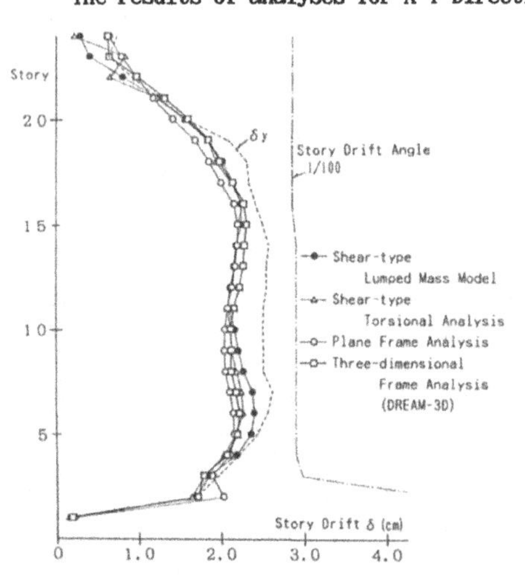

Fig. 2.6 Maximum Story Drift (Y-Direction Input)

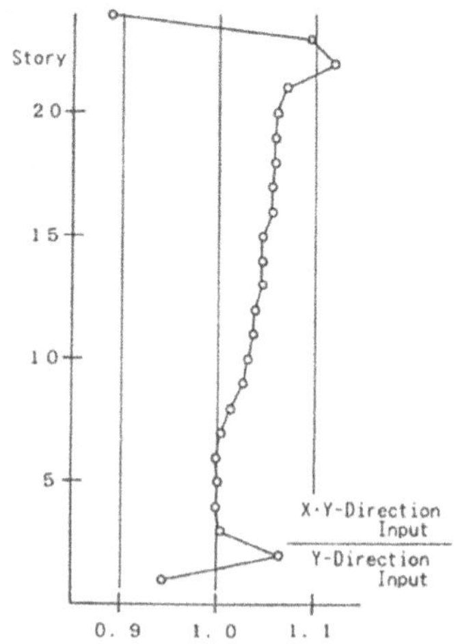

Fig. 2.7 Comparison of Maximum Story Drift

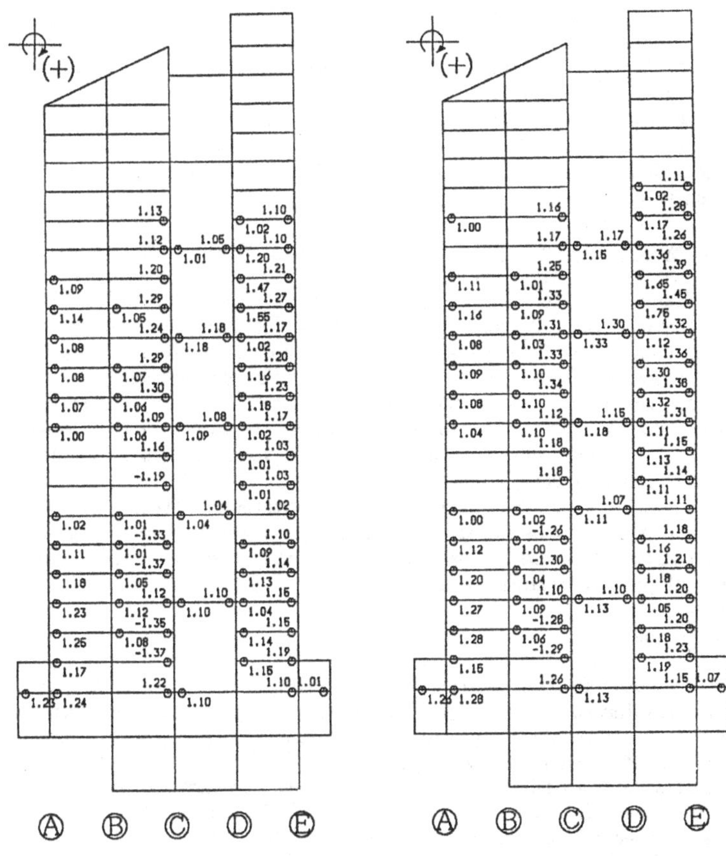

(a) Y-Direction Input (b) X·Y-Direction Input

Fig. 2.8 Ductility Factor of Beams and Columns

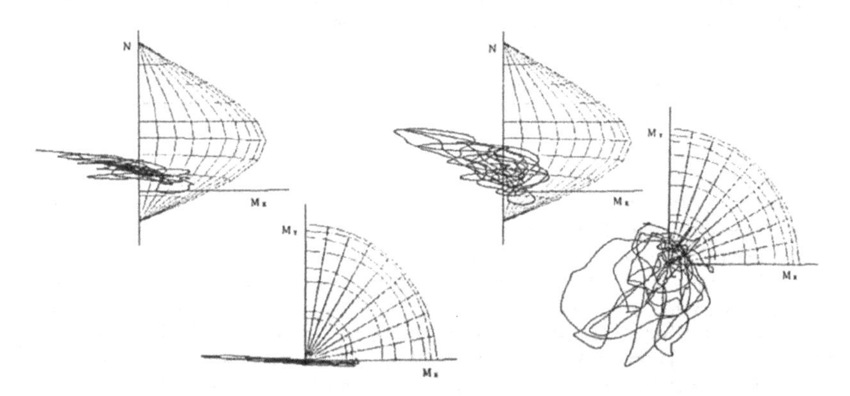

(a) Y-Direction Input (b) X·Y-Direction Input

Fig. 2.9 N-Mx-My Orbital Plots of Corner Column
(Column Ⓑ–⑦ of the first story)

Input show ductility factors to be larger than Y-Direction Input at beam ends of upper stories at Ⓒ~Ⓓ and Ⓓ~Ⓔ. Further, yielding hinges have been produced at the bottom portion of first-story column of other frames as evidence of the influence of two-way directions input.

The maximum response values of ductility factors of beams were μ = 1.55 (beam) for Y-Direction Input, and μ = 1.75 (beam), 1.10 (column), and 1.17 (orthogonal beam) for X·Y-Direction Input. These results are also satisfied by the design criteria.

(3) N-Mx-My Orbital Plots of Columns The N-Mx response hysteresis curves seen from the My-axis direction and the Mx-My response hysteresis curves seen from the N-axis direction at a corner column (Ⓑ—⑦) of the first story in analysis results for Y-Direction and X·Y-Direction Input are shown in Fig. 2.9. The strength envelopes proposed by Dr. Abe[5] as yielding interaction curved surfaces are shown in the figures for reference purposes.

On looking at the N-Mx response hysteresis curves, at the corner column, the axial-force variation in two-way directions is added in the analysis results for X·Y-Direction Input, and axial-force variation is larger than in the results for Y-Direction Input.

On examination of the Mx-My response hysteresis curves, all of the columns are seen to have moved in a roughly 45-deg direction on the Mx-My plane at maximum response of bending moment, and there is a trend for the influence of biaxial bending to appear.

2.4 Conclusion

In order to grasp in more detail the three-dimensional nonlinear behavior of a highrise reinforced concrete building under input of earthquake from arbitrary, multiple directions, a nonlinear seismic response analysis program, "DREAM-3D", employing idealized member models was developed, and simulation analysis of a 23-story building by supercomputer was carried out.

In this analysis, the behaviors as a three-dimensional framework which cannot be grasped by conventional analysis of a plane frame, and the nonlinear behavior under triaxial stress of column can be pursued in detail at the member level, and this can be a powerful tool for enhancing the quality of design in earthquake resistant design of a highrise reinforced concrete building of complex plan and elevation configurations.

References in Chapter 2

1) Takizawa, H., "Notes on Some Basic Problems in Inelastic Analysis of Planar R/C Structures (Part 1)", Transactions of the Architectural Institute of Japan, No. 240, pp. 51-62, Feb. 1976.
2) Riahi, A., Row, D.G. and Powell G.H., "ANSR 1:INEL-2 & 3 —Three Dimensional Inelastic Frame Elements for the ANSR-1 Program", Earthquake Engineering Research Center Report, No. 78/06, Univ. of Calif., Berkeley, Aug. 1978.
3) Li, K., Otani, S. and Aoyama H., "Behaviour of Reinforced Concrete Columns under Varying Axial and Earthquake Loads (Part 2)" (in Japanese), Proc. Kanto Dist. Symp., Architectural Institute of Japan, pp. 157-160, May 1986.
4) Takeda, T., Sozen, M.A. and Nielsen, N.N., "Reinforced Concrete Response to Simulated Earthquakes", Journal of the Structural Division, American Society of Civil Engineers, Vol. 96, No. ST12, pp. 2557-2573, Dec. 1970.
5) Abe, K., "A Proposition of Equations to Calculate the Ultimate Bending Strength and Interaction Curved Surface of Reinforced Concrete Columns" (in Japanese), Transactions of the Architectural Institute of Japan, No. 366, pp. 94-105, Aug. 1986.

3. Computational Analyses of Wind Force

3.1 Structures for Analyses

The structures made the objects of analyses were high-grade apartment houses consisting of two high-rise buildings.The architectural design was made by the world famous architect, Arata Isozaki. The feature of this project from the standpoint of architectural designing is that a circular building of diameter, of standard floor,approximately 28 meters and height of approximately 120 meters (Tower A, 33 stories above ground) and a square prism building of length of one side approximately 24 meters and height of approximately 125 meters (Tower B, 35 stories above ground)are connected by a bridge at a height of 100 meters. There is also a heliport on top of Tower A which can be used for rescue activities in an emergency. The towers are shown in Fig. 3.1.

The feature from the standpoint of structural design is that the bridge connecting the two buildings has free support at the Tower B side, so it can be considered that the buildings will exhibit independent behaviors against an external force. Because the buildings are constructed by steel,the natural period of the buildings is long compared with other structural systems. This makes vibration due to wind force a matter of concern in wind-resistant-design. Particularly, since this project is planned for reclaimed land close to the sea in Fukuoka Prefecture, Japan, along a so-called 'typhoon corridor',there is all the more necessity for special study on wind analyses.

Tower A Tower B

Fig.3.1 Computer graphics of towers

3.2 Purpose of Wind Engineering Study

A wind engineering study of a high-rise building is usually made regarding the following items. ① Prediction of strong wind which the building will encounter in the future through an understanding of the characteristics of wind in the site area. ② Study of the aerodynamics and dynamic effects. It is necessary to evaluate the aerodynamic effects as a primary problem, because wind force acting on a structure depends greatly on its configuration. Further, the dynamic characteristics of a structure will be a major factor of resonance with turbulence of wind and with eddies occurring from behind, so that generation of self-excited vibration and dynamic effects must be studied including these characteristics. Furthermore,studies must be made not only regarding the problem of structural safety, but also the following items in order to secure comfort in daily life. ③ Study of comfort of occupants under influence of vibration of a building. ④ Study of strong surface winds at pedestrian level near tall buildings.

These problems of aerodynamics have already been studied through wind tunnel tests using 1/300 ~1/500 scale models. As a result, it was ascertained that the maximum deformation of the building predicted during the service life of the building responds in the elastic region, and that the wind load is adequately small compared with the earthquake load. The problems of comfort of occupants and strong surface winds near the towers have already been examined.

Purpose of numerical simulations in fluid analyses will be to set up a high performance computer as a numerical wind tunnel in place of a conventional wind tunnel test in the future. In this sense, it is firmly believed that to make comparisons of data of wind tunnel experiments and numerical fluid analysis results in this paper will suggest recognition of the present state of simulations and the principles for future research.

3.3 Simulation of Wind Force

The code for fluid analyses is HotflowII developed by Kozo Keikaku Engineering Inc.. This software solves the Navier-Stokes equations which is the motion equation for fluids by the finite difference method. Further, the turbulence characteristics of wind is modeled by the Large Eddy Simulation, so that the field of unsteady flow can be simulated.

In fluid analyses of the flow field around a body, how much of an analytical region can be secured is the object, and how minute a mesh can be made are important problems. Consequently, there is the aspect of depending on the performance of computer hardware. The analytical region and the condition of the mesh is shown in Fig. 3.2. In this case, the mesh division were 99 (wind-direction) x 97 (closs-wind direction) x 46 (vertical direction). However, work done on exterior walls of the actual buildings and variations in configurations at upper stories of the buildings are modified. The bridge connecting the two buildings is omitted to match the wind tunnel model. The wind speed is taken as the expected wind speed of 17m/s for a return period of one year from the meteorological observation data near the site. The mean wind profile and the direction were made to match the wind tunnel experiment. Mean wind profile is defined by following.

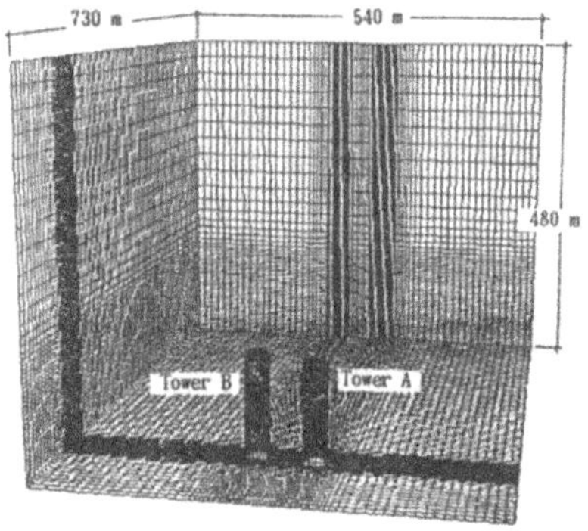

Fig. 3.2 Condition of mesh

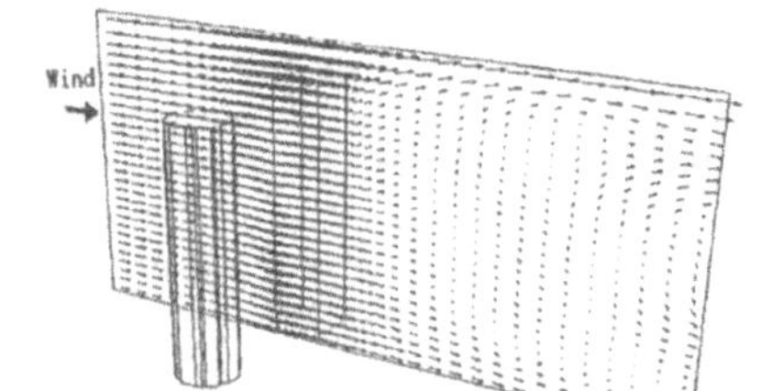

$$U_Z = U_0 (Z/Z_0)^{1/n} \quad , \quad 1/n = 1/7$$

Fig. 3.3 Wind velocity distribution

where, U_Z is mean wind velocity at height Z m (m/s), U_0 is mean velocity of reference wind at reference height Z_0 m (m/s), and 1/n is the roughness parameter of ground in the windward of the site. The wind speed distribution on the vertical plane at an instantaneous condition is shown in Fig. 3.3, and the pressure distribution on the surface of the towers and on the horizontal plane is shown in Fig. 3.4. Figure 3.5 shows the pressure distribution of two-dimensional

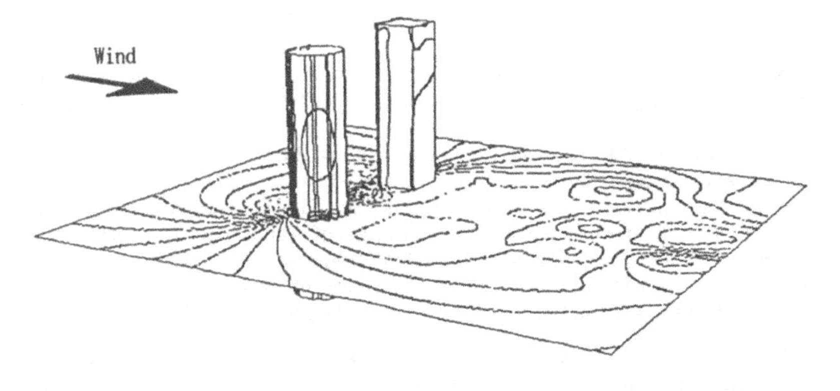

Fig. 3.4 Pressure distribution in three-dimensional analysis

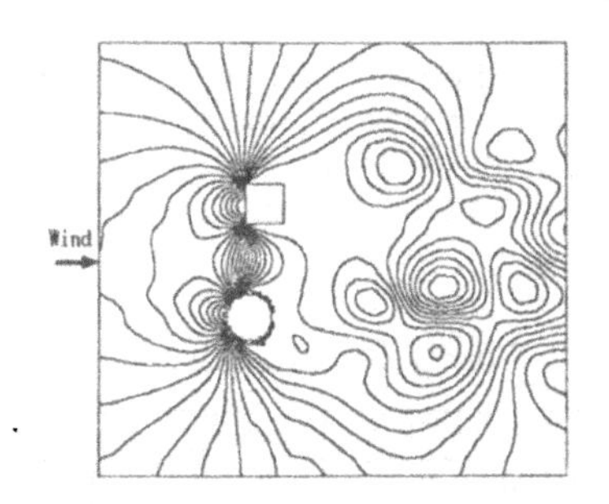

Fig. 3.5 Pressure distribution in two-dimensional analysis

Table 3.1 Coefficient of wind force

Building	Tower A				Tower B			
Coefficient	C_D		C_L		C_D		C_L	
Component	ave	rms	ave	rms	ave	rms	ave	rms
Experiment	0.68	0.04	0.13	0.06	1.13	0.09	0.01	0.20
2-D Simulation	1.17	0.03	0.10	0.01	1.54	0.04	0.03	0.02
3-D Simulation	0.70	0.01	0.06	0.01	0.91	0.01	0.04	0.01

ave : averaged value , rms : value of root mean square

simulation, which is performed extracting a story of standard cross section of towers as the preliminary analysis.

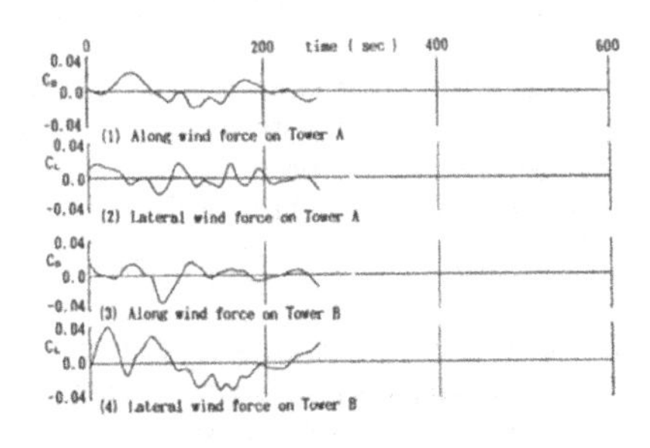

Fig. 3.6 Fluctuating wind force in three-dimensional analysis

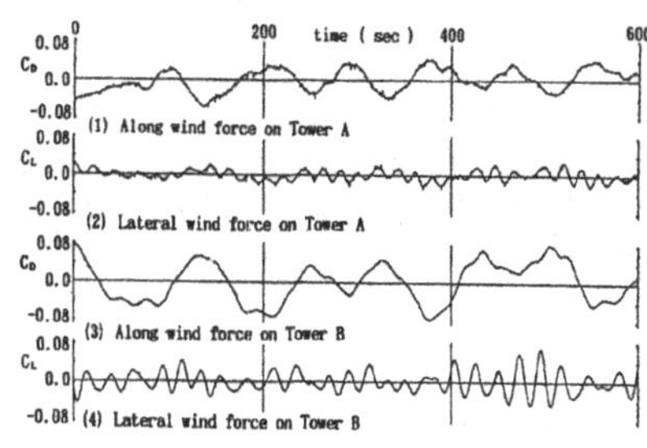

Fig. 3.7 Fluctuating wind force in two-dimensional analysis

The fluctuating wind force acting on the building is calculated with the pressure at the element closest to the building as along-wind force and cross-wind force. The time-history waves, which is wind force acting on the section at 2/3 height of the tower resulted from the three-dimensional analysis, is shown in Fig. 3.6, while the waves resulted from the two-dimensional analysis is shown in Fig. 3.7. The statistical value of wind force is given in Table 3.1. The coefficient of wind force in the figure and the table is defined by the following.

$C_D=Q_D/qA$, $C_L=Q_L/qA$, $q=1/2\cdot\rho U^2$

where, C_D and C_L are coefficients of wind force along-wind and cross-wind, and are respectively called drag coefficient and lateral coefficient. As for Q_D and Q_L, they are wind force (base shear) acting along-wind and cross-wind, q is reference dynamic pressure at height of building top, and A is vertical configuration area of building. ρ is the air density. Further, ave and rms in the table are corresponded averaged wind force and root mean square of fluctuating wind force, respectively. On comparison of analytical and experimental results, two-dimensional simulation results of average coefficients show almost greater than the other, while rms coefficients of deviation are smaller for analytical results than for experimental results both C_D and C_L.

The tendency, which wind force averaged in two-dimensional analysis is larger than that in three-dimensional analysis, is generally similar to that of results in wind tunnel tests. It is an important problem that fluctuating characteristics of wind in the simulations do not grow enough like the natural wind within the computed time span reported herein.

Next, comparisons will be made concerning eddies which occur (call Karman vortex when periodic) from behind the structure. In general, the frequency for eddy occurrence depends on the configuration of the building, and if of the same configuration, this can be defined by the Strouhal number.

$S=nD/U$

where, S is Strouhal number, n is frequency of full cycles of vortex shedding (Hz). D is characteristic dimension of structure projected on a plan normal to the flow and U is wind speed of the oncoming flow (m/s). Strouhal number calculated from experiments and analyses is given in Table 3.2.

Table 3.2 Strouhal number

Building	Tower A	Tower B
Experiment	0.111	0.104
2-D Simulation	0.120	0.106
3-D Simulation	0.111	0.104

The Strouhal number is in all a good agreement and it may be judged that simulation of qualitative state of eddy occurrence is possible even in a complex situation. However, the Vortex behind the towers in three-dimensional simulation occurs weakly and the wind force varies slowly in comparison with the period defined by Strouhal number. In the future, it must be studied to simulate vortex occurred behind a structure for the numerical analyses.

3.4 Conclusion

Numerical simulations of wind forces have been described against a project. Fluid analyses can be simulated roughly well to the results of mechanical wind tunnel tests for averaged field. The characteristics of fluctuating field, however, are not reproduced enough to the wind tunnel condition. This is the important problem, which must be solved to be a numerical wind tunnel in the future. As a conclusion, it may be presumed that the focus of results of numerical fluid analyses is how turbulence characteristics of natural wind are to be simulated. Consequently, in development of a numerical wind tunnel, what is looked forward to is the modeling of turbulence as an analytical technique in the aspect of software, and improvement of memory capacity and computing speed in the aspect of hardware, especially for three-dimensional fluid analyses.

Dynamic Finite Element Analysis of Viscoelastically Damped Composite Structures

H.H. Hilton, S. Yi
Aeronautical and Astronautical Engineering Department, University of Illinois at Urbana-Champaign, Urbana, IL 61801, U.S.A.

Abstract

Numerical procedures for analyzing transient responses of composite structures with time and temperature dependent linear viscoelastic damping have been developed in the time domain using variational principles and the Newmark average acceleration method. Recursion formulas also have been obtained in order to reduce computer storage and only the previous time solution is required to compute the next time solution. Verification studies were conducted to evaluate the accuracy and convergence of the present numerical algorithm and they show that the numerical solutions agree well with the analytical ones. As examples, the dynamic transient responses of long cantilevered T300/934 composite Euler beams subjected to unit step and impulse loads were calculated using the presently developed procedure.

Introduction

The damping ability of viscoelastic materials is of major importance since it is useful in dissipating undesirable sound and vibration energies and in delaying structural fatigue. Furthermore, the replacement of metals by advanced composite materials leads to the development of more weight efficient aerospace structures. Polymer matrix composite materials exhibit viscoelastic behavior [1,2] and thus dynamic analysis of anisotropic viscoelastically damped structures must be performed during the design phase. However, since the system equations for linear viscoelastic materials contain hereditary integrals, it is not possible to analytically solve many dynamic viscoelastic problems, and numerical approaches requiring high speed computers must be used.

Analyses of dynamic responses of viscoelastic structures by finite element methods were developed in References [3-5]. Johnson *et al.* [3] obtained approximate viscoelastic responses by using modal frequency response methods. Bagley *et al.* [4] and Golla *et al.* [5] developed finite element procedures in the transform and time domains respectively, but their methods are computationally expensive since their system equations are much larger and complicated than corresponding elastic ones. Zabaras *et al.* [6] evaluated dynamic responses in laminated composite plates using viscous damping approximations. The present authors [7,8] previously developed finite element procedures to analyze the quasi-static responses of anisotropic viscoelastic solids under time dependent mechanical and hygrothermal loads. These procedures are based on the use of Laplace transforms rather than direct time integrations and they significantly improve solution accuracy and also appreciably reduce computational times and storage needs. However, this method is limited to static analyses since the dynamic system equations require the simultaneous diagonalization of three matrices as opposed to two matrices in the quasi-static case. Otherwise, the system equations must be solve for each specified value of the transform variable and the solutions must then be transformed back into the time domain which may require storage of all solutions in the transform domain.

In the present study, a numerical procedure has been developed in the real time domain to evaluate the transient dynamic responses of viscoelastically damped composite structures. The approach is based on variational principles and the direct time integration method such as the Newmark average acceleration method [9] which has been successfully used for evaluating transient responses of elastic dynamic systems with or without viscous damping.

Governing Equations for Linear Viscoelasticity

The strain-displacement equations for linear viscoelastic materials are in Cartesian tensor notation

$$\epsilon_{ij}(\mathbf{x},t) = \frac{1}{2} \cdot [u_{i,j}(\mathbf{x},t) + u_{j,i}(\mathbf{x},t)] \tag{1}$$

and the constitutive relationship for linear anisotropic hygrothermoviscoelastic composites can be written as [10]

$$\sigma_{ij}(T,\ M,\ \mathbf{x},\ t) = \int_{-\infty}^{t} C_{ijkl}(T,\ M,\ t,\ \tau)\frac{\partial}{\partial \tau}[\epsilon_{kl}(\mathbf{x},\tau) - \epsilon^{*}_{kl}(\mathbf{x},\tau)]\ d\tau \tag{2}$$

where $C_{ijkl} = C_{klij}$, C_{ijkl} are relaxation moduli for the principal material coordinates, σ_{ij} are stress tensors and ϵ_{kl} and ϵ^{*}_{kl} are respectively the total strain tensors and the hygrothermally induced strain tensors.

Due to the postulate of hygrothermo-rheologically simple materials, the pseudo times ζ_{ijkl} associated with the shift functions b_{ijkl} can be defined as

$$\zeta_{ijkl}(\mathbf{x},t) = \int_{0}^{t} b_{ijkl}[T(\mathbf{x},t'),\ M(\mathbf{x},t')]\ dt' \tag{3}$$

Then the relaxation moduli can be expressed in terms of reference temperatures and moistures T_r and M_r as

$$C_{ijkl}(T,\ M,\ t) = C_{ijkl}[T_r,\ M_r,\ \zeta_{ijkl}(\mathbf{x},t)] \tag{4}$$

Variational Principles for Dynamic Viscoelastic Problems

Gurtin [11] has developed variational principles for quasi-static viscoelastic problems and Schapery [12] has proposed variational principles for steady-state dynamic viscoelastic solutions with and without thermomechanical coupling. In the present study, the variational principles analogous to Hamilton's principle are formulated for dynamic anisotropic hygrothermo-viscoelasticity as

$$\pi = \int_V \left[\int_{s=-\infty}^{s=t} \frac{1}{2} \rho \dot{u}_i(\mathbf{x}, t-s) \dot{u}_i(\mathbf{x}, s) ds - \int_{s=-\infty}^{s=t} \int_{\tau=-\infty}^{\tau=t} \left\{ \frac{1}{2} \bar{C}_{ijkl}[\zeta_{ijkl}(\mathbf{x}, t) - \zeta'_{ijkl}(\mathbf{x}, \tau), \zeta_{ijkl}(\mathbf{x}, t) - \zeta''_{ijkl}(\mathbf{x}, s)] \frac{\partial \epsilon_{ij}(\mathbf{x}, \tau)}{\partial \tau} \cdot \frac{\partial \epsilon_{kl}(\mathbf{x}, s)}{\partial s} - \bar{C}_{ijkl}[\zeta_{ijkl}(\mathbf{x}, t) - \zeta'_{ijkl}(\mathbf{x}, \tau), \zeta_{ijkl}(\mathbf{x}, t) - \zeta''_{ijkl}(\mathbf{x}, s)] \frac{\partial \epsilon^*_{ij}(\mathbf{x}, \tau)}{\partial \tau} \frac{\partial \epsilon_{kl}(\mathbf{x}, s)}{\partial s} \right\} d\tau ds \right] dV + \int_{S_T} \int_{s=-\infty}^{s=t} \bar{T}_i(\mathbf{x}, t-s) \frac{\partial u_i(\mathbf{x}, s)}{\partial s} ds dS_T \tag{5}$$

where $\bar{C}_{ijkl}$ are transformed moduli in the lamination directions, ρ is the material density, V is the volume of the viscoelastic solid, S_T is the surface on which tractions $\bar{T}_i$ are applied and u_i are the displacement vectors.

Formulation of Viscoelastic Finite Element Equilibrium Equations

In this section, the general finite element formulation for the dynamic analysis of anisotropic viscoelastic materials is derived using the displacement method. The displacement fields within each element are taken as a function of the nodal displacements

$$\{U(\mathbf{x},\ t)\} = [N(\mathbf{x})]\{\mathbf{q}(t)\} \tag{6}$$

where $[N(\mathbf{x})]$ is the displacement interpolation matrix within an element and $\{\mathbf{q}(t)\}$ are displacement vectors at the element nodes. The strain vector in Eq.(1) can then be described in terms of nodal displacements by differentiating Eq.(6) with respect to x_i as

$$\{\epsilon(t)\} = [\ \mathbf{B}(\mathbf{x})\ ]\{\mathbf{q}(t)\} \tag{7}$$

where $[\mathbf{B}(\mathbf{x})]$ is the element strain-displacement matrix.

The finite element equilibrium equations are obtained by substituting Eqs. (6) and (7) into Eq. (5) and by taking the first variation of the functional π. Then assembling the finite element matrices into global structure matrices provides the following integral system equations

$$M_{mn}\,\ddot{U}_n(t) + \int_{\tau=-\infty}^{\tau=t} K_{mn}[\zeta_{ijkl}(\mathbf{x},t) - \zeta'_{ijkl}(\mathbf{x},\tau)] \cdot \frac{\partial U_n(\tau)}{\partial \tau}\,d\tau = F_m(t) + F_m^{th}(t) \tag{8}$$

$$(m,\ n = 1,\ 2,\ \ldots,\ NT)$$

where $F_m(t)$ and $F_m^{th}(t)$ are nodal force vectors due to prescribed surface tractions and hygrothermal gradients respectively, M_{mn} and K_{mn} are nodal masses and stiffnesses.

Direct integration of Eqs. (8) demands enormous computing time and memory storage, since all the previous solutions would have to be stored in order to evaluate displacements at any nth time step. To avoid these large storage requirements, a numerical algorithm is developed here for the solution of Eqs. (8) using the Newmark average acceleration method [7] and Prony series representations for relaxation moduli.

For computational purposes, the relaxation spectra in Eq. (4) can be expanded in terms of Prony series summations and can be rewritten in the abbreviated form [15]

$$C_{ijkl}(\zeta_{ijkl}) = \sum_{r=1}^{R_{ijkl}} \left[C_r(\zeta_r) = C_r^0 + \sum_{h=1}^{N_r} C_{rh}^t \cdot \exp(-\zeta_r/\lambda_{rh}) \right] \tag{9}$$

where R_{ijkl} is the number of independent relaxation moduli, (i.e.,R_{ijkl}=9 for 3-D orthotropic composite materials and R_{ijkl} can rage up to 21), λ_{rh} are distribution times and N_r is the number of series terms.

In this procedure, the nodal displacements U_n and their time derivatives at time $t + \Delta t$ are approximated as follows:

$$U_n(t+\Delta t) = U_n(t) + \dot{U}_n(t)\Delta t + [\ddot{U}_n(t) + \ddot{U}_n(t+\Delta t)] \cdot \left(\frac{\Delta t}{2}\right)^2$$

analogous to Hamilton's principle are formulated for dynamic anisotropic hygrothermo-viscoelasticity as

$$\pi = \int_V \Bigg[\int_{s=-\infty}^{s=t} \frac{1}{2}\rho \dot{u}_i(\mathbf{x}, t-s)\dot{u}_i(\mathbf{x}, s) ds - \int_{s=-\infty}^{s=t} \int_{\tau=-\infty}^{\tau=t} \Bigg\{ \frac{1}{2}\bar{C}_{ijkl}[\zeta_{ijkl}(\mathbf{x}, t) - \zeta'_{ijkl}(\mathbf{x}, \tau), \zeta_{ijkl}(\mathbf{x}, t) - \zeta''_{ijkl}(\mathbf{x}, s)] \frac{\partial \epsilon_{ij}(\mathbf{x}, \tau)}{\partial \tau} \cdot \frac{\partial \epsilon_{kl}(\mathbf{x}, s)}{\partial s} - \bar{C}_{ijkl}[\zeta_{ijkl}(\mathbf{x}, t) - \zeta'_{ijkl}(\mathbf{x}, \tau), \zeta_{ijkl}(\mathbf{x}, t) - \zeta''_{ijkl}(\mathbf{x}, s)] \frac{\partial \epsilon^*_{ij}(\mathbf{x}, \tau)}{\partial \tau} \frac{\partial \epsilon_{kl}(\mathbf{x}, s)}{\partial s} \Bigg\} d\tau ds \Bigg] dV + \int_{S_T} \int_{s=-\infty}^{s=t} \bar{T}_i(\mathbf{x}, t-s) \frac{\partial u_i(\mathbf{x}, s)}{\partial s} ds dS_T \quad (5)$$

where $\bar{C}_{ijkl}$ are transformed moduli in the lamination directions, ρ is the material density, V is the volume of the viscoelastic solid, S_T is the surface on which tractions $\bar{T}_i$ are applied and u_i are the displacement vectors.

Formulation of Viscoelastic Finite Element Equilibrium Equations

In this section, the general finite element formulation for the dynamic analysis of anisotropic viscoelastic materials is derived using the displacement method. The displacement fields within each element are taken as a function of the nodal displacements

$$\{U(\mathbf{x},\ t)\} = [N(\mathbf{x})]\{\mathbf{q}(t)\} \quad (6)$$

where $[N(\mathbf{x})]$ is the displacement interpolation matrix within an element and $\{\mathbf{q}(t)\}$ are displacement vectors at the element nodes. The strain vector in Eq.(1) can then be described in terms of nodal displacements by differentiating Eq.(6) with respect to x_i as

$$\{\epsilon(t)\} = [\ \mathbf{B}(\mathbf{x})\ ]\{\mathbf{q}(t)\} \quad (7)$$

where $[\mathbf{B}(\mathbf{x})]$ is the element strain-displacement matrix.

The finite element equilibrium equations are obtained by substituting Eqs. (6) and (7) into Eq. (5) and by taking the first variation of the

$$\dot{U}_n(t+\Delta t) = \dot{U}_n(t) + [\ddot{U}_n(t) + \ddot{U}_n(t+\Delta t)] \cdot \frac{\Delta t}{2} \tag{10}$$
$$\ddot{U}_n^{av}(t+\Delta t) = \frac{\ddot{U}_n(t) + \ddot{U}_n(t+\Delta t)}{2}$$

Using the above approximations in Eq. (10), Eq. (8) can be described in a recursive form as

$$\begin{aligned}
&\sum_{r=1}^{R} \left\{ \left(\frac{2}{\Delta t}\right)^2 \cdot M_{mn} + K^0_{mnr} + \sum_{h=1}^{N_r} K^t_{mnrh} \cdot \exp[-\zeta_r(t_p)/\lambda_{rh}] \right. \\
&\left. \cdot [S_{rh}(t_p) \cdot \Delta t - SS_{rh}(\Delta t_p)] \frac{2}{\Delta t^2} \right\} U_n(t_p) = F_m(t_p) + F^r_m(t_p) \\
&+ M_{mn} \left\{ \left(\frac{2}{\Delta t}\right)^2 [U_n(t_{p-1}) + \dot{U}_n(t_{p-1}) \cdot \Delta t] + \ddot{U}_n(t_{p-1}) \right\} \\
&+ \sum_{r=1}^{R} \sum_{h=1}^{N_r} \left[K^t_{mnrh} \cdot \exp(-\zeta_r(t_p)/\lambda_{rh}) \left\{ -SS_{rh}(\Delta t) \frac{2}{\Delta t^2} \cdot [U_n(t_{p-1}) \right. \right. \\
&+ \dot{U}_n(t_{p-1}) \cdot \Delta t] + S_{rh}(t_p) \cdot [U_n(t_{p-1}) \cdot \frac{2}{\Delta t} + \dot{U}_n(t_{p-1})] \\
&\left. \left. - U_n(t_0) + S_{rh}(t_0) \cdot \dot{U}_n(t_0) \right\} + GR_{mrh}(t_p) \right]
\end{aligned} \tag{11}$$

where

$$\begin{aligned}
S_{rh}(t_p) &= \int^{t_p} \exp[\zeta_r'(\tau)/\lambda_{rh}] d\tau \\
SS_{rh}(\Delta t_j) &= \int_{t_{j-1}}^{t_j} S_{rh}(\tau) d\tau
\end{aligned} \tag{12}$$

$$\begin{aligned}
GR_{mrh}(t_p) &= \exp[-\zeta_r(t_p)/\lambda_{rh}] \cdot SS_{rh}(\Delta t_{p-1}) \cdot \ddot{U}_n^{av}(t_{p-1}) \\
&+ \exp[-\Delta\zeta_r(t_p)/\lambda_{rh}] \cdot GR_{mrh}(t_{p-1})
\end{aligned}$$

Eqs. (11) then become the governing system relations to be solved numerically for as many time steps Δt as are needed to achieve the final solution at time t. The size of Δt determines the accuracy of any numerical solution and its value can only be determined by trial and error for specific materials and problems. Because of the need for many repetative solutions for each time step, it is desirable to use a high speed computer to achieve solutions fast. The computational storage requirements are determined by the number of the independent relaxation moduli, the Prony series terms, nodes and nodal degrees of freedom and, of course, increase with the complexibility of the problem, such as nonhomogeneous materials, complicated geometries, etc. In this pilot study, to determine the effectiveness and accuracy of this new method, an IBM 3090 was used. It, like its comparable main frames, offers the advantages of speed, storage, vectorization and parallel processing, although the latter was not used for these computations.

Numerical Examples

A few numerical examples are presented to verify the accuracy and convergence and to demonstrate the use of the present finite element procedure for the transient responses of anisotropic viscoelastically damped structures. First, consider the nonconservative system shown in Fig. 1 and the system is excited by a unit step load. For the sake of simplicity in this example, the time-dependent stiffness of the bar is described by the following Prony series

$$K(t) = 100. + 100. \cdot \exp(-10t) \tag{13}$$

and the exact longitudinal response of a system to unit step excitation is

$$\begin{aligned} X(t) = & 1. \times 10^{-2} - 1.02 \times 10^{-6} \cdot \exp(-9.9t) \\ & - \exp(-4.9995 \times 10^{-2} t) \Big[9.99898 \times 10^{-3} \\ & \cdot \cos(1.00379305t) + 5.0817142 \times 10^{-4} \cdot \sin(1.00379305t) \Big] \end{aligned} \tag{14}$$

The dynamic responses calculated by the present numerical procedure are compared with the closed form solution in Fig. 2. The plots show that the two solutions are in good agreement for a time step $\Delta t = 4 \times 10^{-3}$. As another demonstration example, the transverse responses of a long cantilevered T300/934 composite beam (Fig. 3) are calculated. Twenty two-node beam elements with two degrees of freedom for each node are used. The elastic materials properties for T300/934 composites are $E_{11} = 21 \times 10^6$ psi, $E_{22} = 1.5 \times 10^6$ psi, $\nu_{12} = 0.3$, $G_{12} = 0.25 \times 10^6$ psi and $\rho = 0.05419$ lb/in.^{3} and the viscoelastic time variations and shift factors for T300/934 composites are taken from Ref. [2].

In the principal material directions, the fiber dominated modulus E_{11} is elastic, but the transverse modulus E_{22} and shear modulus G_{12} are both time-dependent since those are controlled by the epoxy matrix. The lamina thickness is taken as 0.005 in. Composite beams with $(0_{50}/90_{50})_s$ and $(90_{50}/0_{50})_s$ laminate orientations are analyzed at temperatures of 25 ° C and 93 °C and the composite beam section dimensions are 0.5 in.(height) × 1. in. (width) × 20 in.(length). Fig. 4 illustrates the maximum transverse deflections of $(0_{50}/90_{50})_s$ at the beam tip at temperatures of 25 ° C and 93 °C. The results indicate that the temperature does not have much affect on the solutions since the 0° layers, which are not time/temperature dependent, rule the flexibility of the beam. In Fig. 5, dynamic responses of the maximum transverse deflections for the $(90_{50}/0_{50})_s$ beams are plotted. The response spectra from t=5 sec. to t=10 sec. are also depicted in Fig. 6. The results show that the temperature influences these solutions since the 90° layers are significantly time/temperature dependent and degrade faster at higher temperatures . In addition, phase angle and amplitude change effects can also be observed from these numerical results. Temperature induced material degradation results in smaller amplitudes and larger vibration frequencies. At the same temperature, the amplitudes for the $(0_{50}/90_{50})_s$ beam are smaller than those of $(90_{50}/0_{50})_s$ but the frequencies for the $(0_{50}/90_{50})_s$ beam are much higher than those of $(90_{50}/0_{50})_s$. Another numerical example was carried out for the cantilever beam of Fig. 3 with a unit impulse load at the end. The results are presented in Figs. 7 and 8. It can be readily seen from Fig. 7 that changes in temperature affect both amplitude and frequency responses. This is due to larger damping material abilities with increasing temperatures increases. Fig. 8, which is an enlargement of the initial time intervals of Fig. 7, clearly shows the transient responses which die out in the first cycle for the case under consideration.

The numerical examples considered by the authors in this paper and in References 7 and 8 consist of pilot real world problems with a moderate number of elements (NOE) and degrees of freedom (DOF). Their purpose is to formulate the analyses and to evaluate numerical procedures and accuracy. The number of elements and DOF considered are 128 and 243 for the plates and 20 and 40 for the beams. For the quasi static analyses, the time dependent bending plate responses are evaluated up to 2×10^4 secs. and time increments of 0.002 secs. (no. of iteration $= 2 \times 10^4$) are used for the dynamic viscoelastic analyses of the beams. The CPU times are displayed in Table 1. It is evident that compiler optimization of the FORTRAN programs leads to significant CPU execution time savings on any of the three IBM mainframes. The effects of vectorization are not as pronounced in the dynamic problems due to their inherent nature which requires a substantially large number of sequential computations. For quasi-static analyses based on Laplace transforms, the running time decreases when using parallel processing for numerical Laplace inversions since each degree of freedom is uncoupled by orthogonal transformations. The use of recursive formuli substantially reduces computer storage in this dynamic analysis, but solution accuracy primarily depends on the time step size and the dynamic analysis for transient responses requires many high speed iterations. Thus, in these analyses, the use of supercomputers leads to an advantage by reducing total CPU times. In problems with considerably larger numbers of elements and degrees of freedom and/or complicated shapes as well as boundary conditions, the use of a supercomputer becomes mandatory. This need is dictated by expanded storage requirements, although kept to a minimum under the present analyses, as well as larger CPU time usage.

Conclusions

Variational principles for dynamic analysis of anisotropic hygrothermoviscoelasticity have been proposed and the finite element formulation for analyzing transient responses of viscoelastically damped structures has been developed in the time domain using variational principles and the Newmark average acceleration method. The recurrence procedure, similarly obtained by Refs. [13-15], permits the new time solutions to be evaluated using only the previous time solutions. The present dynamic algorithm demands the same computer memory storage as the quasi-static methods of Refs. [13-15]. Verification studies show that the numerical solutions agree well with analytical ones. As examples, the dynamic responses of T300/934 composite beams subjected unit step and impulse loads are calculated using the presently developed procedure and $(0_{50}/90_{50})_s$ and $(90_{50}/0_{50})_s$ stacking sequences are considered. The numerical results show that temperature has a large effect on the solutions of $(90_{50}/0_{50})_s$ laminated composite beams since the transverse moduli E_{22} degrade faster at higher temperatures and thus the time-temperature dependence of composite materials can be both an advantage and drawback.

Acknowledgement

This work was carried out under a grant from the IBM Palo Alto Scientific Center. The authors gratefully thank the staff for their support.

References

[1] Mantena, R.M., Gibson, R.F. and Place, T.A., Damping Capacity Measurements of Degradation in Advanced Materilas, SAMPE Quarterly, Vol. 17, No. 3, pp. 20-31, April 1986.

[2] Crossman, F.W. and Mauri, R.E. and Warren, W.J., Moisture Altered Viscoelastic Response of Graphite/Epoxy Composite, Advanced Composite Materials-Environmental Effects, ASTM STP 658, J.R. Vinson, Ed., American Society for Testing and Materials, pp. 205-220, 1978.

[3] Johnson, C.D. and Kienholz, D.A., Finite Element Prediction of Damping in Structures with Constrained Viscoelastically Damped Structures, AIAA Journal, Vol. 20, No. 9, pp. 1284-1290, September 1982.

[4] Bagley, R.L. and Torvik, P.J., Fractional Calculus-A different Approach to the Finite Element Analysis of Viscoelastically Damped Structures, AIAA Journal, Vol. 21, No. 5, pp. 741-748, May 1983.

[5] Golla, D.F. and Hughes, P.C., Dynamics of Viscoelastic Structures-A Time Domain, Finite Element Formulation, Journal of Applied Mechanics, Vol. 52, pp. 897-906, December 1985.

[6] Zabaras, N. and Pervez, T., Viscous Damping Approximation of Laminated Anisotropic Composite Plates Using the Finite Element Method, Computer Methods in Applied Mechanics and Engineering, No. 81 pp. 291-316, 1990.

[7] Hilton, Harry H. and Yi, S., Finite Element Formulation for Thermo-Viscoelastic Analysis of Composite Structures Subjected to Mechanical and Hygrothermal Loading, Proceeding of First NCSA Conf. on Finite Element Applications in Computational Mechanics, University of Illinois at Urbana-Champaign, May 14-15, 1990.

[8] Hilton, Harry H. and Yi, S., Bending and Stretching Finite Element Analysis of Anisotropic Viscoelastic Composite Plates, Proceedings of Third Air Force / NASA Symposium on Recent Advances in Multidisciplinary Analysis and Optimization, San Francisco CA, pp. 488-494, Sept. 24-26, 1990.

[9] Newmark, Nathan M., A Method of Computation for Structural Dynamics, Proceedings of the American Society of Civil Engineers, pp. 67-94, July 1959.

[10] Hilton, Harry H., Viscoelastic Analysis, Engineering Design for Plastics, Reinhold Pub. Corp., New York, pp. 199 - 276, 1964.

[11] Gurtin, M.E., Variational Principles in the Linear Theory of Viscoelasticity, Archive Rat. Mech. and Analysis, Vol. 13, pp. 179-185, 1963.

[12] Schapery, R.A., On the Time Dependence of Viscoelastic Variational Solutions, Quart. Appl. Math., Vol. 22, pp. 207-215, 1964.

[13] Zak, A.R., Structural Analysis of Realistic Solid Propellant Materials, Journal of Space Craft and Rockets, Vol. 5, pp. 270-275, 1967.

[14] Taylor, R.L., Pister, K.S., and Goudreau, G.L., Thermomechanical Analysis of Viscoelastic Solids, International Journal for Numerical Methods in Engineering, Vol. 2, pp. 45-59, 1970.

[15] Lin, K.Y. and Yi, S., Analysis of Interlaminar Stresses in Viscoelastic Composites, International Journal of Solids and Structures, Vol. 27, No. 7, pp. 929-945, 1991.

Table 1. CPU Execution Times (Secs) for Viscoelastic Problems

Mainframe & Fortran Compiler Optimization	4 Sides Simply Supported Plate Static [7,8]	4 Sides Clamped Plate Static [7,8]	SP impulse Beam	DP impulse Beam
IBM4381				
0	2972.25	1040.71	1210.20	1299.50
2	1275.93	449.28	429.01	463.54
3	1267.16	445.18	428.80	461.55
IBM3081				
0	969.22	338.17	741.08	791.28
2	380.61	128.10	329.09	358.69
3	377.88	127.93	220.82	271.13
IBM3090				
0	651.56	99.07	183.16	204.68
2	359.14	36.29	58.57	68.60
2V	252.83	26.72	64.34	72.65
3	99.24	29.16	56.41	65.27
3V	78.05	23.64	63.00	70.35

SP Single Precision, DP Double Precision

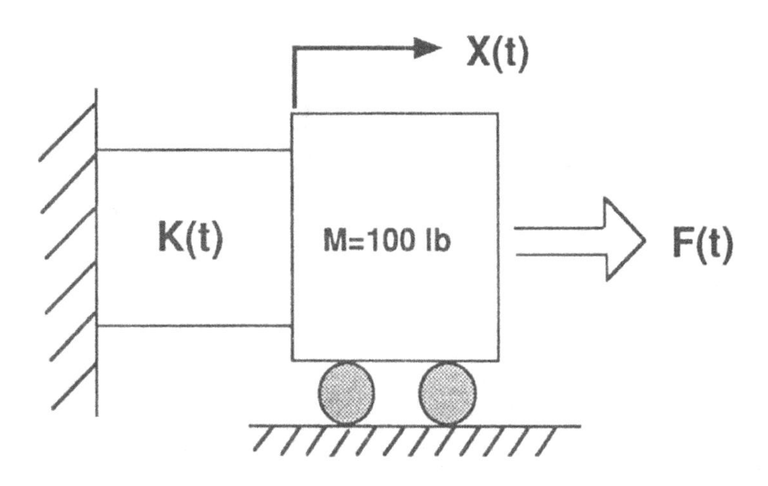

Fig. 1 Single degree of freedom non-conservative system.

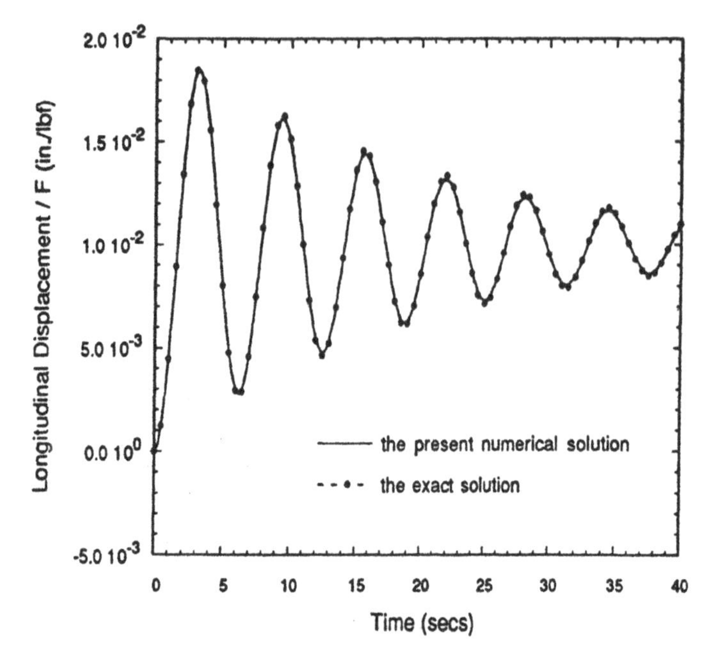

Fig. 2 The displacemnt responses of the system in Fig.1.

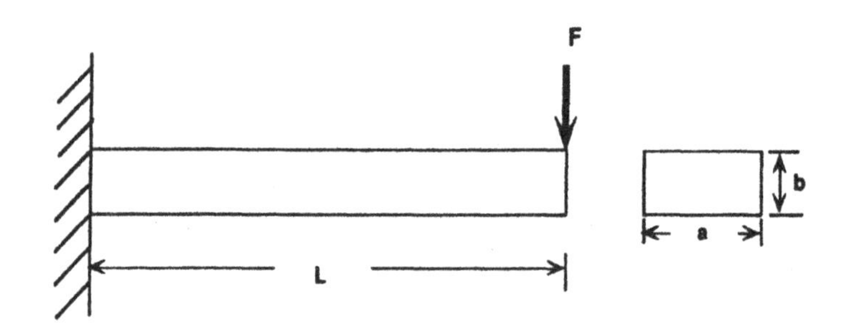

Fig. 3 A long cantilevered Euler beam with rectangular cross section.

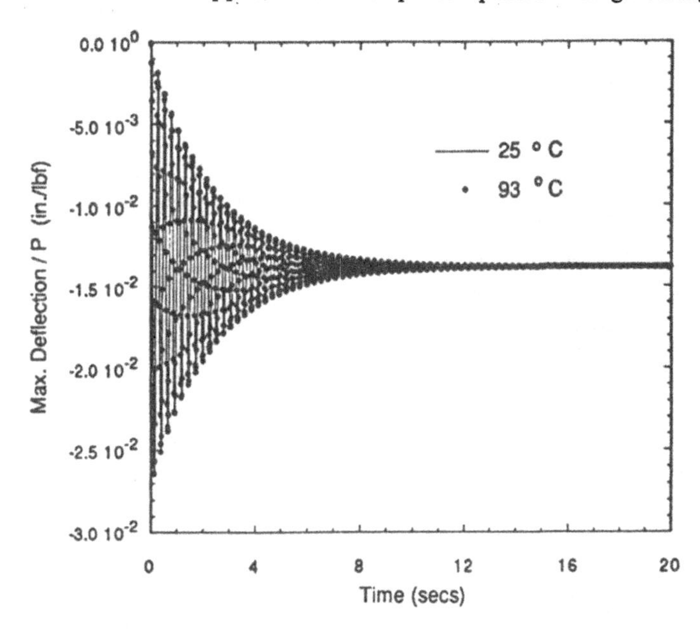

Fig. 4 The deflection responses of cantilevered T300/934 composite beams with (0_{50} / 90_{50}), laminate orientations.

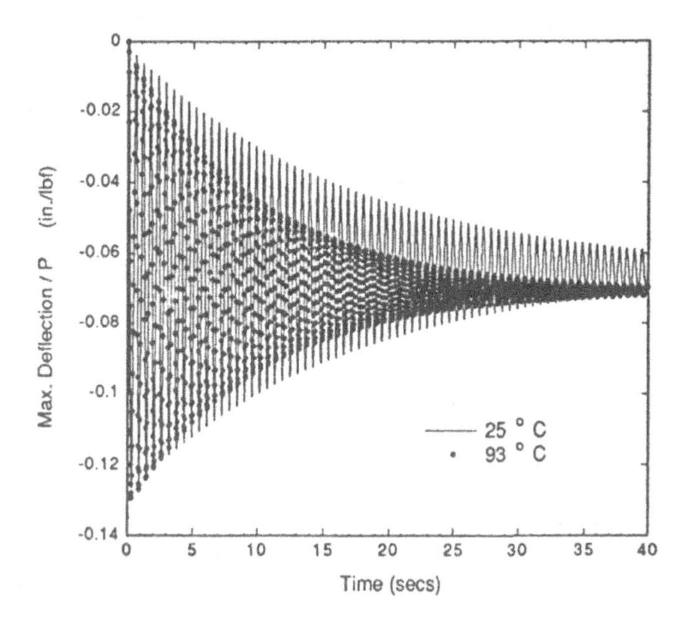

Fig. 5 The deflection responses of cantilevered T300/934 composite beams with (90_{50} / 0_{50}), laminate orientations (unit setp load).

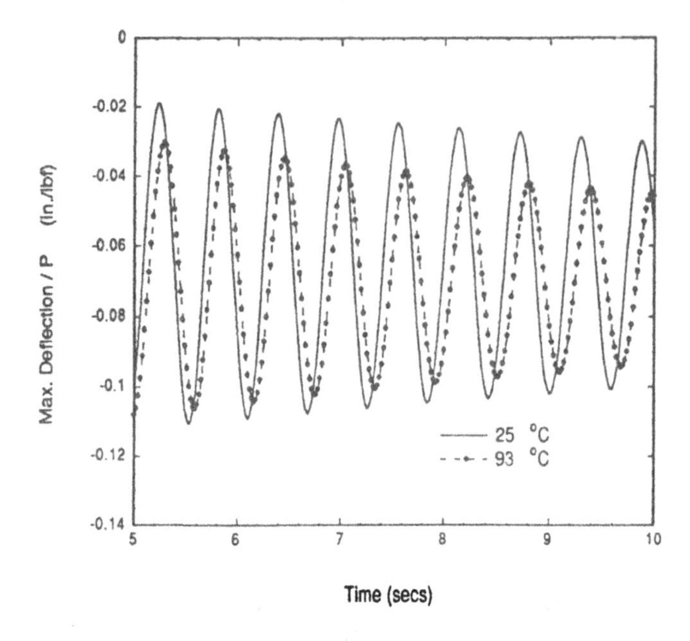

Fig. 6 The magnified deflection responses in Fig. 5.

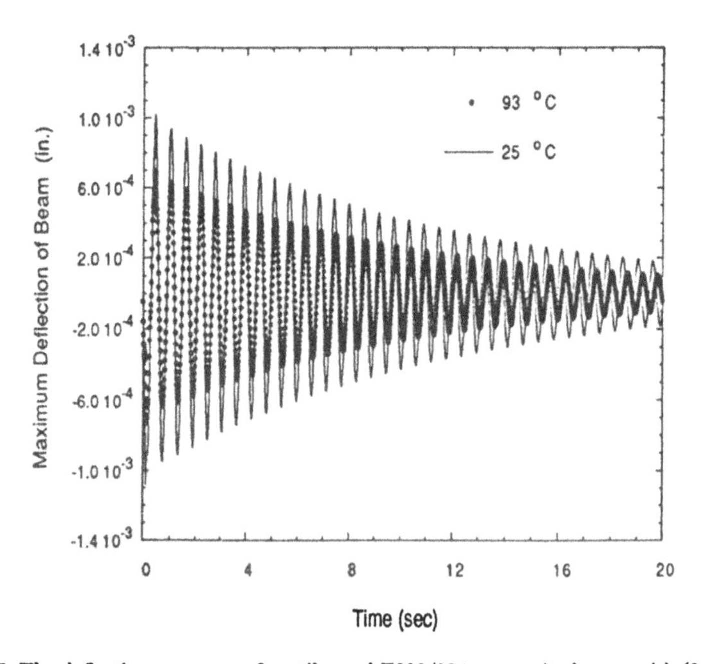

Fig. 7 The deflection responses of cantilevered T300/934 composite beams with (90_{50} / 0_{50})$_s$ laminate orientations (impulsive load).

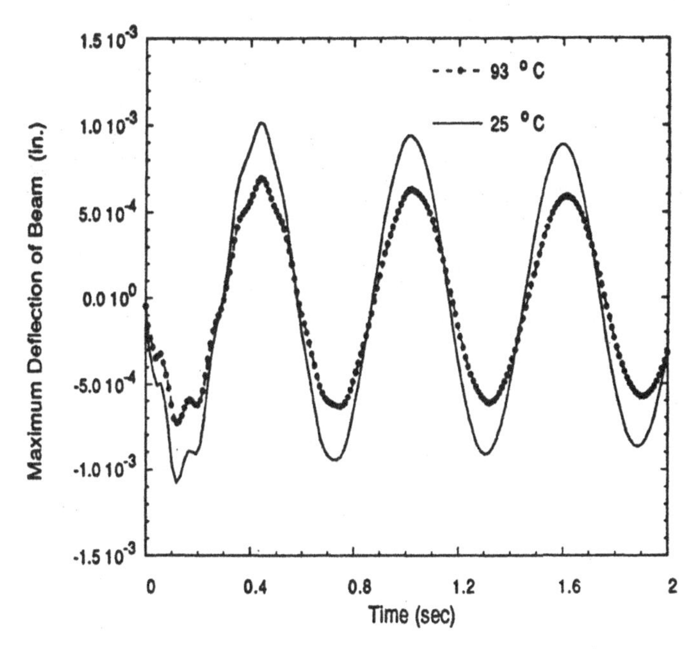

Fig. 8 The magnified deflection responses in Fig. 7.

Thickness-Shear Mode Shapes and Mass-Frequency Influence Surface of a Circular and Electroded At-Cut Quartz Resonator

J.T. Stewart, Y.-K. Yong

Department of Civil Engineering, Rutgers University, P.O. Box 909, Piscataway, NJ 08855-0909, U.S.A.

Abstract

Finite element solutions for the fundamental thickness shear mode and the second anharmonic overtone of a circular, 1.87 Mhz AT-Cut quartz plate with no electrodes are presented and compared with previously obtained [1] results for a rectangular plate of similar properties. The edge flexural mode in circular plates, a new phenomenon not seen in the rectangular plate, is also presented. A 5 Mhz circular and electroded AT-Cut quartz plate is studied and compared with its counterpart with no electrodes. For each plate studied, a portion of the frequency spectrum is constructed in the neighborhood of the fundamental thickness-shear mode. A convergence study is also presented for the electroded 5 Mhz plated. A new two dimensional technique for visualizing the vibration mode solutions is presented. This method departs quite substantially from the three dimensional "wire-frame" plots presented in the previous analysis. The two dimensional images can be manipulated to produce nodal line diagrams and can be color coded to illustrate mode shapes and energy trapping phenomenon. A contour plot of the mass-frequency influence surface for the plated 5 Mhz plate is also presented. The mass-frequency influence surface is defined as a surface giving the frequency change due to a small localized mass applied to the resonator surface.

Introduction

The use of quartz crystal plates in ultrasonics is well studied experimentally. Analytically, however, the problem of a freely vibrating anisotropic elastic plate in general can only be solved on an infinite domain. For a finite domain, solutions may be obtained either approximately, or for certain types of boundary conditions. To perform a practical analysis of the AT-Cut quartz resonator, it is necessary to solve the free vibration problem for a finite plate. In this paper, such a problem is solved using the finite element method.

The complexity of high frequency vibrations as well as the anisotropicity of quartz demands that a general thick plate theory be used. In this study, Mindlin's two dimensional plate equations are truncated to a first order approximation. The first order approximation is chosen because it is the lowest order formulation which includes the major modes of interest, namely, the fundamental thickness-shear modes [2]. Real AT-Cut resonators generally contain a thin electrode plating on a portion of each face, across which a time varying voltage is applied to drive the vibration. The effect of this electrode plating was studied analytically by

Mindlin [3] for an infinite domain. Mindlin's equations for the electroded plate are incorporated into the present analysis so that the effects on frequencies and mode shapes of this electrode plating can be studied. The formulation for the plated problem is also useful for assessing certain performance parameters of a given plate. Yong and Vig [4] proposed that surface contamination may be a contributor to random frequency fluctuations of a thickness shear resonator and a useful parameter in predicting the magnitude of these frequency fluctuations is the net frequency change due to a unit mass loading on the plate's surface. This mass loading effect is highly position dependent. The variation of the change in frequency due to a unit mass loading is illustrated in the frequency influence surface. This surface is calculated for an electroded 5 Mhz plate with a width to thickness ratio of 24.0.

The finite element formulation for the first order plate theory is derived using a non-conforming bilinear quadrilateral element. This element is chosen because its formulation is relatively simple and it performs well as a thick plate element [5]. A finite element program is written to solve the resulting eigenvalue problem. The program includes an efficient eigenvalue solver employing the Lanczos algorithm [6],[7] which allows the user to find resonant frequencies within a specified bandwidth. Using this program, three plates are studied. For each case, the frequency spectrum and mode shapes in the region around the fundamental thickness-shear mode are studied. The computer work was performed on a CRAY-2 supercomputer at the National Center for Supercomputing Applications (NCSA).

Finite Element Formulation of Mindlin's First Order Plate Equations

Figure 1 shows the general plate problem considered: a circular plate of diameter 2a and thickness 2b with optional electrode plating of diameter 2a' and thickness 2b'.

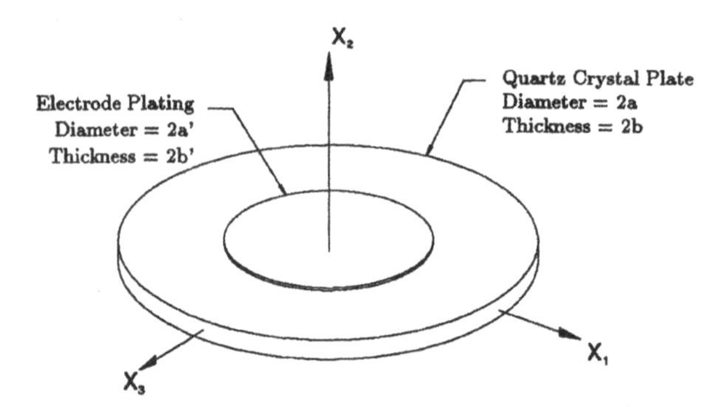

Figure 1 Orientation and dimensions of a circular electroded plate.

Mindlin's two dimensional plate equations are obtained by expanding the three dimensional equations of motion for an elastic continuum into an infinite series of powers of the thickness coordinate (x_2). The resulting equations are then integrated through the thickness, eliminating the x_2 dependence. The approximate plate equations are then obtained by truncating this series to the desired power [2] (in this case, one). In the following description, the results of the finite element formulation are given. For a detailed derivation of these equations, the reader is referred to reference [1].

In the finite element formulation, the displacement vector **u** is interpolated using shape functions of a four-node, bilinear quadrilateral element,

$$\mathbf{u} = \mathbf{N}\mathbf{q} \tag{1}$$

or

$$\begin{bmatrix} u_2 \\ \psi_1 \\ \psi_3 \end{bmatrix} = [\mathbf{N}_1 \quad \mathbf{N}_2 \quad \mathbf{N}_3 \quad \mathbf{N}_4] \begin{bmatrix} \mathbf{q}_1 \\ \mathbf{q}_2 \\ \mathbf{q}_3 \\ \mathbf{q}_4 \end{bmatrix} \tag{2}$$

In equation (2), the components of displacement, u_2, ψ_1, and ψ_3 represent, respectively, the transverse displacement in the x_2 direction, the shear rotation about the x_1 axis, and the shear rotation about the x_3 axis. The components of the shape function matrix, **N**, are given as:

$$\mathbf{N}_i = \begin{bmatrix} N_i & 0 & 0 \\ 0 & N_i & 0 \\ 0 & 0 & N_i \end{bmatrix} \tag{3}$$

where N_i are bilinear shape functions for a quadrilateral element, namely,

$$\begin{aligned} N_1 &= \frac{1}{4}(1-\xi)(1-\eta) \\ N_2 &= \frac{1}{4}(1+\xi)(1-\eta) \\ N_3 &= \frac{1}{4}(1+\xi)(1+\eta) \\ N_4 &= \frac{1}{4}(1-\xi)(1+\eta) \end{aligned} \tag{4}$$

which are represented in local element coordinates, (ξ,η). The nodal displacement vector, $\mathbf{q}_i$, $(i = 1,2,3,4)$ in equation (2) is given as:

$$\mathbf{q}_i = \begin{bmatrix} u_{2i} \\ \psi_{1i} \\ \psi_{3i} \end{bmatrix} \tag{5}$$

The finite element matrix equations are obtained by using the interpolated displacement components of equation (2) in the variational form of the first order plate equations of motion. By applying the standard reasoning of the variational method, the following equation is obtained:

$$\mathbf{Kq} = \mathbf{P} - \mathbf{M}\ddot{\mathbf{q}} \tag{6}$$

where

$$\mathbf{K} = \int_A \mathbf{B}^T \mathbf{D} \mathbf{B} \, dA \tag{7}$$

is the element stiffness matrix,

$$\mathbf{P} = \int_C \mathbf{N}^T \mathbf{p} \, dC \tag{8}$$

is the consistent load vector, and

$$\mathbf{M} = \int_A \mathbf{N}^T \mathbf{m} \mathbf{N} \, dA \tag{9}$$

is the consistent mass matrix. In equation (7), the **B** matrix represents the strain-displacement and curvature-rotation relations in terms of the shape functions, N_i. The matrix **D** in equation

(7) represents the elastic plate stiffness tensor which is a function of the elastic constants for AT-Cut quartz. These constants exhibit monoclinic symmetry when the x_1 axis coincides with the digonal axis of the crystal [2]. The values of the elastic constants used in the present study are those supplied by Bechmann, Ballato, and Lukaszek [8]. The integration over A in equations (7) and (9) represents an area integral over the domain of the element. In equation (8), the vector **p** is the nodal load vector and the integration over C is interpreted as the line integral taken around the boundary of element. The matrix **m** in equation (9) is a diagonal matrix representing the effective plate mass terms which multiply $\ddot{u}_2$, $\ddot{\psi}_1$, and $\ddot{\psi}_3$ in the plate equations of motion. The explicit forms of these matricies are given in reference [1], and will not be pursued here. The global stiffness, mass, and load vector are assembled in the standard way [5] to give:

$$\mathbf{K_g} = \mathbf{P_g} - \mathbf{M_g}\ddot{\mathbf{q}}_\mathbf{g} \tag{10}$$

where the subscript **g** represents the global set of equations. For the free vibration problem, $\mathbf{P_g} = 0$. For simple harmonic motion with a radial frequency ω,

$$\ddot{\mathbf{q}}_\mathbf{g} = -\omega^2 \bar{\mathbf{q}}_\mathbf{g} e^{i\omega t} \tag{11}$$

where $\bar{\mathbf{q}}_\mathbf{g}$ are the magnitudes of the nodal displacements. Using equation (11) in equation (10) results in the following symmetric, positive definite eigenvalue problem:

$$\left(\mathbf{K_g} - \omega^2 \mathbf{M_g}\right) \bar{\mathbf{q}}_\mathbf{g} = 0 \tag{12}$$

In the formulation of the stiffness matrix, extra shape functions

$$N_5 = 1 - \xi^2 \text{ and } N_6 = 1 - \eta^2 \tag{13}$$

are included to prevent "shear locking". The extra degrees of freedom associated with these shape functions are removed by static condensation at the elemental level [5].

To solve equation (12), an efficient finite element program is written. In most engineering applications, the modes of interest occur at the lower end of the spectrum and for these problems it is sufficient to solve for the first few eigen-pairs. In the present study, the modes of interest occur at very high frequencies. The Lanczos algorithm is well suited for the problem at hand because it allows for the calculation of eigenvalues within a user specified range. The particular routines used are found in the LASO2 package at NCSA. The LASO2 package is an efficient implementation of David S. Scott's block Lanczos algorithm [7]. To achieve accurate solutions at these high frequencies, an extremely fine finite element mesh is required, generating a large number of degrees of freedom. For this reason, an efficient "skyline" storage scheme [5] is employed which exploits the sparse nature of the problem.

Visualization of Solutions

The post-processing phase of our finite element analysis may be as important as the actual solution phase. It is not only necessary to compute the frequency spectrum for a given plate, but also to study the complex vibration mode shapes which result. For resonator design purposes, accurate and concise visual data is required to aid in the siting of device mountings. Some of this data has been obtained experimentally through the use of stroboscope X-ray topography, which requires an elaborate and costly laboratory setup.

To visualize the finite element solutions obtained, a two-dimensional technique is employed. Previously, mode shapes were plotted as three-dimensional surfaces with hidden lines removed [1]. In the present study, use is made of a Silicon Graphics Iris workstation to produce

2–D Gouraud shaded images of the vibration mode shapes. A simple FORTRAN program is written which uses v2f and c3f calls to the Iris GL library [9] to construct 2–D polygons with colors specified at each vertex. The color is supplied to the ith vertex (node) via a 3–D vector (R_i, G_i, B_i), where R,G, and B denote percentages of red, green, and blue that make up the desired color. To produce a particular image, the nodal displacements are normalized to the interval [0,1] and the color proportions are computed by mapping this interval into the proper values of R, G, and B. Using this technique, three very useful effects are created.

To produce a linearly graded color mapping from blue into green into red representing the minimum, zero, and maximum of the data, the following transformation is applied to a value $x \epsilon [0,1]$:

$$\begin{aligned} &\text{for } x \leq 0.25, \begin{cases} R = 0 \\ G = 4x \\ B = 1 \end{cases} \\ &\text{for } 0.25 < x \leq 0.5, \begin{cases} R = 0 \\ G = 1 \\ B = -4x + 2 \end{cases} \\ &\text{for } 0.5 < x \leq 0.75, \begin{cases} R = 4x - 2 \\ G = 1 \\ B = 0 \end{cases} \\ &\text{for } 0.75 < x, \begin{cases} R = 1 \\ G = -4x + 4 \\ B = 0 \end{cases} \end{aligned} \tag{14}$$

To produce a grey scale mapping with black and white representing minimum and maximum of the data, respectively, the following transformation is applied to the same value x:

$$R = G = B = x \tag{15}$$

To produce a nodal line image in which the the curves representing the zeros of the vibration mode are shown in black, and the rest of the data is represented in light grey or white, the following transformation is applied to x:

$$R = G = B = |2x - 1|^p \tag{16}$$

The value of p used for this analysis is 0.3.

The figures presented at the end of this paper are grey-scale and nodal-line images of some of the important vibration modes. The grey scale images are comparable to the experimental X-ray stroboscopic images taken of the resonator. The nodal-line images will be useful in the design process because they locate areas where displacements are minimal for mounting purposes. Also, the nodal-line images may be used to identify quickly the modes in the frequency spectrum.

Frequency Spectrums and Mode Shapes Of Circular Plated And Unplated AT-Cut Quartz Plates

A uniformly impressed electric field over the AT-Cut quartz plate will excite modes that have 1) u_2 odd in x_1 and even in x_3, 2) ψ_1 even in x_1 and even in x_3, and 3) ψ_3 odd in x_1 and odd in x_3. The resonator can thus be modeled using a quarter plate and imposing the essential boundary conditions $u_2 = \psi_3 = 0$ along the x_3 axis, and $\psi_3 = 0$ along the x_1 axis. This results in a substantial saving in storage and computation as well as reducing the number of modes within a specified bandwidth. Figure 2 shows a typical finite element discretization with optional electrode plating and mass loading.

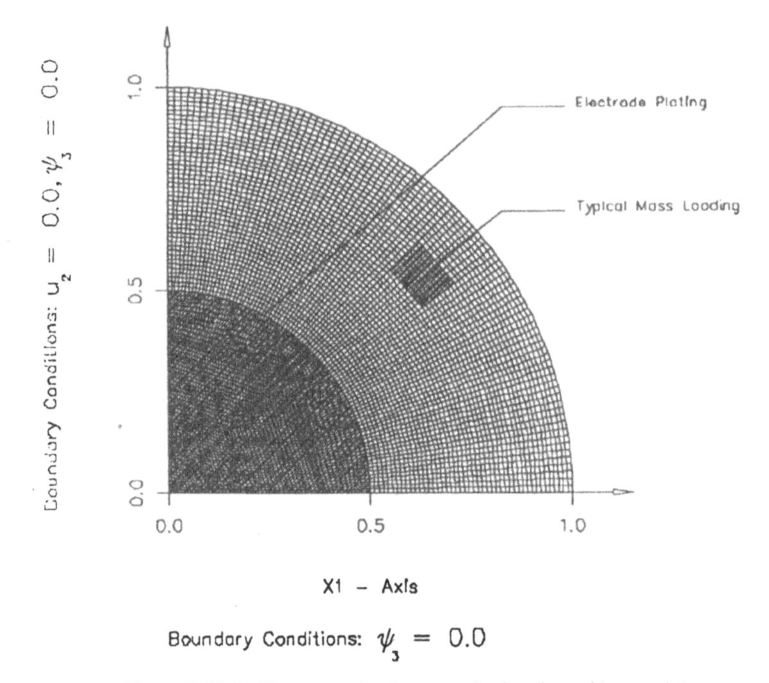

Figure 2 Finite Element mesh of quarter circular plate with essential boundary conditions at plate edges along the coordinate axes

A. 1.87 Mhz Unplated AT-Cut Quartz Resonator

A 1.87 Mhz unelectroded, circular plate is studied for aspect ratios ranging from 15.000 to 16.000 and vibration frequencies in the neighborhood of the fundamental thickness-shear mode. This problem was modeled with 16,707 displacement degrees of freedom. Shown in figure 3 is a portion of the frequency spectrum ranging over the aspect ratios discussed above.

Shown in figures 8 and 10 are nodal-line plots for the fundamental thickness shear mode (TS-1) and the first anharmonic overtone of the fundamental thickness-shear mode (TS-3) for the 1.87 Mhz circular plate. These modes correspond to points Aa and Ab, respectively, in figure 3. Included for comparison in figures 9 and 11 are, respectively, nodal-line plots of the TS-1 and TS-3 modes for a rectangular plate of the same aspect ratio (15.500). We note that even for a rectangular plate, there is no pure straight crested waves traveling in the x_1 direction due to boundary interactions. The assumption for straight crested waves in circular plates is worse than in rectangular plates.

B. 5 Mhz Unplated and Plated AT-Cut Quartz Resonator

A 5 Mhz circular plate is studied with and without electrode plating for aspect ratios from 24.000 to 25.000 in the neighborhood of the fundamental thickness-shear mode (TS-1). A convergence study of a plated circular plate with an aspect ratio of 24.0 was performed for the TS-1 mode (5.07 Mhz) and the TS-3 mode (5.12 Mhz). The results of this study are presented in figure 4. The frequencies converge monotonically from below.

For the 5 Mhz plate, 20,655 displacement degrees of freedom were used. Shown in figures 5 and 6 are, respectively, the frequency spectrums for the unplated and plated cases, ranging over the aspect ratios discussed above.

Figure 12 shows a grey-scale image of the TS-1 mode shape for the unplated 5 Mhz plate for an aspect ratio of 24.375. This mode corresponds to point Ba in figure 5. Shown

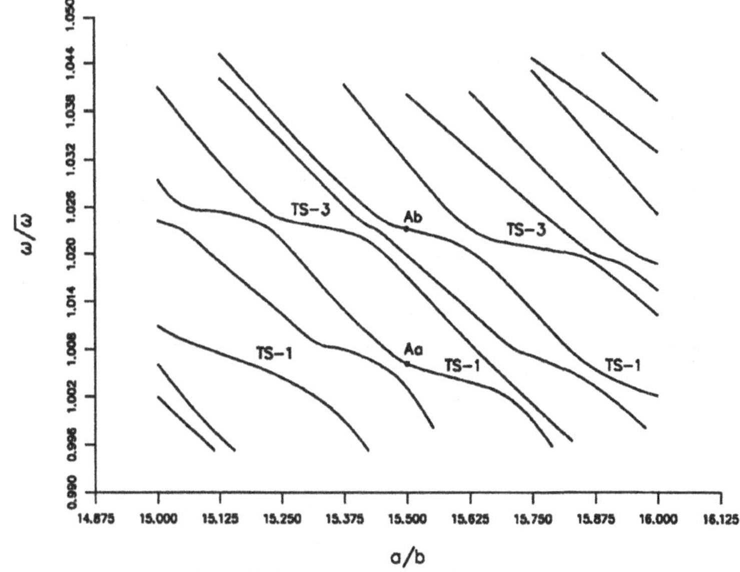

Figure 3 Frequency Spectrum for 1.87 Plate

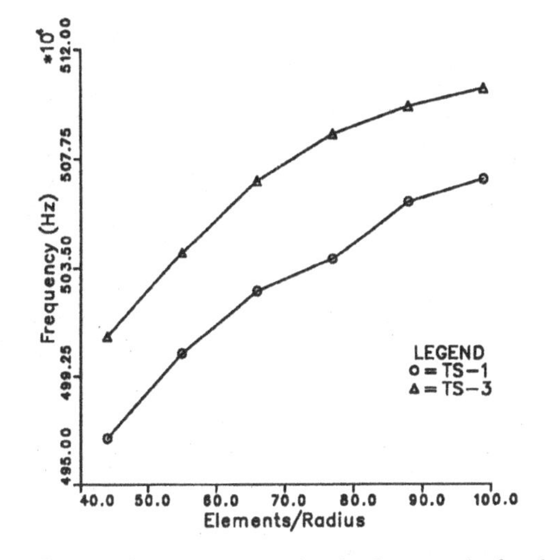

Figure 4 Convergence study for plated quarter circular plate.

in figures 13 and 14 are grey-scale and nodal-line images for the same plate with electrode plating. This mode corresponds to point Ca in figure 6. In comparing figures 12 and 13 it is evident how the electrode plating changes the mode shape of ψ_1. The addition of plating tends to concentrate the displacements to the electroded area. Such a phenomenon is known as energy trapping. Figures 15 and 16 show a grey-scale and nodal-line image for the TS-3 mode for the plated case, corresponding to point Cb in figure 6. This mode is also shown for an aspect ratio of 24.375.

Figure 17 shows a new phenomenon, not previously observed in rectangular plates. This mode is called the edge flexural mode, due to the fact that the predominance of the displacement is contained around the edge of the plate in the u_2 displacement component. Shown in figure 17

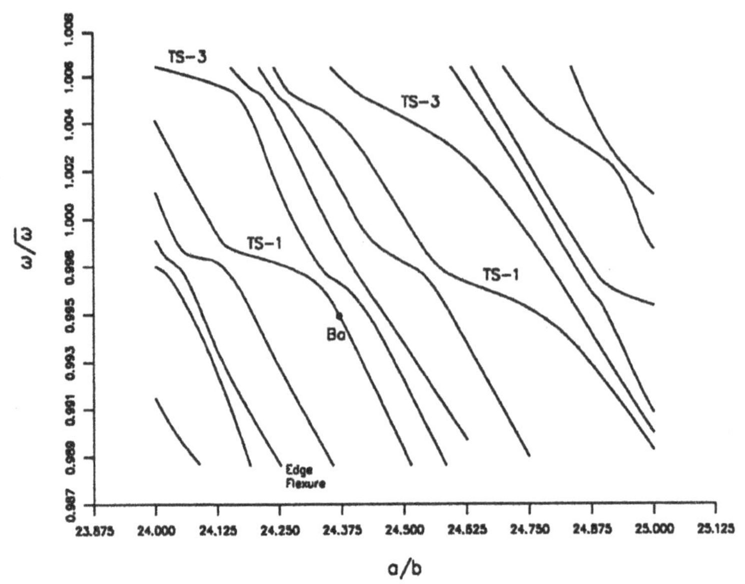

Figure 5 Frequency Spectrum for the unplated 5 Mhz Plate

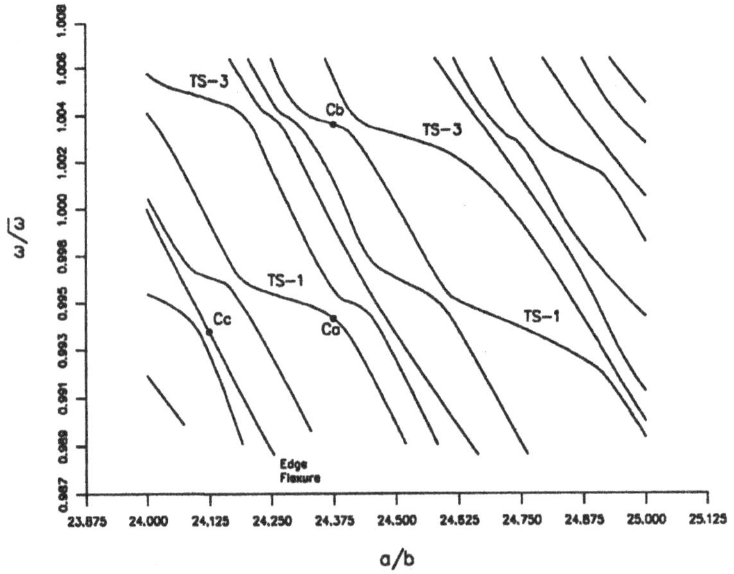

Figure 6 Frequency Spectrum for the plated 5 Mhz Plate

is a grey-scale image of this mode. This image corresponds to point Cc of figure 6. The edge flexural mode may be excited by electrode fingers of alternating voltages and distributed along the circumference. Such a resonator may have useful applications and its resonant frequency is limited only by the number of electrode fingers along the circumference. Hence, very high resonant frequencies may be obtained. The resonator can be mounted at the center.

Mass Frequency Influence Surface of A 5 Mhz Plated Circular AT-Cut Quartz plate

Using the finite element program, the mass frequency influence surface is constructed for the 5 Mhz electroded plate with an aspect ratio of 24.000 and vibrating in the fundamental thickness-shear mode. This surface gives the frequency effect of a small, discrete, mass loading as a function of position on the plate's surface. Suppose the resonator has a frequency influence surface $F(x_1,x_3)$ in units of Hz/μg. If the plate is mass loaded with a mass distribution of $m(x_1,x_3)$ in units of μg/mm^2, the resonant frequency would decrease by the amount:

$$\int_A F(x_1, x_3)\, m(x_1, x_3)\, dA \tag{17}$$

where A is the plate area in mm^2.

Figure 7 shows a contour plot of the mass frequency influence surface. A maximum frequency change of 398 Hz/μg is observed just off center along the x_1 axis. Mass frequency change is concentrated to the electroded area. Comparing this diagram with the mode shape shown in figure 13, it is observed that the mass effect is strongly coupled with both the u_2 and ψ_1 displacement components.

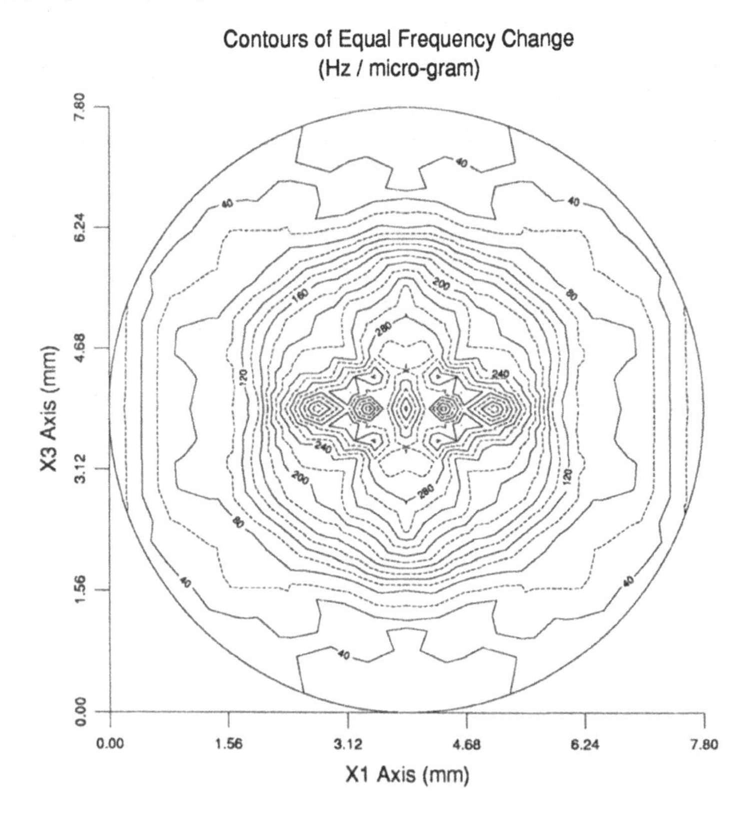

Figure 7 Mass Frequency Influence Surface For A 5 Mhz Electrode Plated Circular Plate.

Acknowledgement

This work was supported by the U. S. Army Research Office contract no. DAAL03–87–K-0107 and the National Center for Supercomputing Applications. We gratefully acknowledge their support.

References

1. "Mass Frequency Influence Surface, Mode Shapes, and Frequency Spectrum of a Rectangular AT-Cut Quartz Plate", Y-K Yong and J. T. Stewart, IEEE Transactions on Ultrasonics, Ferroelectrics, and Frequency Control, Vol. 38, No. 1, 1991, pp. 67–73.

2. An Introduction To the Mathematical Theory of Vibrations of Elastic Plates, R. D. Mindlin, Monograph prepared for The U.S. Army Signal Corps Engineering Laboratories, Fort Monmouth, New Jersey, 1955.

3. "High Frequency Vibrations of Plated Crystal Plates", R. D. Mindlin, Progress In Applied Mechanics, Macmillan, New York, 1963, pp. 73–84.

4. "Resonator Surface Contamination — A Cause of Frequency Fluctuations ?", Y-K Yong and J. R. Vig, IEEE Transactions on Ultrasonics, Ferroelectrics, and Frequency Control, Vol. 36, No. 4, 1989, pp. 452–458.

5. The Finite Element Method, Linear Static And Dynamic Finite Element Analysis, Thomas J. R. Hughes, Prentice-Hall, Englewood Cliffs, New Jersey, 1987.

6. "An Iteration Method for the Solution of the Eigenvalue Problem of Linear Differential and Integral Operators", Lanczos, C., Journal of Research of the National Bureau of Standards, 45, 1950, pp. 255–281.

7. "The Lanczos Algorithm with Selective Orthogonalization", Parlett, B. N., and Scott, D., Mathematics of Computation, 33, No. 145, 1979, pp. 217–238.

8. "Higher-Order Temperature Coefficients of the Elastic Stiffness and Compliances of Alpha-Quartz",Bechmann, R., Ballato, A. D., and Lukaszek, T. J., Proc. IRE, 50, 1962, pp. 1812–1822.

9. Graphics Library Reference Manual, FORTRAN 77 Edition , Ver. 1.0, Silicon Graphics, Inc., Mountain View, California.

Fig.8 Nodal-line plots of (a) u_2, (b) ψ_1 and (c) ψ_3 displacements of the fundamental thickness-shear mode for the circular plate without electrodes, a/b=15.500, frequency=1.89 Mhz:

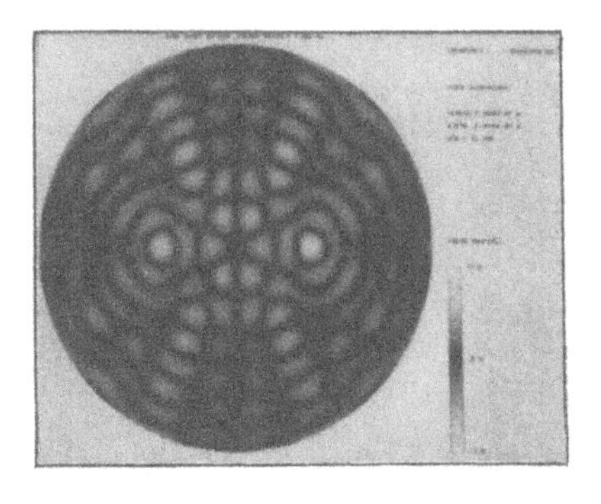

(a) u_2 displacement nodal line plot.

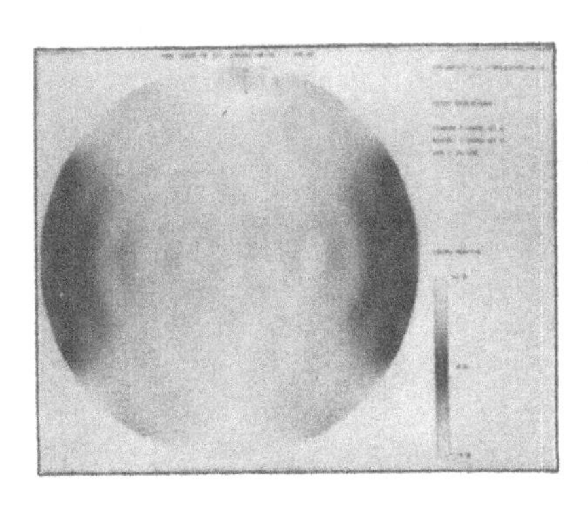

(b) ψ_1 displacement nodal line plot.

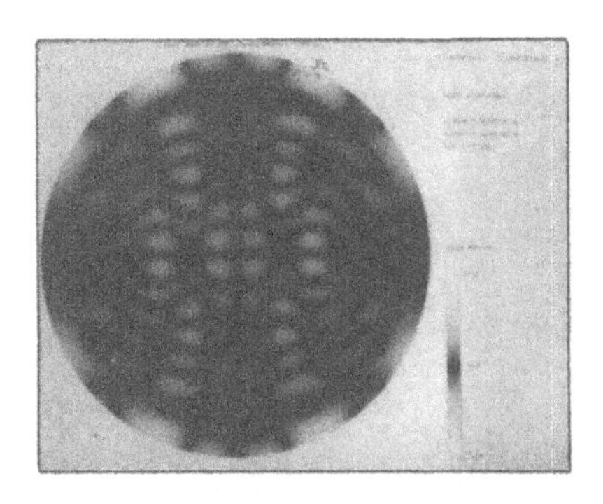

(c) ψ_3 displacement nodal line plot.

Fig.9 Nodal-line plots of (a) u_2, (b) ψ_1 and (c) ψ_3 displacements of the fundamental thickness-shear mode for the rectangular plate without electrodes, a/b=15.500, frequency=1.88 Mhz:

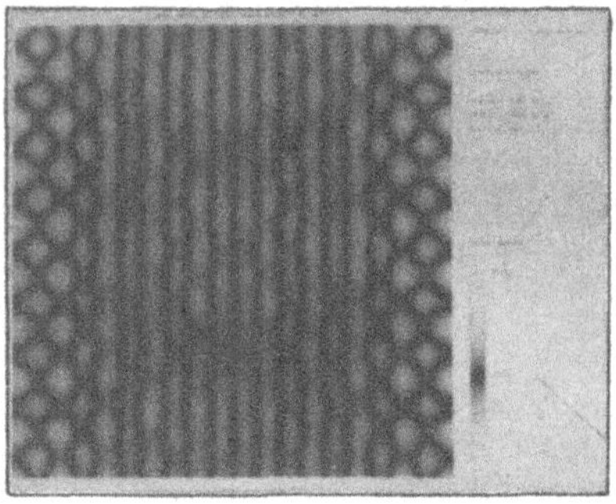

(a) u_2 displacement nodal line plot.

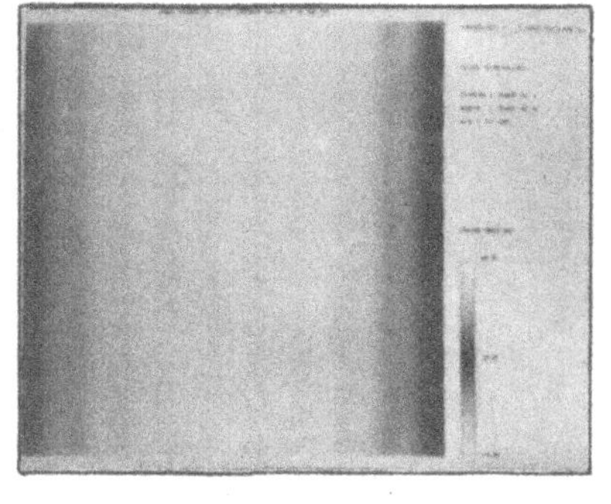

(b) ψ_1 displacement nodal line plot.

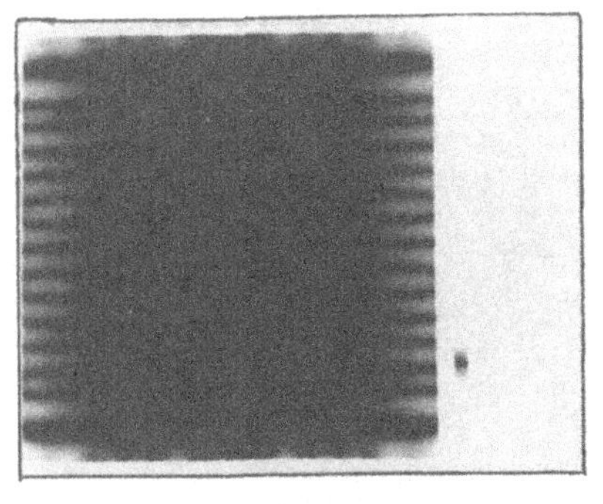

(c) ψ_3 displacement nodal line plot.

Fig.10 Nodal-line plots of (a) u_2, (b) ψ_1 and (c) ψ_3 displacements of the first anharmonic overtone of fundamental thickness-shear mode for the circular plate without electr odes, a/b=15.500, frequency=1.92 Mhz:

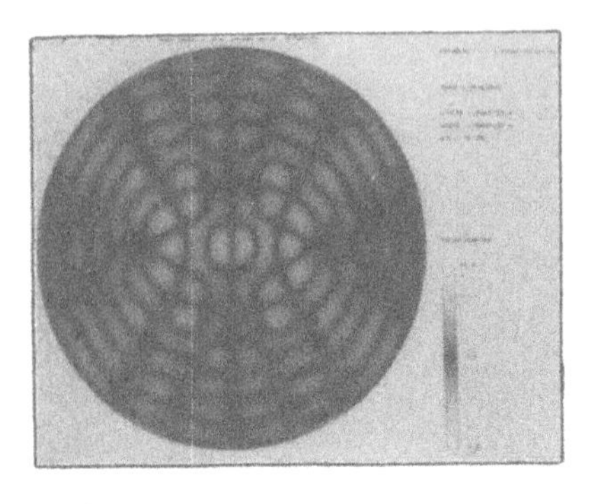

(a) u_2 displacement nodal line plot.

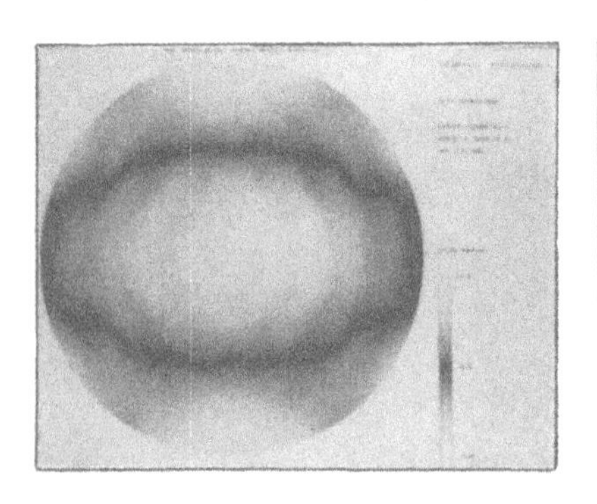

(b) ψ_1 displacement nodal line plot.

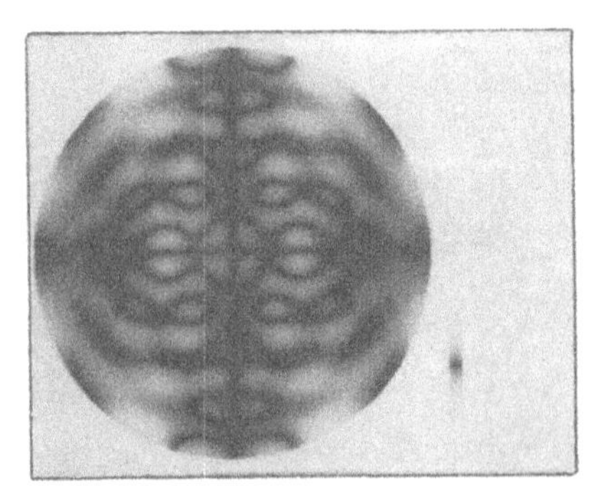

(c) ψ_3 displacement nodal line plot.

Fig.11 Nodal-line plots of (a) u_2, (b) ψ_1 and (c) ψ_3 displacements of the first anharmonic overtone of fundamental thickness-shear mode for the rectangular plate without electrodes, a/b=15.500, frequency=1.91 Mhz,

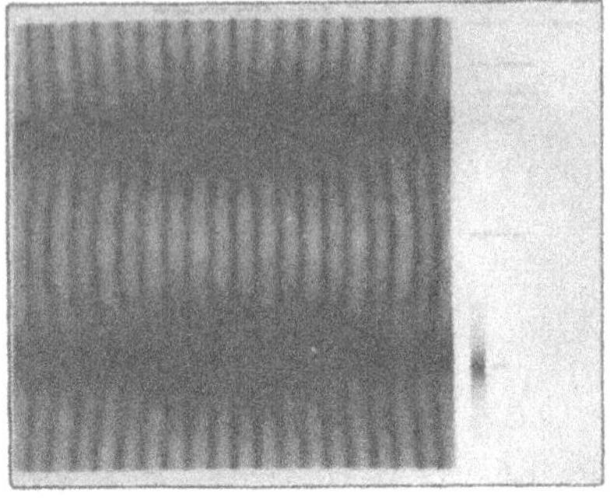

(a) u_2 displacement nodal line plot.

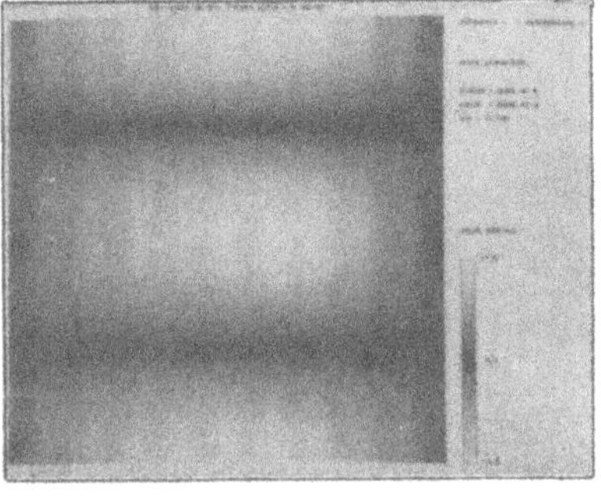

(b) ψ_1 displacement nodal line plot.

(c) ψ_3 displacement nodal line plot.

Fig.12 Grey-scale plots of (a) u_2, (b) ψ_1 and (c) ψ_3 displacements of the fundamental thickness-shear mode for the circular plate without electrodes, a/b=24.375, frequency=5.08 Mhz:

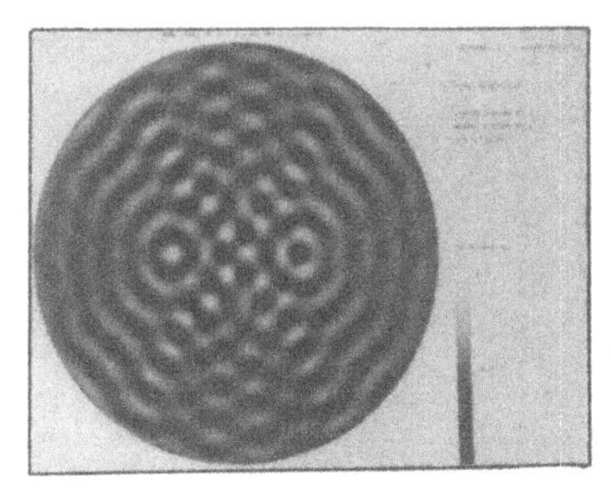

(a) u_2 displacement grey-scale plot.

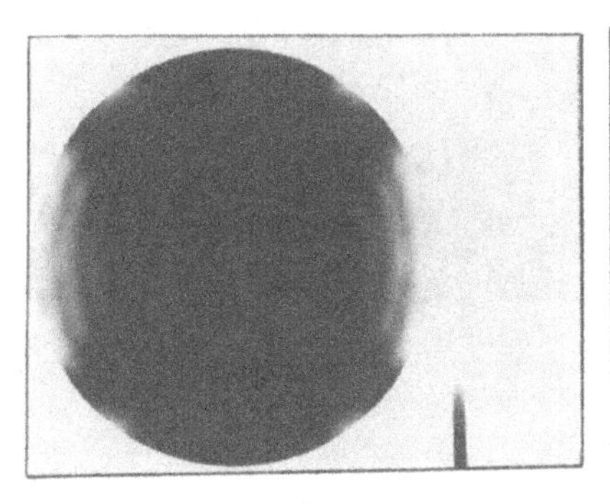

(b) ψ_1 displacement grey-scale plot.

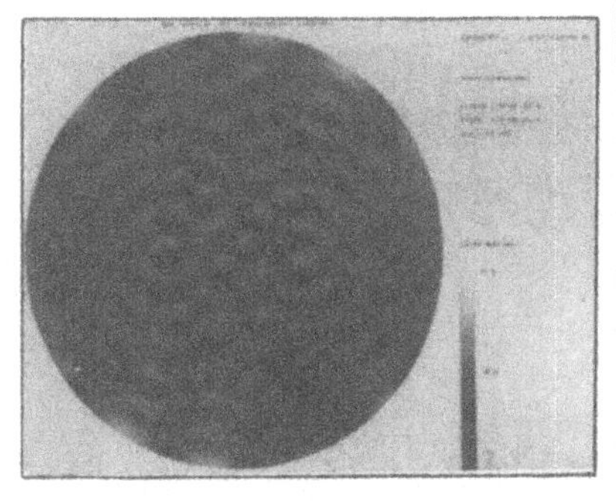

(c) ψ_3 displacement grey-scale plot.

Fig.13 Grey-scale plots of (a) u_2, (b) ψ_1 and (c) ψ_3 of the fundamental thickness-shear mode for the circular plate with electrodes, a/b=24.375, frequency=5.07 Mhz:

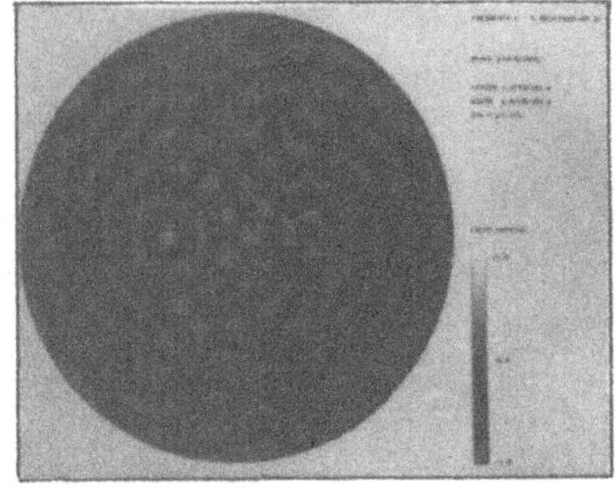

(a) u_2 displacement grey-scale plot.

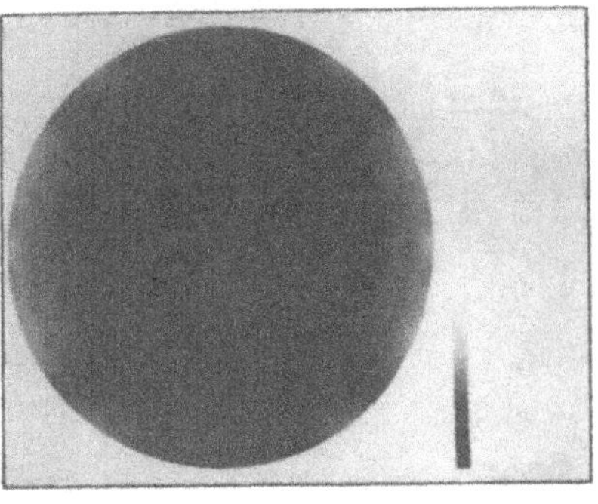

(b) ψ_1 displacement grey-scale plot.

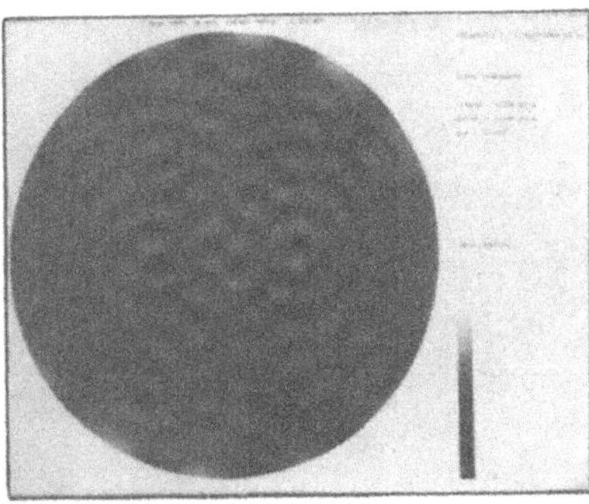

(c) ψ_3 displacement grey-scale plot.

Fig.14 Nodal-line plots of (a) u_2, (b) ψ_1 and (c) ψ_3 displacements of the fundamental thickness-shear mode for circular plate with electrodes, a/b=24.375, frequency=5.07 Mhz:

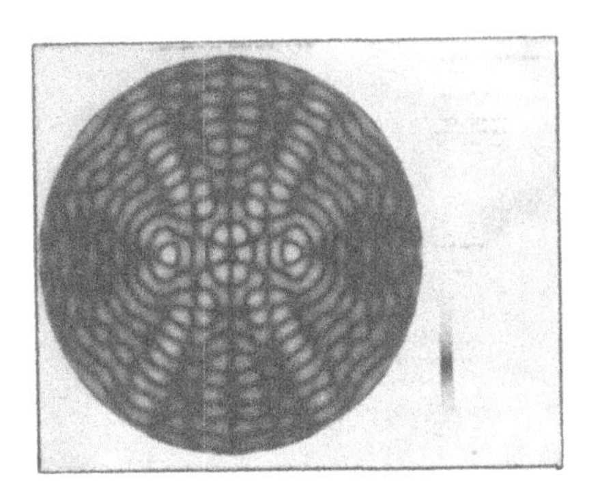

(a) u_2 displacement nodal line plot.

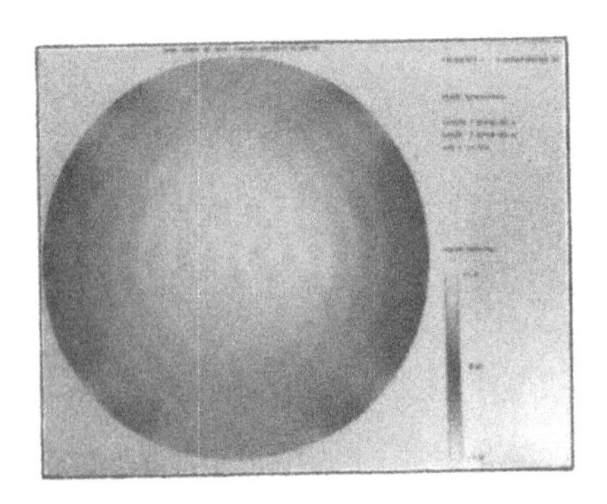

(b) ψ_1 displacement nodal line plot.

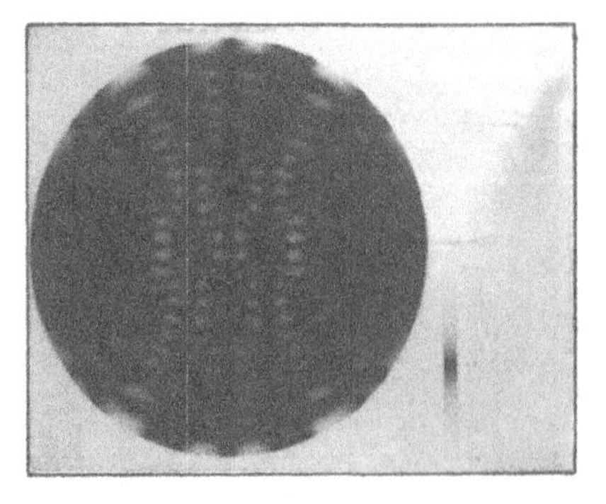

(c) ψ_3 displacement nodal line plot.

Fig.15 Grey-scale plot of (a) u_2, (b) ψ_1 and (c) ψ_3 displacements of the first anharmonic overtone of fundamental thickness-shear mode for the circular plate with electrodes, a/b=24.375, frequency=5.11 Mhz:

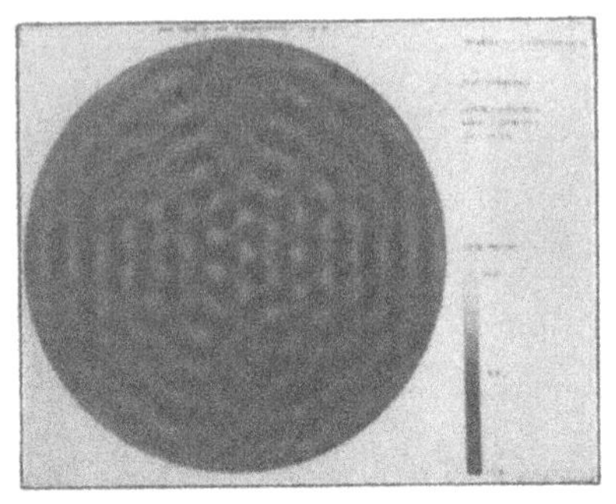

(a) u_2 displacement grey-scale plot.

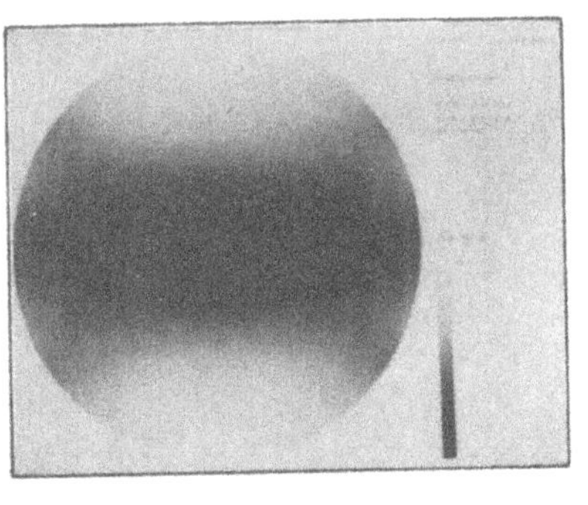

(b) ψ_1 displacement grey-scale plot.

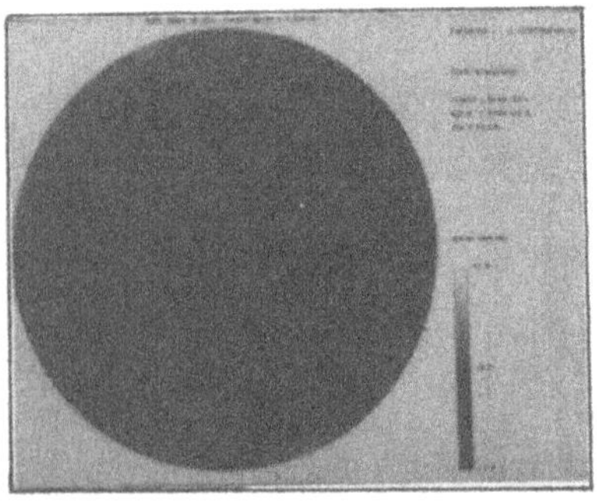

(c) ψ_3 displacement grey-scale plot.

Fig.16 Nodal-line plots of (a) u_2, (b) ψ_1 and (c) ψ_3 displacements of the first anharmonic overtone of fundamental thickness-shear mode for circular plate with electrodes, a/b=24.375, frequency=5.11 Mhz:

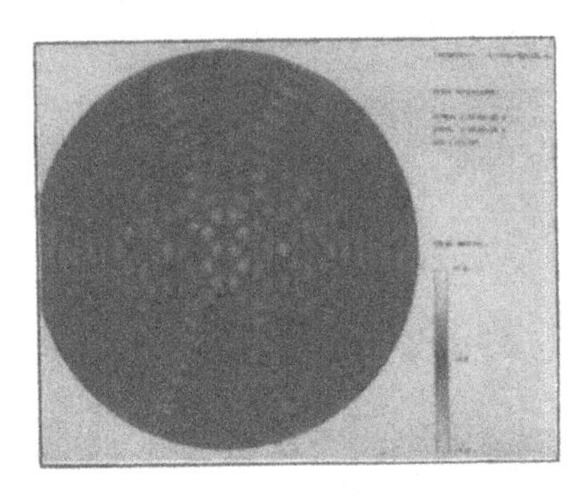

(a) u_2 displacement nodal line plot.

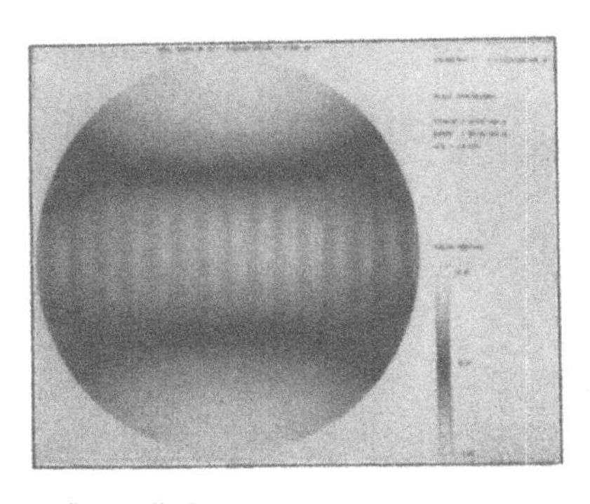

(b) ψ_1 displacement nodal line plot.

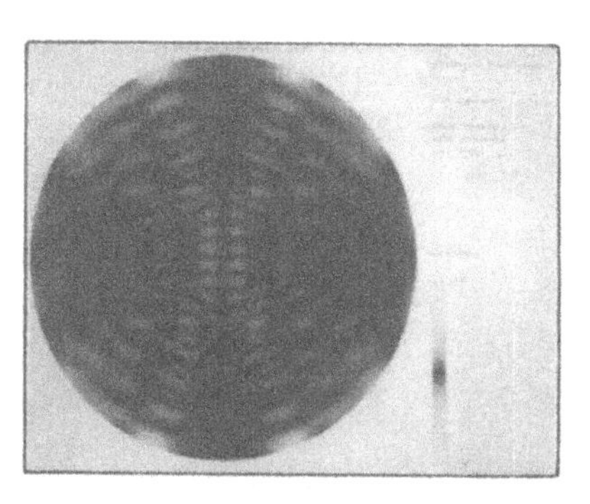

(c) ψ_3 displacement nodal line plot.

Fig.17 Grey-scale plots of (a) u_2, (b) ψ_1 and (c) ψ_3 displacements of the edge flexural mode for the circular plate with electrodes, a/b=24.125, frequency=5.06 Mhz:

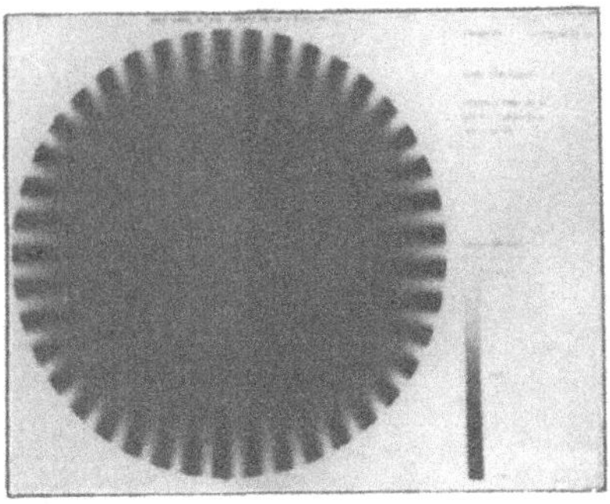

(a) u_2 displacement grey-scale plot.

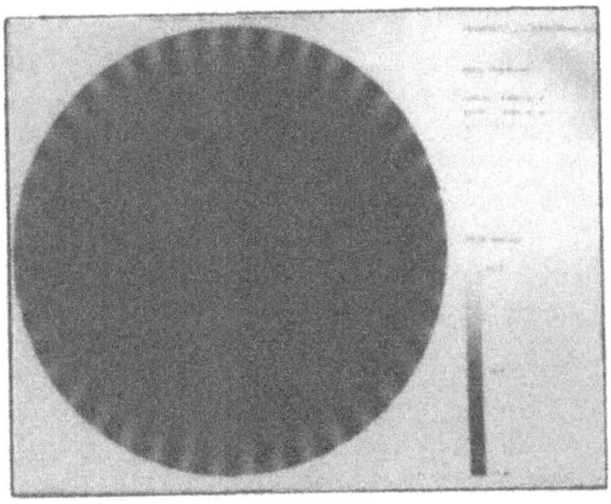

(b) ψ_1 displacement grey-scale plot.

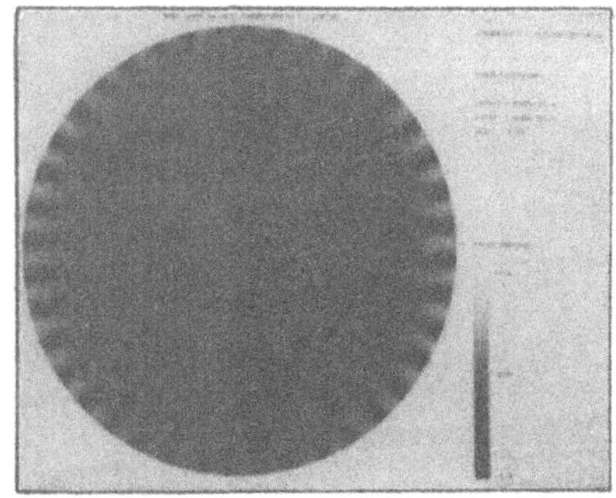

(c) ψ_3 displacement grey-scale plot.

Optimization of Boundary Elements Performance on Supercomputers

A. Elzein

Computational Mechanics BEASY, Ashurst Lodge, Ashurst, Southampton, SO4 2AA, U.K.

Abstract

A boundary element engineering analysis system for stress and heat transfer analyses, is optimized for supercomputers.

The implementation of new integration and data management algorithms and the use of library routines specific to CRAY computers as well as other directives and I/O facilities, results in substantial reductions in CPU, elapsed time as well as memory requirements.

A highly efficient engineering design environment is thus achieved, where the speed with which the analysis can be performed on a supercomputer is matched by a considerable reduction in mesh preparation effort inherent to boundary elements. Furthermore, the additional reduction in memory requirements allows larger and thus more realistic models to be analysed on such an environment.

Results from data models run on both the optimized and non-optimized versions of the program are produced to demonstrate the savings in computer requirements.

I. INTRODUCTION

Fast supercomputers with multiple vector-processors and large amounts of in-core memory, are widely used in a large number of scientific and industrial applications that require speedy and accurate results from numerically intensive analyses. Further speed up can usually be obtained if the analysis computer program is vectorized and/or parallel processing is used. Very often however, such improvements cannot be obtained unless minor or major modifications to the program algorithms are introduced. Vector operations must replace scalar ones and restrictive programming syntax, that may hinder vectorization, removed. When parallel processing is used, different tasks, preferably similar in type, must be assigned to different processors, avoiding at the same time logical, memory etc... conflicts.

The boundary element method (BEM) is a recently developed powerful numerical technique, similar to the finite element method (FEM), for the solution of various engineering problems. Some of the mathematical algorithms widely used in FEM programs are also employed by the BEM e.g. the *Gauss* numerical integration, the solution of linear systems of equations etc... Nevertheless, fundamental differences in the numerical implementation of the two methods exist both in the assembler, where more sophisticated integration techniques are required in BEM, and in the solver, where the BEM left-hand side matrix is, unlike the banded FEM matrix, fully populated.

The optimization of a number of boundary element programs have been reported in the literature. The vectorization of a boundary element code on CYBER computers has been described by *Katnik et al.*[1] and *Hodous et al.*[2] as early as 1982. Parallel processing was later applied by *Kline et al.*[3] to a boundary element solver. *Zucchini* and *Mukherjee*[4] optimized the performance of an axisymmetric single-zone BEM program through vectorial and parallel processing. The performance of both the coefficients assembly and internal points solution was improved. *Drake* and *Gray*[5] presented a parallel implementation of an electro-plating boundary element solution. *Adey*[6] discussed the solver vectorization of a general-purpose boundary element program.

This paper describes an experience in optimizing the performance of a general-purpose boundary element industrial code BEASY, on supercomputers. Although the project was primarily intended for CRAY supercomputers, most of the newly introduced features are also valid for other machines and only a minimal amount of conversion work has been required.

II. THE BOUNDARY ELEMENT METHOD

Introduction

The boundary element method, based on integral equations rather than differential ones, has introduced two useful concepts into the field of computational mechanics: that the shape of the body under consideration can be fully and uniquely be defined, for the purpose of analysis, by its boundary surface rather than its volume and that physically relevant known classical solutions (fundamental solutions) can be used to obtain highly accurate results. In more practical terms, this implies that the user of a program based on boundary elements will only have to discretize the surface of the problem using lines in 2D problems and quadrilaterals or triangles in 3D problems. In addition, a relatively small number of elements will be required to achieve a high degree of accuracy. Clearly, the method's advantages are amplified in a computer-aided design environment where modifications are regularly introduced throughout the design process; the corresponding effort required to modify the mesh is greatly reduced by the fact that the modelling is done on the surface of the model. Furthermore, such an optional advantage can easily become a crucial requirement in 3D analyses where automatic solid mesh generators in Finite Elements are still unable to cope with all possible eventualities of geometries, and where mesh design is one of the most, if not the most, time consuming parts of the design process.

It is rather difficult to review the origins of the boundary element method with exactitude as it is closely linked to the integral equations theory, the beginning of which can be traced back to the late nineteenth century. In addition, such a large number of scientific fields are encompassed by the method that a comprehensive review is well beyond the scope of this paper. It was *Fredholm*[7] in 1908 however, who first derived the integral equations of linear elasticity problems. He determined the conditions of existence and uniqueness, known as *Fredholm* theorems, of the solutions to these equations. In the 1960's, the pioneering work of a number of authors[(8–14)] inaugurated the first boundary element solutions to some engineering problems. The seventies saw an increasing research activity which established boundary elements as a reliable alternative to finite elements in a significant number of scientific fields. Practical[(15–19)] numerical applications and interpretations of the method were presented and problems related to singularities and volume integrals addressed. By the end of the seventies, the method was successfully applied to fields as far apart as plate bending problems, cathodic protection of offshore structures, fracture mechanics etc...

Although the boundary element method is thus relatively recent and some of its applications are still in the research phase, commercial computer programs based on the method are widely used in industrial applications notably for stress and heat transfer analyses.

Boundary Integral Equations of 3D Linear Stress Analysis

Boundary integral equations for 3D linear elastostatics can be written as follows[20]:

$$c_{ij}(P)u_i(P) + \int_s u_j(Q)t^*_{ij}(P,Q)dS = \int_s t_j(Q)u^*_{ij}(P,Q)dS +$$

$$\int_v u^*_{ij}(P,Q)b_j(Q)dV \qquad (1)$$

where S is the surface of the solid under consideration and V its volume. P and Q denote the source and field points, respectively. u_j and t_j refer to the displacement and traction fields (i and j are direction indices). u_i is the displacement at the source point P. u^*_{ij} and t^*_{ij} are the displacement and traction functions of the *Kelvin* point force solution. c_{ij} is a constant depending on the smoothness of the boundary at P and b_j are the body forces.

The volume integral containing the effect of body forces can be transformed into a surface one using a number of available techniques.

The integrals appearing in equations (1) can be numerically evaluated by dividing the boundary of the problem into a finite number of elements and by using standard integration procedures. Either displacements or tractions, in each of the three directions, will be given as boundary conditions to the problem. If equations (1) corresponding to each boundary node are formed, a linear system of equations of the following form is obtained:

$$[A](x)=(b)$$

where [A] is a matrix of coefficients, (b) is a known vector and (x) is a vector containing the unknown tractions and displacements.

When the distance r between the source point P and the field point Q approaches zero, singularities of order 1/r and $1/r^2$ occur in equations (1). The standard gaussian integration scheme would then require a large number of gaussian points to achieve a reasonnable accuracy or, ultimately, it may fail altogether. Special integration techniques, either analytical or numerical, are then resorted to in order to resolve the singularities.

Once the main system of equations has been solved, solutions at selected points inside the volume of the model (internal points) can also be found by using integral equations similar to equations (1).

Clearly, a boundary element analysis program can be divided into 6 main stages, essentially similar to Finite Elements:

1. Reading and data preparation

2. Assembly of coefficient matrices

3. Assembly of coefficients relative to interface degrees of freedom

4. Solution of the system of equations

5. Solution at internal points

6. Writing output

Only when the problem is substructured would stage 3 be required.

One of the most important differences between the performance of a boundary element and a finite element linear program is that, while the latter's most CPU-time intensive part is the solver, the assembler of the former can be equally time consuming because of the integrals singularity.

III. BEASY, GENERAL DESCRIPTION

BEASY is a general-purpose industrial boundary element analysis system. The first version was released in 1982 by Computational Mechanics. Today, in the latest version 4., stress analyses and steady state heat transfer analyses of 2D, axisymmetric and 3D problems can be performed[21]. Linear, elastic conditions are assumed and a large choice of boundary conditions is available. Non linear contact problems featuring gap elements with or without static or dynamic friction can also be analysed. Stress intensity factors of fracture mechanics problems can be automatically calculated for the three crack modes K1, K2 and K3. A hierarchical elements library is available featuring constant, linear and quadratic elements as well as continuous or discontinuous elements with automatic selection of continuity pattern. Discontinuous elements (where nodes are inside the element rather than on its periphery) can simplify the model generation and improve the accuracy in cases of geometrical discontinuities such as corners or physical ones such as the abrupt jump of stress concentration on a crack tip. In fact, the user can employ an unlimited number of adjacent elements having non-aligned edges and/or non-coinciding nodes, which further simplifies the model generation.

Naturally, a problem can be substructured and a large choice of boundary conditions at the interfaces between the various substructures is available. Substructuring can be used as a tool for improving the computational efficiency of the analysis since it introduces a bandwidth into an, otherwise fully-populated, left-hand side matrix. It is also a way of defining components with different material properties in the same structure.

The optimization project initially targeted the 3D stress analysis module of BEASY. But all changes to the program were later ported, quite easily, to the 3D potential analysis module. The 2D and axisymmetric parts of both stress and potential analyses were not modified in this project as the computer requirements of these modules are minimal even for relatively large problems.

IV. CRAY-YMP SUPERCOMPUTERS

While the initial focus of the supercomputer industry has perhaps been on the speed up of the CPU-time, other factors such as I/O efficiency, memory and storage capacities, performance under heavy usage conditions etc... have quickly acquired as much importance. The overall evaluation of a supercomputer technical performance is therefore a rather complicated task which is beyond the scope of this paper. What is proposed instead is a brief presentation of the general attributes of the machine.

The CRAY-YMP/832[22] has up to 8 Central Processing Units (no parallel processing has been used in this project) and a clock cycle of 6 nsec[23]. A maximum of 256 Mbytes of central memory can be delivered. The operating system, UNICOS, is UNIX-based. An optimized scientific subroutine library is available.

V. APPROACH TO OPTIMIZATION

Optimization for Supercomputers

Since most industrial software is produced over long periods of time, conversion work adapting the program to specific machine requirements or new hardware developments is an essential part of the maintenance and development work performed on a computer code such as BEASY. So what is special about supercomputers that requires the modification of the standard version of a program?

In addition to the instantaneous speed of execution and the multiple processing, supercomputers are essentially characterized by a large memory capacity and a potential for further speed up by vectorization. Any free memory available can be used as a temporary storage space to avoid repeated calculations and reduce the I/O operations. The free memory is thus 'converted' into a speed up of both CPU and elapsed time. Such a conversion however requires a change of algorithm. In addition, re-writing of some key loops or subroutines is usually needed to remove any obstacles, such as IF statements or subroutine calls inside a loop, that may hinder vectorization.

In this project, the optimization has targeted three main performance areas: the CPU-time, elapsed time associated with I/O operations and the memory requirements (the storage requirements of intermediate work files have also been reduced but a subsequent BEASY project have concentrated on this particular aspect and have produced more substantial reductions). The purpose of the optimization is essentially to improve the performance as much as possible and, by the same token, to increase the size of the problems that can practically be analysed by, roughly, a factor of 5. The 'size' of the problem, will be based on the number of degrees of freedom.

A Systematic Approach

A systematic approach to the optimization has been found quite useful in maximising the efficiency of the work and even necessary if the supercomputer time is to be used in an economical way. Such an approach can be divided into 5 main stages: profiling, algorithm change, assessment and tuning, special machine features, target problems and final tuning. The reason for this sequence will become apparent in the discussion that follows.

1. Profiling: the profile of a computer program is a computer output chart showing the amount of CPU-time spent in each program subroutine during a complete run. This is an efficient way of identifying the CPU-time intensive parts of the program. Each data problem results in a different profile and a number of models must be run before an accurate picture is obtained. Clearly, target problems cannot be analysed at this stage as they would require a large amount of time to run. Profiles of smaller problems were used instead.

2. Algorithm change: based on the profile results and on a good knowledge of the program theoretical background, decisions on any algorithm changes and programming improvements are made and implemented. This has been the most time consuming part of the project, as major changes to the program have been required.

3. Assessment and tuning: once the previous changes are implemented, a new set of profiles is produced and the effect of stage 2 assessed. Further improvements can be made at this stage, especially by loop re-writing to improve vectorization.

4. Special machine features: special features related to the specific machine are added such as library calls, special directives, special I/O facilities etc... The inclusion of these special features in a separate

stage would facilitate the porting of the optimization work to other machines.

5. Target problems and final tuning: profiles for target problems are produced and further tuning performed.

One particular difficulty has been encountered when the target problems were run for the first time. That is the, sometimes abrupt, shift in the location of CPU-time intensive and I/O-intensive parts of the program as one increases the size of the problem by what amounts to half a degree of magnitude. Clearly, this is due to the fact that initial profiles were based on much smaller models. Moreover, "upsurges" in CPU-time occured in some parts of the program, when running the large target problems, that were almost completely "CPU-time free" for smaller models. The first analogy that comes to mind is that of a famous cartoon character using all the energy he can spare to lock the water tap through which heavy drops of water are falling noisily, only to realise that the spill has started through a neighbouring tap... As you may expect, a less artistic but more rational approach to solve the problem has been used in our case. In fact, these difficulties had to be tackled systematically during the final tuning of stage 5 and sometimes in an iterative way by running the target problem, locating the CPU-time intensive areas, modifying the program to reduce the CPU-time, re-running and so on... until the performance is either satisfactory or optimal. The special CRAY utility tools such as profiling and measuring the MFlops performance proved of utmost importance and usefulness for such a task.

Clearly, the above described systematic approach cannot be applied unconditionally to any optimisation effort of a numerically intensive program. It appears to be particularly valid however for programs, such as boundary element ones, which have more than one CPU-time intensive location.

VI. DESCRIPTION OF MODIFICATIONS

CPU-time Reduction

A new integration algorithm was implemented which has reduced the overall number of operations performed, converted some scalar operations into vector ones, increased the size of inner loops to improve the vectorization efficiency and eliminated the need for any subroutine calls inside them. The new algorithm made use of a number of integration and programming techniques to ensure that the numerical accuracy of the integration and the computational efficiency of the program are both satisfactory. These techniques are essentially the standard gaussian integration scheme[24], the *RISP* method for the efficient management of data within the assembly routines[25] and the self-adaptive integration scheme presented by *Telles*[26] where a polynomial transformation is used to reduce the number of required gaussian points. In addition, a special scheme for the calculation of body forces, based on the homogenization of the differential equation[27], has almost completely eliminated the time cost associated with standard body forces. Further loop re-writing of other integration routines and their subsequent vectorization as well as the use of some library routines and directives specific to CRAY computers has been performed. By the time this project had started, the solver in BEASY was already well vectorized[6]; both optimized and non-optimized versions, referred to in the results section of this paper, are thus using a vector solver.

Elapsed Time Reduction

The elapsed time resulting from I/O operations was also greatly reduced as a new data management system has been introduced throughout BEASY which has, among other things, substantially reduced the number of I/O operations by using whatever free memory is available. This scheme has proven to be of particular efficiency in the solver and the assembler where a large amount of I/O is required. In addition, the scheme requires a relatively small amount of free memory to achieve most of its potential savings. Figure 1 illustrates this behaviour.

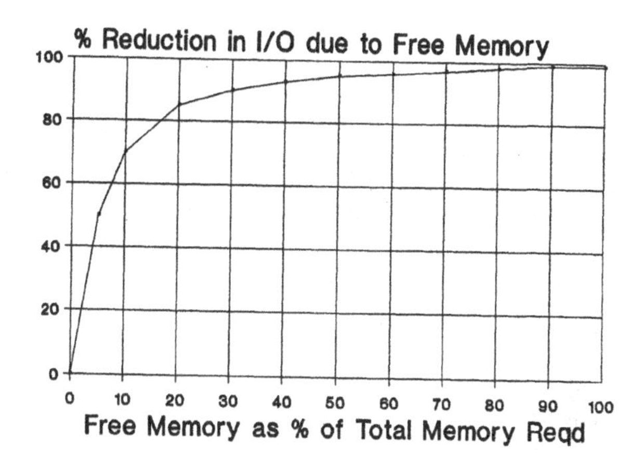

Figure 1. General trend of I/O reduction in the solver versus free memory available.

In this highly typical case, 70% of I/O reduction can be achieved using only 10% of the total free memory required to run the solver in core. This is due to the fact that in the solver, based on a Gaussian elimination scheme, I/O can be greatly reduced by just keeping a small number of rows in memory. In the case of the assembler, the required free memory to achieve all potential reduction is almost invariably very small.

In addition, the CRAY word addressable I/O system has been used for the storage of work files which resulted in more efficient I/O operations. Furthermore, a series of systematic parametric tests have been performed to obtain the optimum configuration, in respect of elapsed and CPU time, of the various program variables associated with I/O.

Memory Reduction

The above mentioned modification of the data management system has provided the opportunity to modify some array structures in order not only to reduce the memory requirements for any particular problem but also to avoid a steep growth of those requirements with the size of the problem. A parallel effort of another BEASY project for reducing the memory in other parts of the program, was also added to that.

VII. RESULTS

Profiles

We have discussed earlier the considerable change in the profile of the program when the large target problems are run. This profile variation is illustrated in figure 2. The general distribution of the CPU-time is shown for 4 problems, of varying sizes, analysed by the optimized version of BEASY. While the CRAY profile output gives a comprehensive description of the CPU-time distribution for all program subroutines, only the overall time shares of the Solver and Assembler are given here.

It is clear from figure 1, that the solver's share in the total CPU-time becomes more important as the size of the problem is increased.

The item "Others" refers mainly to the coefficients assembly corresponding to the interface between the substructures, the solution at internal points and the data processing. Its contribution to the total CPU-time varies from 7% to 21% and appears to be unrelated to the size of the problem. In fact, it depends on other factors such as the number of substructures, the number of boundary condition cards, the number of internal points etc...

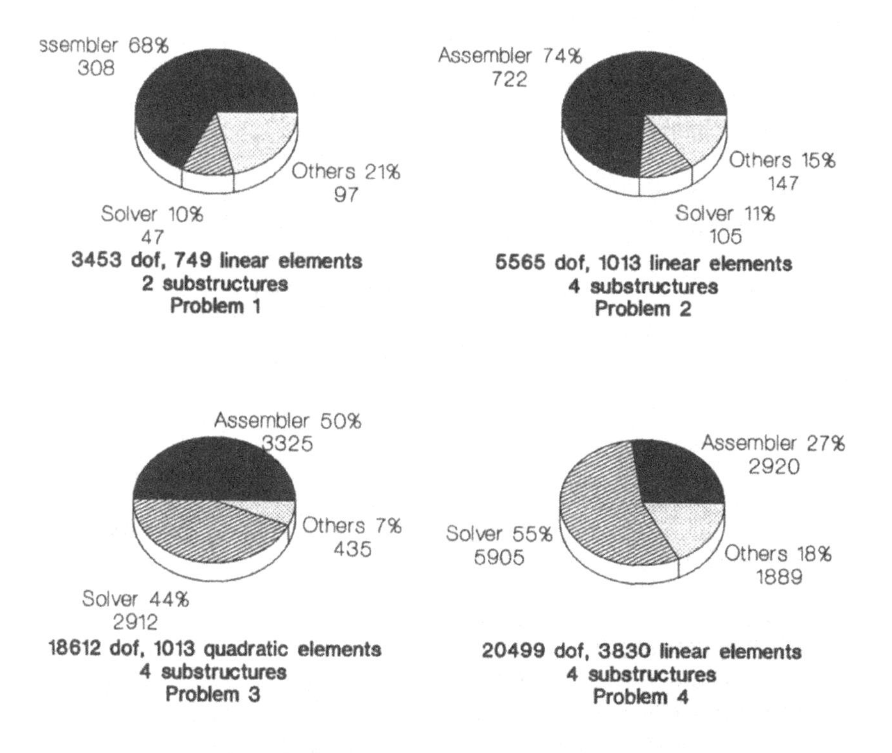

Figure 2. Main CPU time distribution in seconds for 4 problems run on CRAY-YMP using the optimized version of BEASY.

Improvements due to Optimization.

The effect of the optimization is assessed by running on a CRAY-YMP machine, 3 sample problems which include the first two problems of figure 1. A comparison of the CPU-time, I/O data tranferred and memory required from both the optimized and non-optimized versions of BEASY, is made. Tables 1, 2 and 3 describe these results.

	Non-Optimized Version	Optimized Version	% Reduction
Substructures	2	2	
Elements Number	749	749	
D.O.F.	3453	3453	
Memory Reqd Mwords	1.22	0.8	-34 %
Data Transf Mbytes	6.87	2.83	-59 %
CPU-time Assembler	9 mins 25 secs	5 mins 8 secs	-45 %
CPU-time Total	12 mins 28 secs	7 mins 32 secs	-40 %

Table 1. BEASY Optimization on CRAY-YMP: problem 1.

	Non-Optimized Version	Optimized Version	% Reduction
Substructures	4	4	
Elements Number	1202	1202	
D.O.F.	5373	5373	
Memory Reqd Mwords	1.79	1.17	-35 %
Data Transf Mbytes	9.55	4.61	-52 %
CPU-time Assembler	11 mins 33 secs	6 mins 32 secs	-43 %
CPU-time Total	16 mins 12 secs	10 mins 15 secs	-37 %

Table 2. BEASY optimization on CRAY-YMP: problem 2.

	Non-Optimized Version	Optimized Version	% Reduction
Substructures	4	4	
Elements Number	1013	1013	
D.O.F.	5565	5565	
Memory Reqd Mwords	1.76	1.12	-37 %
Data Transf Mbytes	11.34	7.44	-34 %
CPU-time Assembler	19 mins 30 secs	12 mins 2 secs	-38 %
CPU-time Total	25 mins 40 secs	16 mins 14 secs	-37 %

Table 3. BEASY optimization CRAY-YMP: problem 3.

'D.O.F' denotes the number of degrees of freedom. 'Data Transferred' is a CRAY output information on the amount of data transferred between the Central Processing Unit and the Disc, in Mbytes. This, in fact, should be taken as a measure of the number of I/O operations made. All the elements used are linear.

The memory requirements have clearly been reduced by an average of 35%. The average reduction of the assembler CPU-time is about 42%. This can be seen as a substantial reduction if we bear in mind that the operations in the assembler, unlike the solver, are essentially scalar. The average reduction in the total CPU-time is about 38%.

The reduction in data transferred reflects the new data management system which has reduced the overall number of I/O operations. It does not show however all the improvements due to the use of Word Addressable Files and the parametric optimization of I/O variables. These modifications have reduced the elapsed time by speeding the I/O without necessarily reducing the actual number of I/O operations and thus the data transferred. When the reduction in CPU-time is added to that, it becomes clear that the reduction in the elapsed time of the analysis is expected to be much higher than the reduction in the data transferred. This was confirmed by the results as a minimum reduction in elapsed time by a factor of 3.5 was observed. No values for elapsed time were reproduced here because, unless run in dedicated time, they are highly dependent on the overall load on the machine.

Finally, large target problems, containing around 20,000 degrees of freedom or more, were solved using the optimized version only.

VIII. CONCLUSIONS

A systematic approach has been used to speed up and optimize the performance of the assembler of a boundary element code. A new data management scheme has also been introduced to reduce the number of I/O operations by using any available free memory. A substantial reduction in CPU and elapsed time has been obtained in addition to a reduction in memory requirements. In fact, since the end of this project, further reductions in elapsed time and storage requirements have been made and have increased yet again the size of the problems that can practically be analysed by BEASY.

Further reduction in the elapsed time of a BEASY analysis on a Supercomputer can be obtained by using parallel processing. Two approaches can be particularly efficient: the first would assign one substructure for each processor, the second would assign a group of field elements instead.

Acknowledgements

The author wishes to thank HONDA R&D for supporting this project and Mr Junichi Sugita and his team for their keen interest. Thanks are also extended to CRAY Research in Eagan, Minneapolis for the support provided during the final stages of this project.

REFERENCES

1. Katnik R.B., Hodous M.F., Bozek D.G., Ciarelli K.J. and Kline K.A., 'Boundary integral method of structural analysis on a vector processor', *Proc. Symp. CYBER 205 Application*, Colorado State University, 1982.

2. Hodous M.F., Bozek D.G., Ciarelli D.M., Ciarelli K.J., Katnik R.B. and Kline K.A., 'Vector processing applied to boundary element algorithms on the CDC Cyber 205', *Bulletin de la Direction des Etudes et des Recherches, 1983, Serie C No 1*, pp 87-94.

3. Kline, Tsao N.K and Friedlander C.B., 'Parallel processing and the solution of boundary element equations', *Advanced Topics in Boundary Element Analysis*, Cruse et al. (eds.), ASME AMD Publication, ASME, 1985, New York.

4. Zucchini A. and Mukherjee S., 'Vectorial and parallel processing in stress analysis with the boundary element method', *Intern. Journ. Numer. Meth. Eng'g, Vol 31, 1991*, pp. 307-317.

5. Drake J.B. and Gray L.J., 'Parallel implementation of the boundary element method on the IPSC2 hypercube for electro-plating applications', *Oak Ridge National Laboratory Technical Report ORNL-5615*, Oak Ridge, Tennesse, 1988.

6. Adey R.A., 'Computational aspects and applications of boundary elements on supercomputers', *Supercomputers in Engineering Structures*, Melli and Brebbia (eds), Computational Mechanics Publications and Springer-Verlag, 1989.

7. Fredholm I., 'Sur une classe d' équations fonctionelles', *Acta Math., 1963, Vol. 27*, pp.365-390.

8. Kupradze V.D., '*Potential Methods in the Theory of Elasticity*', Oldbourne Press, 1965.

9. Massonet C., 'Résolution graphomécanique des problèmes généraux de l'élasticité plane', *Bull. Centre des Etudes et Recherches, Essais Sc. Génie Civil, 1949, Vol. 4*, pp. 169-180.

10. Hess J.L. and Smith A.M.O., 'Calculation of potential flow about arbitrary bodies', *Prog. Aero. Sci. 8, 1967*, Kuchemann (eds), Pergamon Press.

11. Jaswon M.A., 'Integral equation methods in potential theory I', *Proc. Royal Soc., A, 1963*, p 273.

12. Symm G.T., 'Integral equation methods in potential theory II', *Proc. Royal Soc., A, 1963*, p 275.

13. Rizzo F.J., 'An Integral equation approach to boundary value problems of classical elastostatics', *Applied Mathematics, 25, 1967*, pp. 83-95.

14. Cruse T.A., 'Numerical solutions in three dimensional elastostatics', *Intern. J. Solids Structures, 5, 1969*, pp. 1259-1274.

15. Brebbia C.A. '*The Boundary Element Method for Engineers*', 1978, Pentech Press, London.

16. Lachat J.C., '*A Further Development of the Boundary Integral Technique for Elastostatics*', PhD Thesis, University of Southampton, 1975.

17. Lachat J.C. and Watson J.O., 'Effective numerical treatment of boundary integral equations', *intern. Journ. Numer. Meth. Eng'g, Vol 10, 1976*, pp. 991-1005.

18. Cruse T.A., 'An improved boundary-integral equation method for three dimensional elastic stress analysis', *Computers and Structures, 1974, 4*, pp. 741-754.

19. Symm G.T., 'Treatment of singularities in the solution of laplace's equation by an integral equation method', *NPL Report NAC 31, 1973.*

20. Brebbia C.A. and Dominguez J., *'Boundary Elements: an Introductory Course'*, Computational Mechanics Publications, Southampton and Boston, 1989.

21. '*BEASY User's Guide*', Computational Mechanics Publications, 1990.

22. 'Introducing the CRAY Y-MP computer system', *CRAY Channels, Spring 1988*, CRAY Research Inc. Publication, pp. 2-5.

23. Levesque J.M. and Williamson J.W., '*A Guidebook to Fortran on Supercomputers*', Academic Press, 1989.

24. Zienkiewics O.C., '*The Finite Element Method*', 3rd Edition, McGraw-Hill, London, 1977.

25. Kane J.H., Gupta A. and Saigal S., 'Reusable intrinsic sample point (RISP) algorithm for the efficient numerical integration of three dimensional curved boundary elements', *Intern. Jour. Numer. Meth. Eng'g, Vol. 28, 1989*, pp. 1661-1676.

26. Telles J.C.F., 'A self-adaptive coordinate transformation for efficient numerical evaluation of general boundary element integrals.', *Intern. Journ. Numer. Meth. Eng'g, Vol. 24, 1986*, pp. 3-15.

27. Gründemann H., 'A general procedure transferring domain integrals onto boundary integrals in BEM', *Engineering Analysis with Boundary Elements, Vol. 6, No 4, 1989*, pp. 214-222.

Supercomputer Application in Inelastic Stability of 22-Story Steel Pino-Suarez Complex

F.Y. Cheng, J.-F. Ger

Department of Civil Engineering, University of Missouri-Rolla, Rolla, MO 65401, U.S.A.

ABSTRACT

This paper presents a modular computer program, INRESB-3D-SUP II, which utilizes the vectorization and parallelization resources of IBM 3090 600J supercomputer to investigate the inelastic behavior of a 22-story Pino-Suarez steel building collapsed during the September 19, 1985, Mexico earthquake. INRESB-3D-SUP II is a general program and is developed for analyzing elastic and inelastic building systems subjected to static loads, static cyclic loads and multi-component earthquake motions. The program can also be used to calculate natural frequencies and buckling loads. Dynamic allocation of memory and compact matrix storage are employed in the program for achieving computational efficiency. The structural elements may be 3D beam-columns, reinforced concrete shear walls, springs, and bracing members. The nonlinear behavior of a structural element is based on the assigned material hysteresis model for the element. Modular systems are used in this program for the easy addition of elements, hysteresis models, and solution algorithm. The output solutions include static member forces and joint displacements, dynamic member forces, joint displacements, ductility factors, excursion ratios, damage indices, input energies, and dissipated energies, as well as the option of saving data for plotting.

INTRODUCTION

When a high-rise steel building is subjected to strong earthquake excitations, the structural members may be damaged and become highly nonlinear. For the nonlinear dynamic analysis of large structural systems, excessive computing time and cost may be expected. The computing cost is a function of the number and types of structural elements, the hysteresis models each element uses, the number of degrees of freedom, the time increment for numerical integrations, and the amount of storage required for the structural stiffness, geometric stiffness, and mass matrices.

Such cost, however, could be minimized if the computer program has an efficient memory management and good numerical techniques for the system analysis.

Here, a modular computer program INRESB-3D-SUP II (Inelastic Analysis of Reinforced Concrete and Steel Building Systems for 3-Dimensional Ground Motions) is presented. The dynamic memory management and compact matrix storage algorithm used in this program as well as utilizing the vectorization and parallelization of a supercomputer can achieve the computational efficiency. The program consists of five primary blocks. The first block (STRUCT) defines the structural model and the remaining four (SOL01, SOL02, SOL03, and SOL04) are solution blocks for static loading, seismic loading, natural frequency or buckling load, and static cyclic loading, respectively. The material hysteresis models for open-web-joist girders, bracing members, and box columns to be used in the inelastic analysis are first developed and stored in the material library in the block STRUCT. Several different types of elements which represent 3D beam-columns, reinforced concrete shear walls, and bracing members are also developed and stored in the element library in the block STRUCT for the inelastic study.

The configuration of the 22-story Pino-Suarez building is shown in Figure 1. It contains four bays in the building's long direction (E-W direction) and two bays in the building's short direction (N-S direction). The frames were constructed of welded box columns and specially fabricated open-web-joist girders. An auxiliary bracing system is in one of the bays. The bracing members are build-up section with three plates welded together into an H shape. To investigate the inelastic behavior of this building, the failure ductility of open-web-joist girders and the local buckling of columns as well as the geometric matrix for large deformations are considered in the dynamic analysis.

DESCRIPTION OF INRESB-3D-SUP II COMPUTER PROGRAM

The computer program INRESB-3D-SUP II is capable of analyzing elastic and nonlinear 3D structures subjected to static and seismic loadings. The modular computer program consists of five primary blocks for which the detail of each block is described as follows:

STRUCT- definition of the structural model

The structural model consists of an assemblage of elements. The point where two or more elements connect is called a joint. A structure is modelled by first defining the location and orientation of each joint, then the elements that connect the joints, and the orientations of the elements are defined. The materials that describe the behavior of the elements are given. For dynamic analysis, the lumped mass at each joint is also defined. The flow chart for STRUCT is shown in Figure 2 and briefly described in the following three major steps:

Step 1. Define joints and determine the DOF's The coordinates of the joints and their orientation are defined by the user. The coordinates are in the global coordinate system (GCS). The orientation of each joint defines it joint coordinate system (JCS). Each joint initially has six degrees of freedom (dof) in the JCS. The user then defines the joint's dof that are restrained, constrained, and condensed out. The program generates the structural degrees of freedom.

Step 2. Define material properties The material properties are input and initialized. There are eight different material behaviors available, which constitute the material library, and are discussed as follows

a. Elastic 3D prismatic beam material: This material consists of the elastic section properties of a 3D prismatic beam, A_x, J, I_y, I_z, E, and G. A_x, J, I_y, I_z, represent the cross sectional area, torsional moment of inertia, moment of inertia about the strong axis, and moment of inertia about the weak axis, respectively. E and G represent the Young's modulus and shear modulus, respectively.

b. R/C axial hysteresis model: An axial hysteresis model is available for reinforced concrete shear walls (1,2).

c. Cheng-Mertz B1 bending hysteresis model: A bending hysteresis model is used to model the bending deformations in low-rise reinforced concrete shear wall (2).

d. Cheng-Mertz S1 shear hysteresis model: A shear hysteresis model is included to model the shear deformations in low-rise reinforced shear wall (2).

e. Takeda hysteresis model: A bending hysteresis model is for the bending deformations of slender reinforced concrete members (3).

f. Long-direction open-web-joist hysteresis model (LONG OWJG): A hysteresis model is for the bending deformations in the long-direction open-web-joist girder in the Pino-Suarez building. This hysteresis model (4,5) is sketched in Figure 3.

g. Short-direction open-web-joist hysteresis model (SHORT OWJG): A hysteresis model shown in Figure 4 is developed (5) to model the bending deformations in the short-direction open-web-joist girder in the Pino-Suarez building.

h. Bracing member hysteresis model (BRACE): This model can be applied to the box, angle, and wide flange members. The hysteresis model is sketched in Figure 5.

i. Bilinear hysteresis model (BILN): The hysteresis models are sketched in figures 6a and 6b to model the bending and axial deformations of box-shape columns. The maximum axial force P_{cr} and bending capacity M_{cr} can be determined by detecting local buckling criteria for box columns (5). $\beta' P_{cr}$ and $\beta(M_y)_{cr}$ or $\beta(M_z)_{cr}$ represent the residual axial strength and residual bending strengths of the column in the post-buckling region.

Step 3. Define elements The element data is input. The transformation matrices, initial element structural stiffness and the initial element geometric stiffness are calculated. There are six different elements available in the program, which constitute the element library, and discussed as follows

a. Elastic 3D prismatic beam element: The beam element has axial, torsional, and bending deformations. Warping torsion and shear deformation are not considered. The elastic 3D prismatic beam material is used with the beam element.

b. Spring elements: The spring element consists of an isolated spring that connects the start and end joints. The spring element can be axial, shear, or rotaional spring. The elastic 3D prismatic beam material, R/C axial hysteresis model, bracing member hysteresis model, or bilinear hysteresis model can be used with the axial spring element.

c. Reinforced concrete shear wall element: Bending, shear, and axial deformations are considered in the shear wall element (6,7). The bending, shear, and axial deformations are lumped into three springs. Different materials are used to describe the stiffness of the bending, shear, and axial springs. Typically the bending stiffness is defined by the B1 bending hysteresis model, the shear stiffness is defined by the S1 shear hysteresis model and the axial stiffness is defined by the R/C axial hysteresis model.

d. Inelastic 3D beam-column element: The element has axial, bending, and torsional deformations. Warping torsion and shear deformations are not considered. The LONG OWJG, SHORT OWJG, and BILN hysteresis models can be used for bending deformations. The BILN hysteresis model is also used for axial and torsional deformations.

e. Bracing element: The bracing hysteresis model is used to represent the hysteresis behavior of axially loaded box, angle, or wide flange members.

f. Finite-segment element: The finite-segment element (4,5) considers axial, bending, and torsional deformations. The member is divided into several segments; the cross section is divided into many sectional elements. The material stress-strain relationship for each sectional element is assumed to be elasto-plastic model. A substructural technique is applied to the element for which the internal degrees of freedom are condensed out by Gauss elimination and only the degrees of freedom at both ends of the member are maintained so that a computational efficiency can be achieved.

SOL01 - Elastic static analysis with multiple load cases

This block performs the elastic static analysis with multiple load cases. The flow chart for SOL01 is shown in Figure 7. The joint loads and imposed displacements (support settlements) are input for each load case. The displacements for each load case are calculated by Gauss elimination.

SOL02 - Elastic/nonlinear seismic time history response

This block performs the elastic or nonlinear analysis of a structure subjected to multiple ground accelerations. The flow chart for SOL02 is shown in Figure 8. The hysteresis models in the material library are called to calculate the incremental element forces for the given incremental displacements and previous loading history. For nonlinear analysis, if the element stiffness changes during the incremental displacement, the element's unbalanced forces are calculated. Then the system unbalanced force vector is assembled from the element's unbalanced forces and applied as a load in the next time step.

SOL03 - Elastic natural frequency/elastic buckling load

This block calculates either the natural frequencies and mode shapes of an elastic structure, or the buckling load and mode shapes of an elastic structure. The flow chart for SOL03 is shown in Figure 9.

SOL04 - Nonlinear static cyclic response

This block calculates the nonlinear static cyclic response for a given loading pattern. A loading pattern consists of joint loads, imposed displacements and element loads. The loading pattern is multiplied by positive and negative load factors to generate cyclic loading cycles. The flow chart for SOL04 is shown in Figure 10.

DYNAMIC MEMORY MANAGEMENT

Dynamic memory allocation

The computer program stores all of the joint, material, mass, stiffness, response, and other data in a large linear array, Z. The Z array consists of real numbers. An integer array NZ is set equal to the Z array, by a Fortran EQUIVALENCE statement, to provide storage for integer variables. Index variables beginning with IZ, such as IZxxxx, are used to store the address of the information in the Z array. The variable IZ is used to track the next available space in the Z array. The exact amount of space required to solve a specific problem is reserved in the Z array during the execution of the program. This dynamic allocation of memory allows the program to use the computer memory efficiently.

For example, assume that Z array is empty and IZ = 1. The joint ID numbers are integer variables that are stored in the NZ array. Thus the index for the joint ID numbers is IZID = IZ = 1 and NUMJOI joint numbers are stored in the NZ array. Once the storage for the joint numbers is reserved, the next available storage location is IZ = IZ + NUMJO1. Next the joint coordinates are stored in the Z array, beginning at IZCORD = IZ. The joint coordinates require 3 x NUMJOI storage locations. Thus once the storage of the joint coordinates is reserved, the next available storage location is IZ = IZ + 3*NUMJOI. The Z array with joint numbers and coordinates is shown in Figure 11.

Compact matrix storage

The mass, stiffness, geometric stiffness and dynamic stiffness matrices are sparse symmetric matrices. Because the matrices are symmetric, only half (the upper triangular matrix) of each matrix needs to be stored. Additionally, many of the terms in the upper triangular matrix are zero. The nonzero terms, in a given column, are typically found near the main diagonal. The level of the upper nonzero term in each column is the skyline. Thus only the terms below the skyline of the upper triangular matrix are stored in the Z array.

Capacity of the dynamic memory

The program's capacity is a function of the amount of memory available in the computer. The program compiled in double precision on IBM 3090 600J supercomputer requires about 0.8 megabytes of memory. Assuming 8 megabytes of total memory is available, the Z array may occupy 8-0.8 = 7.2 megabytes of memory. For double precision, on an IBM computer, each variable is 8 bytes long. Thus for 8 megabytes of memory, the Z array may contain up to 900,000 variables.

The number of variables in the Z array required for a specific problem is a function of 1) the number and orientation of the joints, 2) the number of restraints, 3) the number of constraints, 4) the number and types of materials, 5) the number and types of elements and materials each element uses, 6) the number of joint masses, 7) the amount of storage required by the structural stiffness, geometric stiffness, and mass matrices, which is a function of the band width, 8) the type of solution, and 9) the number of degrees of freedom. The capacity of the program is modified by increasing or decreasing the length of the Z array for a specific problem to avoid wasting the unnecessary memory.

SAMPLE STUDIES--INELASTIC DYNAMIC ANALYSIS OF PINO-SUAREZ BUILDING

The structural model used for the nonlinear analysis of this building is shown in Figure 12. The origin of the global coordinate system (GCS) is located at the ground level. It is assumed that the slab of each floor is rigid in the plane of the floor and is flexible in the out-of-plane direction. The dead load for floors 1 to 21 is 67.63 psf and for roof is 99.3 psf. The live load for floors 1 to 21 is 71.72 psf and for roof is 20.49 psf. The total static load on a floor is equal to the total dead load plus the total live load on that floor. The total mass on a floor is equal to the total weight on that floor divided by the acceleration of gravity. The mass is lumped at each joint on the floor. The damping ratio is assumed to be 2 percent of its critical damping. The static load of live and dead load, is first applied to the structure and the structural deformations and member internal forces are treated as initial conditions for the dynamic analysis. The actual Mexico earthquake acceleration records with E-W, N-S, and vertical components are input to this building. The Wilson-θ numerical integration approach is used for the dynamic analysis with θ = 1.4 and the time increment is assumed to be 0.01 seconds. The structural geometric stiffness and the unbalanced force are included in the analysis. The failure ductilities for long-direction girders and short-direction girders are assumed to be 4. The failure ductility is defined as the maximum member-end rotation to the end rotation of a member when its end moment reaches the critical moment M_{cr}. After evaluation of several elastic analyses, it is determined that the bilinear bending and axial hysteresis models with consideration of column local buckling are assumed for all the columns from floors 1 to 7; 50 percent of residual axial strength (β' = 0.5) and 50 percent of residual bending strengths (β = 0.5) are assigned for columns with considering of local buckling; the columns on the floors 8 to 22 are assumed to be elastic; and the bracing hysteresis model is used for all the bracing members. The results based on solution SOL02 are shown in Figures 13a, 13b, and 13c corresponding to the translational responses in the GCS' X and Y directions and the torsional response in the GCS' Z direction at the top floor's mass center, respectively. It can be seen that the displacement in the positive X direction increases dramatically from 10 seconds to 12 seconds, then followed by the large displacement increase in the negative Y direction from 13 seconds to 15 seconds, and from 16 seconds to 20 seconds, large rotation is developed in the negative Z direction with maximum value about 14.5 degrees. The displacement in the positive X direction has the maximum value about 235 inches. The local buckling of the columns and failure of the girders result in significant story drift, increase P-Δ effects, then cause the failure mechanism and the building collapse.

CONCLUSIONS

The computer program INRESB-3D-SUP II is a modular program consisting of five primary blocks. The first block (STRUCT) defines the structural model. The remaining four blocks (SOL01, SOL02, SOL03, AND SOL04) are solution procedures for static loading, seismic loading, natural frequency or buckling load, and static cyclic loading, respectively. The modular form of this program allows for the easy addition of material, elements, and solutions. The program uses dynamic allocation of memory and compact matrix storage and it stores all of the joint, material, element, mass, stiffness, response, and other data in a linear array Z. The capacity of the program can be modified by increasing or decreasing the length of the Z array for a specific problem to manage memory efficiently.

Utilizing the vectorization and parallelization resources of the IBM 3090 600J supercomputer, significant computing time can be saved for the inelastic analysis of large structural system such as the 22-story Pino-Suarez Building. From the inelastic analysis of this building, it shows that the failures of girders combined with the local buckling of columns in the lower part of the building result in significant story drift, building tilt, increasing P-Δ effect, and the failure mechanism.

ACKNOWLEDGEMENTS

This research work is supported by the National Science Foundation under the grant CES 8706531. The work has been conducted on the Cornell National Supercomputer Facility (CNSF). Support for the work is gratefully acknowledged.

REFERENCES

1. Kabeyasawa, T., Shiohara, H., Otani, S., and Aoyama, H., Analysis of the Full-Scale Seven-Story Reinforced Concrete Test Structure, Journal of the Faculty of Engineering, The University of Tokyo, Vol. XXXVII, No 2, 1983, pp. 431-478.

2. Cheng, F.Y. and Mertz, G.E., Inelastic Seismic Response of Reinforced Concrete Low-Rise Shear Walls and Building Structures, NSF Report, the U.S. Dept. of Commerce, National Technical Information Service, Virginia, NTIS PB90-123217, (442 pages), 1989.

3. Takeda, T., Sozen, M.A., and Nielsen, N.N., Reinforced Concrete Response to Simulated Earthquakes, Journal of the Structural Division, ASCE, Vol. 96, No ST12, Dec 1970, pp. 2557-2573.

4. Cheng, F.Y. and Ger, J.F., Post-Buckling and Hysteresis Rules of Truss-Type Girders, Proceedings of Structural Stability Research Council, 1990, pp. 207-218.

5. Ger, J.F., Inelastic Response and Collapse Behavior of Steel Tall Buildings Subjected to 3-D Earthquake Excitations, Ph.D. Dissertation, University of Missouri-Rolla, 1990.

6. Cheng, F.Y., and Mertz, G.E., Recent Studies of Nuclear Power Plant Auxiliary Buildings Subjected to Seismic Excitations, Proceedings of the NSF CCNAA-AIT Joint Seminar on Research and Application for Multiple Hazard Mitigation, Taipei, Taiwan, 1988, pp. 257-271.

7. Cheng, F.Y., and Mertz, G.E., Hysteresis Models of Low-Rise Shear Walls with Coupled Bending and Shear Deformations, Proceedings of 10th International Conference on Structural Mechanics in Reactor Technology, 1989, Vol. K2, pp. 457-472.

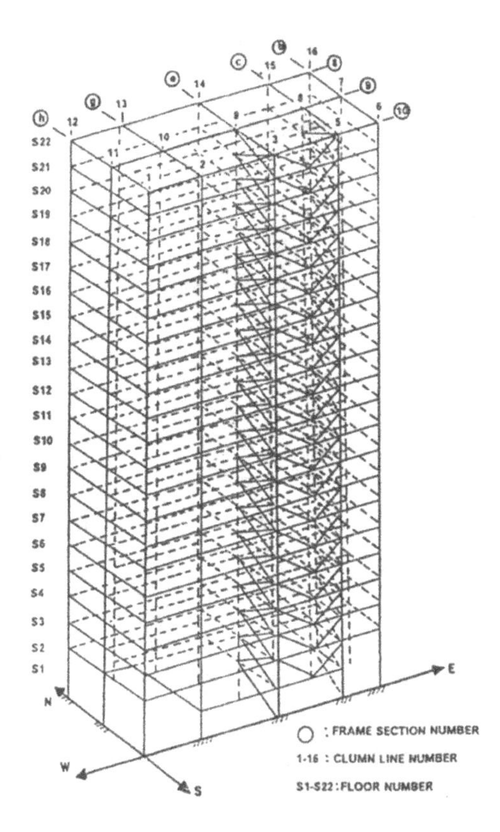

Figure 1. Configuration of Pino-Suarez Building

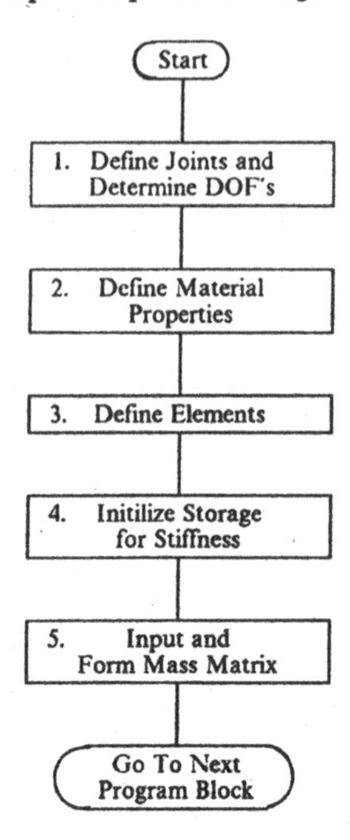

Figure 2. Block STRUCT - Define the Structural Model

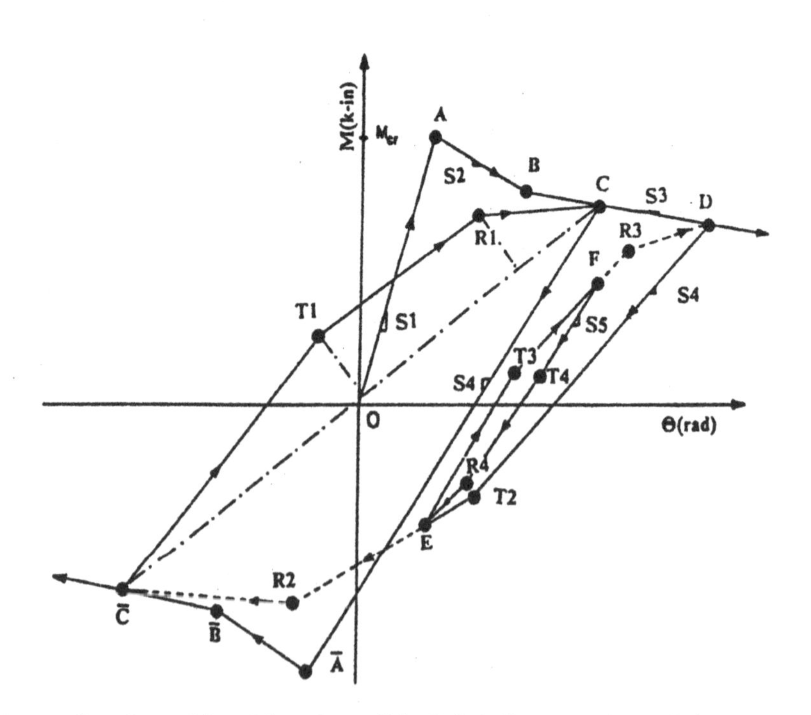

Figure 3. Long-Direction Open-Web-Joist Hysteresis Model

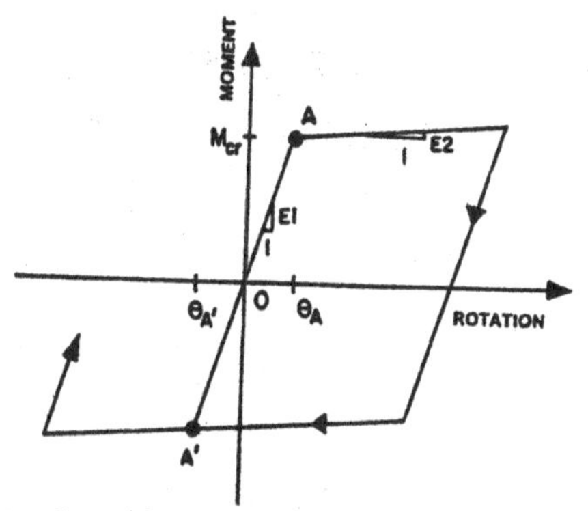

Figure 4. Short-Direction Open-Web-Joist Hysteresis Model

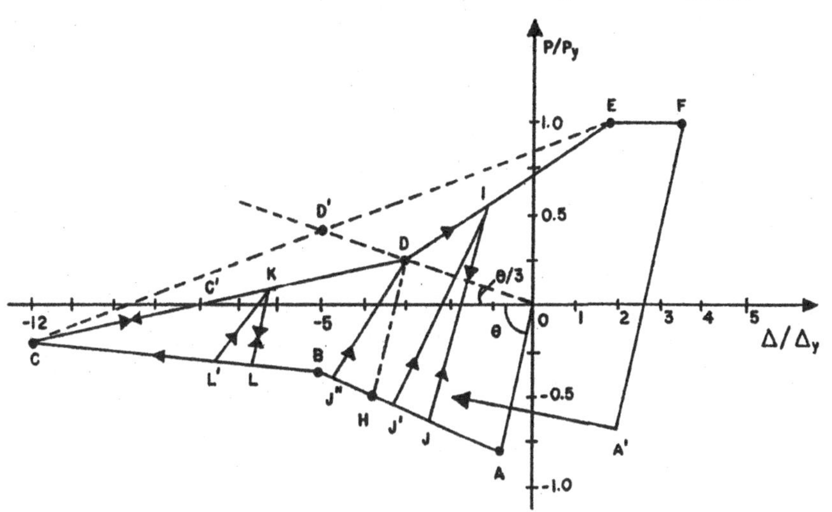

Figure 5. Bracing Member Hysteresis Model

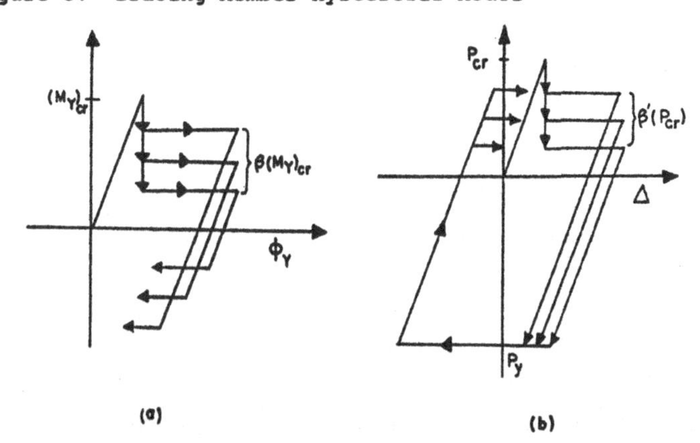

Figure 6. Bilinear Hysteresis Model, a) For Bending Deformation, b) For Axial Deformation

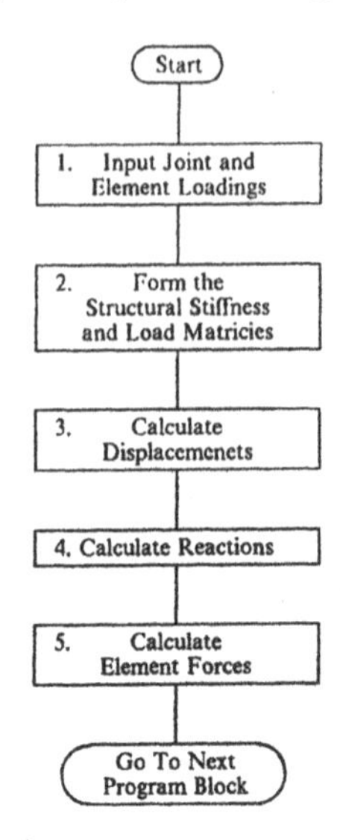

Figure 7. Block SOL01 - Static Analysis

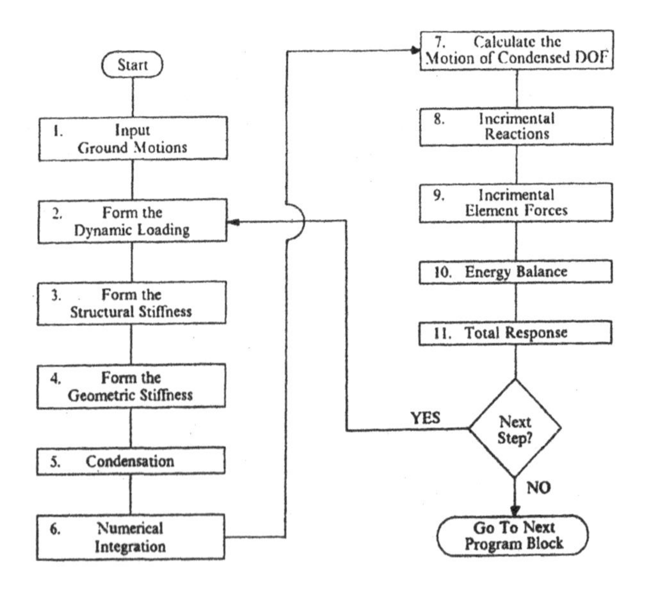

Figure 8. Block SOL02 - Elastic/Nonlinear Dynamic Analysis

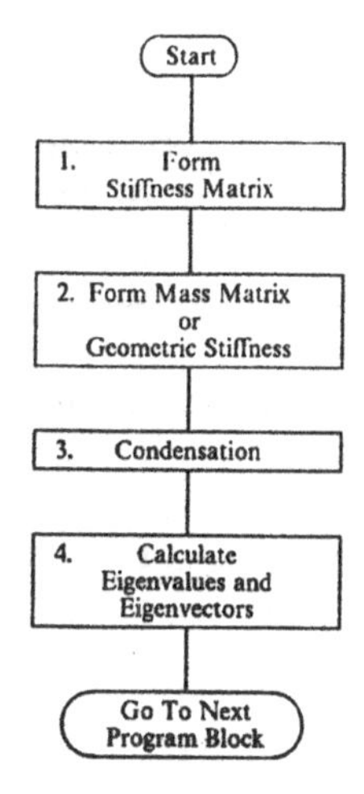

Figure 9. Block SOL03 - Natural Frequency/Elastic Buckling Load

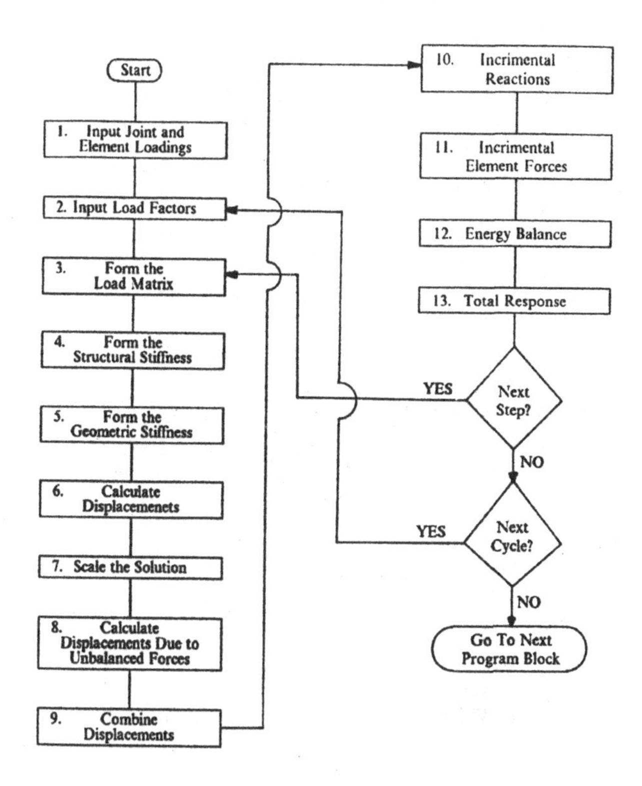

Figure 10. Block SOL04 - Nonlinear Static Cyclic Response

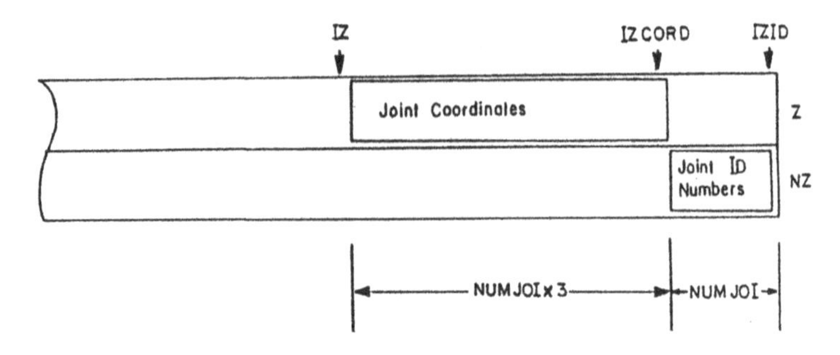

Figure 11. Dynamic Memory Example

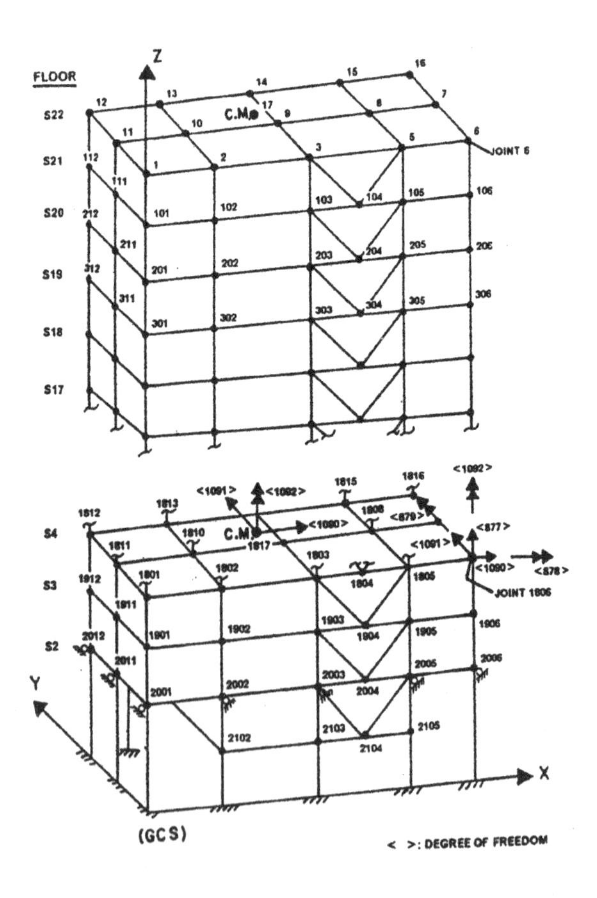

Figure 12. Structural Model of 22-Story Pino-Suarez Building

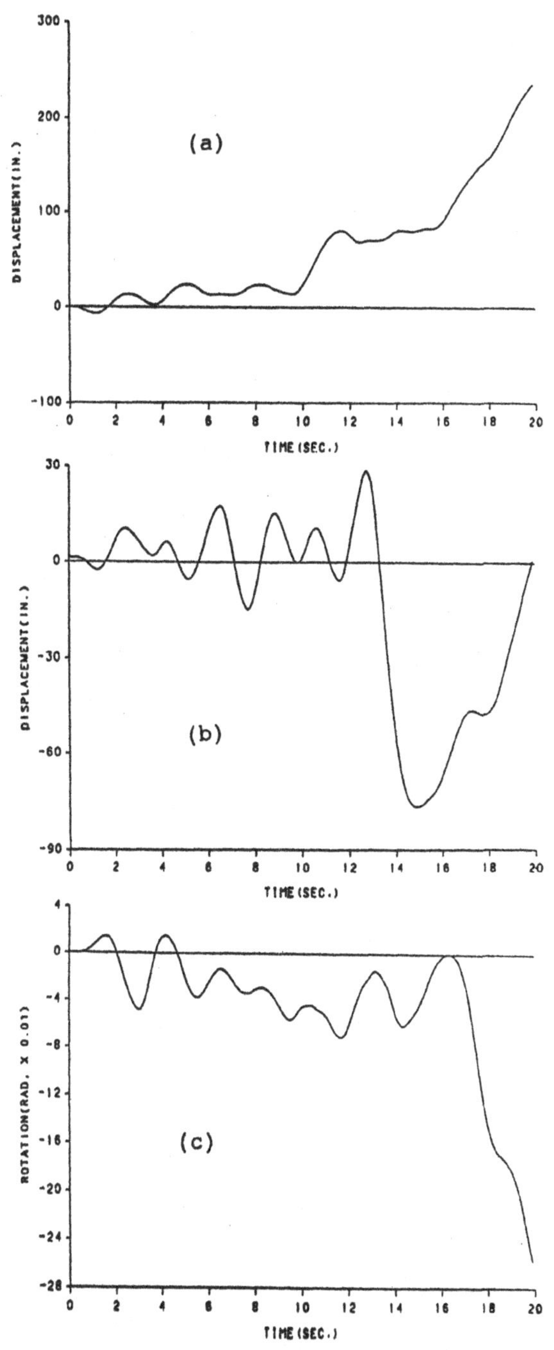

Figure 13. Responses at Top Floor's Mass Center, a) Translational Response in the GCS' X-Direction, b) Translational Response in the GCS' Y-Direction, c) Torsional Response in the GCS' Z-Direction

AUTHORS' INDEX